Fuzzy Systems and Soft Computing in Nuclear Engineering

Studies in Fuzziness and Soft Computing

Editor-in-chief
Prof. Janusz Kacprzyk
Systems Research Institute
Polish Academy of Sciences
ul. Newelska 6
01-447 Warsaw, Poland
E-mail: kacprzyk@ibspan.waw.pl

Vol. 3. A. Geyer-Schulz
Fuzzy Rule-Based Expert Systems and Genetic Machine Learning, 2nd ed. 1996
ISBN 3-7908-0964-0

Vol. 4. T. Onisawa and J. Kacprzyk (Eds.)
Reliability and Safety Analyses under Fuzziness, 1995
ISBN 3-7908-0837-7

Vol. 5. P. Bosc and J. Kacprzyk (Eds.)
Fuzziness in Database Management Systems, 1995
ISBN 3-7908-0858-X

Vol. 6. E. S. Lee and Q. Zhu
Fuzzy and Evidence Reasoning, 1995
ISBN 3-7908-0880-6

Vol. 7. B. A. Juliano and W. Bandler
Tracing Chains-of-Thought, 1996
ISBN 3-7908-0922-5

Vol. 8. F. Herrera and J. L. Verdegay (Eds.)
Genetic Algorithms and Soft Computing, 1996
ISBN 3-7908-0956-X

Vol. 9. M. Sato et al.
Fuzzy Clustering Models and Applications, 1997, ISBN 3-7908-1026-6

Vol. 10. L. C. Jain (Ed.)
Soft Computing Techniques in Knowledge-based Intelligent Engineering Systems, 1997
ISBN 3-7908-1035-5

Vol. 11. W. Mielczarski (Ed.)
Fuzzy Logic Techniques in Power Systems, 1998, ISBN 3-7908-1044-4

Vol. 12. B. Bouchon-Meunier (Ed.)
Aggregation and Fusion of Imperfect Information, 1998
ISBN 3-7908-1048-7

Vol. 13. E. Orłowska (Ed.)
Incomplete Information: Rough Set Analysis, 1998
ISBN 3-7908-1049-5

Vol. 14. E. Hisdal
Logical Structures for Representation of Knowledge and Uncertainty, 1998
ISBN 3-7908-1056-8

Vol. 15. G. J. Klir and M. J. Wierman
Uncertainty-Based Information, 2nd ed. 1999
ISBN 3-7908-1242-0

Vol. 16. D. Driankov and R. Palm (Eds.)
Advances in Fuzzy Control, 1998
ISBN 3-7908-1090-8

Vol. 17. L. Reznik, V. Dimitrov and J. Kacprzyk (Eds.)
Fuzzy Systems Design, 1998
ISBN 3-7908-1118-1

Vol. 18. L. Polkowski and A. Skowron (Eds.)
Rough Sets in Knowledge Discovery 1, 1998
ISBN 3-7908-1119-X

Vol. 19. L. Polkowski and A. Skowron (Eds.)
Rough Sets in Knowledge Discovery 2, 1998
ISBN 3-7908-1120-3

Vol. 20. J. N. Mordeson and P. S. Nair
Fuzzy Mathematics, 1998
ISBN 3-7908-1121-1

Vol. 21. L. C. Jain and T. Fukuda (Eds.)
Soft Computing for Intelligent Robotic Systems, 1998
ISBN 3-7908-1147-5

Vol. 22. J. Cardoso and H. Camargo (Eds.)
Fuzziness in Petri Nets, 1999
ISBN 3-7908-1158-0

Vol. 23. P. S. Szczepaniak (Ed.)
Computational Intelligence and Applications, 1999
ISBN 3-7908-1161-0

Vol. 24. E. Orłowska (Ed.)
Logic at Work, 1999
ISBN 3-7908-1164-5

continued on page 481

Da Ruan (Editor)

Fuzzy Systems and Soft Computing in Nuclear Engineering

With 233 Figures
and 69 Tables

Physica-Verlag

A Springer-Verlag Company

Dr. Da Ruan
Belgian Nuclear Research Centre (SCK*CEN)
Boeretang 200
2400 Mol
Belgium
Email: druan@sckcen.be

ISBN 3-7908-1251-X Physica-Verlag Heidelberg New York

Cataloging-in-Publication Data applied for
Die Deutsche Bibliothek – CIP-Einheitsaufnahme
Fuzzy systems and soft computing in nuclear engineering / Da Ruan (ed.). – Heidelberg; New York: Physica-Verl., 2000
    (Studies in fuzziness and soft computing; Vol. 38)
    ISBN 3-7908-1251-X

© Physica-Verlag Heidelberg 2000
Printed in Germany

The use of general descriptive names, registered names, trademarks, etc. in this publication does not imply, even in the absence of a specific statement, that such names are exempt from the relevant protective laws and regulations and therefore free for general use.

Hardcover Design: Erich Kirchner, Heidelberg

SPIN 10745440   88/2202-5 4 3 2 1 0 – Printed on acid-free paper

Preface

Fuzzy systems and soft computing are new computing techniques that are tolerant to imprecision, uncertainty and partial truths. Applications of these techniques in nuclear engineering present a tremendous challenge due to its strict nuclear safety regulation. The fields of nuclear engineering, fuzzy systems and soft computing have nevertheless matured considerably during the last decade.

This book presents new application potentials for *Fuzzy Systems and Soft Computing in Nuclear Engineering*. The root of this book can be traced back to the series of the first, second and third international workshops on Fuzzy Logic and Intelligent Technologies in Nuclear Science (FLINS), which were successfully held in Mol, September 14-16, 1994 (FLINS'94), in Mol, September 25-27, 1996 (FLINS'96), and in Antwerp, September 14-16, 1998 (FLINS'98). The conferences were organised by the Belgian Nuclear Research Centre (SCK•CEN) and aimed at bringing together scientists, researchers, and engineers from academia and industry, at introducing the principles of fuzzy logic, neural networks, genetic algorithms and other soft computing methodologies, to the field of nuclear engineering, and at applying these techniques to complex problem-solving within nuclear industry and related research fields.

This book, as its title suggests, consists of nuclear engineering applications of fuzzy systems (Chapters 1-10) and soft computing (Chapters 11-21). Nine pertinent chapters are based on the extended version of papers at FLINS'98 and the other 12 chapters are original contributions with up-to-date coverage of fuzzy and soft computing applications by leading researchers written exclusively for this book.

The first chapter analyses selected structures for model-based measuring methods using fuzzy logic. Chapter 2 describes a set of fuzzy systems to automate the manual procedures for reactor power level changes. Chapter 3 presents a fuzzy controller for the real time supervision of nuclear power reactors. Chapter 4 provides a comparative study of fuzzy control, PID control, and advanced fuzzy control for a simulation of hydraulic analogy of a nuclear reactor. In Chapter 5 a fuzzy adaptation of the set of output membership functions is incorporated into a fuzzy control algorithm applied to a point-kinetic model of a TRIGA Mark-III research reactor. Chapter 6 deals with the utilisation of fuzzy logic related artificial intelligent methods in an implementation of various subsystems and their plant modifications in nuclear power plants. Chapters 7-8 apply fuzzy set theory to the fields of radioactive waste processing systems and management diagnostics. Chapter 9 reviews the general neuro-fuzzy control methods and their applications in presurized water reactors. Chapter 10 discusses various techniques, mainly a combination of fuzzy clustering and artificial neural networks to approach the problem of identifying events in dynamic processes of nuclear power plants.

Chapters 11-12 review applications of neural networks in signal processing, reactor diagnostics and monitoring. Chapter 13 presents the use of regularization methods for inferential sensing in nuclear power plants. Chapters 14-15 apply genetic algorithms to the nuclear reactor design optimization and the nuclear power plant operation. Chapter 16 presents a digital control system for an ITU TRIGA Mark-II reactor using genetic algorithms with simulated annealing. Chapter 17 discusses logic-based hierarchies for modelling behaviour of complex dynamic systems with applications at pressurized water reactors. Chapter 18 provides a rapid prediction method in time series forecasting, which may have a potential application in the domain of nuclear safeguards. Chapter 19 introduces a possibilistic approach to target classification in civilian and military observation systems. Chapter 20 presents tools, developed from a rule-based expert system to a genetic-algorithm-based optimization, for fuel reload pattern design. The last chapter demonstrates diagnosis of unanticipated plant component faults in portable expert systems.

This volume highlights the advantages of applying *Fuzzy Systems and Soft Computing in Nuclear Engineering*, which can be viewed as complementary to traditional methods; as a result, fuzzy systems and soft computing provide a powerful tool for solving intricate problems pertaining in nuclear engineering. Each chapter of this book is self-contained and also indicates the future research direction on this topic of applications of *Fuzzy Systems and Soft Computing in Nuclear Engineering*.

The 21 chapters of the book are co-authored by 48 experts in nuclear engineering, fuzzy systems and soft computing from 14 countries in the world. Special thanks are due to all contributors for their kind co-operation in helping to prepare this book; to Professor Kacprzyk for his kind advice and his stimulating role in the editing of this book; to Gert Van den Eynde for his reformatting of almost all chapters in this book; to Greet Ruan for her proof-reading of part of the manuscript, and to the Belgian Nuclear Research Centre (SCK•CEN) for its support.

Da Ruan
Mol, May 1999

Contents

Preface   v

Da Ruan

1 Analysis of Selected Structures for Model- Based Measuring
  Methods Using Fuzzy Logic   1

  R. Hampel, W. Kästner, A. Fenske, B. Vandreier, and S. Schefter

  1 Introduction   1
  2 Overview - Problems and Solution Steps   3
  3 Fuzzy-Based Algorithm for Adaptation of Model Matrix
    Elements   9
  4 Fuzzy-Based Algorithm for the Adaptation of the Observer
    Gain Matrix   22
  5 Fuzzy-Based Algorithm for the Description of Process
    Disturbances   28
  6 Conclusions   35
  References   38

2 A Set of Fuzzy Systems to Automate the Manual Procedures for
  Reactor Power Level Changes   40

  Byung Soo Moon

  1 Introduction   40
  2 A Fuzzy System to Represent Spline Interpolation   41
  3 Examples of Fuzzy Systems   45
  4 Fuzzy System to Compute Total Power Defect   47
  5 Fuzzy System to Evaluate Integral Rod Worth   48
  6 Fuzzy System to Compute the Sum and the Difference   48
  7 Simulation Results   49
  8 Conclusion   51
  References   51

3 A Fuzzy Controller for the Real Time Supervision of Nuclear Power Reactors   **52**

Mohand Si Fodil, François Guly, Patrick Siarry, and Jean Luc Tyran

    1 Introduction   52
    2 Control Methods   53
    3 Working Method   53
    4 The Fuzzy Logic Controller   54
    5 Control Board of the Reactor   60
    6 Conclusion   61
    References   63

4 Adaptive Fuzzy Control for a Simulation of Hydraulic Analogy of a Nuclear Reactor   **65**

Da Ruan, Xiaozhong Li, and Gert Van den Eynde

    1 Introduction   65
    2 Fuzzy Control   67
    3 PID Control   69
    4 Advanced Fuzzy Control   69
    5 Comparative Study   75
    6 Conclusion   76
    References   80

5 Controlling Neutron Power of a TRIGA Mark III Research Nuclear Reactor with Fuzzy Adaptation of the Set of Output Membership Functions   **83**

Jorge S. Benítez-Read and Daniel Vélez-Díaz

    1 Introduction   84
    2 Adaptation of Output Membership Functions in a Fuzzy
      Controller   86
    3 Application of the SOMF Fuzzy Generation Method to a Research
      Nuclear   Reactor   97
    4 Results and Conclusions   108
    References   111

6 NPP Operator Support in Decision Making - Diagnostics of the Operation Failures Using Fuzzy Logic   **115**

Ivan Petruzela

1 Use of Artificial Intelligence Methods in the Control of Complex Technological Systems   115
2 A Proposal for a System to Support the Operator during NPP Operation in Frequency Control   119
3 Measuring Errors and Their Minimisation Using Estimation Methods   124
4 Fuzzy System of Automatic Failure Classification   130
5 Conclusion   134
References   134

7 Optimal Operation Planning of Radioactive Waste Processing System by Fuzzy Theory   **135**

Jin Yeong Yang and Kun Jai Lee

1 Introduction   135
2 Fuzzy Algorithm   136
3 Goal Programming   138
4 System Modeling   142
5 Results and Discussion   147
6 Conclusion   151
References   152

8 A Fuzzy Inference System for the Economic Calculus in Radioactive Waste Management   **153**

Pierre Kunsch, Antonio Fiordaliso, and Philippe Fortemps

1 Introduction   153
2 The Conventional Approach of Economic Calculus   154
3 The Nature of Contingency Factors in Radioactive Waste Management   155
4 A Fuzzy Approach for Radioactive Waste Management Funding   157
5 Results   162
6 Conclusions   170
References   171

9 Neuro-Fuzzy Control Applications in Pressurized Water Reactors **172**

Man Gyun Na

1 Introduction 172
2 Neuro-Fuzzy Control Methods 174
3 Application to the Steam Generator Level Control 184
4 Application to the Power Distribution Control 193
5 Conclusions 204
References 205

10 Neural and Fuzzy Transient Classification Systems: General Techniques and Applications in Nuclear Power Plants **208**

Davide Roverso

1 Introduction 208
2 Related Work 209
3 Neural Models for Transient Classification in ALADDIN 211
4 Case Study 1: Simulated BWR Events 221
5 Case Study 2: PWR Plant Islanding Events 229
6 Future Developments 232
References 233

11 Neural Networks in Signal Processing **235**

Rekha Govil

1 Introduction 235
2 The Use of ANN for Signal Processing 239
3 Dynamic Modeling 241
4 Model Based ANN's for Image Processing 242
5 Statistical Learning Networks 243
6 ANN for Eigen-Structure-Based Signal Processing 245
7 The Functional Link ANN 246
8 Active Learning 248
9 Optical Signal Processing 248
10 Superconducting Neural Networks 250
11 Neural Networks in Signal Processing Applications 250
12 Future Challenges in Neural Networks Signal Processing 252
References 254

12 Application of Neural Networks in Reactor Diagnostics and
   Monitoring 258

   I. Pázsit, N. S. Garis, and P. Lindén

   1 Introduction 258
   2 Determination of Axial Control Rod Elevation from the Measured
      Flux Shape with Neural Networks 260
   3 Localisation of a Vibrating Control Rod from Neutron Noise
      Measurements 264
   4 Choice of Input Data Representation 270
   5 Investigation of the Erroneous Identifications of the Trained
      Network 271
   6 Determination of Mass Flow of Water with Pulsed Neutron
      Activation and Neural Networks 272
   7 Conclusions 282
   References 283

13 Regularization Methods for Inferential Sensing in Nuclear Power
   Plants 285

   J. Wesley Hines, Andrei V. Gribok, Ibrahim Attieh, and
   Robert E. Uhrig

   1 Introduction 285
   2 Methodology 289
   3 Results 294
   4 Conclusions 308
   References 310

14 Genetic Algorithms Applied to Nuclear Reactor Design
   Optimization 315

   C.M.N.A. Pereira, R. Schirru, and A.S. Martinez

   1 Introduction 315
   2 The Genetic Algorithm Search Paradigm 316
   3 A Simple Nuclear Reactor Design Optimization Problem 317
   4 The Genetic Algorithm Implementation 318
   5 Application of the Genetic Algorithm to the Simple Problem 320
   6 The Robustness of the GA 326
   7 Conclusions 332
   References 333

15 Genetic Algorithms Applied to the Nuclear Power Plant Operation   335

R. Schirru, C.M.N.A. Pereira, and A.S. Martinez

1 Introduction   335
2 PWR's Fuel Management Optimization Using Genetic Algorithm   336
3 Genetic Algorithm in the Optimization of a Centroids Based
   Diagnosis System   340
References   349

16 Reactor Controller Design Using Genetic Algorithms with Simulated Annealing   351

Kadir Erkan and Erhan Bütün

1 Introduction   351
2 Discrete-Time Model of the Reactor   352
3 Controller Design   353
4 Trajectory   355
5 Genetic Algorithms for Parameter Estimation   355
6 Simulated Annealing   359
7 Simulation Results   360
8 Conclusion   362
References   362

17 Logic-Based Hierarchies for Modeling Behavior of Complex Dynamic Systems with Applications   364

Y.-S. Hu and M. Modarres

1 Hierarchy – the Nature of Complexity   364
2 The GTST-DMLD Hierarchy Modeling   367
3 Application of DMLD to Nuclear Plant Direct Containment
   Heating Phenomenon   377
4 Analysis of DCH Using DML-US 98 Software   388
5 Other Areas of GTST-DMLD Applications   391
6 Conclusions   392
References   394

18 Continued Fractions in Time Series Forecsting   396

    Andrew Zardecki

    1 Introduction   396
    2 Continued Fractions   397
    3 Ergodic Properties of Continued Fractions   400
    4 Time Series Coding   405
    5 Numerical Results   406
    6 Conclusions   411
    References   411

19 A Possibilistic Approach to Target Classification   413

    Albert G. Huizing and Frans C.A. Groen

    1 Introduction   413
    2 Problem Formulation   414
    3 Target Classification   416
    4 Performance Criterion   420
    5 A Case Study   421
    6 Conclusions   429
    References   430

**20 From FUELCON to FUELGEN: Tools for Fuel Reload Pattern
  Design**   432

    E. Nissan, A. Galperin, J. Zhao, B. Knight, and A. Soper

    1 A Description of the Domain and of the Task   432
    2 A Taxonomy   438
    3 Further Considerations on FUELGEN   445
    References   446

**21 Diagnosis of Unanticipated Plant Component Faults in a Portable
  Expert System**   449

    Jaques Reifman and Thomas Y. C. Wei

    1 Introduction   449
    2 Diagnostic Methodology   451
    3 Knowledge Bases   457

4 Simulation Tests   467
5 Discussion   474
References   475

Subject Index   477

1 Analysis of Selected Structures for Model-Based Measuring Methods Using Fuzzy Logic

R. Hampel, W. Kästner, A. Fenske, B. Vandreier, and S. Schefter

Institute for Process Technique, Process Automation and Measuring Technique
(IPM) at the University of Applied Sciences Zittau/Görlitz
D - 02763 Zittau, Theodor-Körner-Allee 16, Germany
r.hampel@htw-zittau.de

Monitoring and diagnosis of safety-related technical processes in nuclear engineering can be improved with the help of intelligent methods of signal processing such as analytical redundancies. This chapter gives an overview about combined methods in form of hybrid models using model based measuring methods (observer) and knowledge-based methods (fuzzy logic). Three variants of hybrid observers (fuzzy-supported observer, hybrid observer with variable gain and hybrid non-linear operating point observer) are explained. As a result of the combination of analytical and fuzzy-based algorithms a new quality of monitoring and diagnosis is achieved. The results will be demonstrated in summery for the example water level estimation within pressure vessels (pressurizer, steam generator, and Boiling Water Reactor) with water-steam mixture during the accidental depressurization.

1 Introduction

For the validation of the process state of safety-related systems during as well as after transients and accidents the knowledge of measurable and non-measurable variables is of fundamental importance. This knowledge is a precondition for the accident management and analysis. For improving the quality of information analytical redundancies in form of Model-based Measuring Methods (MMMs) are applied. With the help of them additional information about the actual process state can be generated which exceeds the information available from pure measurements.

The quality of the process information generated by means of MMMs depends on the quality and validity of the MMMs themselves as well as the implemented models they based on. The strong change of parameters in result of the accidental conditions leads to the following problems:

− non-linearities and as a result a limited validity of the model,

− uncertain and fuzzy mathematical description of physical effects,

− influence of process disturbances, which cannot be considered in the model.

The global aim of the investigation is to improve the quality and reliability of the information about the actual process state during strong parameter changing and to determine non-measurable process variables. The aspired quality improvement is realized by the application of MMMs including methods of signal processing of fuzzy information by fuzzy logic.

The efficiency and the verification of the developed methods and algorithms are demonstrated by the determination of the collapsed level and the mixture level in pressure vessels with water-steam mixture (pressurizer, steam generator, and reactor pressure vessel).

The chosen example is predestinated for the methodical investigations and for testing the developed methods to describe strong non-linear thermodynamic and thermohydraulic effects within the pressure vessel. Furthermore the process is characterized by the combination of a lot of parameters which have an influence on the process state (pressure, temperature, steam content, water level, mass flow, and heat flux). An exact information about the actual process state is very important to the safety-related character of the water level as a process parameter.

Most of all, the application of fuzzy logic for specific problems of power plants was realized in the field of control [10], especially the control of plant components (power control [2, 21, 17], feed water control [12, 13], speed control, turbine power control [6, 8, 4], steam generator control [18, 14]), and for fault detection [20, 7, 11, 15]. There are some parallels in international literature for using the described methodology of the combination of model- and fuzzy-based algorithms. In [19] a fuzzy-supported extended Kalman-estimator was proposed for the on-line state estimation of biological processes. An adaptation of the gain matrix was explained in [1]. In [3] a model-based fuzzy observer for the state estimation of an inverted pendulum was presented. A rule- and model-based controller was applied in [16]. As it is shown by the references, the trend to combine model- and knowledge-based algorithms for the realization of an exact and robust state monitoring is confirmed.

In Section 2 the specific problems of the investigated process and the developed methods for their solution are explained. The detailed description of the designed methods follows in Sections 3, 4 and 5. Section 6 contains the conclusions.

2 Overview - Problems and Solution Steps

2.1 Strategy of Problem Solution

Starting point of the methodical investigations is the assumption that the combination of analytical and knowledge-based methods leads to a new quality of signal processing methods for the monitoring of safety-related systems.

In Figure 2.1, the methods representing the current state of art such as MMMs on the basis of analytical modelling (observer, Kalman estimator) and knowledge-based algorithms like fuzzy logic and neural networks are illustrated.

The combination of the advantages of both approaches leads to hybrid methods, which combine model- and knowledge-based elements.

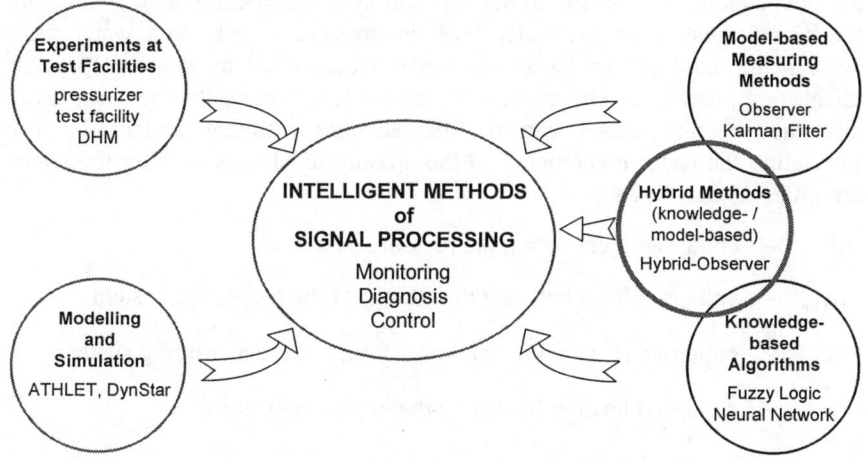

Fig. 2.1. Components for the design of intelligent signal processing methods

Experiments at test facilities (pressurizer test facility DHM) and simulations by means of complex simulation tools (thermal-hydraulic code ATHLET) were used to generate the data basis for the model- and knowledge-based algorithms.

The simulation of the hybrid methods was carried out by means of the simulation tool DynStar, which allows the realization of model- and knowledge-based algorithms [5].

The test facility allows the investigation of special single effects. Further information about the process state can be obtained by using additional measuring techniques. Based on this information a comprehensive data basis can be generated.

The realized experiments were post-calculated by means of the complex thermohydraulic code ATHLET. The code reproduces all interactions between the physical parameters of the real process. Furthermore it enables to verify the functionality range of the methods and the models used by the MMM for such parameter ranges or process states (accidents) which are not realizable at the test facility.

By the ATHLET-code all measurable and non-measurable variables and parameters are provided which characterize the process state. The ATHLET-code substitutes the real process and was used to verify the developed hybrid methods.

2.2 Application Problems and Their Solution

The subject of investigation is the water level measuring system in connection with pressure vessels (pressurizer, steam generator, and reactor vessel) under two-phase flow conditions. The aim of the application of analytical redundancies is the exact state estimation of the water level in pressure vessels with water-steam mixture in the case of accidental transients characterized by negative pressure gradients (e.g., leaks). To characterize the complete process state, the water levels within the pressure vessel can be divided into different collapsed levels (representing the water inventory) and the mixture level (representing the water-steam mixture) as follows:

$\Rightarrow$ hc   -   collapsed level within the pressure vessel

$\Rightarrow$ hc_{bF} -   collapsed level between the fittings of the measuring system

$\Rightarrow$ hc_{lF} -   collapsed level below the lower fitting of the measuring system

$\Rightarrow$ hc_i -   collapsed level indicated by the measuring system

$\Rightarrow$ hm   -   mixture level within the pressure vessel

Figure 2.2 shows the pressure vessel in connection with the hydrostatic water level measuring system.

Based on the measurable variables collapsed level hc within the pressure vessel and collapsed level hc_{bF} between the fittings of the narrow range measuring system the MMM was applied to estimate the collapsed level hc_{lF} below the lower fitting of the narrow range measuring system. The variable hc_{lF} characterizes the evaporation in this zone. It has an important influence on the mixture level hm.

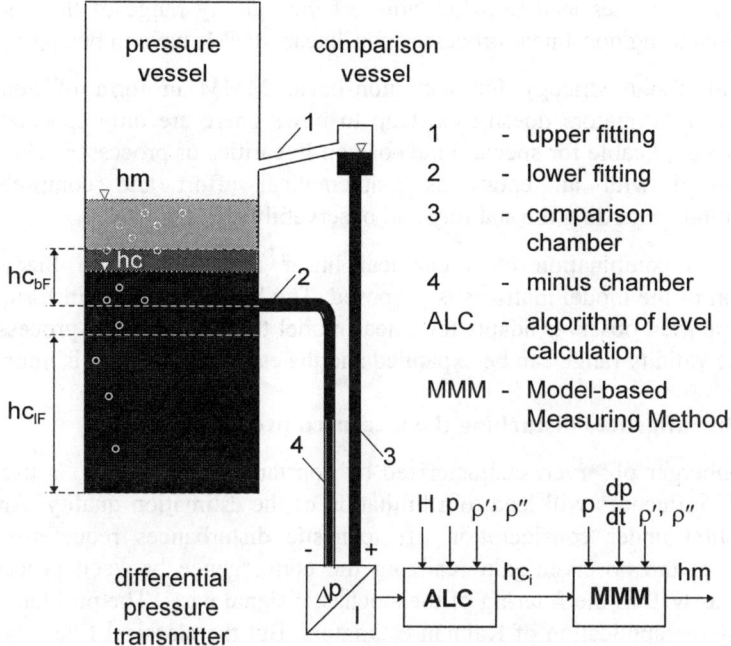

Fig. 2.2. Pressure vessel in connection with a hydrostatic level measuring system

The following specific problems have an influence on the model- and fuzzy-based description of the process state:

Non-linearities

The thermohydraulic and thermodynamic processes in pressure vessels with water-steam mixture are characterized by non-linearities which can be classified as follows:

⇒ **non-linear structure of the model**

products of state variables and input variables within the state equations

e.g.   product of the state variable collapsed level hc and the input variable pressure gradient dp/dt

⇒ **time-dependent coefficients in the model equations**

non-linear change of thermodynamic properties

e.g.   dependence of density and enthalpy on pressure

⇒ **non-linear behaviour as a result of the change of process state**

e.g.   disturbances like feed-in, bleed-off, spray

These non-linearities lead to a limitation of the validity range of the linearized model. For strong non-linear processes non-linear MMMs have to be applied.

A general design strategy for such non-linear MMM in form of non-linear observers or estimators doesn't exist up to now. There are only special design algorithms applicable for special kinds of non-linearities or processes. The design is connected with an enormous mathematical effort and comprehensive investigations regarding the stability and observability.

Therefore a combination of a classical linear observer with a fuzzy-based adaptation of the model matrices is proposed. The hybrid observer in form of the fuzzy-supported observer adapts the linear model to the non-linear process. As a result, the validity range can be expanded and the estimation quality is improved.

Stochastic influences disturbing the measured process signals

If a Luenberger observer, characterized by constant gain elements, is used then stochastic influences will lead to a limitation of the estimation quality. An exact reproduction under consideration of stochastic disturbances requires variable elements of the observer gain realizing the convergence between process and observer as well as the filtering of the stochastic signal parts. The problem can be solved by the application of Kalman estimators. But the classical filter algorithm has to be expanded to guarantee the numerical stability and to improve the robustness of the covariance matrix.

For adapting the observer gain elements of the classical observer in dependence on the stochastic properties of the measuring signals the implementation of a fuzzy-based algorithm is proposed (Section 4).

Process disturbance

Process disturbances lead to estimation errors which cannot be compensated by the classical observer in all cases. Therefore the influence of process disturbances has to be modelled by a mathematical or rule-based description.

The fuzzy-based modelling of a process disturbance is investigated on the example of the reproduction of the heat flux between the pressure vessel wall and the water-steam mixture during the depressurization. In the presented example a non-linear operating point observer is expanded by a dynamic fuzzy controller for the description of the heat flux (Section 5).

2.3   Fuzzy-Based Adaptation of the Observer Model Parameters- Hybrid Observer

The design of the classical observer is based on a state space model which consists of state equations for the description of the behaviour of process and measuring system. By means of these equations the correlations between the vector of input variables **u**, the vector of state variables **q** and the vector of output variables **x** are defined. Both, process as well as measuring system, can be influenced by disturbances (process disturbance **v**, disturbances on measuring system **z**) which can be reflected by the state space model only in a limited way.

Furthermore, the structure of the classical observer is characterized by the feed-back of the estimation error $\mathbf{e_x} = [\mathbf{x} - \hat{\mathbf{x}}]$ beyond the observer gain matrix **K**.

The equations (2-1) to (2-4) illustrate the state equations for the process and for the observer.

State space model for process description:

$$\frac{d\,\mathbf{q}(t)}{dt} = \overbrace{\mathbf{A_p} \cdot \mathbf{q}(t) + \mathbf{B_p} \cdot \mathbf{u}(t)}^{\text{process model}} + \overbrace{\mathbf{v}(t)}^{\text{process disturbance}} \qquad (2\text{-}1)$$

$$\downarrow$$

adaptation of
process disturbances

$$\mathbf{x}(t) = \underbrace{\mathbf{C_p} \cdot \mathbf{q}(t)}_{\text{model of the measuring system}} + \underbrace{\mathbf{z}(t)}_{\text{measuring disturbance}} \qquad (2\text{-}2)$$

State space model of Observer:

$$\frac{d\,\hat{\mathbf{q}}(t)}{dt} = \overbrace{\mathbf{A_b} \cdot \hat{\mathbf{q}}(t) + \mathbf{B_b} \cdot \mathbf{u}(t)}^{\text{process model}} + \overbrace{\mathbf{K} \cdot [\mathbf{x}(t) - \hat{\mathbf{x}}(t)]}^{\text{correction term}} \qquad (2\text{-}3)$$

adaptation of
model matrices

adaptation of the
observer gain matrix

$$\hat{\mathbf{x}}(t) = \underbrace{\mathbf{C_b} \cdot \hat{\mathbf{q}}(t)}_{\text{model of the measuring system}} \qquad (2\text{-}4)$$

The general structure of a hybrid observer is shown in Figure 2.3. The hybrid observer consists of a classical observer and a fuzzy-based adaptation of selected parameters and is connected in parallel to the real process as an analytical redundancy.

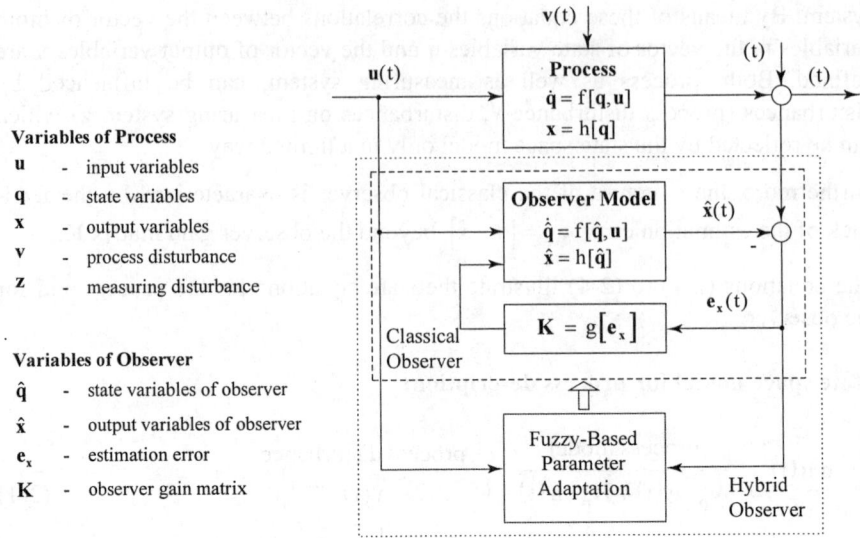

Variables of Process

u   -   input variables

q   -   state variables

x   -   output variables

v   -   process disturbance

z   -   measuring disturbance

Variables of Observer

$\hat{q}$ -   state variables of observer

$\hat{x}$ -   output variables of observer

e_x -   estimation error

K   -   observer gain matrix

Fig. 2.3.   Structure of a hybrid observer

Based on the analysis of the reasons for deviations between process variables and estimated observer variables the following possibilities of fuzzy-based parameter adaptation and corresponding applied methods summerized in Figure 2.4. can be considered [9]:

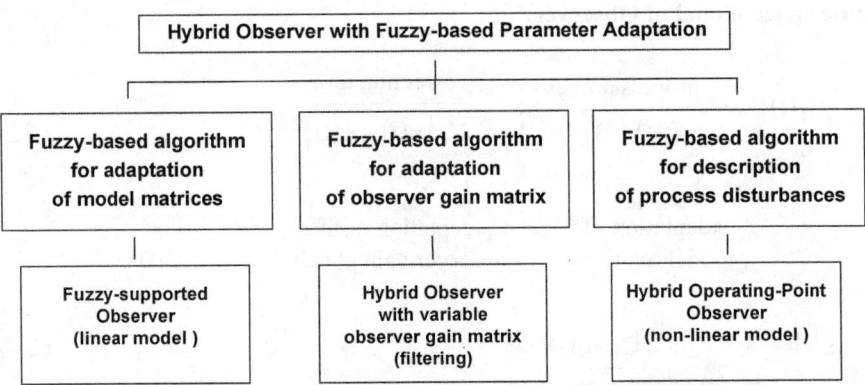

Fig. 2.4.   Variants of fuzzy-based parameter adaptation within observer structures

The fuzzy-based algorithm for the adaptation of the elements of the model matrices A_b, B_b in form of the fuzzy-supported observer enables the compensation of non-linearities and an improvement of the estimation quality.

The fuzzy-based algorithm for the adaptation of the observer gain matrix **K** in form of the hybrid observer with variable gain improves the robustness of the observer regarding to stochastic disturbances **z** of the measuring signals.

The fuzzy-based algorithm for the description of process disturbances **v** in form of the hybrid operating point observer with the dynamic fuzzy controller makes it possible to describe the heat flux between the pressure vessel wall and the water-steam mixture during the depressurization.

In the following sections the developed hybrid methods for the process monitoring are explained.

3 Fuzzy-Based Algorithm for Adaptation of Model Matrix Elements

3.1 State Space Model

The state equations describe the deviation of collapsed level within the following zones of the pressure vessel:

- **zone below the lower fitting of the measuring system with a constant height of the zone** z_{lF}
  (volume of the pressure vessel below the lower fitting of the measuring system, characterized by the non-measurable state variable steam content φ_{lF} or collapsed level hc_{lF} of the zone 1)

- **zone between the fittings of the measuring system with variable height of the zone** Δz_{bF}
  (volume of the water-steam mixture between the fittings of the measuring system, characterized by the non-measurable state variable steam content φ_{bF} and by the measurable state variable collapsed level hc_{bF} of zone 2).

Figure 3.1 shows the nodalization scheme of the pressure vessel, where the zone below the lower fitting and the zone between the fittings are defined as separate zones. The nodalization allows a one-dimensional description of the thermodynamic effects and an axial determination of the steam content distribution. This model of second order was favoured because of its simple structure and the minimal number of elements which have to be adapted.

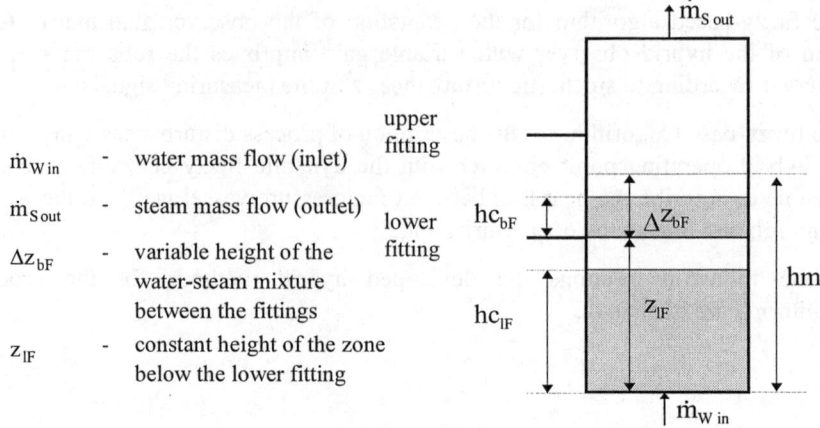

Fig. 3.1. Nodalization scheme of pressure vessel

The state equations (3-1) and (3-2) describe the state variables depending on the input variable pressure gradient $d\Delta p/dt$:

Δhc_1 - deviation of the collapsed level within the zone below the lower fitting of the measuring system (zone 1)

Δhc_2 - deviation of the collapsed level within the zone between the fittings of the measuring system (zone 2)

$$
\begin{bmatrix} \Delta \dot{h}c_1 \\ \Delta \dot{h}c_2 \end{bmatrix} = \begin{bmatrix} a_{11} & 0 \\ a_{21} & a_{22} \end{bmatrix} \begin{bmatrix} \Delta hc_1 \\ \Delta hc_2 \end{bmatrix} + \begin{bmatrix} b_1 \\ b_2 \end{bmatrix} \begin{bmatrix} \dfrac{d\Delta p}{dt} \end{bmatrix} \tag{3-1}
$$

$$
\Delta hc_2 = \begin{bmatrix} 0 & 1 \end{bmatrix} \begin{bmatrix} \Delta hc_1 \\ \Delta hc_2 \end{bmatrix} \tag{3-2}
$$

Using this state space model the matrix elements a_{11}, a_{21}, a_{22} of the system matrix and b_1, b_2 of the input matrix have to be adapted by separate controllers.

3.2 Fuzzy-Supported Observer

The aim of the fuzzy-based adaptation of the model matrix elements is the expansion of the validity range of the linearized observer model. The model matrix elements are adapted to the change of process state, that means the influence of the non-linearities is compensated by the fuzzy-based parameter adaptation. This aim is realized by use of the developed fuzzy-supported observer.

Assuming that

– the depressurization has no influence on the response characteristics of the measuring system respectively the influence will be compensated by correction algorithms $\mathbf{C}_b \neq f(p,\ dp/dt)$ (equation 2-4) and

– the observer gain matrix $\mathbf{K}$ (equation 2-3) is designed in such a way, that the Observer realizes a sufficient convergence within the defined parameter range,

only the model matrices $\mathbf{A}_b$ and $\mathbf{B}_b$ of the linearized observer model (equations 2-3 and 2-4) have to be adapted. For each element of these model matrices a separate fuzzy controller for adaptation has to be designed.

3.3 Design of the Fuzzy-Supported Observer

3.3.1 Fuzzy Controller for Adapting the Elements of the System Matrix $\mathbf{A}_b$

Preliminary remarks

As essential variables which have an influence on the matrix elements a_{11}, a_{21}, a_{22} the actual values of pressure p and pressure gradient dp/dt were analyzed. In the following section the design of the matrix element a_{11} of the system matrix is exemplarily explained.

$$a_{11} = f\left(p(t), \frac{dp}{dt}(t)\right) \tag{3-3}$$

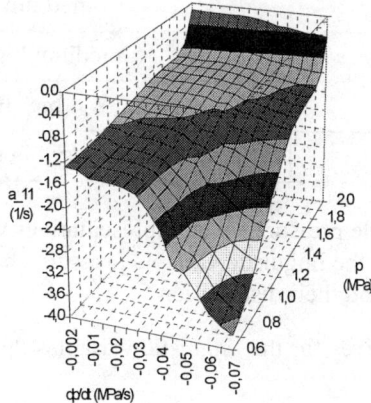

Fig. 3.2. Non-linear characteristic field of the element a_{11} of the system matrix depending on the pressure and the pressure gradient

Figure 3.2 illustrates the dependence of the absolute value of the element a_{11} on the pressure and the pressure gradient. The parameter range of pressure reduction was defined on the basis of the realized experiments at the test facility by $p = [0.6 \ldots 2.0]\,\text{MPa}$. The pressure gradient is changing in the range of $dp/dt = [-0.002 \ldots -0.07]\,\text{MPa}/\text{s}$.

Basic rule

The basic rule defines the correlation between the input variables of pressure and pressure gradient and the output variable of absolute value of system matrix element a_{11}:

$$\text{IF} \quad p \qquad \text{AND} \qquad dp/dt \qquad\qquad \text{THEN} \qquad |a_{11}| \qquad (3\text{-}4)$$

Linguistic variables

The input and output variables pressure, pressure gradient and matrix element are defined as linguistic variables. The dimensioning of the matrix element was realized for selected working points of pressure (Table 3.1).

Table 3.1. Representatives for the linguistic variable pressure

Representatives	Values of the linguistic variable
p = 0,6 MPa	very small (VS)
p = 1,0 MPa	small (S)
p = 1,2 MPa	medium small (MS)
p = 1,5 MPa	medium (M)
p = 1,6 MPa	medium big (MB)
p = 1,8 MPa	big (B)
p = 2,0 MPa	very big (VB)

For the linguistic variable pressure seven representatives were defined. Figure 3.3 shows the fuzzy sets for the linguistic variable pressure characterized by triangular membership functions and their distribution.

The defined representatives for the linguistic variables dp/dt and $|a_{11}|$ are listed in Table 3.2.

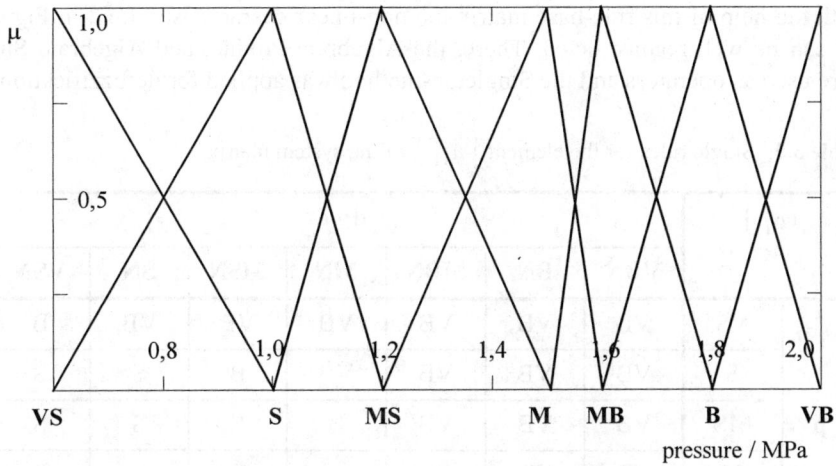

Fig. 3.3. Fuzzy-set for the linguistic variable pressure

Table 3.2. Linguistic values of the linguistic variables pressure gradient dp/dt and matrix element $\left| a_{11} \right|$

dp/dt	$\left| a_{11} \right|$
VSN (very small negative)	VS (very small)
SN (small negative)	S (small)
MSN (medium small negative)	B (big)
MN (medium negative)	VB (very big)
MBN (medium big negative)	
BN (big negative)	
VBN (very big negative)	

Rule base

The correlation between the linguistic values is defined on the basis of equation (3-4). In this way the single rule for the first element of the rule base is defined as:

IF   ' p ' = 'VS'   AND   ' dp/dt ' = 'VBN'

THEN   '$\left| a_{11} \right|$ ' = 'VB'   (3-5)

All single rules are represented in the rule base matrix (Table 3.3).

With the help of this rule-base matrix the non-linear characteristic field in Figure 3.2 can be well reconstructed. There, the Algebraic Product and Algebraic Sum were used as operators and the Singleton-method was applied for defuzzification.

Table 3.3. Single rules for the element $|a_{11}|$ of the system matrix

$|a_{11}|$		dp/dt						
		VBN	**BN**	**MBN**	**MN**	**MSN**	**SN**	**VSN**
	VS	VB	VB	VB	VB	VB	VB	B
	S	VB	VB	VB	VB	B	S	S
	MS	VB	VB	VB	B	S	S	S
p	**M**	VB	B	S	S	S	S	S
	MB	B	S	S	S	S	S	S
	B	S	S	S	S	S	S	S
	VB	VS	VS	VS	VS	VS	VS	VS

3.3.2 Fuzzy Controller for Adapting the Elements of the Input Matrix B_b

Preliminary remarks

The elements of the input matrix B_b depend on the process variables pressure (actual pressure p(t), initial pressure p_0 at the beginning of depressurization) and collapsed level (initial collapsed level hc_0 at the beginning of depressurization):

$$\mathbf{B} = \begin{bmatrix} b_1 \\ b_2 \end{bmatrix} = f\{p(t), p_0, hc_0\} \tag{3-6}$$

In Table 3.4 the experiments which were carried out and post-calculated with the help of the ATHLET-code are listed. The initial values of pressure and collapsed level were varied.

Table 3.4. Experiments with different initial values of pressure p_0 and collapsed level hc_0 at the beginning of depressurization (R - experiments with the character of reference points, O - experiments for verifying the algorithm)

Experiment	Initial Pressure		
Initial Collapsed Level	$p_0 = 22$ bar	$p_0 = 18$ bar	$p_0 = 14$ bar
$hc_0 = 195$ cm	R	O	R
$hc_0 = 170$ cm	O	O	O
$hc_0 = 155$ cm	R	O	R
$hc_0 = 135$ cm	O	O	O
$hc_0 = 115$ cm	O	O	O
$hc_0 = 105$ cm	R	O	R

Figure 3.4 shows the non-linear dependence of the element b_1 on the actual pressure calculated by the ATHLET-code and validated by experiments for three depressurizations with the initial pressure $p_0 = 22$ bar and different initial collapsed levels hc_0. For the input matrix element b_2 qualitatively similar relations can be found.

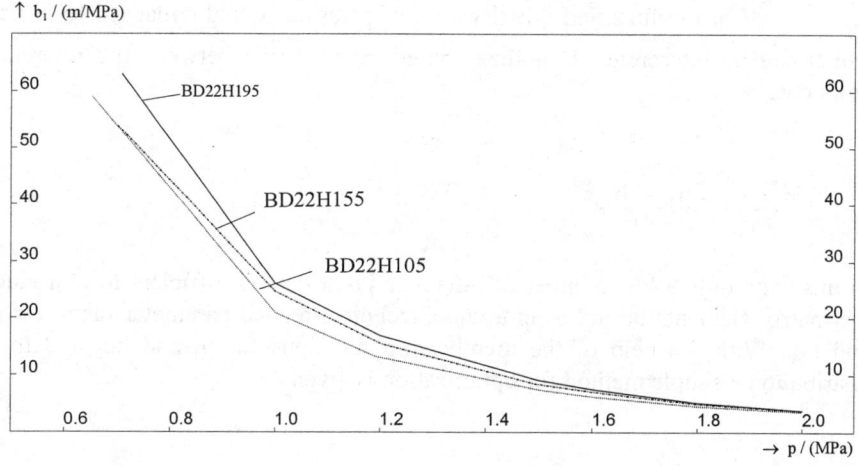

Fig. 3.4. Dependence of the input matrix element b_1 on the actual pressure p for Blow Down experiments with the initial pressure $p_0 = 22$ bar and the initial collapsed levels $hc_0 = 105; 155; 195$ cm

There are two possibilities for the design of the fuzzy controller for adapting the elements b_1 and b_2 of the input matrix $\mathbf{B_b}$:

Variant a)   one 3D-Controller for each matrix element with the input variables actual pressure $p(t)$, initial pressure p_0 and initial collapsed level hc_0

Variant b)   a higher number of 1D-Controllers for selected reference points p_0 and hc_0 with the input variable $p(t)$ and an adaptation in dependence on the real initial values of pressure and collapsed level

The advantage of variant a) is the compact solution in form of one fuzzy controller for each matrix element. However, depending on the defined fuzzy sets for the input and output variables a high number of single rules is necessary to achieve a sufficient reproduction. The overview about the influences regarding the adaptation is insufficient.

Variant b) is illustrated in Figure 3.5. It is based on the experimental data base of the six reference points (R) (see Table 3.4). The basic idea of the algorithm is to weight the fuzzy controller output signals $b_{i_{p0,hco}}$ with the membership values μ_{p0} , μ_{hc0} of the fuzzified initial values of pressure p_0 and collapsed level hc_0, which can be interpreted as a fuzzy-based interpolation between the reference point data:

$$b_i = \sum_{(R)} \mu_{p0} \cdot \mu_{hc0} \cdot b_{i_{p0,hc0}} \qquad i = 1,2 \qquad\qquad (3\text{-}7)$$

In this way, only a low number of reference point data is sufficient to reproduce the matrix elements b_1 and b_2 in a wide, technical-related parameter range for p_0 and hc_0. With the help of the membership functions for p_0 and hc_0 and their distribution a simple method for optimization is given.

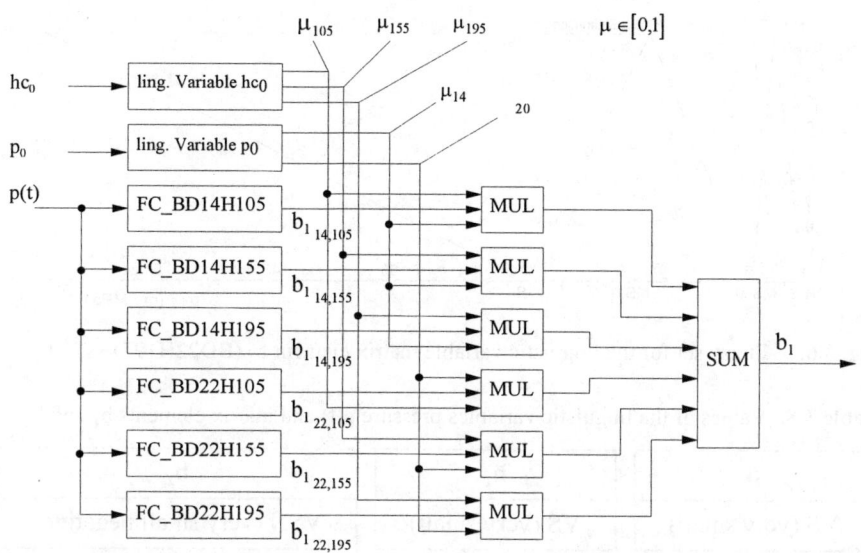

Fig. 3.5. Determination of the matrix element b_1 by a combination of weighted fuzzy controllers

Basic rule

The basic rules of the fuzzy controllers in Figure 3.5 for the matrix elements b_1 and b_2 are defined as follows:

$$\text{IF} \quad p(t) \qquad\qquad \text{THEN} \qquad b_1 \tag{3-8}$$

$$\text{IF} \quad p(t) \qquad\qquad \text{THEN} \qquad b_2 \tag{3-9}$$

Starting from experimental data the 1D-Controllers for the reference points (R) were designed.

Linguistic variables

The representatives for the linguistic variable pressure $p(t)$ were defined analogous to section 3.3.1. The fuzzy sets for the matrix element b_1 are shown in Figure 3.6 for the reference experiment BD22H195. Table 3.5 summarizes the values of the used linguistic variables.

18

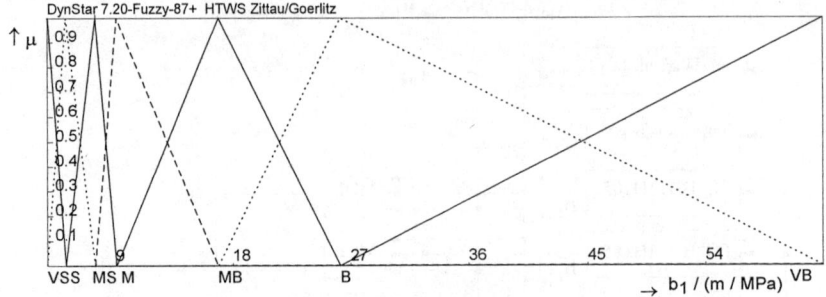

Fig. 3.6. Fuzzy set for the linguistic variable matrix element b_1 (BD22H195)

Table 3.5. Values of the linguistic variables pressure p(t) and matrix elements b_1 and b_2

p	b_1	b_2
VS (very small)	VS (very small)	VSN (very small negative)
S (small)	S (small)	SN (small negative)
MS (medium small)	MS (medium small)	MSN (medium small negative)
M (medium)	M (medium)	MN (medium negative)
MB (medium big)	MB (medium big)	MBN (medium big negative)
B (big)	B (big)	BN (big negative)
VB (very big)	VB (very big)	VBN (very big negative)

The fuzzy sets for the initial values of pressure p_0 and collapsed level hc_0 are illustrated in Figures 3.7 and 3.8.

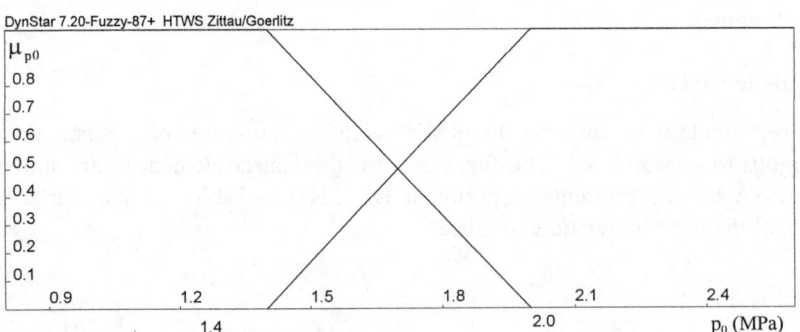

Fig. 3.7. Fuzzy set of the linguistic variable initial pressure (p_0) to assess the fuzzy controllers for the reference points

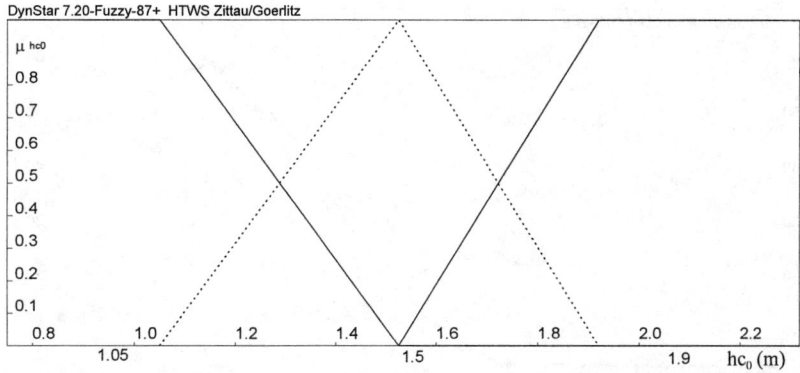

Fig. 3.8. Fuzzy set for the linguistic variable initial collapsed level (hc₀) to assess the fuzzy controllers for the reference points

Rule base

The single rules which represent the correlation between the linguistic variables are summarized in the rule base in form of a rule table (Table 3.6).

Table 3.6. Rule table for the input matrix element b_1 and b_2

IF p	THEN b_1		IF p	THEN b_2
VS	VB		VS	VBN
S	B		S	BN
MS	MB		MS	MBN
M	M		M	MN
MB	MS		MB	MSN
B	S		B	SN
VB	VS		VB	VSN

In Figure 3.9, the result of the fuzzy-based interpolation described here is shown exemplarily for the element b1f interpolated for a process state characterized by parameters for the initial values of pressure p_0 and collapsed level hc_0 between the defined reference points (Table 3.4).

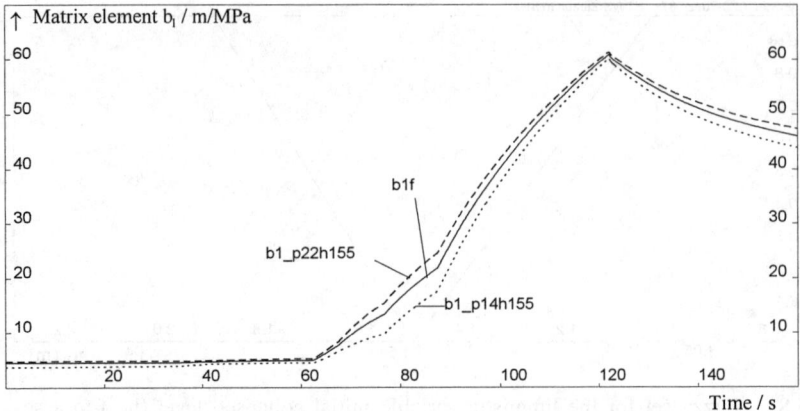

Fig. 3.9. Interpolation of the element b_1 for the process state characterized by the initial values $p_0 = 18$ bar and $hc_0 = 155$ cm

As shown in Section 3.4, the sufficient quality of the calculated values for b1f leads to a good correspondence between state observation and experimental results.

So it could be shown, that the matrix elements can be reproduced by the fuzzy-based interpolation between the output variables of the fuzzy controllers designed for the reference points in a sufficient quality. For that, only a low number of reference points and a low number of single rules are necessary.

3.4   Results of the State Estimation

Figures. 3.10-3.12 represent the results of state estimation for the collapsed levels within the zones by means of the fuzzy-supported Observer. The elements of the system matrix $\mathbf{A_b}$ as well as the input matrix $\mathbf{B_b}$ were adapted by the fuzzy algorithms described here. For that, to demonstrate the estimation quality, such experiments were chosen which did not serve as the basis of the controller design (Table 3.4). For the state variables of the second order state space model the time characteristics between process and fuzzy-supported observer were compared.

In all cases, the measurable state variable $\Delta hc2$ (deviation of collapsed level between the fittings of the measuring system) was estimated very well. The estimated non-measurable state variable $\Delta hc1$ (deviation of collapsed level below the lower fitting of the measuring system) was also in a good correspondence with the process behaviour. So, the estimation quality can be assumed as sufficient for all experiments.

These figures demonstrate the functionality and universality of the developed algorithms within the defined parameter range.

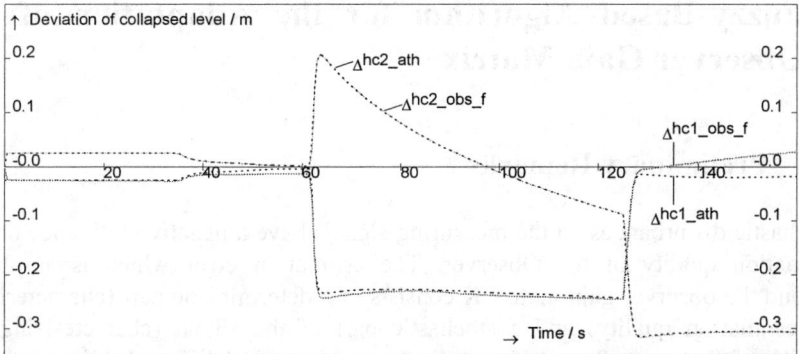

Fig. 3.10. State estimation (experiment BD18H170) with interpolation between the output variables of the fuzzy controllers

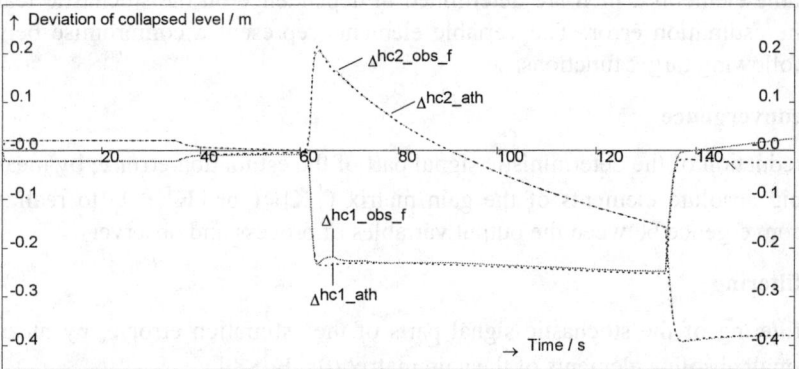

Fig. 3.11. State estimation (experiment BD22H170) with interpolation between the output variables of the fuzzy controllers

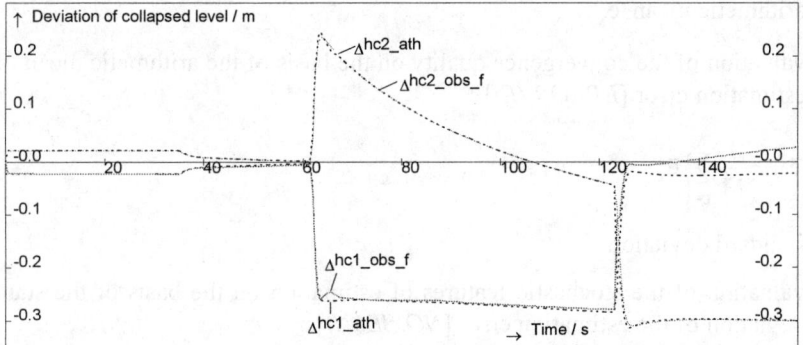

Fig. 3.12. State estimation (experiment BD22H135) with interpolation between the output variables of the fuzzy controllers

4 Fuzzy-Based Algorithm for the Adaptation of the Observer Gain Matrix

4.1 Preliminary Remarks

Stochastic disturbances on the measuring signals have a negative influence on the estimation quality of the Observer. The estimation error which is amplified beyond the observer gain matrix **K** consists of a deterministic part (characterizing the estimation quality) and a stochastic part of the signal (characterizing the quality of filtering). The different signal parts require different values of gain matrix elements.

The constant gain elements of the classical observer have to be replaced by variable elements which are determined in dependence on the stochastic features of the estimation error. The variable elements represent a compromise between the following target functions:

$\Rightarrow$ **convergence**

reduction of the deterministic signal part of the estimation error $\mathbf{e}_x$ by means of big absolute elements of the gain matrix ($|\mathbf{K}| \approx 1$ or $|\mathbf{K}| \gg 1$) to realize the convergence between the output variables of process and observer

$\Rightarrow$ **filtering**

filtering of the stochastic signal parts of the estimation error $\mathbf{e}_x$ by means of small absolute elements of the gain matrix ($0 < |\mathbf{K}| \ll 1$)

The valuation of the estimation error features was realized with the help of the following statistics:

- arithmetic mean $\overline{e}_x$

valuation of the convergence quality on the basis of the arithmetic mean of the estimation error [*ERROR (E)*]

$$\overline{e}_x = \frac{1}{M} \sum_{i=1}^{M} \overline{e}_{x_i} \tag{4-1}$$

- standard deviation δ

valuation of the stochastic features of estimation on the basis of the standard deviation of the estimation error [*NOISE (N)*]

$$\delta = \sqrt{\frac{1}{M-1} \sum_{i=1}^{M} \left(e_{x_i} - \overline{e}_x \right)^2} \tag{4-2}$$

The aim of the investigation was to prove the functionality of the fuzzy algorithm, but not a detailed analysis regarding to the choice or variation of the statistics. That is why the elements of the observer gain matrix were calculated based on these statistic values.

4.2 Design of the Hybrid Observer with Variable Gain Matrix

The fuzzy-based adaptation of the observer gain matrix is realized in form of the hybrid observer with variable gain. For that, the classical observer is combined with a fuzzy controller for the determination of the observer gain elements.

Basic rule

The fuzzy controller for the calculation of the gain elements is characterized by the following basic rule:

IF „*NOISE*" AND „*ERROR*"

THEN „*GAIN ELEMENT*" (4-3)

Linguistic variables

For the input and output variables of the fuzzy controller linguistic variables were defined presented in Table 4.1.

Table 4.1. Values of linguistic variables noise, error and gain matrix element

NOISE (N) (standard deviation of estimation error δ)	ERROR (E) (arithmetic mean of estimation error $\bar{e}_x$)	GAIN MATRIX ELEMENT (K) (absolute value)
small (S)	small (S)	small (S)
medium (M)	medium (M)	medium (M)
big (B)	big (B)	big (B)

Figure 4.1 demonstrates the chosen membership function of the linguistic variable gain matrix element.

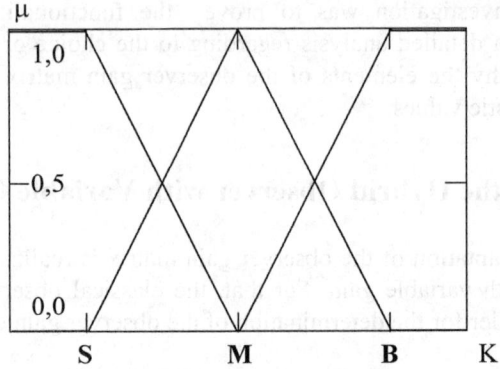

Fig. 4.1. Fuzzy set of the linguistic variable element of gain matrix K

The fuzzy sets for the input variables were created in the same way.

Rule base

Based on experiences single rules can be defined such as follows:

IF   *NOISE = 'Small'*   AND   *ERROR = 'Big'*
                    THEN   *GAIN MATRIX ELEMENT ='Big'*

IF   *NOISE = 'Big'*   AND   *ERROR = 'Small'*
                    THEN   *GAIN MATRIX ELEMENT ='Small'*

$$(4\text{-}4)$$

Starting from the basic rule and the experiences the complete rule base presented in Table 4.2 can be generated.

Table 4.2. Rule base to determine the gain matrix element K depending on the variables error E (arithmetic mean) and noise N (standard deviation) of the estimation error

K		N		
		S	M	B
	S	M	S	S
E	M	B	M	S
	B	B	B	M

4.3   Results of State Estimation

The test of the hybrid observer was carried out for the nodal pressure vessel model to simulate the deviation of collapsed level during a blow down experiment. The measurable output variable was falsified by a stochastic disturbance.

The fuzzy-supported observer described in Section 3 was expanded by an additional fuzzy controller for the calculation of the variable gain matrix elements.

The state equations (3-1) and (3-2) were used to represent the state space model. For the test the following variants of observers were compared:

- Variant V1:   Constant, *small* value of gain element, $\left| K\_V1_n \right| = 0.1$

- Variant V2:   Constant, *big* value of gain element, $\left| K\_V2_n \right| = 5.0$

- Variant V3:   *Variable* value of gain element, $\left| K\_V3_n \right| = \left[0.1 \ldots 5.0 \right]$
  (fuzzy-based determined depending on the variables *Noise* and *Error*)

Figure 4.2 represents a part of the time characteristics of the undisturbed and stochasticly disturbed measurable output variable of the collapsed level in the simulation period t= 0 ... 200 s.

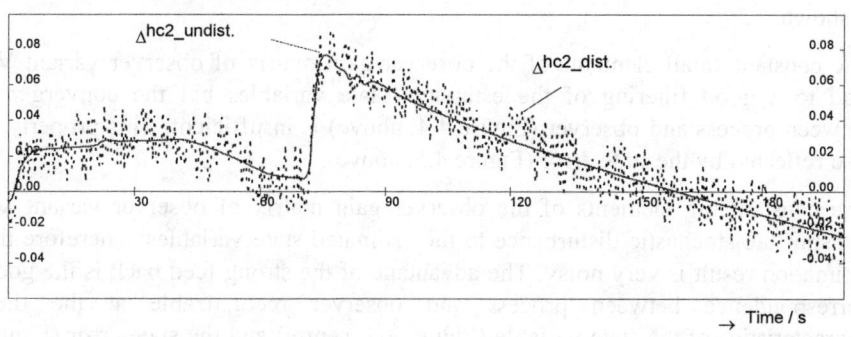

Fig. 4.2.   Time characteristics of the undisturbed (Δhc2_undist.) and the stochasticly disturbed (Δhc2_dist.) output variable of the process

During the simulation period $t = 0 \ldots 60$ s the heaters of the pressurizer are in operation. Therefore a deviation of the collapsed level (Δhc2) can be observed as a result of the evaporation. The heating elements switch off. The depressurization starts at t= 70 s. As a result of the evaporation a rapid increasing of the collapsed level can be indicated. The loss of water leads to a continuous decrease of the collapsed level during the rest of the simulation time.

Figure 4.3 represents the defined constant elements of the observer gain (K_V1, K_V2) and the time characteristics of the variable gain element calculated by the fuzzy controller (K_V3).

↑ Element of observer gain matrix

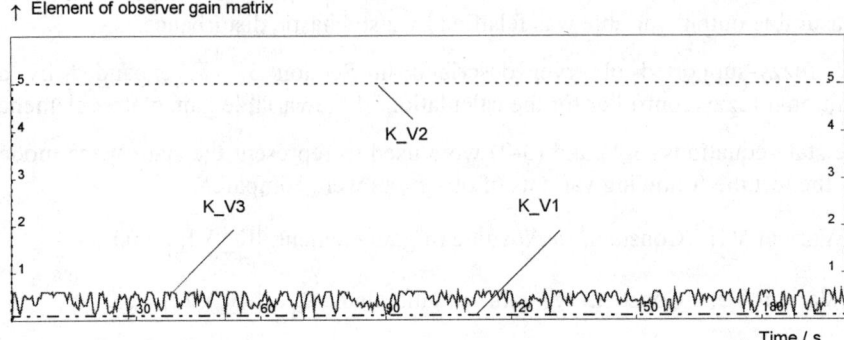

Fig. 4.3. Time characteristics of the constant and variable values of the observer gain matrix element K

Figure 4.4 demonstrates the estimation results of the different observer variants for the non-measurable state variable Δhc1 (deviation of collapsed level below the lower fitting of the measuring system). In Figure 4.5 the state error of this variable is shown.

The constant small elements of the observer gain matrix of observer variant V1 lead to a good filtering of the estimated state variables but the convergence between process and observer (Figure 4.4, above) is insufficient. This property is also reflected by the state error (Figure 4.5, above).

The constant big elements of the observer gain matrix of observer variant V2 transmit the stochastic disturbance to the estimated state variables. Therefore the estimation result is very noisy. The advantage of the strong feed back is the good correspondence between process and observer recognizable at the time characteristics of the state variable (Figure 4.4, centre) and the state error (Figure 4.5, centre).

With the help of the fuzzy algorithm of observer variant V3 elements of the observer gain matrix are calculated which vary within the range $\left|K\_V3_n\right| = \left[\,0.1 \ldots 0.8\,\right]$. The estimated state variable is filtered (Figure 4.4 below) and the state error is reduced (Figure 4.5 below).

By the fuzzy-based adaptation a good compromise between the target functions convergence and filtering could be found.

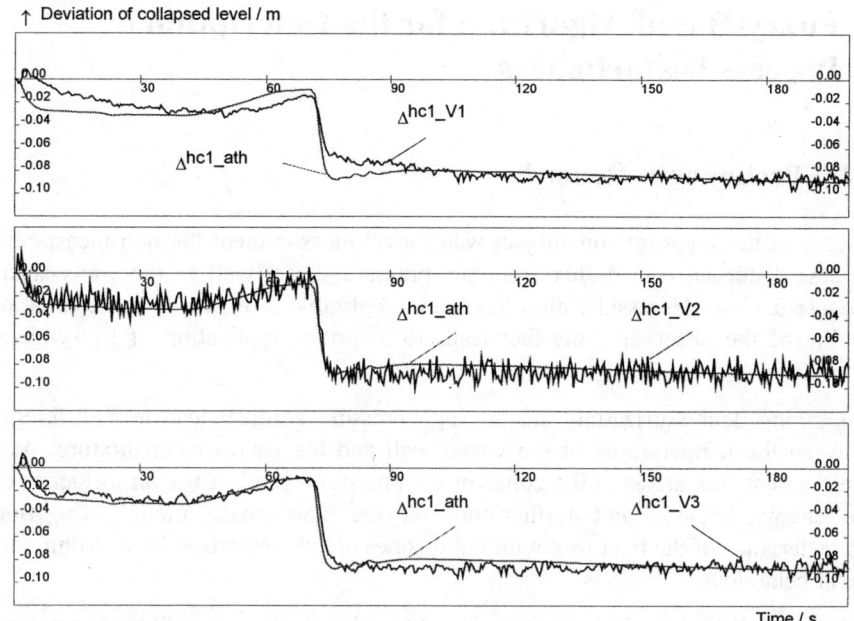

Fig. 4.4. Time characteristics of the estimated state variable Δhc1 for the different observer variants compared with the process behaviour Δhc1_ath

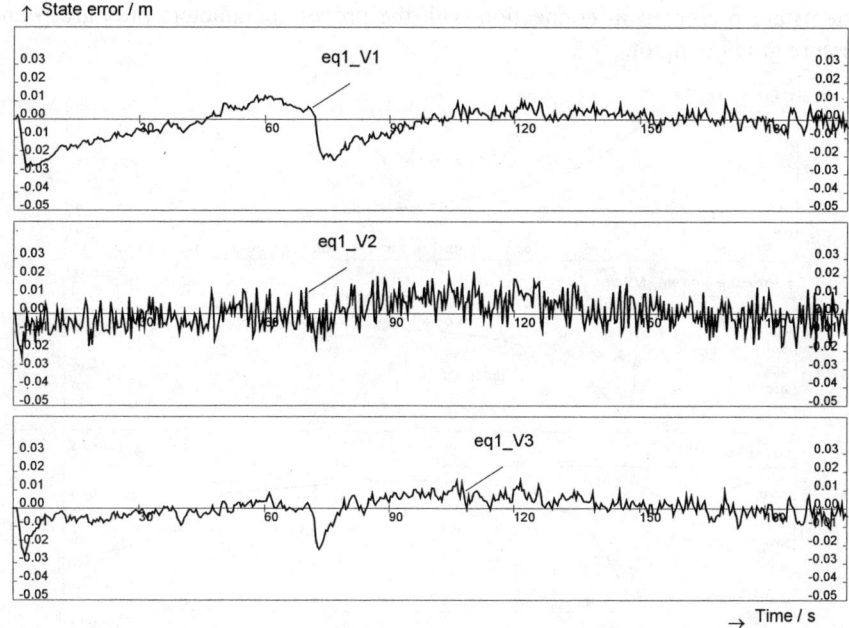

Fig. 4.5. Time characteristics of the state error eq1 of the non-measurable state variable for the different observer variants

28

5 Fuzzy-Based Algorithm for the Description of Process Disturbances

5.1 Preliminary Remarks

In case of the investigation subject water level measurement the non-measurable process disturbance heat flux from the pressure vessel wall to the water-steam mixture during depressurization has a non-negligible influence to the estimation quality of the observer. This fact leads to a further application of fuzzy-based algorithms.

The strong depressurization and the high pressure gradient lead to a difference between the temperatures of the vessel wall and the water-steam mixture. As a result a heat flux arises in the zones of the pressure vessel. It has an influence on the energy balance and furthermore on the state space model. The time characteristics of the heat flux within the zones are characterized by a strong non-linear behaviour.

In Figure 5.1 this property is shown for a blow down experiment at the pressurizer test facility. The post-calculated time characteristics of the heat flux $\dot{Q}_1$ (ATHLET-simulation) of zone 1 (zone below the lower fitting of the measuring system) are presented in connection with the process parameters pressure p and pressure gradient dp/dt.

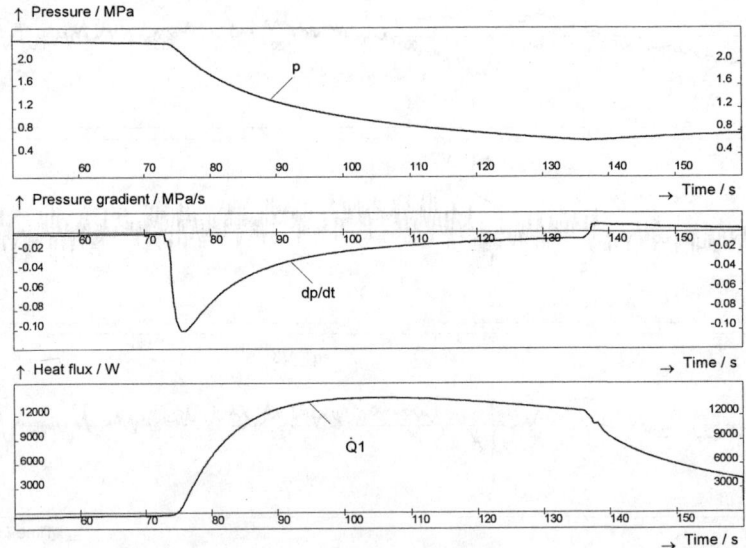

Fig. 5.1. Time characteristics of the parameters pressure (p), pressure gradient (dp/dt) and heat flux of zone 1 ($\dot{Q}_1$) for a depressurization within the period t= 75 s till 135 s

For including such process disturbances the model has to be expanded by a submodel for describing the heat flux. This can be realized in the following two ways:

Variant a)   analytical model for the description of the heat flux in form of a submodel which is implemented into the state space model of the Observer

Variant b)   knowledge-based model for the description of the heat flux by means of fuzzy logic and implementation into the existing Observer structure

The main emphasis was focused on the reproduction of the heat flux by fuzzy controllers (variant b)).

There, starting from experiments and ATHLET-code calculations, a comprehensive data basis is available for the generation of knowledge-based models. With the help of the fuzzy controller a fuzzy characteristic field was generated, which represents the dependence of the heat flux in the zone from the measurable variables pressure and pressure gradient. The developed fuzzy controller was implemented within a non-linear operating point observer.

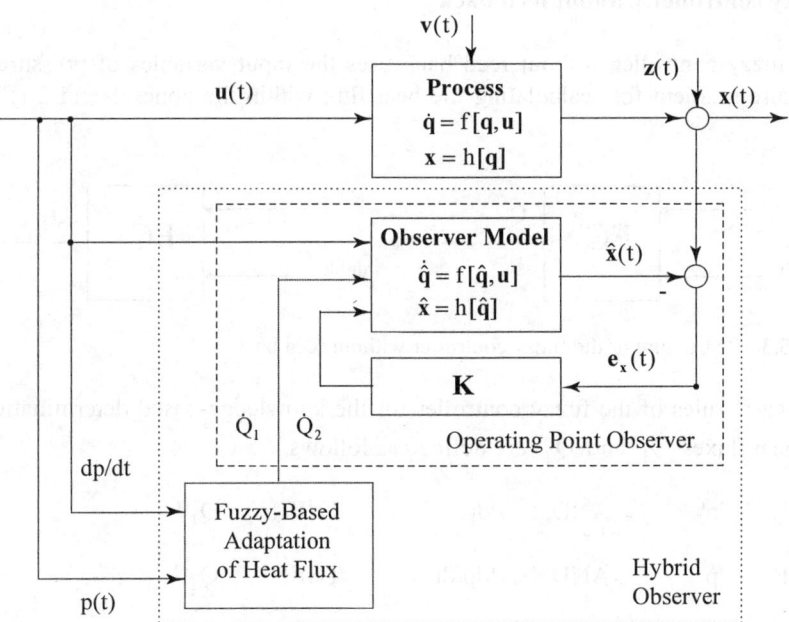

Fig. 5.2.   Hybrid observer based on a non-linear operating point observer in connection with a fuzzy-based adaptation of the input variable heat flux

The hybrid observer structure (Figure 5.2) consists of:

⇒ a non-linear process model for the Observer with the state variables $\hat{q}$ (estimated collapsed level hc1 and hc2) and the input variables **u** (pressure gradient dp/dt and heat flux $\dot{Q}_1$ and $\dot{Q}_2$),

⇒ a constant observer gain matrix **K,**

⇒ a fuzzy-based adaptation of the heat fluxes $\dot{Q}_1$ and $\dot{Q}_2$, which are interpreted as additional input variables of the observer.

5.2   Description of the Heat Flux by Fuzzy Algorithm

As input variables for the fuzzy controller in Figure 5.2 the process parameters pressure and pressure gradient were used. The parameter range was defined based on the experiments which were carried out at the test facility. The design of the fuzzy controllers was realized in the two zones of the described state space model of second order. Using the simulation system DynStar with Fuzzy-Shell the fuzzy controllers and the observer structures could be simulated [5].

Fuzzy controller without feed back

The fuzzy controller without feed back uses the input variables of pressure and pressure gradient for calculating the heat flux within the zones 1 and 2 (Figure 5.3).

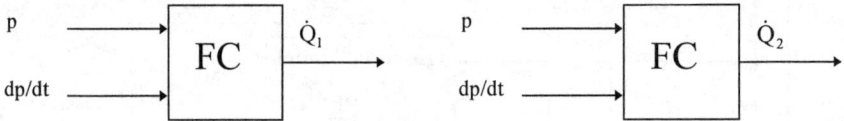

Fig. 5.3.   Structure of the fuzzy controller without feed back

The basic rules of the fuzzy controller for the knowledge-based determination of the heat fluxes $\dot{Q}_1$ and $\dot{Q}_2$ are defined as follows:

$$\text{IF} \quad \text{'p'} \quad \text{AND} \quad \text{'dp/dt'} \quad \text{THEN} \quad \text{'}\dot{Q}_1\text{'} \tag{5-1}$$

$$\text{IF} \quad \text{'p'} \quad \text{AND} \quad \text{'dp/dt'} \quad \text{THEN} \quad \text{'}\dot{Q}_2\text{'} \tag{5-2}$$

Extensive pre-investigations have shown that during depressurization the input variables repeatedly pass the same parameter ranges. Furthermore the time characteristics of the heat flux are characterized by high dynamics (Figure 5.4).

Therefore an exact reproduction of the heat flux is not achievable by a fuzzy controller of simple structure like in Figure 5.3.

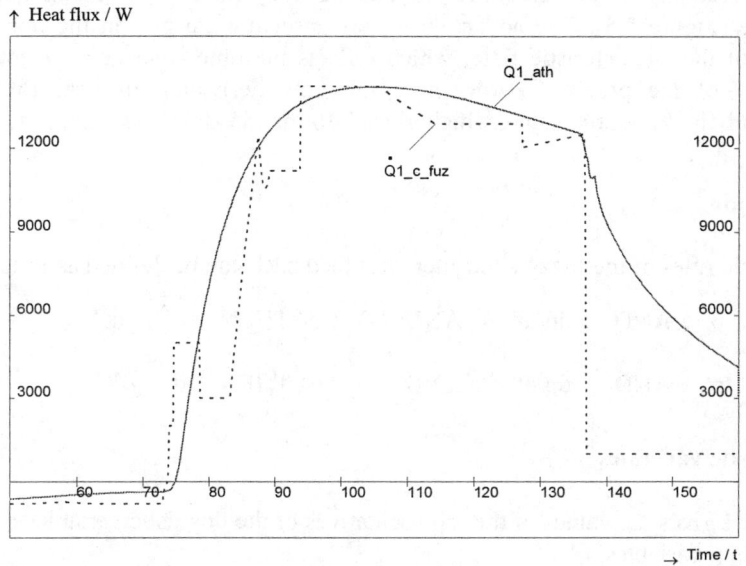

Fig. 5.4. Time characteristics of the heat flux within zone 1 calculated by the ATHLET-code and by the fuzzy controller without feed back

As a result of sensibility analysis concerning stability, time characteristics, fuzzy set optimization and parameter ranges the classical fuzzy controller was extended by a dynamic structure. The so-called the dynamic fuzzy controller [4] is characterized by a feed back.

Fuzzy controllers with feed back (dynamic fuzzy controller)

Pressure and pressure gradient are the global input and the heat fluxes the output variables of the controller with feed back as well.

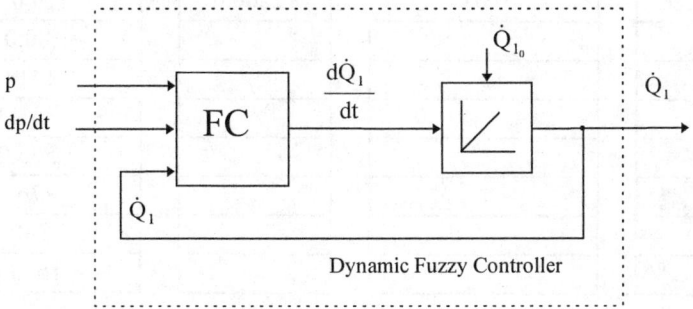

Fig. 5.5. Structure of the dynamic fuzzy controller

The structure of the classical controller is expanded by an integration of the time derivation of heat flux which is the direct output variable of the fuzzy controller and the feed back of the integrated value to the fuzzy controller as additional input variable (Figure 5.5). The advantage of this structure consists in the non-linear statics of the characteristic field, which reflects the time constants and the non-linearity of the process. Furthermore the time derivation of heat flux is a physically interpretable value, which gives information about the process state.

Basic rule

The basic rules of the fuzzy controller with feed back can be defined as follows:

$$\text{IF} \quad \text{'p'} \quad \text{AND} \quad \text{'dp/dt'} \quad \text{AND} \quad \text{'}\dot{Q}_1\text{'} \quad \text{THEN} \quad \text{'}d\dot{Q}_1/dt\text{'} \tag{5-3}$$

$$\text{IF} \quad \text{'p'} \quad \text{AND} \quad \text{'dp/dt'} \quad \text{AND} \quad \text{'}\dot{Q}_2\text{'} \quad \text{THEN} \quad \text{'}d\dot{Q}_2/dt\text{'} \tag{5-4}$$

Linguistic variables

Table 5.1 gives the values of the representatives of the linguistic variables of input and output variables.

Table 5.1. Representatives of the linguistic variables of the dynamic fuzzy controller for the determination of the heat flux $\dot{Q}_1$

Pressure [MPa]	Pressure gradient [MPa/s]	Heat flux zone 1 [W]	Time derivation of heat flux of zone 1 [W/s]
0.5	-0.1	-500.0	-2200.0
0.7	-0.05	0.0	-500.0
1.0	-0.017	2200.0	-140.0
2.4	0.0	7300.0	-70.0
	0.01	12200.0	0.0
			80.0
			141.0
			353.3
			565.0
			776.7
			1200.0
			1600.0

The definition of the linguistic variables, the sets and the generation of the rule base were realized on the basis of three experimental data sets, which are characterized by different initial pressure values.

As membership functions triangular functions were used illustrated on the fuzzy sets of the linguistic variable pressure (Figure 5.6).

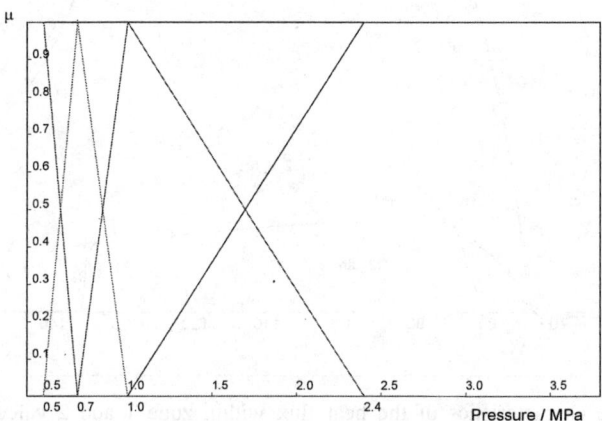

Fig 5.6. Membership function of the linguistic variable pressure

Rule base

The correlations between the linguistic variables are defined with the help of single rules which are summarized in form of the rule base. For illustration some selected single rules are presented:

IF 'p'='2.4' AND 'dp/dt'='0.0' AND '$\dot{Q}_1$'='-500.0' THEN '$d\dot{Q}_1/dt$'='0.0'

IF 'p'='1.0' AND 'dp/dt'='-0.05' AND '$\dot{Q}_1$'='2200.0' THEN '$d\dot{Q}_1/dt$'='776.7'

IF 'p'='0.7' AND 'dp/dt'='-0.017' AND '$\dot{Q}_1$'='7300.0' THEN '$d\dot{Q}_1/dt$'='-140.0'

IF 'p'='0.5' AND 'dp/dt'='0.01' AND '$\dot{Q}_1$'='12200.0' THEN '$d\dot{Q}_1/dt$'='-70.0'

$$(5-5)$$

The achieved quality of heat flux reproduction within zone 1 and zone 2 is demonstrated in Figure 5.7. The time characteristics of the heat flux determined by the dynamic fuzzy controller correspond with the process behaviour calculated by the ATHLET-simulation very well.

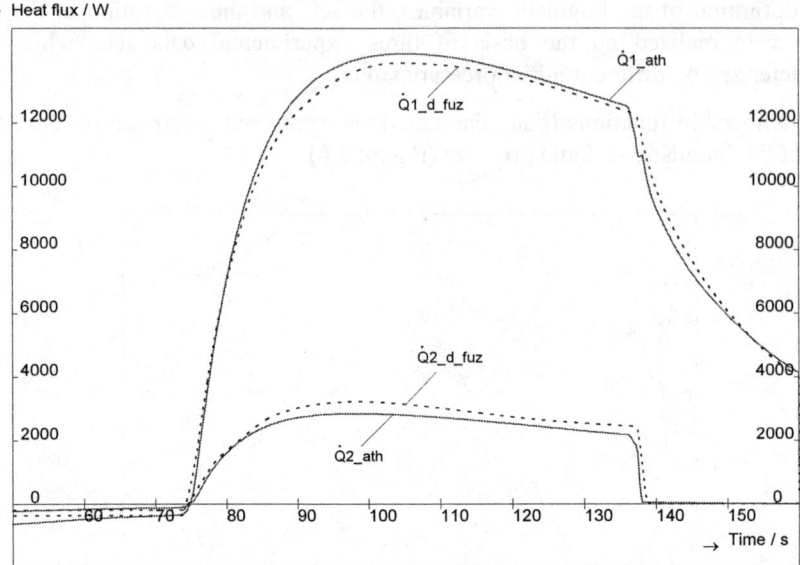

Fig. 5.7. Time characteristics of the heat flux within zone 1 and 2 calculated by the ATHLET-code and by the fuzzy controller with feed back

5.3 Results of State Estimation Using the Fuzzy-Based Description of Heat Flux

The high quality of heat flux reproduction by the fuzzy controller with feed back leads also to an improvement of the state estimation of the collapsed levels by the hybrid non-linear operating point observer. Figure 5.8 shows the estimation results for the state variables collapsed level of zones 1 and 2. The process behaviour calculated by the ATHLET-code was compared with the following observer variants:

– classical non-linear operating point observer (hc1_obs, hc2_obs)
(model without consideration of heat flux)

– Hybrid non-linear operating point observer (hc1_h_obs, hc2_h_obs)
(determination of heat flux by the fuzzy characteristic field)

Using the hybrid operating point observer it can be achieved a better estimation quality regarding the non-measurable state variable hc1_ath (collapsed level in zone 1) than by the classical observer. The investigation demonstrates that the dynamic behaviour of physical variables can be reproduced exactly by dynamic fuzzy controllers. The implementation of these fuzzy structures within classical MMMs leads to an improvement of the estimation quality.

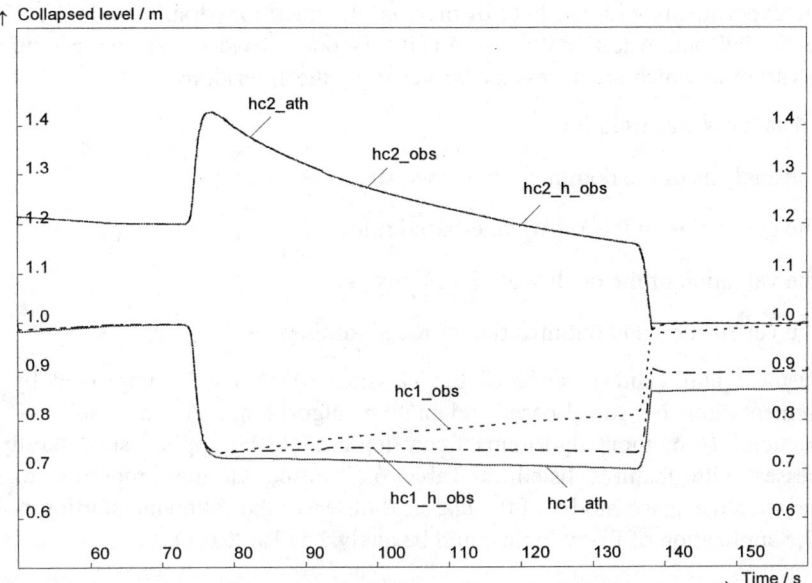

↑ Collapsed level / m

Fig. 5.8. Time characteristics of the state variables hc1, hc2 (collapsed level within zone 1 and 2) calculated by ATHLET (hc1_ath, hc2_ath), by the classical non-linear operating point Observer (hc1_obs, hc2_obs) and by the hybrid non-linear operating point observer (hc1_h_obs, hc2_h_obs)

6 Conclusions

For the realization of the process monitoring of safety-related processes characterized by

⇒ non-linearities

⇒ process disturbances

⇒ stochasticly disturbed measuring signals

a general conception was developed which combines analytical and knowledge-based components.

The design, test and proof of functionality of the developed algorithms were realized exemplarily by the determination of collapsed level within pressure vessels with water-steam mixture (pressurizer) during accidental transients as a result of negative pressure gradients (leak in the steam dome).

As preposition for the generation of the data and knowledge basis and for the development of the algorithms experiments at test facilities and post-calculations of the experiments with the help of the complex thermohydraulic code ATHLET were carried out. Additionally the ATHLET-code provided the non-measurable state variables which are necessary for verifying the algorithms.

The data basis was used for:

⇒ the analysis of the dominant process variables

⇒ the generation of the knowledge-based rules

⇒ the valuation of the quality of algorithms

⇒ the verification and optimization of the algorithms

Robustness and validity range of the classical MMMs were improved by the implementation of fuzzy-based adaptation algorithms. As a result of the reproducibility of their algorithms fuzzy logic could be applied successfully to processes with features listed in Table 6.1. Based on the properties of the linearized state space model of the classical observer the following starting-points for the application of fuzzy logic could be analyzed (Table 6.1).

Table 6.1. Starting-points for the application of fuzzy logic within classical observers

Feature of the classical observer model	Starting-point for the application of fuzzy logic
constant elements of model matrices of the observer model	adaptation of the elements of the model matrices to the changing process state
constant elements of observer gain matrix	adaptation of the elements of the observer gain matrix to the stochastic features of the measuring signals
process disturbances, the influence of which could not be considered by the model	knowledge-based description of process disturbances

The following variants for fuzzy-based parameter adaptations were investigated and applied in form of hybrid observers to the subject of investigation (pressure vessel with water-steam mixture):

⇒ Fuzzy-based algorithm for the adaptation of model matrices A_b, B_b (fuzzy-supported observer)

  – improvement of the estimation quality by means of the compensation of non-linearities

  – efficient application only for low-dimensional state space models because of the adaptation of each element by a separate fuzzy controller

⇒ Fuzzy-based algorithm for the adaptation of the observer gain matrix **K** (hybrid observer with variable gain)

    – improvement of robustness of the observer regarding stochasticly disturbed measuring signals by variable elements of the observer gain matrix

    – alternative of the use of Filter-algorithms

⇒ Fuzzy-based algorithm for the description of process disturbances (hybrid operating point observer with dynamic fuzzy controller)

    – description of the heat flux between the pressure vessel wall and the water-steam mixture during depressurization)

    – improvement of the estimation quality by additional process information

    – application recommendable if the analytical description is difficult or not applicable to the state space model

    – the fuzzy-based description requires a comprehensive data basis

The functionality and universality of the developed methods were proved.

The design of the hybrid observer is a contribution to the development of a new class of methods of analytical redundancy.

In principle the developed methods are applicable to other problems of the determination of safety-related non-measurable parameters as well.

For the application of fuzzy logic the following methodical conclusions could be found:

⇒ the reproduction of strong non-linear high-transient process variables with the help of fuzzy algorithms can be realized by dynamic fuzzy controllers

⇒ the influence of additional input variables on existing fuzzy controllers can be considered by a fuzzy-based interpolation of these Controllers

⇒ the characteristic field represents the correlation defined by the chosen degrees of freedom between the input and output variables of the fuzzy controller and is very suitable for the verification of the fuzzy algorithm

⇒ the map-based concept allows a real-time application of fuzzy algorithm

References

1 Adjallah, K., Non-linear Observers using Fuzzy Gain Adaptation, International Workshop on Fuzzy Technologies in Automation and Control, Duisburg 1994, 73-85

2 Akin, H.L.; Altin, V., Rule-based Fuzzy Logic Controller for a PWR-type nuclear power plant, IEEE Transactions on Nuclear Science, Vol. 38, No. 2, April 1991

3 Berger, M.; Jelali, M., Robust Model-Based Fuzzy Observer for an Inverted Pendulum, IEEE Transactions 1996, 118 - 122

4 Chaker, N.; Wagenknecht, M.; Fenske, A.; Hampel, R., Fuzzy Controller Structure Transformation, Proceedings of 3rd International FLINS Workshop, Antwerp, Belgium, September, 1998, 99-110

5 DynStar- Ein Simulationsprogramm für Automatisierungstechniker, Programm-beschreibung 12/97, HTWS Zittau/Görlitz (FH), 1997, Regelungstechnik 9, 10, 11/1984

6 Frank, P. M.; Kiupel, H.; Bux, O, Fuzzy Control of Steam Turbines, Fuzzy Sets and Systems 1994 (63), 319

7 Frank, P.M., Fuzzy Supervision - Einsatz der Fuzzy Logic in der Prozeßüberwachung, VDI-Bericht 1113, 181, VDI-Verlag, 1994

8 Hampel, R., Investigation in Utilization of Fuzzy Logic in NPP, ICM on Advanced Control and Instrumentation Systems in NPP, Espoo Helsinki, June 1994

9 Hampel, R., Meß- und Automatisierungstechnik zur Störfall-beherrschung - Methoden der Signalverarbeitung, Simulation und Verifikation, Abschlußbericht zum BMBF-Projekt 150 10 15, HTWS Zittau/Görlitz (FH), Januar 1999

10 Handschin, E.; u. a., Einsatz von Fuzzy-Reglern in der Kraftwerks-technik, VDI-Berichte, 151, 1113 VDI-Verlag 1994,

11 Holbert, K.E.; Sharif Heger; A.; Nahrul K. Alang-Rashid, Redundant Sensor Validation by Using Fuzzy Logic, Nuclear Science and Engineering 118, 54 - 64 (1994)

12 Iijima, T.; Nakajima, Y.; Sakurai, N., Fuzzy Logic Control System for Reactor Feedwater Control of The Fugen Nuclear Power Station, International Symposium on Instrumentation and Control, Tokyo, May 1992

13 Iijima, T.; Nakajima, Y.; Nishiwaki, Y., Application of Fuzzy Logic, Control System for Reactor Feed-Water Control, Proc. of 1st International FLINS Workshop, Mol, Belgium, 1994

14 Jung, C.H.; Ham, C.S.; Lee, K.L., A real time self tuning Fuzzy Controller for the steam generator through Scaling Factor Adjustment, Proc. of 1st International FLINS Workshop, Mol, Belgium, 1994

15 Kim, Byung-Kook; et. al., Fuzzy Logic utilization for the diagnosis of metallic loose part impact in nuclear power plant, Proc. of 2nd International FLINS Workshop, Mol, Belgium, 1996

16 King, P.J.; Burnham, K.J.; James, D.J.G., A combined rule-based and model-based adaptive control scheme, Proc. IEE International Conference CONTROL'94, Warwick, 1994

17 Liu, Z.; Ruan, D., Experiments of Fuzzy Logic Control on a Nuclear Research Reactor, Proc. of 2nd International FLINS Workshop, Mol, Belgium, 1996

18 Na, N.; Kwon, K.; Ham, C.; Bien, Z., A study on water level control of PWR steam generator at low power and the self-tuning of ist Fuzzy Controller, Proc. of 1st International FLINS Workshop, Mol, Belgium, 1994

19 Simutis,R.; Havlik,I.; Lübbert, A., A fuzzy-supported Extended Kalman Filter: a new approach to state estimation and prediction exemplified by alcohol formation in beer brewing, Journal of Biotechnology, 24 (1992), 211-234

20 Worlitz, F., Anwendung klassischer Verfahren und Fuzzy-Logik zur Verbesserung der hydrostatischen Höhenstandsmessung, Dissertation, Technische Hochschule Zittau, 1992

21 Yung Joon Hah, Byong-Whi Lee, Fuzzy Power Control Algorithm for a Pressurized Water Reactor, Nuclear Technology, Vol. 106, May 1994

2 A Set of Fuzzy Systems to Automate the Manual Procedures for Reactor Power Level Changes

Byung Soo Moon

Korea Atomic Energy Research Institute
P.O. Box105, Taeduk Science Town, Taejon, Korea 305-600
bsmoon@nanum.kaeri.re.kr

In this chapter, we describe a set of fuzzy systems which automate the manual part of procedures being used in the reactor operations for PWR-type nuclear power plants. One of the fuzzy systems evaluates the total power defect as a function of the boron concentration and the reactor power level. Others are used to compute the amount of step changes for the control rod as a function of the current rod position and the increment of the total power defect needed for the power level change. Each fuzzy system is either an exact or an approximate representation of the cubic spline interpolation for the corresponding function provided as graphs. The resulting set of fuzzy systems not only improves the efficiency in performing the curve readings but also reduces the interpolation errors involved.

1 Introduction

A fuzzy system can be considered as a representation of a function which interpolates the points provided in the form of fuzzy rules. We will show in the next section that one can design a fuzzy system so that it is an exact representation of the cubic spline interpolation function. In cases of smooth functions of two variables such as low order polynomials, the fuzzy system approximates the function very accurately with as few rules as of size 7×7. This justifies that a fuzzy system based on a set of linguistic rules or by a knowledge base can represent an unknown function very accurately and conveniently.

There are lots of works reported on applications of fuzzy algorithms for the reactor power control. Some use knowledge bases either to improve the power system stability [1] or to automate manual part of the operation procedures [2]. Others use fuzzy algorithms to automate the optimizing process for the reactor power control [3] or to replace the mathematical modelling problem [4]. All of these studies utilize fuzzy systems of some kind and the interpolation properties are essential for these applications.

Consider the problem of automating the manual part of procedures being used during the low power operations of PWR-type nuclear power plants. As an example, we can take the case where the reactor power level is at 25% and is to be reduced to 10%. The reactor operators can either insert the control rods or increase the boron concentration to reduce the power to the desired level. We will consider only the cases where the control rods are used to change the reactor power level.

The manual steps for changing the reactor power level can be summarized as follows. First, the operators compute the total power defect values corresponding to the two different reactor power levels by using graphs in Figure 1. The difference $\triangle$PCM of the two values is computed next. The third step is to read the integral rod worth corresponding to the current control rod position from the curve in Figure 2. This value is added to $\triangle$PCM computed earlier to obtain the integral rod worth corresponding to the new power level. Finally, the resulting integral rod worth is used to read from Figure 2 the new rod position corresponding to the desired reactor power level.

In the following, we will show that each of the above procedures can be performed by a fuzzy system and hence all the procedures by a set of fuzzy systems.

2 A Fuzzy System to Represent Spline Interpolation

Let $f(x,y)$ be a continuously differentiable function of two variables on the interval $[a,b] \times [c,d]$. We divide the intervals $[a,b]$ and $[c,d]$ into 2n subintervals and let $x_j = a + jh_x$, $y_j = c + jh_y$ with $h_x = \frac{b-a}{2n}$ and $h_y = \frac{d-c}{2n}$. Let $B_i(t)$'s be the cubic B-spline functions defined on $[t_{i-2}, t_{i+2}]$ for i=-1,0,1,...,2n+1, by

$$B_i(t) = \frac{1}{6h^3} \begin{cases} (t - t_{i-2})^3, & t \in [t_{i-2}, t_{i-1}] \\ h^3 + 3h^2(t - t_{i-1}) + 3h(t - t_{i-1})^2 - 3(t - t_{i-1})^3, & t \in [t_{i-1}, t_i] \\ h^3 + 3h^2(t_{i+1} - t) + 3h(t_{i+1} - t)^2 - 3(t_{i+1} - t)^3, & t \in [t_i, t_{i+1}] \\ (t_{i+2} - t)^3, & t \in [t_{i+1}, t_{i+2}] \\ 0, & \text{otherwise} \end{cases}$$

for i=-1,0,1,...,2n+1. Then the spline interpolation function for $f(x,y)$ can be written as $S(x,y) = \sum_{i,j=-1}^{2n+1} c_{ij} B_i(x) B_j(y)$, where $B_i(x)$'s are defined on $[x_{i-2}, x_{i+2}]$ and $B_j(y)$'s are defined on $[y_{j-2}, y_{j+2}]$.

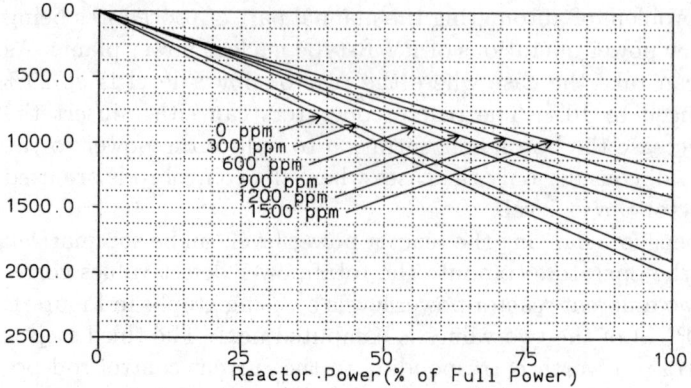

Fig. 1. Reactor power vs total power defect

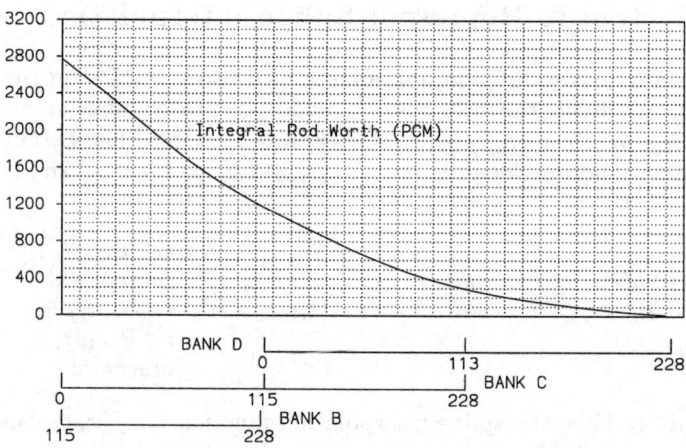

Fig. 2. Control rod position vs integral rod worth

For simplicity, we assume in the following that both of the intervals $[a, b]$ for x and $[c, d]$ for y are the same as $[-1, 1]$ and that $h_x = h_y = \frac{1}{n}$. Then the interpolation constraints that $S(x, y)$ must satisfy are:

$$\frac{\partial^2 S}{\partial y \partial x}(x_0, y_0) = \frac{\partial^2 f}{\partial y \partial x}(x_0, y_0)$$

$$\frac{\partial S}{\partial x}(x_0, y_j) = \frac{\partial f}{\partial x}(x_0, y_j), \qquad 0 \leq j \leq 2n$$

$$\frac{\partial^2 S}{\partial y \partial x}(x_0, y_{2n}) = \frac{\partial^2 f}{\partial y \partial x}(x_0, y_{2n})$$

$$\frac{\partial S}{\partial y}(x_i, y_0) = \frac{\partial f}{\partial y}(x_i, y_0), \qquad 0 \leq i \leq 2n$$

$$S(x_i, y_j) = f(x_i, y_j), \qquad 0 \leq j \leq 2n, \quad 0 \leq i \leq 2n$$

$$\frac{\partial S}{\partial y}(x_i, y_{2n}) = \frac{\partial f}{\partial y}(x_i, y_{2n}), \qquad 0 \leq i \leq 2n$$

$$\frac{\partial^2 S}{\partial y \partial x}(x_{2n}, y_0) = \frac{\partial^2 f}{\partial y \partial x}(x_{2n}, y_0)$$

$$\frac{\partial S}{\partial x}(x_{2n}, y_j) = \frac{\partial f}{\partial x}(x_{2n}, y_j), \qquad 0 \leq j \leq 2n$$

$$\frac{\partial^2 S}{\partial y \partial x}(x_{2n}, y_{2n}) = \frac{\partial^2 f}{\partial y \partial x}(x_{2n}, y_{2n})$$

The above equations form a set of $(2n+3)^2$ linear equations with $(2n+3)^2$ unknowns c_{ij}, $-1 \leq i, j \leq 2n + 1$. When the equations are ordered as in the above, the coefficient matrix is invertible without any pivoting algorithm even for a fairly large n. In the following, we use these c_{ij}'s to set up a fuzzy system for representing the cubic spline interpolation function $S(x, y) = \sum_{i,j=-1}^{2n+1} c_{ij} B_i(x) B_j(y)$.

(1) Fuzzy Sets for Input Fuzzification

Let $x_i = -1 + ih$, $y_j = -1 + jh$ with $h = \frac{1}{n}$ and define $B_i(x)$ and $B_j(y)$ as described above. Then the B-spline function $B_i(x)$ has support $[x_{i-2}, x_{i+2}]$ with the maximum value of $\frac{2}{3}$ at x_i, and $\frac{1}{6}$ at both x_{i-1} and x_{i+1}. We take these functions $B_i(x)$ and $B_i(y)$ for $i = -1, 0, 1, \ldots, 2n + 1$ as fuzzy sets for input variables x and y respectively. If t is an arbitrary point in $[-1, 1]$, then $t \in [x_{k-1}, x_k]$ for some $1 \leq k \leq 2n + 1$. If t is fuzzified by $\{B_i(x) | i = -1, 0, 1, \ldots, 2n + 1\}$ to obtain $\lambda_i = B_i(t)$, then we have $\sum_{i=-1}^{2n+1} \lambda_i = 1$ since $\sum_{i=-1}^{2n+1} B_i(x) = 1$ for all x. Note that $\lambda_i = 0$ for $i < k - 2$ or for $i \geq k + 2$

and hence t has nonzero membership in $B_i(x)$ only when $k - 2 \leq i \leq k + 1$ so that $\sum_{i=k-2}^{k+1} \lambda_i = 1$.

(2) Generation of Fuzzy Rules

We sort the $(2n+3)^2$ cubic spline interpolation coefficients c_{ij} in an increasing order and delete the duplicate ones, i.e., delete c_{kl} for example when $|c_{ij} - c_{kl}| \leq 10^{-7}$. Assign the ordinal number 1 to the smallest c_{ij} and 2 to the second smallest, and so forth. The largest c_{ij} will have ordinal number N which is less than or equal to $(2n + 3)^2$. We then form the rule matrix R so that (i, j)th entry R_{ij} is the ordinal (fuzzy set) number corresponding to the coefficient c_{ij}. For the fuzzy inferences, we use the Larsen's product rule so that if x belongs to $B_i(x)$ with membership value λ_i and y belongs to $B_j(y)$ with μ_j, then (x, y) belongs to the set R_{ij} with membership value $\lambda_i \mu_j$.

(3) Output Fuzzy Sets

Let $\{t_k \mid k = 1, 2, \ldots, N\}$ be the sorted array of c_{ij}'s. For each k, define an output fuzzy set T_k to be triangular set (spike function) whose support is $[t_{k-1}, t_{k+1}]$ and center of support is at t_k. For the first fuzzy set, we use an arbitrary point t_0 such that $t_0 < t_1$ and similarly for the last set, a point t_{N+1} with $t_{N+1} > t_N$ is used. Note that if T_k is the output fuzzy set corresponding to the (i, j)th entry of the rule table, then the center of support for T_k is c_{ij}, i.e., $t_k = c_{ij}$.

(4) Defuzzification

For the defuzzification of the output, we use the center area defuzzification method [5]. When 16 of the fuzzy sets $T_{k(ij)}$'s with nonzero weights $\nu_{k(ij)}$'s are the output from the fuzzy inferences, we compute

$$\frac{\sum_{ij} \nu_{k(ij)} Center(T_{k(ij)})}{\sum_{ij} \nu_{k(ij)}} = \sum_{ij} \nu_{k(ij)} t_{k(ij)}$$

where the sums range for only 16 nonzero ν_{ij}'s and $Center(T_{k(ij)})$ is the center of support $t_{k(ij)} = c_{ij}$ for the fuzzy set $T_{k(ij)}$. Note that the denominator $\sum_{ij} \nu_{k(ij)}$ is equal to $(\sum_{i=-1}^{2n+1} \lambda_i)(\sum_{j=-1}^{2n+1} \mu_j)$ which is 1 since both factors are 1.

The following theorem proves that the fuzzy system described above is an exact representation of the cubic spline interpolation function for a continuously differentiable function of two variables.

Theorem 1 [6] Let $f(x,y)$ be a continuously differentiable function and let $S(x,y) = \sum_{i,j=-1}^{2n+1} c_{ij} B_i(x) B_j(y)$ be the spline interpolation of $f(x,y)$ at $\{(x_i, y_j) \mid 0 \le i, j \le 2n+1\}$, where $x_i = -1 + ih$, $y_j = -1 + kh$, and $B_i(x)$'s and $B_j(y)$'s are the cubic B-splines with supports $[x_{i-2}, x_{i+2}]$, $[y_{j-2}, y_{j+2}]$ respectively. If F is the fuzzy system designed as above using the coefficients c_{ij}'s, then the output $F(x,y)$ of the fuzzy system at $(x,y) \in [-1,1] \times [-1,1]$ is identical to the value of the spline interpolation function, i.e., $F(x,y) = S(x,y)$.

Theorem 2 [6] Let $f(x,y)$ and $g(x,y)$ be continuously differentiable functions in x, y and let $\sum_{i,j=-1}^{2n+1} c_{ij} B_i(x) B_j(y)$, $\sum_{ij=-1}^{2n+1} d_{ij} B_i(x) B_j(y)$ be the corresponding spline interpolation functions. If $\{(i(n), j(n)) \mid n = 1, 2, \ldots, N\}$ are the indices of c_{ij}'s when they are sorted in an increasing order with duplicate ones deleted and if $\{(k(n), l(n)) \mid n = 1, 2, \ldots, N\}$ are the corresponding indices for d_{kl}'s, then the rule table for $f(x,y)$ is identical to that of $g(x,y)$ if and only if $k(n) = i(n)$, $l(n) = j(n)$, for all $n = 1, 2, \ldots, N$.

3   Examples of Fuzzy Systems

In this section, we describe three examples of the fuzzy system designed to represent the cubic spline interpolation of polynomials in x and y. We compare the evaluation results with those of a fuzzy system based on the Lagrangian interpolation using spike functions as the input fuzzy sets. Double precision calculations are used for all cases so that the calculation error is relatively minor compared to the intrinsic system error. For all of the cases in the following examples, we evaluated the fuzzy systems at 10,000 points in $[-1, 1]$ and computed the average and the maximum of the absolute errors.

Example 1. Fuzzy system for $P(x,y) = xy$
When the procedure described in the previous section is applied to $P(x,y) = xy$ with n=2, we obtain 7×7 fuzzy rules in Table 1, with the centers of support for the output fuzzy sets at -2.25, -1.50, -1.00, -0.75, -0.50, -0.25, 0.00, 0.25, 0.50, 0.75, 1.00, 1.50, 2.25. The maximum evaluation error is found to be of order 10^{-15} for both fuzzy systems.

Table 1. Fuzzy rules to compute $P(x, y) = xy$

		\(y\)					
		1	2	3	4	5	6	7
	1	13	12	10	7	4	2	1
	2	12	11	9	7	5	3	2
	3	10	9	8	7	6	5	4
x	4	7	7	7	7	7	7	7
	5	4	5	6	7	8	9	10
	6	2	3	5	7	9	11	12
	7	1	2	4	7	10	12	13

Example 2. Fuzzy system for $P(x, y) = x^4 + y^4$

The fuzzy rules obtained in this case for n=2 are as shown in Table 2 and the centers of support for the output fuzzy sets are at -.083333, -.020833, .041666, .479166, .541666, 1.041666, 3.916666, 3.979166, 4.479166, 7.916666. Table 3 shows a summary of the evaluation errors at 10,000 points, including those by a fuzzy system using triangular fuzzy sets for n=2, n=5, n=10.

Table 2. Fuzzy rules to compute $P(x, y) = x^4 + y^4$

		\(y\)					
		1	2	3	4	5	6	7
	1	10	9	7	8	7	9	10
	2	9	6	4	5	4	6	9
	3	7	4	1	2	1	4	7
x	4	8	5	2	3	2	5	8
	5	7	4	1	2	1	4	7
	6	9	6	4	5	4	6	9
	7	10	9	7	8	7	9	10

Table 3. Comparison of evaluation errors (Spline vs Triangle)

n	Maximum Error		Average Error		No of Rules
	Spline	Triangle	Spline	Triangle	(Triangle)
2	0.007812	0.434699	0.004166	0.162233	49(25)
5	0.000196	0.097400	0.000107	0.026293	169(121)
10	0.000012	0.026187	0.000007	0.006393	529(441)

Example 3. Fuzzy system for $P(x,y) = x^3 - x^2 y$

The fuzzy rules obtained in this case for n=2 are shown in Table 4. A summary of the evaluation results by the fuzzy system at 10,000 points, along with those by a fuzzy system using triangular fuzzy sets for n=2, n=5, n=10 are shown in Table 5.

Table 4. Fuzzy rules to compute $P(x,y) = x^3 - x^2 y$

					y			
		1	2	3	4	5	6	7
	1	23	9	6	4	3	2	1
	2	25	22	12	10	8	7	5
	3	23	22	20	18	16	14	13
x	4	15	16	17	18	19	20	21
	5	23	22	20	18	16	14	13
	6	31	29	28	26	24	14	11
	7	35	34	33	32	30	27	13

Table 5. Comparison of evaluation errors (Spline vs Triangle)

	Maximum Error		Average Error		No of Rules
n	Spline	Triangle	Spline	Triangle	(Triangle)
2	0.15×10^{-7}	.2034	$.21 \times 10^{-8}$	.0632	49(25)
5	0.18×10^{-7}	.0370	$.14 \times 10^{-8}$	.0102	169(121)
10	0.19×10^{-7}	.0093	$.14 \times 10^{-8}$	.0025	529(441)

4   Fuzzy System to Compute Total Power Defect

In this section, we describe a fuzzy system which evaluates the total power defect shown in Figure 1 as a function of two variables; the boron concentration Bor and the reactor power $Pold$ or $Pnew$ as percent of full power. We use cubic B-splines as the input fuzzy sets for the boron concentration, with centers of support x_i, for $0 \le i \le 5$ at 0, 300, 600, 900, 1200, 1500 ppm. To cover the range of values for $Bor \in [0, 1500]$, we need two more fuzzy sets one at each end so that $x_{-1} = -300$, $x_6 = 1800$ are necessary. Similarly for the reactor power, we use eight B-splines with centers of support y_j at -20, 0, 20, 40, 60, 80, 100, 120.

When the cubic interpolation function is computed to interpolate the curves in Figure 1 at (x_i, y_j) for $0 \le i, j \le 5$ and the derivatives at the boundary points, we obtain $\sum_{i,j=-1}^{6} c_{ij} B_i(x) B_j(y)$. It turns out that there are 64 different values of c_{ij}'s when the double precision is used for the calculation. There are, however, 33 different values when only 4 digit accuracy is maintained. We take them as centers of support for the output fuzzy sets. When these values are sorted in an increasing order, we find that they are

-0.459 -0.398 -0.307 -0.257 0.0006 0.265 0.390 0.432 0.489
0.556 0.615 0.692 0.782 0.869 0.906 0.988 0.107 1.104
1.218 1.301 1.405 1.454 1.542 1.603 1.651 1.729 1.791
1.952 1.998 2.197 2.309 2.383 2.596

We use the ordinal numbers of these c_{ij}'s as the output fuzzy sets and form a rule table as shown in Table 6. With the center area defuzzification, the fuzzy system thus formed represents an approximation of the spline interpolation.

Table 6. Fuzzy rules to compute total power defect

		\multicolumn{8}{c}{p}
		1	2	3	4	5	6	7	8
	1	1	2	2	3	3	4	4	4
	2	5	5	5	5	5	5	5	5
	3	9	8	7	6	6	6	6	6
Bor	4	15	13	13	11	11	10	9	8
	5	21	19	19	15	15	13	13	12
	6	27	24	24	19	19	17	16	14
	7	30	28	29	22	21	20	19	18
	8	33	31	32	26	25	23	21	20

5   Fuzzy System to Evaluate Integral Rod Worth

In this section, we descibe two fuzzy systems which represent the cubic spline interpolations of the curve shown in Figure 2 and of its inverse curve. First, we consider the curve itself which represents the integral rod worth as a function of the rod position. The total number of steps for the control rods ranges from 0 to 343, and hence we use interval $[0, 350]$ for the independent variable. We divide the interval into 10 subintervals and interpolate the curve at the 11 points including the deivative estimates at the two end points.

With a scaled variable x for the rod position and y for the rod worth, the interpolation function turns out to be $g(x) = \sum_{i=1}^{13} c_i B_i(x)$, where $(c_i) = (1.175, .989, .804, .596, .45, .332, .218, .132, .077, .043, .014, .0, -.014)$. Since c_j's are in a decreasing order, we take these to be centers of support for the $(14 - j)$th output fuzzy sets and form a set of 13 triangular fuzzy sets. Then the fuzzy rules "if rod is x_i, then $rod\ worth$ is '13-j'" will produce the same value as $g(rod)$ by the center area defuzzification method. A similar method to generate the fuzzy system for the inverse of the curve is omitted.

6   Fuzzy System to Compute the Sum and the Difference

In this section, we describe two fuzzy systems. One is to compute the difference $\Delta T = T_1 - T_0$, where T_0 and T_1 are the total power defect values corresponding to the current reactor power $Pold$ and the target reactor power $Pnew$ respectively. The other is to compute the sum $Tnew = \Delta T + T_2$, where T_2 is the the rod worth at the current rod position.

Table 7. Fuzzy rules to compute $\delta(x, y) = x - y$

					y				
		1	2	3	4	5	6	7	8
	1	8	9	10	11	12	13	14	15
	2	7	8	9	10	11	12	13	14
	3	6	7	8	9	10	11	12	13
x	4	5	6	7	8	9	10	11	12
	5	4	5	6	7	8	9	10	11
	6	3	4	5	6	7	8	9	10
	7	2	3	4	5	6	7	8	9
	8	1	2	3	4	5	6	7	8

For the difference, we use the fuzzy system representation of a cubic spline interpolation of $\delta(x) = x - y$ on the interval $[0, 1] \times [0, 1]$ using eight fuzzy sets for both x and y. When the spine interpolation coefficients are computed, approximated within 1.E-6, and sorted in an increasing order, we obtain 15 values $\frac{j-8}{30}, j = 1, 2, \ldots, 15$. Taking these as centers of support for the output fuzzy sets, we obtain the rule table shown in Table 7.

Similarly, the fuzzy system for $\sigma(x, y) = x + y$ is generated based on its cubic spline interpolation on $[0, 1] \times [0, 1]$ with eight fuzzy sets each and with an 8×8 rule table. The output fuzzy sets are triangular fuzzy sets with centers of support at $\frac{j-8}{15}, j = 1, 2, \ldots, 15$ with the fuzzy rules given as $r_{ij} = i + j - 1$ in this case.

7 Simulation Results

As shown in Figure 3, the input to the fuzzy systems are the current boron concentration Bor, the initial reactor power $Pold$, the currnt rod position $Rold$, and the desired reactor power $Pnew$. The final output from the set of fuzzy systems is the new rod position $Rnew$. The first fuzzy system f computes T_0 from input Bor and $Pold$, and also computes T_1 from Bor and $Pnew$. The computed values T_0 and T_1 are input to the second fuzzy system $\delta(x, y) = x - y$ to compute $\Delta T = T_1 - T_0$. The next fuzzy system g evaluates the rod worth T_2 corresponding to the current rod position $Rold$. The compted value T_2 and ΔT computed previously are summed by the fuzzy system $\sigma(x, y) = x + y$ and is input to the last fuzzy system g^{-1} to compute the new rod position $Rnew$.

With initial conditions of $Bor = 900$ ppm, $Pold = 25\%$, and $Rold = 190$ step, we computed the control rod positions (Bank D) corresponding to

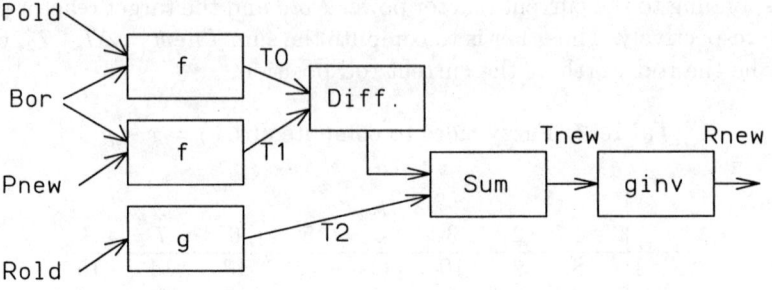

Fig. 3. Layout of the fuzzy systems

possible ranges of reactor power. The result is as shown in Figure 4. Note that the reactor power level can be raised to about 30% with all the control rods in Bank D pulled out fully at 228 step. To the minimum, however, the reactor power level can be decreased to 0% without hitting the rod insertion limit.

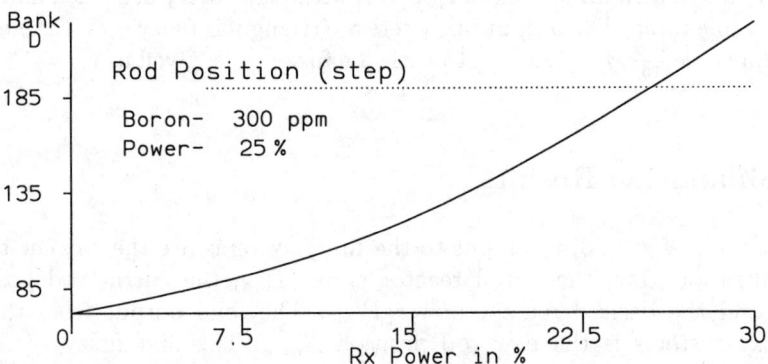

Fig. 4. Simulation results

8   Conclusion

Each of the fuzzy systems we designed is an exact or an approximate representation of the cubic spline interpolation function. The approximation is done in such a way that the coefficients that are close to each other are replaced by one representative value. Thus, the interpolation error is either the same as or close to that of the cubic spline interpolation $O(h^4)$. Through this approximation process, we eliminated unnecessarily many output fuzzy sets and obtained a reasonbly small number of output fuzzy sets. The set of fuzzy systems we designed not only automate the manual procedures being used in the power plants but also reduce the errors involved in reading the curves involved.

References

1. Z. Liu and D. Ruan, Experiments of Fuzzy Logic Control on a Nuclear Research Reactor, *Intelligent Systems and Soft Computing for Nuclear Science and Industry; Proc. 2nd Int. FLINS Workshop*, (1996) 336-348.
2. Y. J. Hah and B. W. Lee, Fuzzy Power Control Algorithm for a Pressurized Water Reactor, *Nuclear Technoloy*, 106(2) (1994) 242-253.
3. P. Ramaswamy, R. M. Edwards, K. Y. Lee, A Fuzzy Logic Controller Design for Nuclear Power Plant, *Proc. 12th triennial World Congress of the Int. Federation of Automatic Control*, Vol.4 Applications II (1994) 703-706.
4. C. Lin and H. W. Lin, Application of Fuzzy Logic Controller to Load-Follow Operations in Pressurized Water Reactors, *J. Nuclear Science and Technoloy*, 31(5) (1994) 407-419.
5. P. Wang, et al., Constructive Theory for Fuzzy Systems, *Fuzzy Sets and Systems*, 88 (1997) 195-203.
6. B. S. Moon, A Practical Algorithm for Representing Polynomials of Two Variables by Fuzzy Systems with Accuracy $O(h^4)$, *Fuzzy Sets and Systems*, To Appear.

3 A Fuzzy Controller for the Real Time Supervision of Nuclear Power Reactors

Mohand Si Fodil[1], Franois Guly[1], Patrick Siarry[1], and
Jean Luc Tyran[2]

[1] Ecole Centrale Paris, Grande Voie des Vignes
    92295 Chtenay-Malabry, France
    sifodilm@cti.ecp.fr
[2] EDF, Direction des Etudes et Recherches, Dpt. CCC
    6, Quai Watier, 78401 Chatou, France
    Jean-Luc.Tyran@der.edfgdf.fr

We describe the development of a real time fuzzy controller aimed at an adequate core axial power distribution inside a pressurized water reactor (PWR). The system has been implemented in simulation using the EDF model on Matlab-Simulink on a Sun workstation. Some experiments tests at different power levels have been performed: several low levels modeling falls in the network consumption and tests modeling falls in the consumption followed by a recovery of the consumption. The results of all these experiments will be discussed.

1   Introduction

Even if most of the tasks are automated in PWR (Pressurized Water Reactor) type nuclear plants, the power axial-offset is still manually controlled. The operation is too complex and requires a permanent human expert evaluation.

The reactivity inside the reactor has to be maintained constant [1]. The neutronic flux is not homogeneous in every place of the reactor. It results principally from the moderating effect of the temperature [2]. Power repartition imbalance between the higher part and the lower part of the reactor has to be evaluated. The power axial-offset $\triangle I$[1] is defined as follows:

$$\triangle I = 100 * \frac{(P_{high} - P_{low})}{(P_{high} + P_{low})_{100}} \,.$$

(1)

$(P_{high} + P_{low})_{100}$ is the power in the heart of the reactor when the electric power is equal to 100 % of the nominal power. The aim of the manual control is to maintain the automatic device as close as possible to the reference. Our aim is to make the fuzzy logic controller work better than the human operator or at least as efficient.

---

[1] subsequently expressed by X_1 in the figures

2 Control Methods

2.1 Short-Term Regulation

Control rods are used. They are made with an absorbent solid material. The insertion[2] induces the power axial-offset. The rods should not be inserted more than a low limit.

2.2 Medium and Long-Term Regulation

They allow a control of the reactivity. Boron concentration in the reactor coolant primary system is modified. Boron is a neutron-absorbing element. According to the situation, boric acid or water will be injected in the circuit. The primary circuit volume being constant, the use of boric acid leads to liquid wastes. The reprocessing of waste volumes is very expensive. Manual operations are performed according to some directives [1]: because of the great difficulty to automate this task by using classical controllers, the fuzzy logic seems well suited; it is easier to describe the operator's behavior than to fully model the process.

3 Working Method

The following work has been done on simulation with a simplified model of the EDF PWR 900 MW power plant [3]. The fuzzy controller has been created and integrated using Matlab-Simulink on a Sun workstation.

Fuzzy logic is used in this approach of the power axial-offset control. The membership functions are parameterized because the reference ΔI and the rods insertion limits are varying according to the electrical power.

The method is based on the regulation of the power axial-offset ΔI, and of the position of the rods, in the same time and with a suitable command. The command to use depends on ΔI and posR according to the Mamdani rules [4]:

IF *condition 1* THEN *action 1*
IF *condition 2* THEN *action 2*
$$\vdots \qquad\qquad \vdots \qquad\qquad \vdots$$
IF *condition n* THEN *action n*

Membership functions have been associated to the input variables ΔI and posR (see Figure 1 and Figure 2).

So the system is multi-variable. Total extraction of the rods during a long time will be so avoided.

Human expertise instructions state for the following behavior [5,6]:

If posR *high* then *dilute*

---

[2] the R rods position posR is subsequently expressed by X_2 in the figures

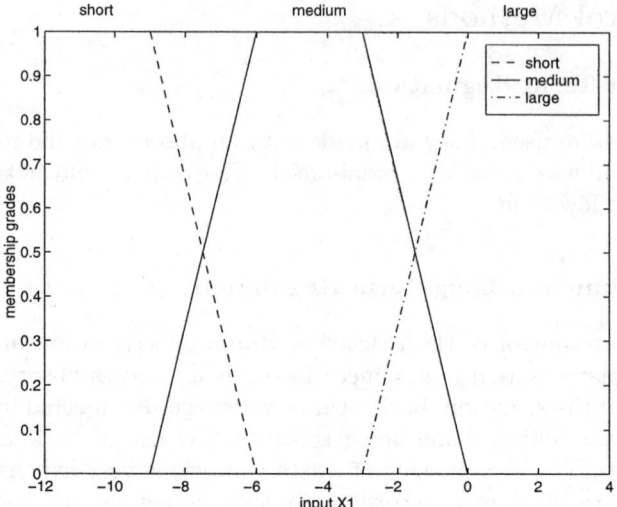

Fig. 1. $\triangle$I input membership functions

If posR *middle* then *do nothing*
If posR *low* then *boricate*
If $\triangle$I *large* then *dilute*
If $\triangle$I *medium* then *do nothing*
If $\triangle$I *short* then *boricate*
If $\triangle$I *large* and posR *low* then *do nothing*
The $\triangle$I and posR values between 0 and 1 are obtained by linear interpolation.

$$\mu_{short}\,(\triangle\mathrm{I}) = 1, \text{ if } \triangle I < -9$$
$$\mu_{medium}(\triangle\mathrm{I}) = 1, \text{ if -6 } \text{¡} \triangle I < -3$$
$$\mu_{large}(\triangle\mathrm{I}) = 1, \text{ if } \triangle I > 0$$
$$\mu_{low}(posR) = 1, \text{if } posR < 189$$
$$\mu_{middle}(posR) = 1, \text{if } 201 < posR < 213$$
$$\mu_{high}(posR) = 1, \text{if } posR > 225$$

4   The Fuzzy Logic Controller

The fuzzy logic controller (FLC) is associated to the system in a closed-loop way as shown in Figure 3.
The device is divided into four blocks (see the "action chain" in Figure 4).

4.1   Fuzzification

In the first block "Fuzzification," the $\triangle$I and posR values are compared with corresponding thresholds, and then membership degrees are associated to the

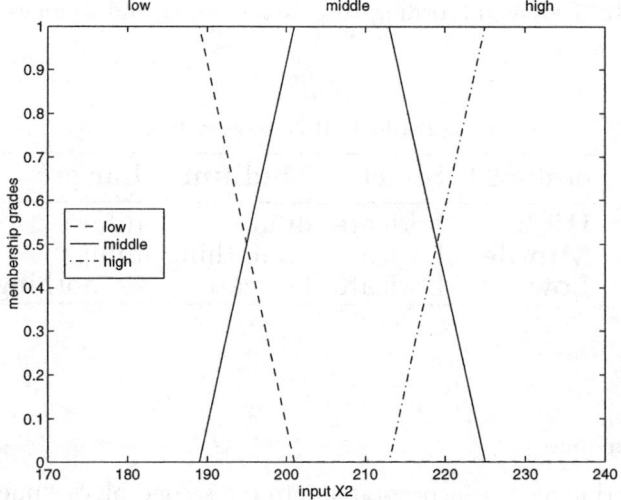

Fig. 2. posR input membership functions

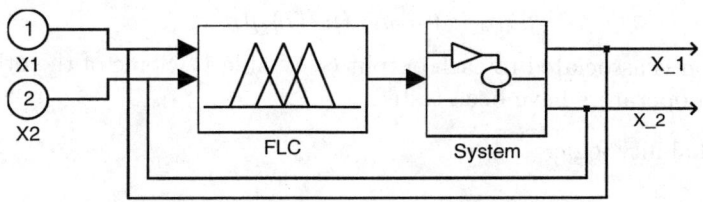

Fig. 3. Closed loop of the FLC

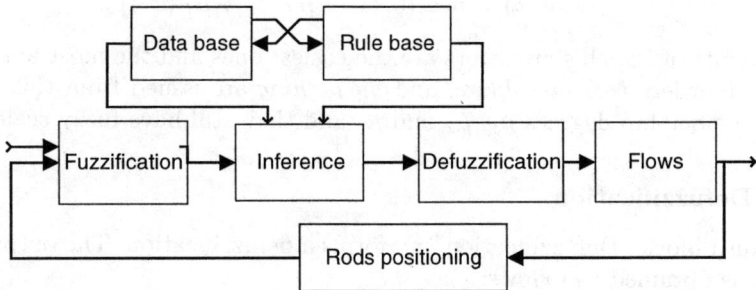

Fig. 4. Blocks of the FLC

membership functions. Fuzzy variables are issued from this block (see Figure 1 and Figure 2). Then it is possible to draw these rules in a plane ($\triangle$I, posR) divided into 9 areas, according to 3 zones of $\triangle I$ and 3 zones of posR (see Table 1).

Table 1. Rule base matrix

posR/$\triangle I$	Short	Medium	Large
High	boricate	dilute	dilute
Middle	boricate	do nothing	dilute
Low	boricate	boricate	do nothing

4.2 Inference

The fuzzy rule base is incorporated into the second block "Inference", using Zadeh's AND and OR operators, as follows:

$$\mu_{A \cup B}(x) = max(\mu_A(x), \mu_B(x)) \ . \tag{2}$$

$$\mu_{A \cap B}(x) = min(\mu_A(x), \mu_B(x)) \ . \tag{3}$$

An action is associated to each output (see Table 1). Some of the other most common operators have been tested:

1. Probability logic:

$$\mu_{A \cup B}(x) = \mu_A(x) + \mu_B(x) - \mu_A(x).\mu_B(x) \ . \tag{4}$$

2. Luckasiewicz:

$$\mu_{A \cap B}(x) = max(0, (1 - (\mu_A(x) + \mu_B(x)))) \ . \tag{5}$$

The results of Zadeh's operators are the easiest ones and the most appropriate. The orders *boricate*, *dilute*, and *do nothing* are issued from this block, with membership degrees μ_1, μ_2 and μ_3 and they still have fuzzy scales.

4.3 Defuzzification

The third block "Defuzzification" performs a defuzzification. The output signal, u, is obtained as follows:

$$u = \frac{\sum_i u_i.\mu_i}{\sum_i \mu_i} \ . \tag{6}$$

where u_i is the command relative to the ith fuzzy subset with the weights: $u_1 = -1$, $u_2 = 0$, $u_3 = 1$, and μ_i is the membership degree associated to the ith fuzzy subset (see Figure 5 and Table 2).

1. $u_{t=10} = \dfrac{-(1)*(0)+1*(0.9771)+0*(0.0229)}{0+0.9771+0.0229} = 0.9771$

2. $u_{t=19} = \dfrac{-(1)*(0)+1*(0.7999)+0*(0.2001)}{0+0.7999+0.2001} = 0.7999$

3. $u_{t=20} = \dfrac{-(1)*(1)+1*(0)+0*(0)}{1+0+0} = -1.0000$

4. $u_{t=60} = \dfrac{-(1)*(0.5456)+1*(0)+0*(0.4544)}{0.5456+0+0.4544} = 0.5456$

5. $u_{t=100} = \dfrac{-(1)*(0.5061)+1*(0)+0*(0.4939)}{0.5061+0+0.4939} = -0.5061$

Table 2. Clear values obtained according to the fuzzy inference outputs

t/deg.	μ_1	μ_2	μ_3	u
10 s	0.000	0.977	0.023	0.977
19 s	0.000	0.799	0.200	0.799
20 s	1.000	0.000	0.000	-1.000
60 s	0.546	0.000	0.454	0.546
100 s	0.506	0.000	0.493	-0.506

The table of control drawn in Figure 6 shows the areas covered by the commands. It allows to know how smooth is the control and how better it will be when we increase the number of rules because of the abrupt threshold.

We also point out a linear behavior when the process moves from a state to another, which justifies the simplicity of the membership functions we used to build the controller.

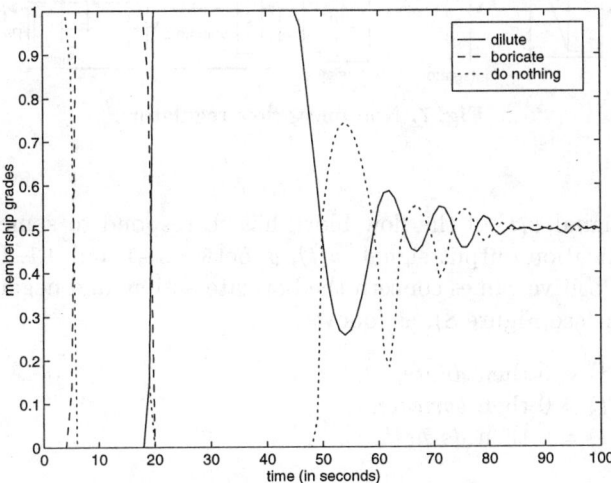

Fig. 5. Output membership functions

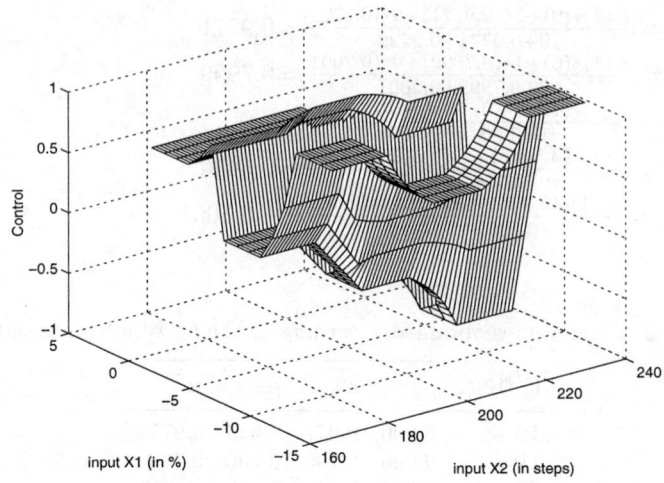

Fig. 6. Table of control according to the command

4.4 Flow Generation

The last block "flow generation" (see Figure 7) indicates what is to do according to the action i.

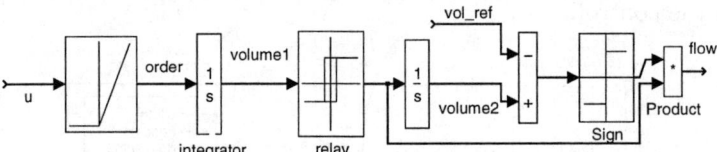

Fig. 7. Non linear flow regulator

The input signal $u(t)$ of the flow block has to respond to some conditions. The defuzzification output signal, $u(t)$, is between -1 and +1. It enters the flow block. Positive values concern the boricate action, and negative ones the dilute action (see Figure 8), as follows:

1. when $u(t) < 0$ then *dilute*,
2. when $u(t) > 0$ then *boricate*,
3. when $u(t) = 0$ then *do nothing*.

$$\begin{cases} F = F_{max} * u(t) \\ u(t) \in [-1, +1], (F = flow) \end{cases} \tag{7}$$

$$F_{max} = \begin{cases} 9 \text{ if } u(t) > 0 \\ 27 \text{ if } u(t) < 0 \end{cases} \qquad (8)$$

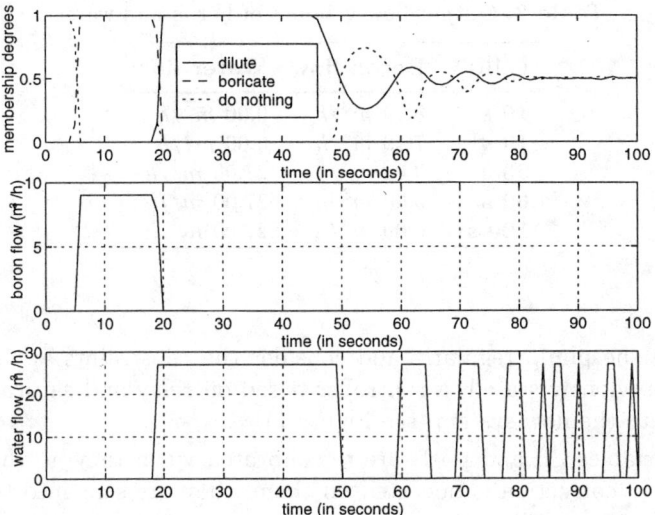

Fig. 8. Output volumes relative to the fuzzy grades

Examples of the flow generation interpretation:

1. $u_{10} = 0.9771 > 0$ then boricate.
2. $u_{19} = 0.7999 > 0$ then boricate.
3. $u_{20} = -1.000 < 0$ then dilute.
4. $u_{60} = 0.5456 > 0$ then boricate.
5. $u_{100} = -0.5061 < 0$ then dilute.

4.5 How the Flow Regulator Works

Let us consider a boron input signal u(t) in the flow block. u(t) is between 0 and +1. An imaginary flow value between 0 and 9 m^3/h corresponds to that input signal, u(t), in a linear way.

1. $u_{10} = 0.9771 \longrightarrow F_{10} = 9 \cdot u_{10} = 8.79 \ m^3/h$.
2. $u_{19} = 0.7999 \longrightarrow F_{19} = 9 \cdot u_{19} = 7.19 \ m^3/h$.
3. $u_{100} = -0.5061 \longrightarrow F_{100} = 27 \cdot u_{100} = -13.66 \ m^3/h$.

If the flow value is under the threshold (1 m^3/h for the boron or flow), then the system integrates this volume and puts it into a relay cycle until it gets a value over the threshold (see Figure 7 and Table 3).

Remark. For a better control, the water volumes are increased up to 27 m^3/h (max. value).

Table 3. Output flow volumes of boron and water

t/flow	Boron flow	Water flow
10 s	8.79 m^3/h	0.00 m^3/h
19 s	7.19 m^3/h	0.00 m^3/h
20 s	0.00 m^3/h	27.00 m^3/h
60 s	0.00 m^3/h	27.00 m^3/h
100 s	0.00 m^3/h	27.00 m^3/h

To request the pump regularly and consider the constraints applied to the flows, as previously stated, a controller based on relay and saturation is integrated into the flow system (see Figure 7).

The variables ΔI and posR are re-calibrated when they go through the FLC; after a correction is operated on them, they are sent into the system for a new loop (see Figure 9 and Figure 10).

The evolution of flows (boron and water) are very important and have to be compared to the evolution of ΔI and posR. Hence, these flows depend on the outputs ΔI and posR (see Figure 6 and Figure 11).

The evolutions of the rods insertion and the power axial offset during a complete cycle of 25000 seconds are shown in Figures 12 and 13.

5 Control Board of the Reactor

A graphical interface has been worked (see Figure 14). Its aim is to warn the operator about the FLC defections. The board allows the operator to know at any time where is the "current point" and where it goes. It describes the working area, shows the convergence of the system, draws the membership functions, the evolution of the variables ΔI and posR and the volumes sent into the coolant primary system. The operator can select, using that board, the state he needs to see. An historical work has been recorded in the memory, the board so allows to go back and follow the previous supervision. With this device, the operator can anticipate the future events. It is interesting to visualize those states of the reactor because the FLC work is displayed, so a manual supervision can replace the FLC one, if there is any divergence. It also allows to know what rule operates, what kind of deal has been created and what rules have been involved.

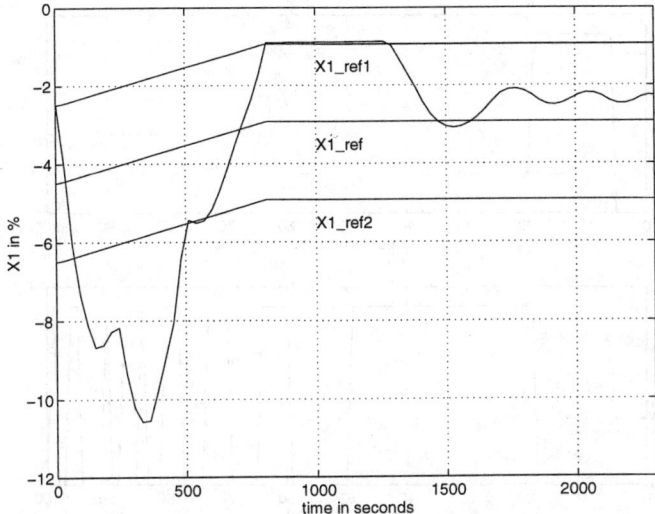

Fig. 9. Calibration of $\triangle$I by the FLC

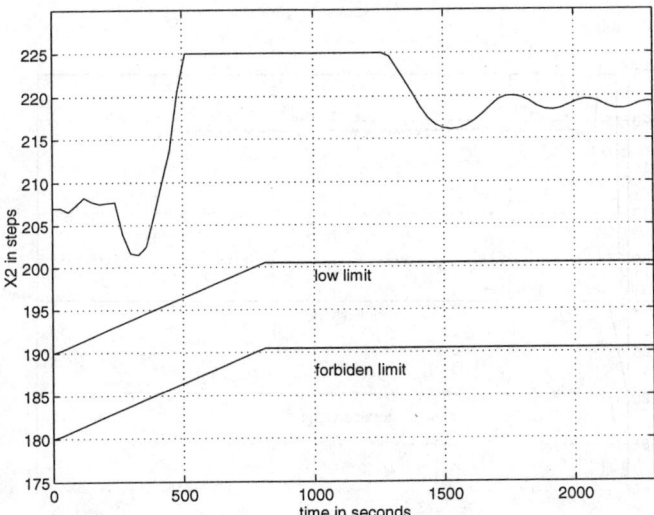

Fig. 10. Calibration of posR by the FLC

6   Conclusion

The fuzzy logic controller is easy to incorporate. The first results are attractive. The more there are rules, the better the result. It is also interesting to integrate new rule bases for more particular cases, as we have seen. The sys-

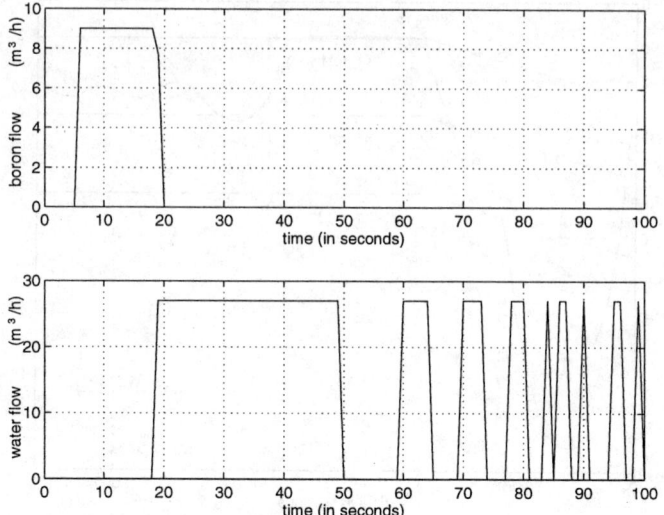

Fig. 11. Output flows according to the relay system

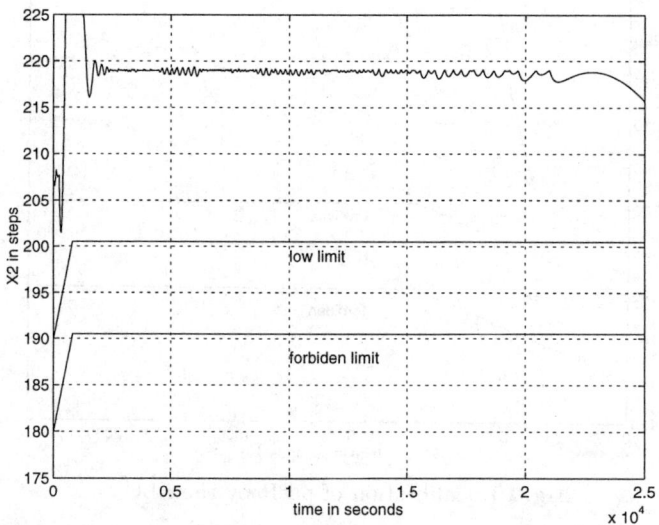

Fig. 12. Evolution of posR during a complete simulation cycle

tem losts more time than we expected to catch the reference. This problem is currently taken in consideration.

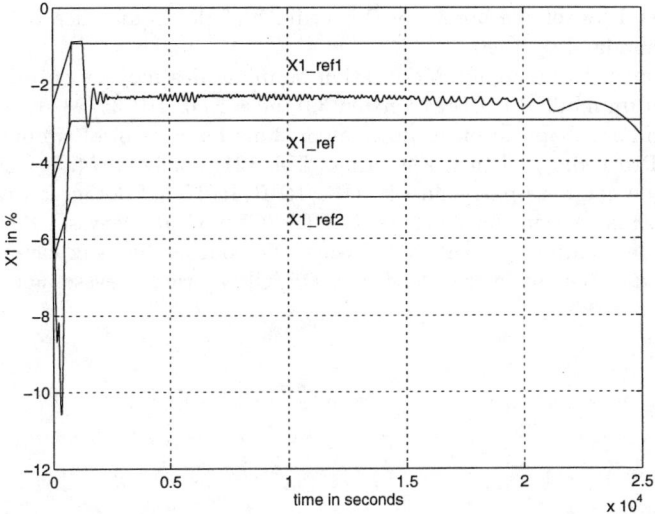

Fig. 13. Evolution of $\triangle$I during a complete simulation cycle

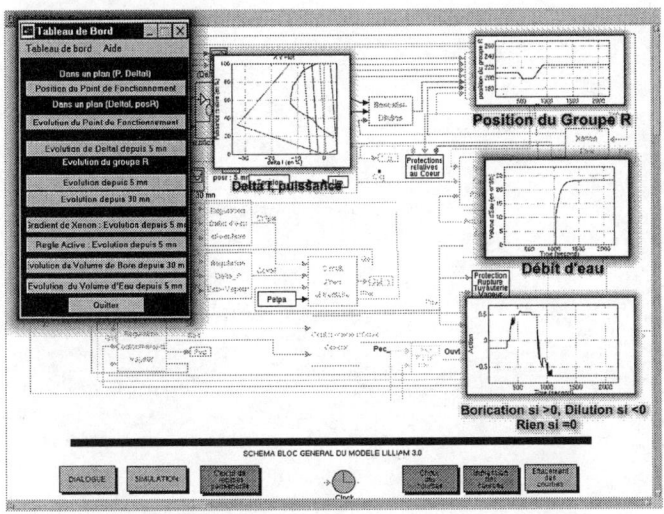

Fig. 14. Board screen for the reactor supervision

References

1. J-L. Tyran. Pilotage des REP 900 mode G au niveau 1, Automate de Borication-Dilution. In note EDF no. HP-35/90-14, February 1991.
2. Y. Pacarin, F. Guly, P. Siarry, Y. Saillard, J-L. Tyran. A contribution of fuzzy logic to the control of pressurized water reactors : automation of the power axial-

offset control by soluble boron. In Proceedings of the Conference IPMU'96, pp 75-79, Granada, July 1996.

3. J.-P. Corfmat, E. Obe-Guy. Modlisation simplifie des moyens de production : conception du modle LILLIAM version 3.0, note EDF HP-35/93/021/A, 1993.

4. E.H. Mamdani. Application of fuzzy algorithms for control of simple dynamic plant. In Proceedings of Inst. Elec. Eng., Vol. 121, no. 12, 1974.

5. Consigne de conduite particulire F RRC/RPR/RPN 1, Pilotage du racteur en mode gris. Instructions for Authors. In EDF/CPN Cruas-Meysse, 1985.

6. Consigne de conduite gnrale GC6, Recueil de courbes lies aux caractristiques physiques du coeur de la tranche 4. In EDF/CPN Cruas-Meysse, Feb. 1991.

4 Adaptive Fuzzy Control for a Simulation of Hydraulic Analogy of a Nuclear Reactor

Da Ruan, Xiaozhong Li, and Gert Van den Eynde

Belgian Nuclear Research Centre (SCK•CEN)
Boeretang 200, B-2400 Mol, Belgium
druan@sckcen.be

In the framework of the on-going R&D project on fuzzy control applications to the Belgian Reactor 1 (BR1) at the Belgian Nuclear Research Centre (SCK•CEN), we have constructed a real fuzzy-logic-control demo model. The demo model is suitable for us to test and compare some new algorithms of fuzzy control and intelligent systems, which is advantageous because it is always difficult and time consuming, due to safety aspects, to do all experiments in a real nuclear environment. In this chapter, we first report briefly on the construction of the demo model, and then introduce the results of a fuzzy control, a proportional-integral-derivative (PID) control and an advanced fuzzy control, in which the advanced fuzzy control is a fuzzy control with an adaptive function that can self-regulate the fuzzy control rules. Afterwards, we present a comparative study of those three methods. The results have shown that fuzzy control has more advantages in term of flexibility, robustness, and easily updated facilities with respect to the PID control of the demo model, but that PID control has much higher regulation resolution due to its integration term. The adaptive fuzzy control can dynamically adjust the rule base, therefore it is more robust and suitable to those very uncertain occasions.

1 Introduction

Fuzzy-logic control techniques are very mature in today's most engineering areas, but not in nuclear engineering, though some research has been done [Bernard and Khedkar, 1988; Hah and Lee, 1994; Lin et al., 1997; Matsuoka, 1990]. The main reason is that it is impossible to do experiments in nuclear engineering as easily as in other industrial areas. For example, a reactor is usually not available to an individual. Even for specialists in nuclear engineering, an official license for doing any on-line test is necessary. That is why we are still conducting the project of "fuzzy control applications at BR1" [Ruan, 1995; Ruan and Li, 1997, 1998; Ruan and van der Wal, 1998]. In this framework, we find that although there are already many fuzzy control applications, it is difficult for us to select the most suitable method for experiments at BR1 and to compare with some other algorithms. Moreover, due to the safety regulations of the nuclear reactor, it is not realistic to perform many experiments at BR1. In this situation, we have to conduct part of the pre-processing experiments outside the reactor, e.g., comparisons of different methods and the preliminary choices of the parameters. One solution is to make a simulation programme in a computer, but this has the disadvantage that the real time property cannot be well reflected. Therefore we adopted another strategy, that is, we designed and made a water-level control system, refered to as the demo model, which is suitable for our testing and experiments. In particular, this demo model (Figure 1) is designed to simulate the power control principle of BR1 [Li and Ruan, 1997].

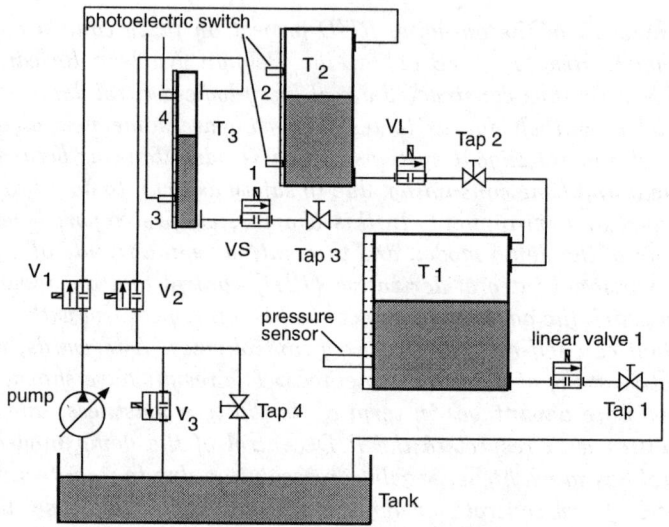

The Demo Model Structure

Fig. 1. The working principle of the demo model

In this demo model, our goal was to control the water level in tower T_1 at a desired level by means of tuning VL (the valve for the large control tower T_2) and VS (the valve for the small control tower T_3). The pump keeps on working to supply water to T_2 and T_3. All taps are manualy set at this time. Valves V_1 and V_2 are used to control the water level in T_2 and T_3 respectively. For example, when the water level in T_2 is lower than the photoelectric switch sensor 1 then the on-off valve V_1 will be opened (on), and when the water level in T_2 is higher than the photoelectric switch sensor 2 then the on-off valve V_1 will be closed (off). The same is true for V_2. Only when both V_1 and V_2 are closed will V_3 be opened, because it can decrease the pressure of the pump and thereby prolong its working life. The pressure sensor is used to detect the height of water level in T_1. So T_1 is a dynamic system with two entrances and one exit for water flow.

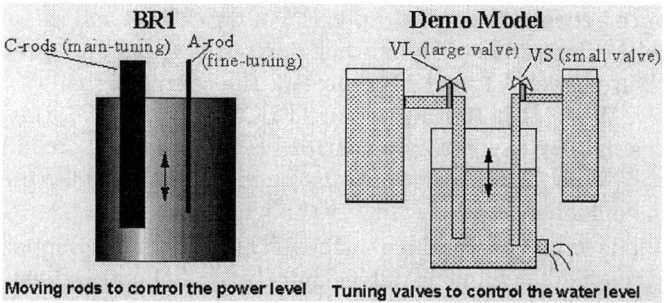

Fig. 2. The demo model is made to mimic the control principle of the BR1 reactor

As shown in Figure 2, the main tower T_1 symbolically represents the nuclear reactor itself. The water level can be interpreted as the power level of the reactor. In the reactor, we control this power level by inserting or withdrawing control rods at different speeds. In the demo model we use two control towers, a big one T_2 (corresponding to the control rods for coarse control) and a small one T_3 (corresponding to the control rods for fine tuning), which are connected to the main tower T_1 with two linear valves, VL and VS. In reality both valves are the same, but in the software we make sure that the maximal flow through VL is twice the maximal flow through VS.

The BR1 reactor is an over 40-year old research reactor, in which the control method is a simple on-off method. Many methods called traditional methods, when compared to fuzzy logic, are still very new to the BR1 reactor. One of these, PID control, has to be tested as well as the fuzzy logic method. So far, we have tested the normal fuzzy control, traditional PID control, and an advanced fuzzy control on this demo model. To obtain a better demonstration, these three approaches have been programmed and integrated into one controller system based on the programmable logic controller (PLC) of the OMRON company. The purpose of this chapter is to report comparative experimental results of these three methods for the demo model. Section 2 introduces a normal fuzzy control and its result. Section 3 introduces a PID control and its result. Section 4 introduces an advanced fuzzy control which is able to self-regulate the fuzzy control rules. Section 5 compares the previous three methods and their results. An Appendix gives an implementation of a computer simulation using adaptive fuzzy control in the demo model.

2 Fuzzy Control

The fuzzy control algorithm in this demo model is a normal algorithm based on the Mamdani model. To simulate the BR1 reactor, we use two fuzzy controllers (FLC1 and FLC2) to control VL and VS separately (note: it is possible to use one fuzzy logic controller with two outputs to control VL and VS and the related result can be referred to [Li and Ruan, 1997]). Let D be

68

the difference between the actual value (P) of water level and the set value (S) and DD be the derivative of D, in other words, the speed and direction of the change of water level. VL and VS represent the control signal to VL and VS, respectively. When D is too big, we use FLC1 to control VL (main-tuning); When D is small, we use FLC2 to control VS (fine-tuning). We choose D and DD as inputs of the fuzzy logic controller, and VL or VS as the output of the fuzzy logic controller.

D and DD must be fuzzified before fuzzy inference. Suppose the universes of discourse (or input variables' intervals) of D and DD are [-d, d] and [-dd, dd], respectively. We use 7 fuzzy sets to partition them, i.e., Negative Large (NL), Negative Middle (NM), Negative Small (NS), Zero (ZE), Positive Small (PS), Positive Middle (PM), and Positive Large (PL). As for VL and VS, because the result of fuzzy reasoning is also a fuzzy linguistic value, the universes of discourse of VL and VS also need to be fuzzified. We use the same 7 fuzzy linguistic terms.

Symmetrical trianglar-shaped functions are used to define the membership functions for input variables [Li et al., 1995], and singletons are used for output variables [Omron, 1992].

Each fuzzy controller has one rule base which contains 49 fuzzy control rules. The i-th rule can be represented as the following form:

if D is A_i and DD is B_i, then VL (or VS) is C_i

where A_i, B_i, and C_i are fuzzy linguistical values, such as NL, PS, and so on. The above rule is sometimes abbreviated as $(A_i, B_i : C_i)$.

Figure 3 shows a control effect of a synthetic control process. It first goes up from 0 to 20 cm then keeps on at 20 cm, next drops down from 20 cm to 10 cm and finally keeps on at 10 cm.

In view of this figure, we know that the fuzzy control has quick responses (quickly approaching the set value) and small overshoot (almost invisible), but with a small steady error (not so smooth in steady state).

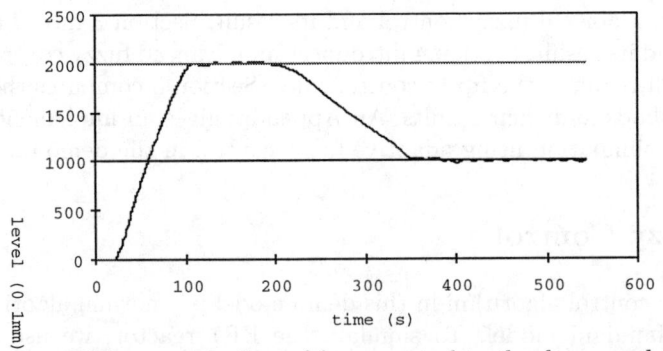

Fig. 3. The control effect of fuzzy control to the demo model

3 PID Control

In the PID control, it is difficult to control VL and VS separately like the previous fuzzy control with a good control result, because the integration term of the PID control needs some time, and this will result in an oscillation when switching the control signal between VL and VS. From this point of view the PID control is worse than the fuzzy control. Therefore, in our tests, VL and VS have to be controlled by the same signal. We use the following formula:

$$U(t) = K_p e + \int \frac{e}{T_i} dt + T_d \frac{de}{dt}$$

By substitution,

$$U(t) = K_p e + \int K_i e dt + K_d \frac{de}{dt}$$

where $U(t)$ is the control value to VL and VS at time t; e is the set value minus the real value at time t; K_p is the proportional parameter $(1/PB) \times 100\%$, PB is the proportional band; K_i is the integration parameter $1/T_i$, T_i is the integration time; K_d is the differential parameter T_d (the differential time).

In practice, a discrete form of the above formula is used:

$$U(t) = K_p e(t) + K_i T_s [e(1) + e(2) + \cdots + e(t)] + \frac{K_d}{T_s} [e(t) - e(t-1)]$$

where T_s is the sample period. Figure 4 shows a result of the PID control, where $PB = 15\%$, $T_i = 30\,s$, $T_d = 10\,s$.

In view of this figure, the PID control is very stable (very smooth in steady state), and has quick responses too, but with visible overshoots.

4 Advanced Fuzzy Control

The kernel part of the fuzzy logic control is the fuzzy rule base with linguistic terms, though the membership functions and scale factors also have an important effect on the fuzzy logic controller. There are some papers which discuss how to adjust membership functions and/or scale factors [Batur and Kasparian, 1991; Chou and Lu, 1994; Tonshoff and Walter, 1994; Zheng, 1992]. This section focuses on rules. Normally the methods of deriving rules can be broadly divided into two types, sourceable and non-sourceable. The term "sourceable method" means that the rules are obtained from some information source, such as human experience or historical input-output data. Experience has been widely used by the fuzzy engineers, especially by the early fuzzy engineers. The problem of using human experience is that it is time-consuming, and to some degree subjective. To overcome these problems, particularly avoiding the subjectivity, historical input-output data if available can be used. To obtain rules from such data, many methods are used, one of

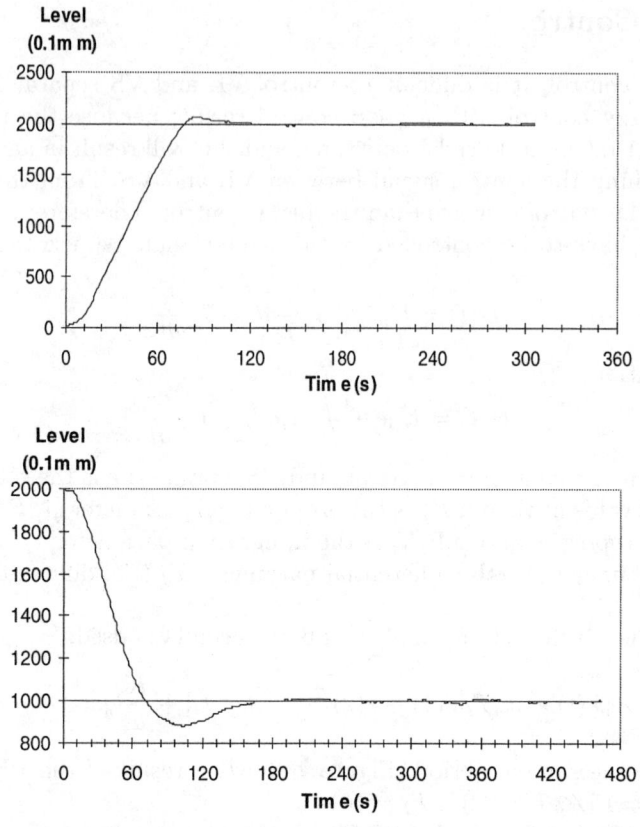

Fig. 4. The trajectory of the water level by the PID control

the popular approaches is neural networks (NN) [Berenji and Khedkar, 1992; Halgamuge and Glesner, 1994; Jang, 1992; Kosko, 1992; Li et al., 1995; Lin et al., 1995; Takagi and Hayashi, 1991; Wang and Mendel, 1992]. One problem of the sourceable method is that it depends strictly on the source which will be transformed into rules. In the case that the source is noisy, then the rules might be biased. Another problem of the sourceable method is that it is usually non-adaptive, i.e., all the rules are fixed, therefore it cannot perform well in a dynamic environment. The non-sourceable methods are source-free and they produce and choose rules according to a performance measurement of the controller, such as genetic algorithms (GAs) [Karr, 1991; Lim et al., 1996; Qi and Chin, 1997] (mostly also generating membership functions and scale factors) and self-organizing controllers (SOC) [He et al., 1993; Li et al., 1996; Lin et al., 1997, Procyk and Mamdani, 1979; Shao, 1988; Tanscheit and Scharf, 1988; Wu et al., 1992]. With a GA it is possible to find integratedly optimal parameters but a GA is very computation rich. Perhaps the SOC is the only method which has the following advantages: objective, adaptive, less computation required, more error-tolerant, and simple.

The general principle of the SOC is that the controller monitors its own performance and adjusts its control rules to improve performance for time-varying and unknown plants. The problem of the SOC is how to perform the performance measurement. The basic way is to design a performance measurement table which looks like a fuzzy control rule table and to use it to assess the performance of the controller (rules) [Procyk and Mamdani, 1979], but to design such a performance measurement table is also very difficult [Chung and Oh, 1993] and it is system-dependent. Based on the SOC, this section will introduce an adaptive method which uses a set of new norms to replace the former performance measurement. The new norms are very simple and system-independent, therefore they can be easily applied to most fuzzy controllers.

In this section, the advanced fuzzy control means the above SOC, in other words, a fuzzy control with an adaptive function, where the adaptive function contains two steps: performance judgement and changing fuzzy control rules. Figure 5 illustrates how an adaptive function is incorporated into the fuzzy control system. At the beginning of each cycle, the controller's last behaviour is judged and then the rule base is changed accordingly. In this cycle, the controller will use the new rule base and output the result to the controlled object. The behaviour of the new rule base will be judged and changed again in the next cycle.

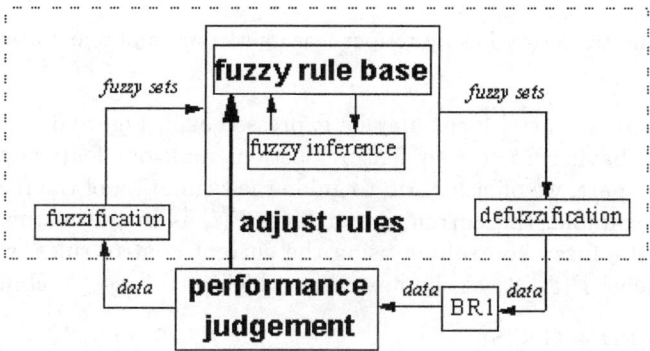

Fig. 5. An adaptive function is incorporated into a fuzzy control system

4.1 The Principle of the Adaptive Function

Let D and DD represent error (the difference between the actual value and the desired value) and change in error, respectively. Let $D(t)$ and $DD(t)$ represent error and change in error at time t, respectively. They are two input variables.

Let U be an output variable, and assume the total number of the rules is n, then every rule has the following form,

if D is A_i and DD is B_i, then U is C_i, $i = 1, 2, \cdots, n$.

where A_i, B_i, and C_i are fuzzy linguistic values and i is an index pointing out each rule's position in the rule table (or the rule data file). Use r[i] to represent the fuzzy control magnitude (conclusion fuzzy set) of the i-th rule, and let simply

$$r[i] = 1, 2, 3, 4, 5, 6, 7$$

where 1=NL, 2=NM, 3=NS, 4=ZE, 5=PS, 6=PM, 7=PL.

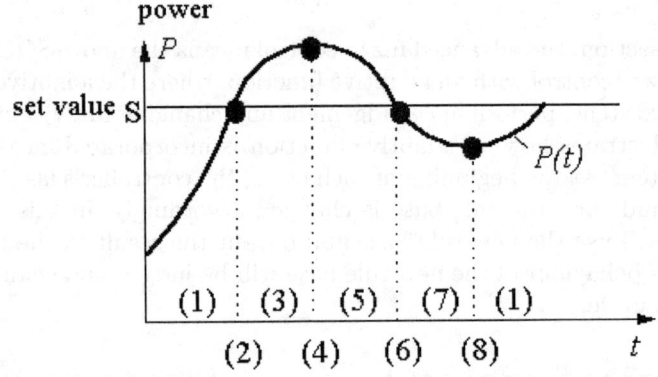

Fig. 6. Any trajectory has up to four feature sections and four feature points

In general, a control locus may be expressed as in Figure 6, and it can be regarded as having up to four feature sections and four feature points. For each feature part, we offer a norm to guide the regulation of the fuzzy control rules. For example, the current water level $P(t)$ is in the feature part (1), then after the fuzzy controlling using the current control rules, we measure the water level $P(t+1)$ at the next time which has three possibilities:

- $P(t) < P(t+1) \leq S$;
- $P(t+1) < S$ and $P(t+1) \leq P(t)$;
- $P(t+1) > S$.

The related norm to guide how to change rules is the following:

(i) if $D(t+1) \leq 0$ and $DD(t+1) > 0$, that is, $P(t) < P(t+1) \leq S$, then $r[i] = r[i]$;

(ii) if $D(t+1) < 0$ and $DD(t+1) \leq 0$, that is, $P(t+1) < S$ and $P(t+1) \leq P(t)$, then $r[i] = r[i] + \alpha$;

(iii) if $D(t+1) > 0$, that is, $P(t+1) > S$, then $r[i] = r[i] - \alpha$.

Where α is a step size and $\alpha = 1, 2, 3, 4, 5, 6$. In case (i), the fuzzy controller makes the water level $P(t+1)$ closer to the set value S, therefore the behaviour of the fuzzy controller is good, no rules should be changed. In case (ii), the fuzzy controller makes the water level $P(t + 1)$ go further away from the set value S, therefore the behaviour of the fuzzy controller is not good, the strength is too weak and the action of the corresponding rules should be stronger. In case (iii), the fuzzy controller makes the water level $P(t + 1)$ overpass the set value S, therefore the behaviour of the fuzzy controller is not good, the strength is too strong and the action of the corresponding rules should be weakened.

Not all rules but some of those that are activated in last cycle should be regulated. We use the following formula to describe which should be adjusted:

$$A_i(D) \bigwedge B_i(DD) = \bigvee_{C_j = C_i} (A_j(D) \bigwedge B_j(DD))$$

which means the i-th rule is changed only if it is the largest activated among those activated rules which have the same conclusion part. For example, (NL, NM: PL) and (NM, NM: PL) are two activated rules and have the same conclusion part, PL. Comparing NL(D)$\bigwedge$ NM(DD) with NM(D)$\bigwedge$ NM(DD), the larger one corresponds to the rule which should be adjusted.

4.2   An Experimental Result

To guarantee no overshoot, the best way is to initialize all rules as the same conclusion part: NL, as shown in Table 1.

Table 1. The initial rule table for both FLC1 and FLC2

DD\D	NL	NM	NS	ZE	PS	PM	PL
NL	NL	NL	NL	NL	NL	NL	NL
NM	NL	NL	NL	NL	NL	NL	NL
NS	NL	NL	NL	NL	NL	NL	NL
ZE	NL	NL	NL	NL	NL	NL	NL
PS	NL	NL	NL	NL	NL	NL	NL
PM	NL	NL	NL	NL	NL	NL	NL
PL	NL	NL	NL	NL	NL	NL	NL

In this table, for example, NL at the row 2 and column 3 means: if D is NM and DD is NL then VL or VS is NL. All rules have the same conclusion part though condition parts are different. Figure 7 illustrates the comparison result between fuzzy adaptive control and fuzzy control with the above rule base. In this example, the set value is 20 cm. Both start from 0 cm. During the first stage, i.e., increasing from zero, some analytic rules manipulate the valves and not fuzzy control. Only after the water level reaches 18 cm does the fuzzy controllers start to operate VL and VS. Apparently, the adaptive fuzzy control has a much better result by self-regulating gradually fuzzy control rules. The normal fuzzy control without adaptive function cannot self-regulate rules, therefore it cannot draw up the water level.

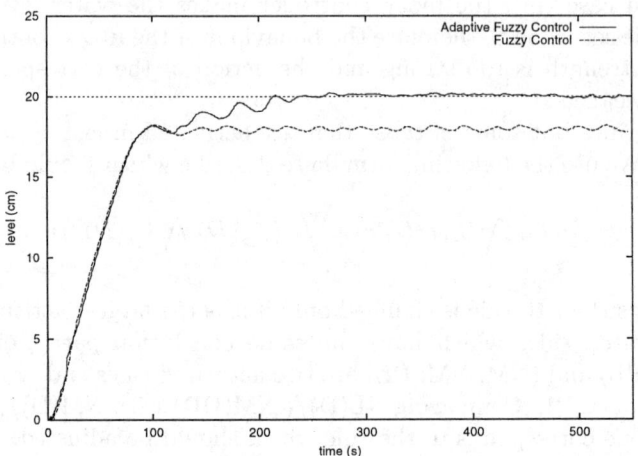

Fig. 7. Comparison between adaptive fuzzy control and fuzzy control

About 10 minutes later, we observe the rule tables on the screen and find both rule tables have changed a lot. Table 2 gives the result of FLC1 and Table 3 gives the result of FLC2, where the regulated rules are marked by bold.

Table 2. The regulated rule table for FLC1

DD\D	NL	NM	NS	ZE	PS	PM	PL
NL	NL	NL	NL	NL	NL	NL	NL
NM	**NM**	NL	NL	NL	NL	NL	NL
NS	NL	**NS**	**PM**	NL	NL	NL	NL
ZE	**NS**	**PL**	**PL**	**ZE**	NL	NL	NL
PS	NL	NL	**NM**	NL	NL	NL	NL
PM	NL	NL	NL	NL	NL	NL	NL
PL	NL	NL	NL	NL	NL	NL	NL

4.3 Some Remarks for the Adaptive Function

- The initial idea about the previously described norms of the adaptive function, which was published in [Li et al., 1996], and where a simulated

Table 3. The regulated rule table for FLC2

DD\D	NL	NM	NS	ZE	PS	PM	PL
NL	NM	PM	PL	PL	NM	NL	NL
NM	PL	PL	PM	PL	PL	NL	NL
NS	PM	PL	PL	PS	PL	NL	NL
ZE	PL	PL	PS	ZE	NL	NL	NL
PS	PL	PL	PM	NS	NL	NL	NL
PM	PM	PL	PL	NM	NL	NL	NL
PL	ZE	ZE	NM	NL	NL	NL	NL

inverted pendulum system and a real industrial heating system were used to make testings, gave satisfactory results.
- The parameter α is influential on the overshoot and response time (rise time). When α is too big, there will be a large overshoot possibly; when α is too small, possibly there will be a long response time [Li et al., 1996].
- The adaptive function considers only the last value, that is, it uses $P(t-1)$ not $P(t-\tau)$ (τ is the delay) to decide $P(t)$, but our experimental results show the effect is good, although the valves of the demo model have a maximum delay of 90 seconds.
- The adaptive function selects only some of the rules according to the formula in Section 4.1 for adjustmnt, not all activated rules like [Lin et al., 1997]. This makes the transition of the rules more smooth, i.e., without or with less resonance.
- Selecting initial rules appropriately will benefit the control effect. For example, if the overshoot is strictly limited, we may initialize all rules with the conclusion part of NL, as was done in the previous experiment. Once some experience has been obtained, it can be transformed into the initial rules of the adaptive function, the advantage being that the rise time will be shorter [Li and Ruan, 1998].
- The rule, "if D is ZE and DD is ZE then U is ZE," should be fixed, and this will help the system to become stable.
- The adaptive function is very helpful in keeping the system stable in a steady state. It cannot guarantee no overshoot if the initial rules are randomly selected.
- The adaptive function cannot adjust membership functions and scale factors.

5 Comparative Study

Each method has both advantages and disadvantages, the details of which are described in Table 4, where "*" is used to represent the degree of a property, and the more "*", the higher the degree. For example, the realisation of an adaptive fuzzy logic controller (FLC) is more difficult than a normal fuzzy controller, but a normal fuzzy controller is more difficult to realize than a PID controller. The PID control has the smallest static error and steady error. The dynamic regulation of the control rules in an adaptive controller can help in reducing the static error and steady error [Li et al., 1996]. As for robustness, it has been accepted that FLC is more robust than PID. Herein we also give one example, as shown as in Figure 8. This experiment was carried out after tuning the Tap 1 (see Figure 1) to make the outflow much smaller. We found that the reaction of FLC was better than that of PID, though the FLC had a small static error.

If we count the total number of "*" for each method, we will find that PID and FLC have the same score, 17. Adaptive FLC has a higher score of 20. This interesting result can be explained by the following facts. PID and FLC have their own strong points, and they compensate each other. Adaptive FLC adds an adaptive function to a normal FLC, therefore its score should be higher than that of FLC. A natural result is that combining FLC and PID should be better than each method alone.

The comparative method above is perhaps a little subjective, but it does reflect some objective properties and relationships among those three methods. In the real world, one may use other ways to evaluate these methods. For example, if robustness is stressed, then it should be highly weighted when the total scores are calculated.

For further descriptions of comparative studies between FLC and PID, readers may refer to [Boverie et al., 1991; Chao and Teng, 1997; Misir et al., 1996; Mizumoto, 1995; Moon, 1995; and Wu and Mizumoto, 1996].

Table 4. Comparative study of FLC, PID and adaptive FLC

	PID	FLC	adaptive FLC
easiness for realisation	***	**	*
objectivity (vs. subjectivity)	***	*	***
easiness for trying parameters	*	**	***
overcoming overshoots	*	***	**
quick response	**	***	***
reducing steady error	***	**	**
reducing static error	***	**	***
robustness	*	**	***
total (*)	17	17	20

6 Conclusion

This chapter gives comparisons between fuzzy control, PID control, and advanced fuzzy control based on the experimental results of a demo model which simulates the control principle of the BR1 reactor. Fuzzy control is

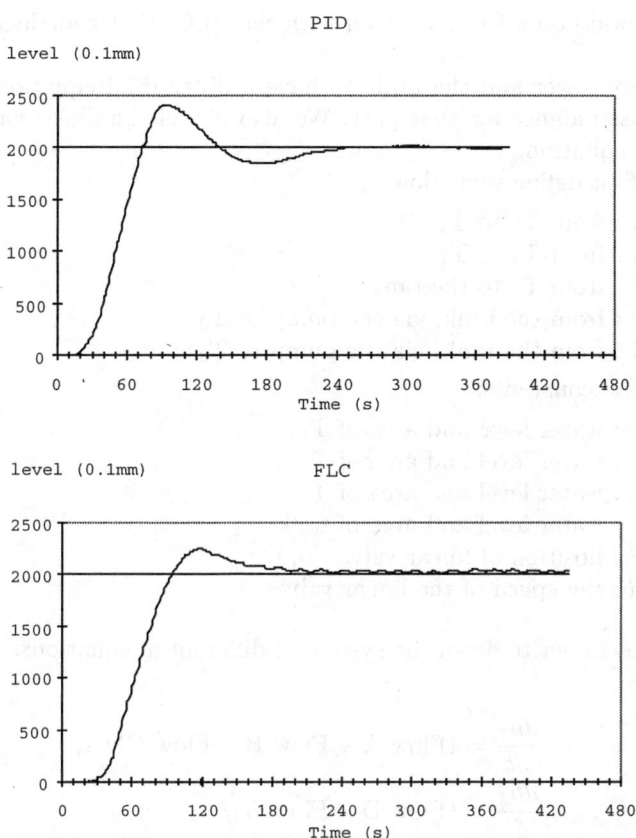

Fig. 8. FLC is more robust than PID

more robust than PID control, but with a well-characterized system, such as a reactor, it should be better to use a hybrid method which inherits the advantages of both methods. Furthermore, the adaptive fuzzy control is able to aid the designer in finding the fuzzy control rules, especially for systems posessing much of dynamical uncertainty.

Appendix: Implementaion of a Computer Simulation with Adaptive Fuzzy Control

All control is done by an early version of a PLC and fuzzy logic unit provided by OMRON, Belgium, both for the nuclear reactor and the demo model. Unfortunately, this early hardware has no facilities for adaptive fuzzy logic control. This is the main reason why we needed to simulate this demo model. The other option was to keep using the real demo model and running the control software on a PC connected with the PLC (that gets degraded to an IO unit).

For every tower and the tank we have a differential equation representing the mass balance for that part. We also model the linear valves with a differential equation.

Let us first define some flows:

Flow A : from T_2 to T_1
Flow B : from T_3 to T_1
Flow C : from T_1 to the tank
Flow D : from the tank, via the pump to T_2
Flow E : from the tank, via the pump to T_3

And some constants:

h_1, A_1 : water level and area of T_1
h_2, A_2 : water level and area of T_2
h_3, A_3 : water level and area of T_3
h_4, A_4 : water level and area of tank
VL, VS : position of linear valves $[0, 1]$
$V speed$: the speed of the linear valves

Now we can write down the system of differential equations:

$$\frac{dh_1}{dt} = (\text{Flow A} + \text{Flow B} - \text{Flow C})/A_1$$

$$\frac{dh_2}{dt} = (\text{Flow D} - \text{Flow A})/A_2$$

$$\frac{dh_3}{dt} = (\text{Flow E} - \text{Flow B})/A_3$$

$$\frac{dh_4}{dt} = (\text{Flow C} - \text{Flow D} - \text{Flow E})/A_4$$

$$\frac{dVL}{dt} = \pm V speed \text{ or } 0$$

$$\frac{dVS}{dt} = \pm V speed \text{ or } 0$$

Note: the formulas for the various flows are omitted here. They are standard hydraulic equations. For example

$$\text{Flow A} = VL * \sqrt{\frac{-\Delta_2^1 h}{K_{dA} * L_A}} * \alpha,$$

with $\Delta_2^1 h$ the difference in water level between T_2 and T_1, VL the position of the big valve, K_{dA} the K-factor of the connection, L_A the length of the connection and α a correction factor obtained by comparing experiments and the simulation. We assume a linear relation for the valve position. This is, of course, an approximation of the real system.

We want to simulate the complete behaviour of the demo model and the control actions that are coordinated by a PLC and a fuzzy logic unit. Every step in our simulation consists of two parts: simulate the evolution of the hydraulic system for a certain time period T_{sample} and simulate the fuzzy logic control action afterwards. The time period T_{sample} resembles the real time needed for the PLC to finish its program (loop time). In our case this is approximately 10 ms.

The evolution of the system is simulated using the system equations given above. We integrate this set of differential equations from a certain moment in time T to $T + T_{sample}$ using numerical techniques for solving ordinary differential equations.

The next flow-chart shows the overall structure of the simulation process.

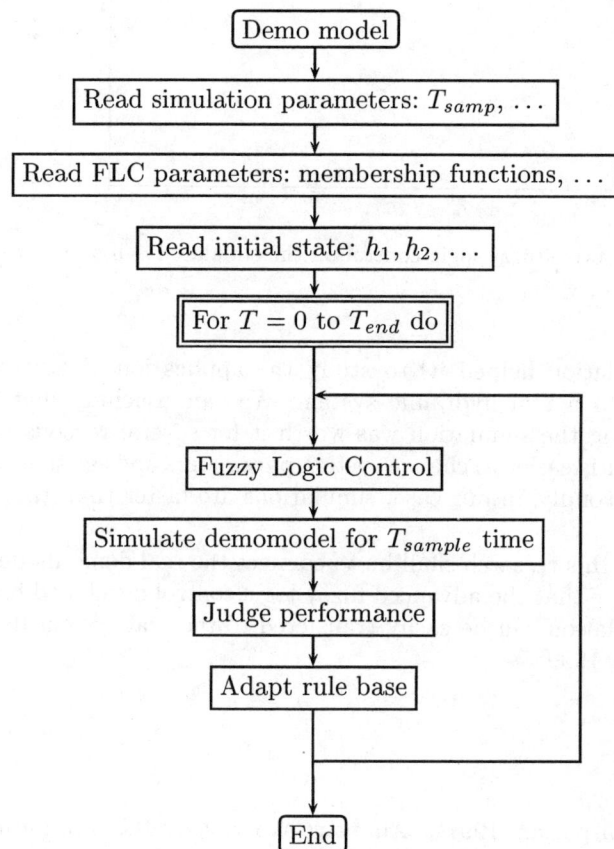

We present here the first results of using adaptive fuzzy logic control on the simulated demo model. We start with an 'empty' rule base, i.e., for every state of the error and the change of error (our two input parameters), we keep both valves completely closed. This means we don't do any control action at al. During the process of simulation and control, the adaptive function tries to change the rule base so the system performs well.

We see in Figure 9 that for both states (steady at 20 cm and steady at 15 cm) we first get a learning phase (oscillations), but after some time the system performs well.

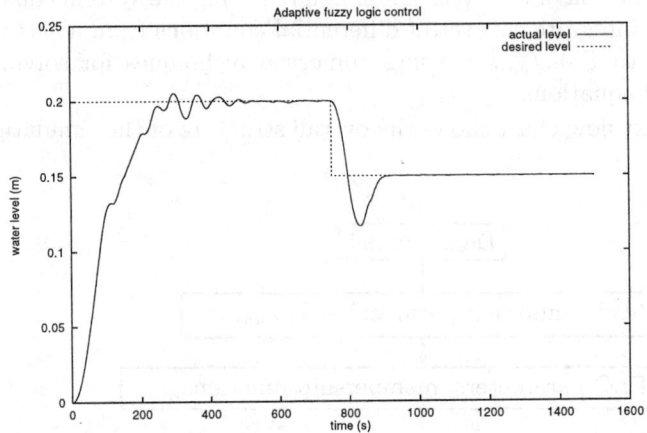

Fig. 9. Using adaptive fuzzy logic control on the simulated demo model

The use of simulation helped us to study the application of advanced fuzzy logic control to a real hydraulic system. We can conclude that the effort of implementing the simulation was worth it for several reasons. The main reason is that it is easier to change system parameters and see the effect of these changes. Secondly, in our case, simulations are faster than the real demo model.

The next step in this research shall be to connect the real demo model to our software. We hope that the advanced fuzzy logic control developed based on a computer simulation can be easily transfered to the real system [Ruan and Van den Eynde, 1999].

References

Batur, C. and V. Kasparian (1991), "Adaptive expert control," *International Journal of Control*, 54(4), pp. 867–881.

Berenji, H.R. and P. Khedkar (1992), "Learning and tuning fuzzy logic controllers through reinforcements," *IEEE Trans. Neural Networks*, 3, pp. 724–740.

Bernard, J.A. (1988), "Use of a rule-based system for process control," *IEEE Control Systems Magazine*, 8(5), pp. 3–13.

Boverie S., B. Demaya and A. Titli (1991), "Fuzzy logic control compared with other automatic control approaches," *Proceedings of the 30th IEEE-CDC conference on decision and control*, Brighton, UK, pp. 1212–1216.

Chao, C.T. and C.C. Teng (1997), "A PD-like self-tuning fuzzy controller without steady-state error," *Fuzzy Sets and Systems*, 87, pp. 141–154.

Chou, C.H. and H.C. Lu (1994), "A heuristic self-tuning fuzzy controller," *Fuzzy Sets and Systems*, 61, pp. 249–264.

Chung, B.M. and J.H. Oh (1993), "Control of dynamic systems using fuzzy learning algorithm," *Fuzzy Sets and Systems*, 59, pp. 1–14.

Hah, Y.J. and B.W. Lee (1994), "Fuzzy power control algorithm for a pressurized water reactor," *Nuclear Technology*, 106, pp. 242–252.

Halgamuge, K. and M. Glesner, "Neural networks in designing fuzzy systems for real world applications," *Fuzzy Sets and Systems*, 65, pp. 1–12.

He, S.Z., S.H., Tan, C.C. Hang and P.Z. Wang (1993), "Control of dynamical processes using an on-line rule-adaptive fuzzy control system," *Fuzzy Sets and Systems*, 54, pp. 11–22.

Jang, J.S.R. (1992), "Self-learning fuzzy controllers based on temporal back propagation," *IEEE Trans. on Neural Networks*, 3(5), pp. 714–723.

Karr, C.L. (1991), "Design of an adaptive fuzzy logic controller using a genetic algorithm," *Proc. 4th ICGA*, pp. 450–457.

Kosko, B. (1992), *Fuzzy Systems and Neural Networks*, Prentice-Hall, Englewood Cliffs, NJ.

Li, X. and D. Ruan (1997), "Constructing a fuzzy logic control demo model at SCK•CEN," *Proceedinds of the 5th European Congress on Intelligent Techniques and Soft Computing*, pp. 1408–1412.

Li, X., S. Bai and Z. Zhang (1996), "An introduction to a fuzzy adaptive control algorithm," *Chinese Journal of Advanced Software Research*, 3(1), pp. 1–11.

Li, X., S. Bai and Z. Liu (1995), "An algorithm for self-learning and self-completing fuzzy control rules," *Informatica* 19, pp. 301–312.

Lim, M.H., S. Rahardja and B.H. Gwee (1996), "A GA paradigm for learning fuzzy rules," *Fuzzy Sets and Systems*, 82, pp. 177–186.

Lin, C., F. Jeng and C. Lee (1997), "Hierarchical fuzzy logic water level control in advanced oiling water reactors," *Nuclear Technology*, 118(3), pp. 254–263.

Lin, C.T., C.J. Lin and C.S. George Lee (1995), "Fuzzy adaptive learning control network with on-line neural learning," *Fuzzy Sets and Systems*, 71, pp. 25–45.

Matsuoka, H. (1990), "A simple fuzzy simulation model for nuclear reactor system dynamics," *Nuclear Technology*, 94, pp. 228–241.

Misir, D., H.A. Malki and G. Chen (1996), "Design and analysis of a fuzzy proportional-integral-derivative controller," *Fuzzy Sets and Systems*, 79, pp. 297–314.

Mizumoto, M. (1995), "Realization of PID controls by fuzzy control method," *Fuzzy Sets and Systems*, 70, pp. 171–182.

Moon, B. S. (1995), "Equivalence between fuzzy logic controllers and PI controllers for single input systems," *Fuzzy Sets and Systems*, 69, pp. 105–114.

Omron (1992), *C200H–FZ001 Fuzzy Logic Unit, Operation Manual*, Omron company.

Procyk, T.J. and E.H. Mamdani (1979), "A linguistic self-organizing process controller," *Automat.*, 15(1), pp. 15–30.

Qi, X.M. and T.C. Chin (1997), "Genetic algorithms based fuzzy controller for high order systems," *Fuzzy Sets and Systems*, 91, pp. 279–284.

Ruan, D. (1995), "Fuzzy logic in the nuclear research world," *Fuzzy Sets and Systems*, 74(1), pp. 5–13.

Ruan, D. and X. Li (1997), "The test of fuzzy logic control with a closed loop at BR1 reactor," *Proceedings of the 7th World Congress IFSA '97*, Prague, Czech Republic, 3, pp. 126–131.

Ruan, D. and X. Li (1998), "Fuzzy-logic control applications to the Belgian reactor 1 (BR1)," *Computers and Artificial Intelligence*, 17(2-3), pp. 127–150.

Ruan, D. and A.J. van der Wal (1998), "Controlling the power output of a nuclear reactor with fuzzy logic," *Information Sciences*, 110, pp. 151–177.

Ruan, D. and G. Van den Eynde (1999), "Adaptive fuzzy control for simulating a nuclear reactor operation," 13th European Simulation Multiconference, Warsaw, Poland, June 1–4, 1999.

Shao, S. (1988), "Fuzzy self-organizing controller and its application for dynamic processes," *Fuzzy Sets and Systems*, 26, pp. 151–164.

Takagi, H. and I. Hayashi (1991), "NN-driven fuzzy reasoning," *Internat. J. Approximate Reasoning*, 5, pp. 191–212.

Tanscheit, R. and E.M. Scharf (1988), "Experiments with the use of a rule based self-organizing controller for robotic applications," *Fuzzy Sets and Systems*, 26, pp. 195–214.

Tonshoff, H.K. and A. Walter (1994), "Self-tuning fuzzy-controller for process control in internal grinding," *Fuzzy Sets and Systems*, 63, pp. 359–373.

Wang, L.X. and J.M. Mendel (1992), "Generating fuzzy rules by learning from examples," *IEEE Trans. Systems Man Cybern.*, 22, pp. 1414–1427.

Wu, Z.Q., P.Z. Wang, H.H. Teh and S.S. Song (1992), "A rule self-regulating fuzzy controller," *Fuzzy Sets and Systems*, 47, pp. 13-21.

Wu, Z.Q. and M. Mizumoto (1996), "PID type fuzzy controller and parameters adaptive method," *Fuzzy Sets and Systems*, 78, pp. 23–35.

Zheng, L. (1992), "A practical guide to tune of proportional and integral (PI) like fuzzy controllers," *Proc. of the 1st IEEE Int. Conf. on Fuzzy Systems*, pp. 633–640.

5 Controlling Neutron Power of a TRIGA Mark III Research Nuclear Reactor with Fuzzy Adaptation of the Set of Output Membership Functions

Jorge S. Benítez-Read and Daniel Vélez-Díaz

Instituto Nacional de Investigaciones Nucleares
Gerencia de Ciencias Aplicadas
Apartado Postal 18-1027, Col. Escandón
11801 México, D.F.
jsbr@nuclear.inin.mx

A fuzzy method is developed to obtain the set of output membership functions (SOMF) to be used in the defuzzification process of a typical fuzzy rule based control system. The universe, domain, and distribution of the controller's output membership functions are defined at the beginning of every control cycle, thus having a dynamic generation of the SOMF. One or more relevant parameters of the controlled process are used as inputs to the SOMF fuzzy adaptation stage. The fuzzy method designed is incorporated into a fuzzy control algorithm applied to a point-kinetic model of a TRIGA Mark III research nuclear reactor. The control objective is to take the neutron power from its source level up to a desired setpoint, avoiding undesirable power excursions, specially at low power levels, in order to maintain the reactor period above the pre-specified lower limit during power ascent. Likewise, once the desired power level is attained in a soft manner, the fuzzy controller should act as a regulator of the power level for long periods of time. Simulation results are analysed and compared with the response provided by a fuzzy controller without SOMF adaptation. A general conclusion drawn from the better performance obtained with adaptation is that knowledge of the dynamic characteristics of the controlled system may be used to provide some kind of knowledge-based adjustment of the controller's parameters to improve the response of the closed-loop system.

1 Introduction

Although productivity has been one of the traditional and most required goals of control systems, the potential risk of radiation exposure to humans in case of accidents in nuclear power plants requires that safety be the dominant concern. Thus, control systems for nuclear reactors have always been designed with a great deal of care, based on well founded theories, and tested first through simulation under different operating conditions. From 1942 to 1960, much of the analysis and design of control systems for nuclear reactors was based on classical methods [Lipinsky and Vacroux 1970]. Several research groups [Weaver and Vanesse 1967; Silvinsky and Schultz 1970] followed this trend, applying modern control theory to nuclear reactors. The concepts of optimal control were first used in the 1970s, developing optimal regulators and time-optimal controllers for nuclear power plants [Raju and Fadra 1973; Tepper 1975; Saif 1989; Benítez et al. 1992]. Different approaches of adaptive control theory, such as model reference, self-tuning, and adaptive linearization, among others, have been proposed for nuclear power control [Bereznai 1973; Benítez and Jamshidi 1992; Jungin et al. 1993; Nah 1995]. Likewise, robust control theory has also been reported [Weng et al. 1994] in feedforward-feedback control designs.

Safety aspects in nuclear power control have always been present, either explicitly or implicitly, in every single design. In this sense, some reactivity constraints may be imposed on the control system [Bernard et al. 1984], or a specified reactor period may be attained by velocity control of control rods' actuator mechanisms [Bernard 1988]. Moreover, controllers that consider the presence of disturbances and modelling uncertainties may reduce the number of unnecessary plant trips [Weng et al. 1994].

The need to use computers for nuclear power plant design, engineering, operation and maintenance has been growing since the inception of commercial nuclear power electricity generation in the 1960s. The rapid advance of computer hardware and software technology in the last two decades has greatly widened the potential of computer applications to plant instrumentation and control of future plants, as well as to those needed for operation of existing plants [Sun 1997]. For instance, the automatic start-up of the Toshiba Training Reactor, a 100 kW swimming pool type reactor, from its source level up to the power level is made possible with a digital computer [Takahashi and Takamatsu 1965]. The International Atomic Energy Agency has also sponsored expert meetings to study the incorporation of personal computers for data monitoring and analysis in research reactors [Leopando 1998].

More recently, the theory of fuzzy sets and fuzzy logic has been the basis of new design approaches for the control of nuclear reactors. This theory has been used to incorporate a human operator's linguistic "rules of thumb" in a HTR NPP controller [Bubak and Moscinski 1983]; another report proposes the construction of a knowledge base and its use in developing a rule-based methodology for power control on the 5

MWt MIT Research Reactor [Bernard 1986]. Several control research projects, involving fuzzy and neural network approaches, are being developed at Pennsylvania State University [Edwards et al. 1992]. Fuzzy logic control algorithms have shown, through computer simulation, equal or better performance than conventional P-I controllers [Moon 1993; Lin 1995]. Likewise, a fuzzy logic control system, aimed to enhance the reactor feedwater control, has been successfully applied, since 1992, to a 165 MWe prototype reactor (heavy water moderated, boiling light water cooled and pressure tube type) [Shibuya and Iijima 1996].

Higher functional reliability and a deeper degree of automation can be achieved in nuclear power plants using automatic control systems and algorithms, which can first be developed in research reactors such as the TRIGA Mark III reactor at the Nuclear Centre of Mexico. In this particular system, an important constraint is the time required by the neutron power, at any instant of time, to be increased by a factor "e," considering that the neutron power rate of change is kept constant [DeGroot 1968]; this parameter is called the reactor period (T). The aim of the control system in a TRIGA Mark III reactor is to attain full power level, starting from a source level (~50 W), and maintaining it for long periods of time, which are commonly around 24 hours. An important safety requirement is to maintain, at all times, the period above a specified lower limit, typically 3 seconds, when increasing power [Nava 1991]. Different algorithms to regulate the reactor's neutron power, based on advanced control theories and knowledge-based expert systems, have been proposed and studied for this reactor [Benítez et al. 1994; Benítez and Vélez 1996]. An important point to consider is the suitability and feasibility of using a particular control architecture for defined regions of the plant's operation. The knowledge base is obtained by simulation of the reactor behaviour [Vélez and Benítez 1995], where the controlled system is modelled by a simple set of nonlinear point kinetic equations [Hetrick 1971]. This model, which includes temperature-based reactivity feedback, has been used to infer a set of fuzzy rules for the reactor's response to different insertions of reactivity [Benítez and Vélez 1996]. A reduction of the response time, using fuzzy rule based controllers on this reactor, has been obtained by a Boolean selection among different sets of the output membership functions, which are defined to operate in their corresponding power regions [Vélez and Benítez 1997].

In this chapter, a fuzzy method is described to obtain the set of output membership functions (SOMF) to be used in the defuzzification process of a typical fuzzy rule based control system. The universe, domain, and distribution of the controller's output membership functions are defined at the beginning of every control cycle, thus having a dynamic generation of the SOMF. One or more relevant parameters of the controlled process may be used as inputs to the SOMF fuzzy adaptation stage; in this particular case, these parameters are the instantaneous neutron power n(t) and the normalised power deviation %PD. The fuzzy method designed is incorporated into a fuzzy control algorithm applied to a point-kinetic model of a TRIGA Mark III research nuclear reactor. The control objective is to take the neutron power from its source level up to a desired setpoint, avoiding undesirable power excursions, specially at low power levels, in order to maintain the reactor period above the pre-specified lower limit during power ascent. Likewise, once the desired power level is attained in a soft manner, the fuzzy controller should act as a regulator of the power level for long periods of time. Simulation results are analysed and compared with the response provided by a fuzzy controller without SOMF adaptation.

2 Adaptation of Output Membership Functions in a Fuzzy Controller

A typical fuzzy control architecture consists of the following four main stages: fuzzification, rule evaluation, aggregation, and defuzzification. The process of fuzzification transforms crisp inputs to fuzzy input variables, using a set of input membership functions (IMF); then the fuzzy If-Then rules are evaluated to yield several output fuzzy sets, which in turn are aggregated to produce an equivalent fuzzy output; finally, the defuzzification process transforms this output back into a crisp value by some weighted average method. The last process is typically made using only one set of output membership functions (SOMF) for each crisp output variable. For some systems, it is possible to improve the response characteristics of the closed-loop system if several SOMFs are related to a crisp output; each SOMF is then used in a predetermined operational region of the controlled system. This technique is referred to as *"Boolean selection of a SOMF based on system's operational regions"* [Vélez and Benítez 1997; Benítez and Vélez 1998]. Furthermore, the use of different SOMFs may be extended towards a refreshing or re-definition process of the SOMF, performed every time the fuzzy controller uses the SOMF. This re-definition will be called a *SOMF dynamic generation*. Different schemes can be devised to carry out this process; here, a fuzzy structure is presented.

The SOMF dynamic generating process is accomplished using the typical blocks of a fuzzy controller, and this similarity leads to the label *SOMF fuzzy generation method*. The principal idea of this method is to generate, in a fuzzy manner, the best-suited SOMF at the beginning of every control cycle. Figure 2.1 shows the main blocks of a fuzzy control system with the SOMF dynamic generating process integrated.

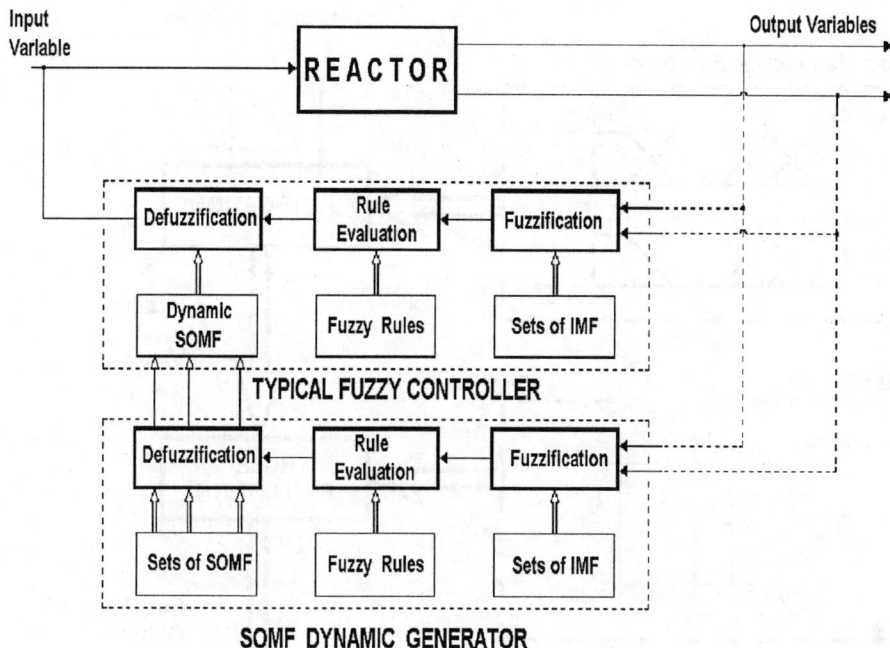

Fig. 2.1. Block diagram of a fuzzy control system using SOMF dynamic generation

2.1 Structure of the SOMF Dynamic Generator

The SOMF dynamic generator is structured in a similar manner as a typical fuzzy controller, meaning that, in addition to a knowledge base, it contains fuzzification, inference, aggregation, and defuzzification stages. A detailed block diagram of the generator, interacting with the fuzzy controller, is shown in Figure 2.2.

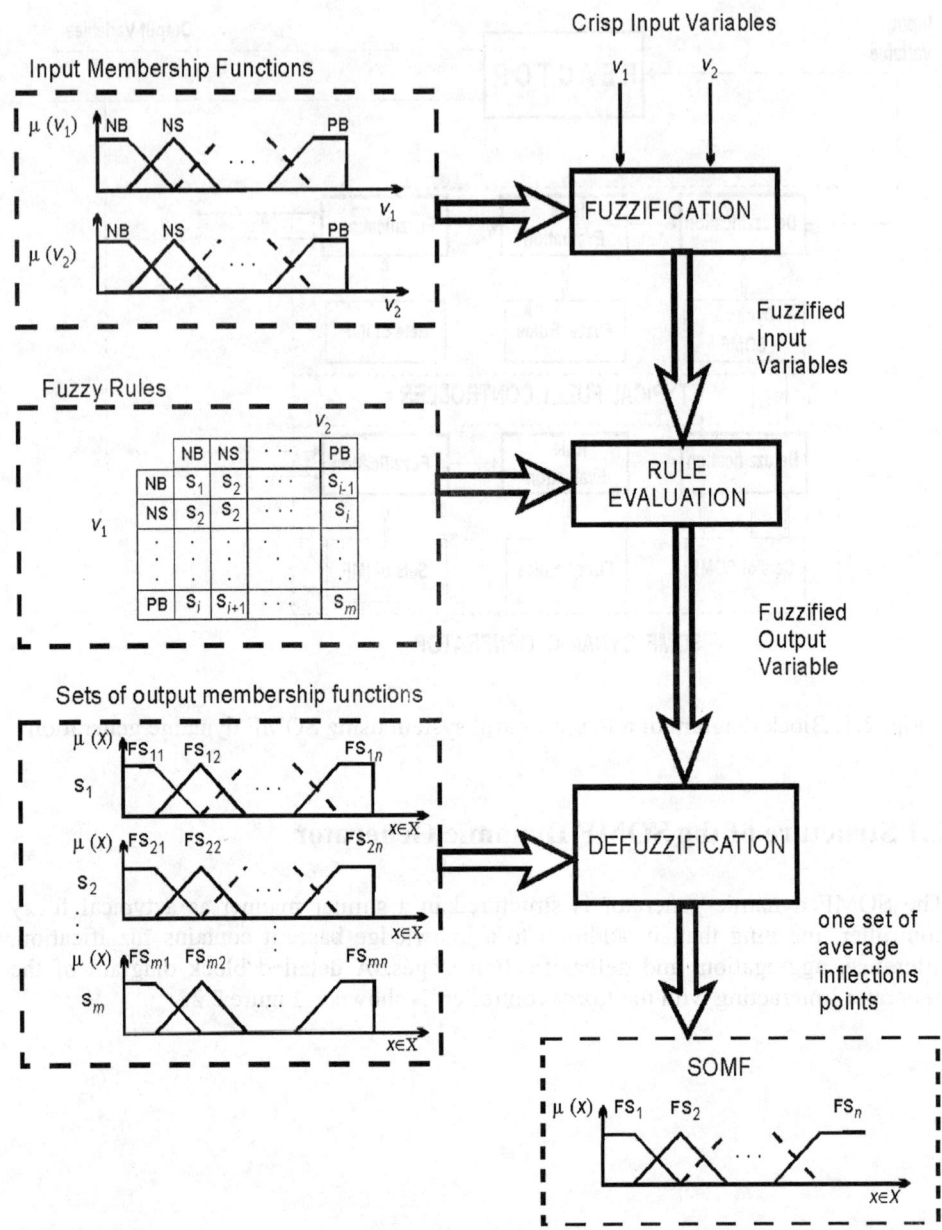

Fig. 2.2. SOMF dynamic generator

The crisp input variables reflect the operating conditions of the controlled system. These variables are fuzzified using the group of fuzzy sets previously defined for each crisp input. The fuzzified input variables are evaluated by the set of fuzzy rules of the generator (If-Then type rules), thus assigning certain weights to the consequents of the fuzzy associative memory (FAM). When several rules activate the same consequent, only one of them is selected, according to the inference method utilised. The outcome of the inference process is a fuzzified variable

$$\psi = \mu_{S1}(x) / S_1 + \mu_{S2}(x) / S_2 + \cdots + \mu_{Sm}(x) / S_m \qquad (2.1)$$

where $\left\{ S_1, \cdots, S_m \right\}$ is the set of different consequents contained in the FAM, and $\mu_{S1}(x)$ to $\mu_{Sm}(x)$ represent the weights assigned to the corresponding consequents, after rule evaluating the fuzzified input variables. The inference method is determined by a specific type of implication, such as the Mamdani type, which is commonly used in control applications.

The generation of the set of inflection points that finally defines the fuzzy controller's SOMF, is the most elaborated stage within the SOMF dynamic generator. For now, this process can be very briefly described as a defuzzification of weighted singleton functions, which are directly related to the weighted consequents of the generator's FAM. In order to describe the details of this process, some concepts will first be introduced in the next section.

2.2 Preliminary Concepts

Any fuzzy set A is completely defined on a universal set X by its membership function μ_A, such that

$$\mu_A : X \rightarrow \left[0,1 \right].$$

The most common membership function shapes are singleton (I), triangular (Λ), and trapezoidal (Π) (Figure 2.3). Although other shapes may be more representative of naturally occurring phenomena, they require more complicated equations.

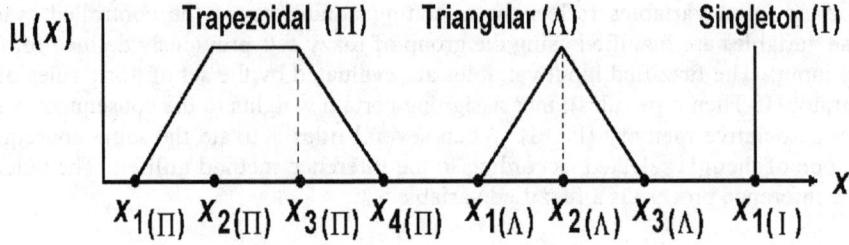

• **X** value where the functions present inflection points

Fig. 2.3. Common membership function shapes

Inflection Set of a Fuzzy Set

Let A be a normal fuzzy set of singleton, triangular, or trapezoidal shape, defined on a universal set X. Then, the membership function that defines A can be easily described by the values where the function presents inflection points. For example, the trapezoidal function in Figure 2.3 can be described as

$$\mu_\Pi(x) = \begin{cases} 0, & x \leq x_1 \ or \ x_4 \leq x \\ \dfrac{x - x_1}{x_2 - x_1}, & x_1 \leq x \leq x_2 \\ \dfrac{x_4 - x}{x_4 - x_3}, & x_3 \leq x \leq x_4 \\ 1, & x_2 \leq x \leq x_3 \end{cases} \qquad (2.2)$$

In general, the membership functions, such as those shown in Figure 2.3, can be described by other functions, defined on certain intervals; that is,

$$\mu_A(x) = \begin{cases} f_1(x), & x \leq x_1 \\ f_2(x), & x_1 \leq x \leq x_2 \\ \vdots & \vdots \\ f_k(x), & x_{k-1} \leq x \end{cases} \qquad (2.3)$$

where $x \in X$ and $f_i \in F = \{f_1, f_2, \cdots, f_k\}$.

The set $\left\{ x_1, x_2, \cdots, x_{k-1} \right\}$ contains the x values where the membership function $\mu_A(x)$ presents inflection points. For the sake of simplicity, these x values will be called "inflection points." Such set of values is called the inflection set of A.

Definition 1. Inflection set.

Let A be a fuzzy set of triangular or trapezoidal shape, defined on the universal set X. The inflection set of A is a crisp set, denoted by *inflect* (A), defined as

$$inflect\ (A) = \left\{ x_i \ \middle|\ \mu_A(x_i) = f_i(x_i) = f_{i+1}(x_i), \ \ i = 1, \cdots, k-1 \right\} \tag{2.4}$$

where $x_i \in X$ and $f_i, f_{i+1} \in F$, $i = 1, \cdots, k-1$.

For the case of a singleton fuzzy set, I, the inflection operation is defined as

$$inflect\ (I) = \left\{ x_1 = x \ \middle|\ \mu_I(x) = 1, \ \ x \in X \right\} \tag{2.5}$$

The notation $x_{i(A)}$ can also be used in the previous definitions to identify, by means of the subscript (A), the fuzzy set to which the i-th inflection point belongs.

Example. The inflection sets of the fuzzy sets shown in Figure 2.3 are:

$$inflect\ (I) = \{ x_{1(I)} \}$$
$$inflect\ (\Lambda) = \{ x_{1(\Lambda)}, x_{2(\Lambda)}, x_{3(\Lambda)} \}$$
$$inflect\ (\Pi) = \{ x_{1(\Pi)}, x_{2(\Pi)}, x_{3(\Pi)}, x_{4(\Pi)} \}$$

A given fuzzy set A cannot, in general, be defined by the set *inflect* (A), since different fuzzy sets may have the same inflection set.

Singleton Generator

The generation of a singleton fuzzy set (or an array of singleton fuzzy sets), from a scalar (or a set of scalars), is possible through the following function:

Definition 2. Singleton(s) from scalar(s).

Let A be a set of scalars, $A = \{a_1, ..., a_n\}$. The array of singleton fuzzy sets related to A, denoted by *singen (A)*, is defined as

$$singen\ (A) = \{\ I_{1(A)}\ ...\ I_{n(A)}\ \} \tag{2.6}$$

where $I_{i(A)} = 1 / a_i$ (singleton fuzzy set located at a_i).

2.3 Dynamic Generation of the SOMF

Referring again to the SOMF dynamic generator of Figure 2.2, the output of the fuzzy rule evaluation stage is the fuzzified variable ψ, given by Eq. (2.1). Each consequent S_i, $i = 1, ... , m$, of the fuzzy rule table (FAM), contains n fuzzy sets (*FS*) defined over the universe of discourse of the output variable of the fuzzy controller. Thus, each S_i, $i = 1, ..., m$, can be denoted by

$$S_i = \{\ FS_{ij}\ |\ j \in N_n,\ FS_{ij} \in \sim\!\mathcal{P}(X)\ \} \tag{2.7}$$

where $i \in N_m$, $\sim\!\mathcal{P}(X)$ is the power set of all possible fuzzy sets defined on the same universal set X, and FS_{ij} can be defined as

$$FS_{ij} = \int_{x \in X} \mu_{FS_{ij}}(x)/x \tag{2.8}$$

Steps to generate the fuzzy controller's set of output membership functions (SOMF dynamic):

Step 1. Array of singleton fuzzy sets.

An *array of the fuzzy sets* contained in the FAM's consequents, can be defined on the same universal set X as,

$$
S = \left\{ \begin{array}{c} S_1 \\ S_2 \\ \vdots \\ S_m \end{array} \right\} = \left\{ \begin{array}{cccc} FS_{11} & FS_{12} & \cdots & FS_{1n} \\ FS_{21} & FS_{22} & \cdots & FS_{2n} \\ \vdots & \vdots & \ddots & \vdots \\ FS_{m1} & FS_{m2} & \cdots & FS_{mn} \end{array} \right\} . \tag{2.9}
$$

Since $inflect\,(FS_{ij}) = \{\, x_{1(ij)}, x_{2(ij)}, \ldots, x_{k\text{-}1(ij)} \,\}$, where the subscript $(i\,j)$ is a short notation for FS_{ij}, then the application of the $inflect$ operation to each element of S produces the following array of "inflection points"

$$
IP = \left\{ \begin{array}{cccc} x_{1\,(11)} & x_{2\,(11)} & \cdots & x_{3\,(1n)} \\ x_{1\,(21)} & x_{2\,(21)} & \cdots & x_{3\,(2n)} \\ \vdots & \vdots & \ddots & \vdots \\ x_{1\,(m1)} & x_{2\,(m1)} & \cdots & x_{3\,(mn)} \end{array} \right\} . \tag{2.10}
$$

The operation $singen$ is then applied to IP, resulting in the following *array of singleton fuzzy sets* (Figure 2.4)

$$
AS = \left\{ \begin{array}{cccc} I_{1\,(11)} & I_{2\,(11)} & \cdots & I_{3\,(1n)} \\ I_{1\,(21)} & I_{2\,(21)} & \cdots & I_{3\,(2n)} \\ \vdots & \vdots & \ddots & \vdots \\ I_{1\,(m1)} & I_{2\,(m1)} & \cdots & I_{3\,(mn)} \end{array} \right\} . \tag{2.11}
$$

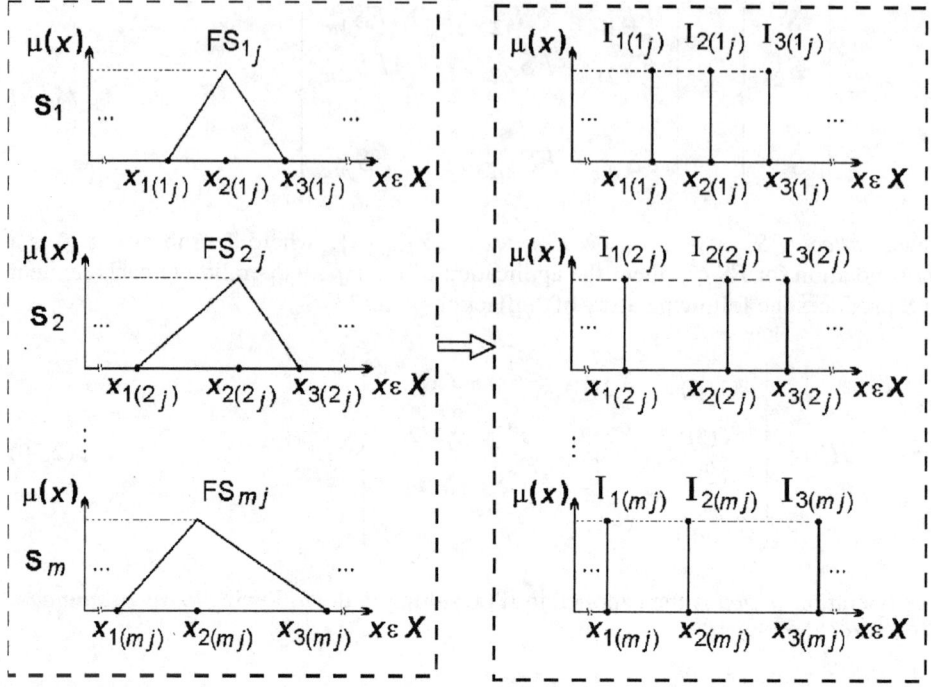

Fig. 2.4. Array of singleton sets corresponding to the *j*-th set in every row of *AS*

Step 2. Array of weighted singleton fuzzy sets.

Each row of singletons in the array *AS* is weighted by the corresponding membership grade μ_{S_i}, $i = 1, \ldots, m$, obtained from the FAM's fuzzified output variable ψ (Eq. (2.1)). The resulting *array of weighted singleton fuzzy sets WS* is (left block of Figure 2.5)

$$
WS = \left\{ \begin{array}{cccc}
\mu_{S1} I_{1(11)} & \mu_{S1} I_{2(11)} & \cdots & \mu_{S1} I_{3(1n)} \\
\mu_{S2} I_{1(21)} & \mu_{S2} I_{2(21)} & \cdots & \mu_{S2} I_{3(2n)} \\
\vdots & \vdots & \ddots & \vdots \\
\mu_{Sm} I_{1(m1)} & \mu_{Sm} I_{2(m1)} & \cdots & \mu_{Sm} I_{3(mn)}
\end{array} \right\}
\tag{2.12}
$$

where, for instance, $\mu_{Sm} I_{1(m1)}$ is a one-element fuzzy set equivalent to $\mu_{Sm} / x_{1(m1)}$.

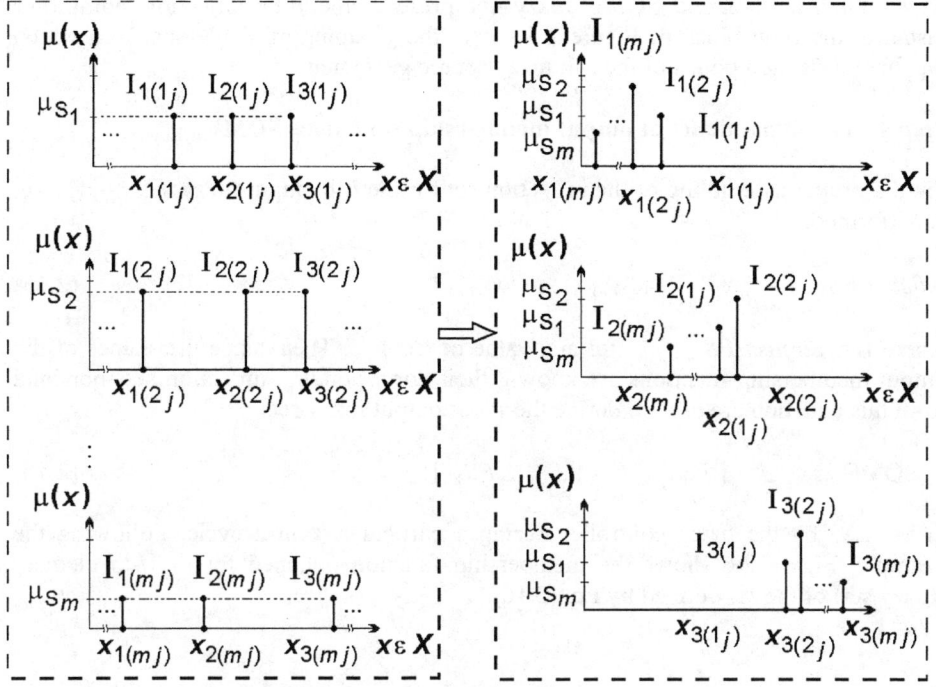

Fig. 2.5. Array of weighted singleton fuzzy sets

Step 3. Column wise defuzzification.

This step produces the array of scalar elements,

$$IP_{SOMF} = \{ x_{1(1)} \quad x_{2(1)} \quad \dots \quad x_{1(j)} \quad x_{2(j)} \quad \dots \quad x_{1(n)} \quad x_{2(n)} \quad x_{3(n)} \} \tag{2.13}$$

where $x_{p(j)} \in X$, and

$$x_{p(j)} = \frac{\displaystyle\sum_{i=1}^{m} x_{p(ip)} \times \mu_{Si}(x_{p(ij)})}{\displaystyle\sum_{i=1}^{m} \mu_{Si}(x_{p(ij)})} \tag{2.14}$$

where for every $j = 1, \dots, n$, the index p takes integer values from 1 to the cardinality of FS_{ij}, $| \, inflect(FS_{ij}) \, |$. The "inflection point" $x_{p(j)}$ represents the value $x \in X$ for which the j-th dynamic SOMF fuzzy set presents the p-th inflection point. For instance, the right block of Figure 2.5 shows the grouping of singletons from which the three inflection points of the j-th fuzzy set are generated.

Step 4. The dynamic set of output membership functions, SOMF $_{dynamic}$.

First, a proper assembling of the inflection set for the j-th output fuzzy set, $j = 1,..., n$, is performed,

$$inflect (FS_j) = \{ x_{1(j)} , x_{2(j)} , ... , x_{L(j)} \} \tag{2.15}$$

where $L = |\ inflect(FS_{ij}) |$, for any value of $i \in N_m$. Then, since the shapes of the output membership functions are known, their corresponding inflection sets obtained from this equation are used to define the set of output fuzzy sets

$$SOMF_{dynamic} = \{ FS_1 , FS_2 , ... , FS_n \} \tag{2.16}$$

to be used by the fuzzy controller during a particular control cycle. Following the example, Figure 2.6 shows the membership function obtained for the j-th element (fuzzy set) of the set defined by Eq. (2.16).

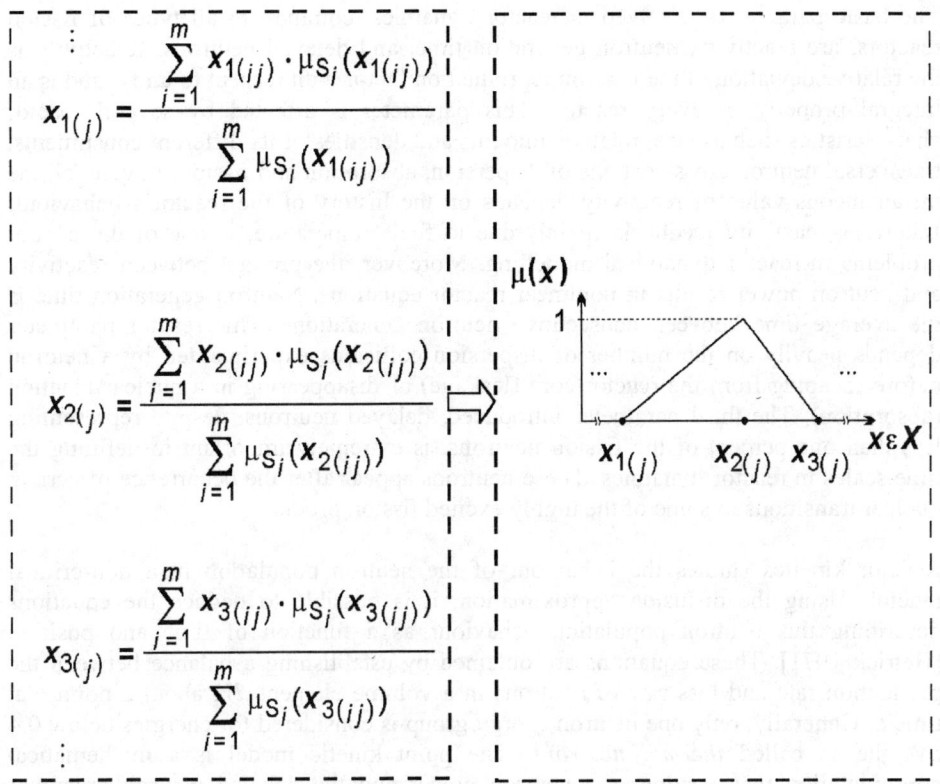

$$x_{1(j)} = \frac{\sum\limits_{i=1}^{m} x_{1(ij)} \cdot \mu_{S_i}(x_{1(ij)})}{\sum\limits_{i=1}^{m} \mu_{S_i}(x_{1(ij)})}$$

$$x_{2(j)} = \frac{\sum\limits_{i=1}^{m} x_{2(ij)} \cdot \mu_{S_i}(x_{2(ij)})}{\sum\limits_{i=1}^{m} \mu_{S_i}(x_{2(ij)})}$$

$$x_{3(j)} = \frac{\sum\limits_{i=1}^{m} x_{3(ij)} \cdot \mu_{S_i}(x_{3(ij)})}{\sum\limits_{i=1}^{m} \mu_{S_i}(x_{3(ij)})}$$

Fig. 2.6. Membership function obtained for the j-th fuzzy set of the SOMF$_{\text{dynamic}}$

3 Application of the SOMF Fuzzy Generation Method toa to a Research Nuclear Reactor

3.1 Point Kinetic Equations

The dynamic behaviour of the neutron population in the core of a nuclear reactor is the main factor to determine its power. In fission nuclear reactors, the essential phenomenon is the fuel's isotope fission induced by neutrons present in the core. This process liberates neutrons of the fissile nuclei, which in turn, if available in sufficient quantities, may attain a self-sustained fission chain reaction.

The basic parameters involved in reactor dynamics, common to all types of fission reactors, are reactivity, neutron generation time, and delayed neutrons. Reactivity is the relative deviation of the neutron reproduction factor with respect to unity, and is an integral property of every reactor. This parameter is affected by several reactor characteristics such as size, relative amounts and densities of its different constituents, transversal neutron cross-sections of dispersion, absorption and fission. In general, the instantaneous value of reactivity depends on the history of the reactor's behaviour. Likewise, reactivity feedback, mainly due to fuel temperature, is one of the central problems in reactor dynamical modelling. Moreover, the product between reactivity and neutron power results in nonlinear reactor equations. Neutron generation time is the average time between consecutive neutron generations. This reactor parameter depends heavily on the number of dispersion collisions experimented by a neutron before escaping from the reactor core (leakage) or disappearing in a nuclear reaction (absorption). The third parameter introduced, delayed neutrons, despite representing less than one percent of the fission neutrons, is extremely important in defining the time scales in reactor dynamics. These neutrons appear after the occurrence of certain nuclear transitions in some of the highly excited fission products.

Reactor kinetics studies the behaviour of the neutron population in a non-critical reactor. Using the diffusion approximation, it is possible to deduce the equations describing this neutron population behaviour as a function of time and position [Hetrick 1971]. These equations are obtained by establishing a balance between the production rate and loss rate of neutrons in a volume element dV, about a point r at time t. Generally, only one neutron energy group is considered for energies below 0.4 eV, the so called *thermal neutrons*. The point kinetic model is a mathematical approximation that assumes the neutron power and the delayed neutron precursor concentration as functions separable in a product of time and space functions. By considering a simple linear feedback model involving exclusively temperature changes, the following point kinetic equations are obtained:

$$\frac{d\,n\,(t)}{dt} = \frac{1}{\Lambda}\,\rho_{\text{int}}(t)\,n\,(t) - \frac{\beta}{\Lambda}\,n\,(t) + \frac{1}{\Lambda}\,\rho_{ext}(t)\,n\,(t) + \lambda\,C\,(t)$$

$$\frac{d\,C\,(t)}{dt} = \frac{\beta}{\Lambda}\,n\,(t) - \lambda\,C\,(t)$$

$$\frac{d\,\rho_{\text{int}}(t)}{dt} = -\,\alpha\,\kappa\left[n\,(t) - n_0\right] - \gamma\,\rho_{\text{int}}(t) \ ,$$

$$(3.1)$$

where $n(t)$ is the neutron concentration or neutron power, $\rho_{ext}(t)$ is the reactivity amount due to external events such as fuel or control rod insertion or withdrawal, $\rho_{int}(t)$ is the reactivity quantity due to changes in the internal characteristics of the reactor components, β is the delayed neutron fraction, Λ is the prompt neutron generation time, $C(t)$ is the one equivalent group delayed neutron precursor concentration, λ is the one equivalent group decay constant, α is the negative of the temperature reactivity coefficient, κ is the reciprocal of the reactor heat capacity (°C/W · s), γ is the reciprocal of the average time of fuel to coolant heat transfer (s^{-1}), and n_0 is a stable system's initial neutron power. Typical values of the point kinetic parameters for a nuclear reactor of the TRIGA Mark III type can be found in [Pérez 1994].

3.2 Design of the Fuzzy Control System

The operation of the reactor is monitored by the continuous measuring of two parameters, the reactor period and the normalised deviation of the actual power level from the setpoint. Taking into consideration the safety requirement of maintaining the reactor period above the 3 seconds level, the fuzzy controller brings the neutron power from its source level to the desired setpoint. A block diagram of the closed-loop fuzzy control system is shown in Figure 3.1. The reactor's measured variables are neutron power, $n(t)$, and reactor period, T. The control input $\rho_{ext}(t)$ represents the external reactivity applied to the reactor by the insertion or withdrawal of the control rod [DeGroot 1968]. The crisp inputs to the fuzzy controller are T and the normalised percentage of the neutron power deviation from its setpoint, %PD. The block diagram of the closed-loop fuzzy control system is shown in Figure 3.1, where the reactor behaviour (controlled system) is simulated using the system's equations (3.1), $\rho_{ext}(t)$ is the system input variable (control variable) and $n(t)$ is the system output variable (controlled variable).

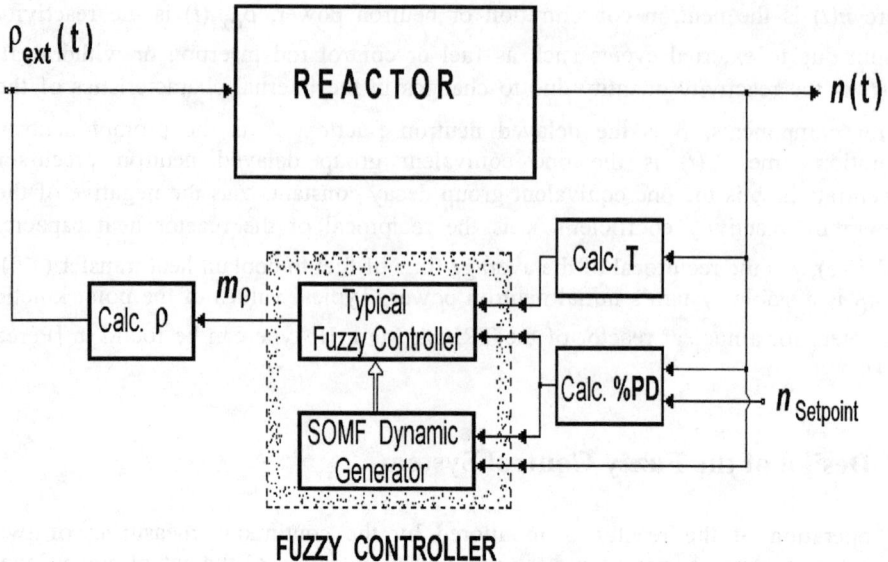

Fig. 3.1. Block diagram of reactor fuzzy control system

One of the crisp input variables to the fuzzy controller is the reactor period (T), which represents the time required by the power, at any instant of time, to be increased by a factor "*e*," considering that the neutron power rate of change is kept constant. The reactor period has a universe of discourse from 0 to infinite seconds, and is evaluated for simulation purposes as

$$T = \frac{t - t_i}{\ln\left(\dfrac{n(t)}{n(t_i)}\right)},$$

(3.2)

where $n(t) \neq n(t_i)$. The other crisp input to the fuzzy controller is the normalised percentage of the neutron power deviation from its setpoint, called %PD, which is computed by the expression

$$\%PD = \frac{n(t) - n_{setpoint}}{n_{setpoint} - n_0} \times 100\%$$

(3.3)

for which the universe of discourse takes on values from -110% to +10%.

Each pair of crisp input variables to the fuzzy controller produces one crisp output variable, m_ρ, representing the rate of change of external reactivity. Once m_ρ is defined, it is used in the following expression, to obtain the external reactivity, $\rho_{ext}(t)$ (control variable):

$$\rho_{ext}(t_i) = \rho_{ext}(t_{i-1}) + m_\rho(t_i - t_{i-1}) \qquad (3.4)$$

Two blocks can be identified within the fuzzy controller: the upper block is the typical fuzzy controller, whereas the lower one represents the SOMF dynamic generator.

Typical Fuzzy Controller

The control of neutron power is effected by the typical fuzzy controller, for which its main components are described below.

Fuzzification of crisp inputs. The crisp input variable T is fuzzified using five input membership functions: μ_{CR}, μ_{CC}, μ_{NR}, μ_{BG}, and μ_{TI}, which define the fuzzy sets shown in Figure 3.2(a). The fuzzy labels assigned to the different fuzzy sets represent the following: CR, critical; CC, close to critical; NR, normal; BG, big; and TI, tends to infinite. In a similar manner, the other crisp input variable %PD is fuzzified by means of the following input membership functions: μ_{NB}, μ_{NM}, μ_{NS}, μ_{ID}, and μ_{PS}, which define the fuzzy sets shown in Figure 3.2(b). The fuzzy labels assigned to the different fuzzy sets represent the following: NB, negative big; NM, negative medium; NS, negative small; ID, ideal; and PS, positive small.

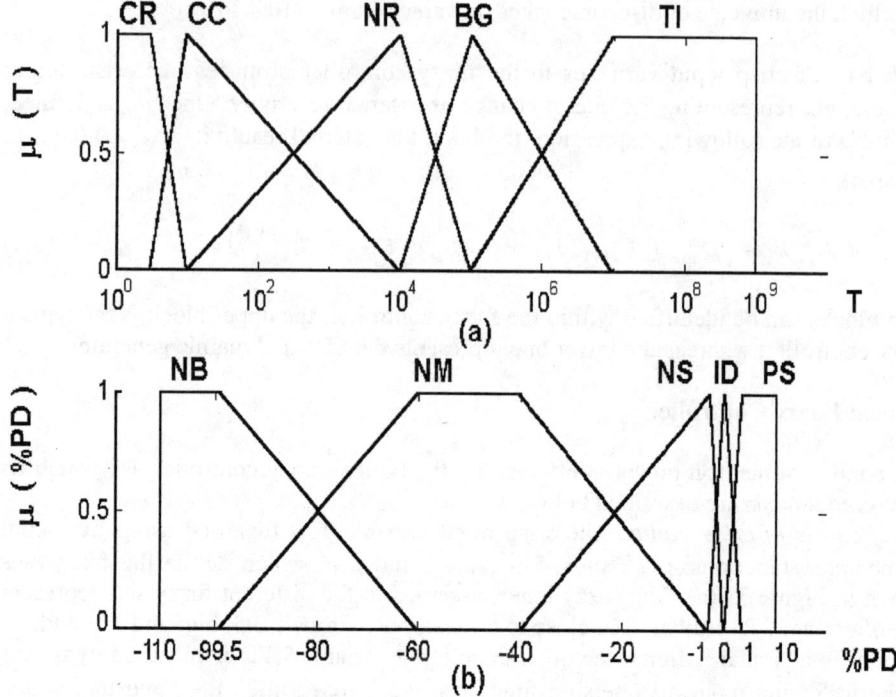

Fig. 3.2. Input membership functions used by the typical fuzzy controller

Fuzzy output sets. These sets, defined over the crisp variable m_ρ by the membership functions μ_{NS}, μ_{ZE}, μ_{PS}, and μ_{PB}, are shown in Figure 3.3. The fuzzy labels assigned represent the following: NS, negative small; ZE, zero; PS, positive small; and PB, positive big. The exact definition of these sets, over the universe of m_ρ, is carried out in every control cycle by the SOMF dynamic generator.

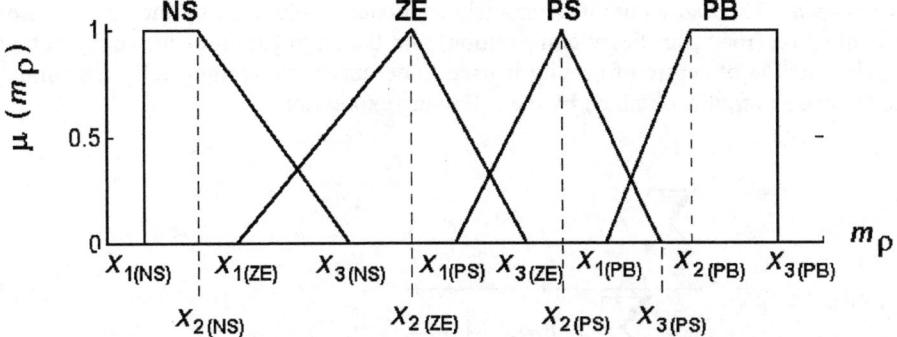

Fig. 3.3. Output membership functions used by the typical fuzzy controller

Fuzzy rules. The operation of the reactor generally considers the reactor period and the normalised deviation of the actual power level from the setpoint. An important safety requirement is to maintain, at all times, the period above a value of three seconds, when increasing power. Based on these parameters, and also on the analysis of the dynamic open-loop response [Vélez and Benítez 1995], a set of twenty five fuzzy rules is proposed for the implementation of the fuzzy control algorithm. The set of fuzzy rules is shown in Figure 3.4.

T

	CR	CC	NR	BG	TI
NB	ZE	PS	PS	PS	PS
NM	ZE	PS	PB	PB	PB
% PD **NS**	ZE	PS	PS	PS	PS
ID	NS	ZE	ZE	ZE	ZE
PS	NS	NS	NS	NS	NS

m_ρ

Fig. 3.4. Fuzzy rule table (FAM)

Crisp Output. The fuzzy output membership function is obtained by the processes of rule evaluation (max-min fuzzy composition) and the aggregation of the fuzzy output sets. The method of centre of gravity is used to defuzzify the composite fuzzy output; thus, the crisp output is obtained by the following expression:

$$
m_{\rho(COG)} = \frac{\sum_{m_\rho} \mu(m_\rho) \cdot m_\rho}{\sum_{m_\rho} \mu(m_\rho)}
\tag{3.5}
$$

The universe of discourse of the crisp output is a dynamic range.

SOMF Dynamic Generator

The SOMF dynamic generator determines the set of the fuzzy controller's output membership functions in every control cycle. The main characteristics of this generator are described below.

Crisp inputs. Two parameters from the controlled system, neutron power $n(t)$ and normalised power deviation %PD, are taken as the crisp inputs to the generator.

Fuzzification of crisp inputs. The fuzzy sets for the crisp input $n(t)$ are shown in Figure 3.5(a): small neutron power, S; medium neutron power, M; and big neutron power, B. On the other hand, five fuzzy sets are used to fuzzify the crisp input %PD. These sets are shown in Figure 3.5(b), and the labels and meaning used for them are the same as those used in the fuzzification stage of the typical fuzzy controller.

Fuzzy rules. These rules consider several regions of operation of the reactor, represented by the fuzzified variables obtained from the fuzzification of the parameters $n(t)$ and %PD. The consequents of the rule table have been devised to generate an adequate distribution of the output membership functions (SOMF), according to the reactor operating conditions. The rule table (FAM) is shown in Figure 3.6, where four different consequents can be identified. As mentioned in Section 2.3, the consequents of this FAM are composed of a series of fuzzy sets.

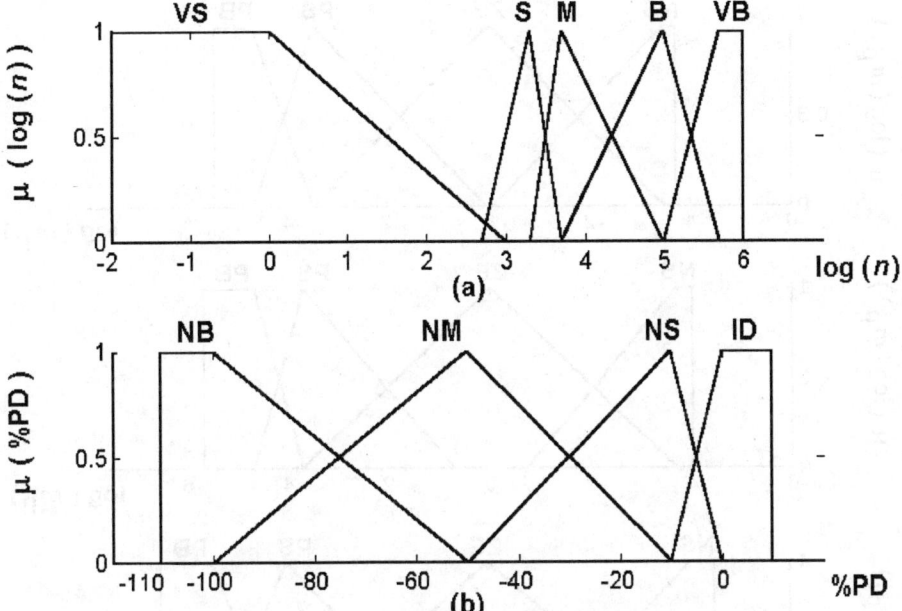

Fig. 3.5. Input membership functions used by the SOMF dynamic generator

$n(t)$

% PD	VS	S	M	B	MB
NB	S_1	S_1	S_2	S_3	S_4
NM	S_1	S_2	S_3	S_4	S_4
NS	S_2	S_3	S_4	S_4	S_4
ID	S_2	S_3	S_4	S_4	S_4

Fig. 3.6. Fuzzy rule table (FAM) of the SOMF dynamic generator

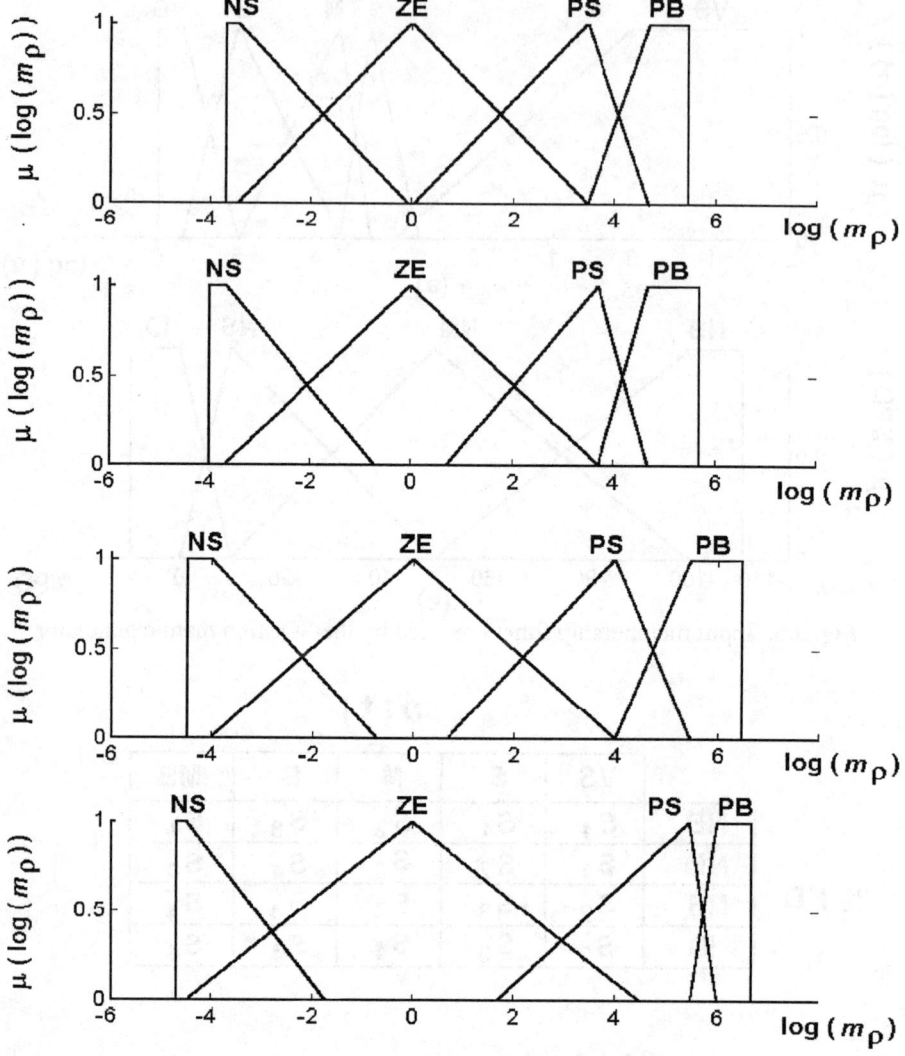

Fig. 3.7. Series of SOMF

Series of SOMF. Each of the four consequents is composed of four fuzzy sets. The four groups of four fuzzy sets each are similar in the sense of sharing the same labels and shapes, but differing in their domains over the universe of discourse of m_ρ. The sets of output membership functions associated to the consequents are shown in Figure 3.7, where the fuzzy labels represent the following: NS, negative small; ZE, zero; PS, positive small; and PB, positive big.

Final SOMF per control cycle. The output membership functions to be used by the typical fuzzy controller in a given cycle are obtained following the four-step procedure of the SOMF dynamic generation method described in Section 2.3. A general outcome of this process for this application is

$$\text{SOMF}_{dynamic} = \{NS, ZE, PS, PB\} \tag{3.6}$$

where the inflection set of each fuzzy set of the SOMF $_{dynamic}$ is

$$inflect(NS) = \{m_{\rho 1(NS)}, m_{\rho 2(NS)}, m_{\rho 3(NS)}\} \tag{3.7}$$

$$inflect(ZE) = \{m_{\rho 1(ZE)}, m_{\rho 2(ZE)}, m_{\rho 3(ZE)}\} \tag{3.8}$$

$$inflect(PS) = \{m_{\rho 1(PS)}, m_{\rho 2(PS)}, m_{\rho 3(PS)}\} \tag{3.9}$$

$$inflect(PB) = \{m_{\rho 1(PB)}, m_{\rho 2(PB)}, m_{\rho 3(PB)}\} \tag{3.10}$$

4 Results and Conclusions

The external reactivity (control variable) required to control the neutron power is shown in Figure 4.1, where $\rho_{ext2}(t)$ is obtained by using the fuzzy controller with SOMF dynamic generation, $\rho_{ext1}(t)$ is obtained with a typical fuzzy controller (with only one set of output membership functions).

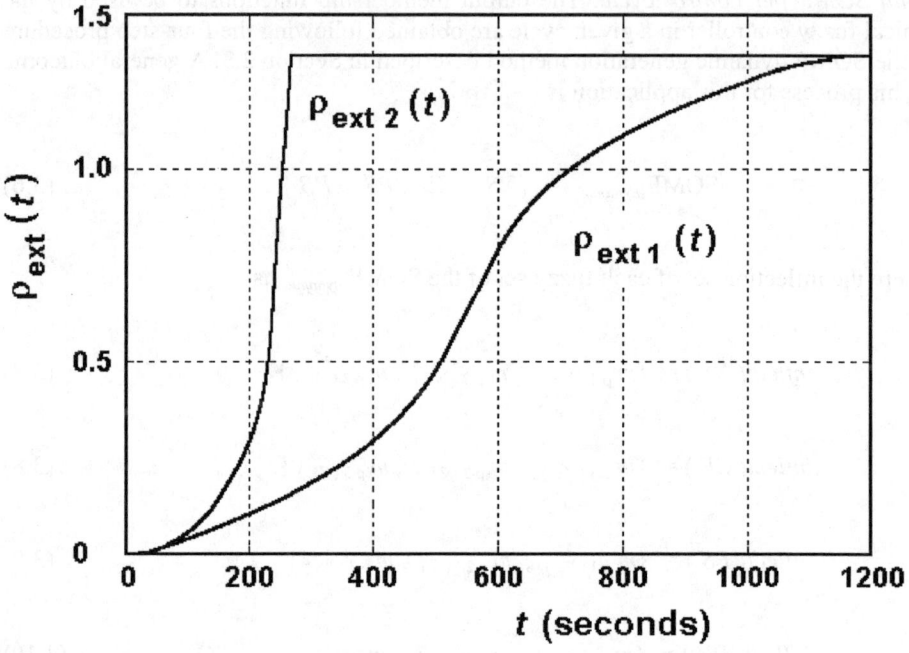

Fig. 4.1. External reactivity applied to the reactor

The results of the neutron power response (controlled variable) are shown in Figure 4.2, where $n_2(t)$ and $n_1(t)$ are the responses obtained when the fuzzy controller with SOMF dynamic generation and the typical fuzzy controller are used, respectively.

The amounts of external reactivities inserted, $\rho_{ext1}(t)$ and $\rho_{ext2}(t)$, are small and very close to each other, when the reactor is operating at low levels of neutron power. As a consequence, the modelled plant responds in a very similar fashion for both $\rho_{ext1}(t)$ and $\rho_{ext2}(t)$. Another characteristic to be inferred from the result graphics is the fact that the same levels of neutron power are attained for the same levels of external reactivity inserted, where the difference is the growing rate of the external reactivity determined by each controller. In both cases, the interrelation of reactor power, period, and power difference with respect to the desired power level produces a rate of change of reactivity that prevents power overshoots.

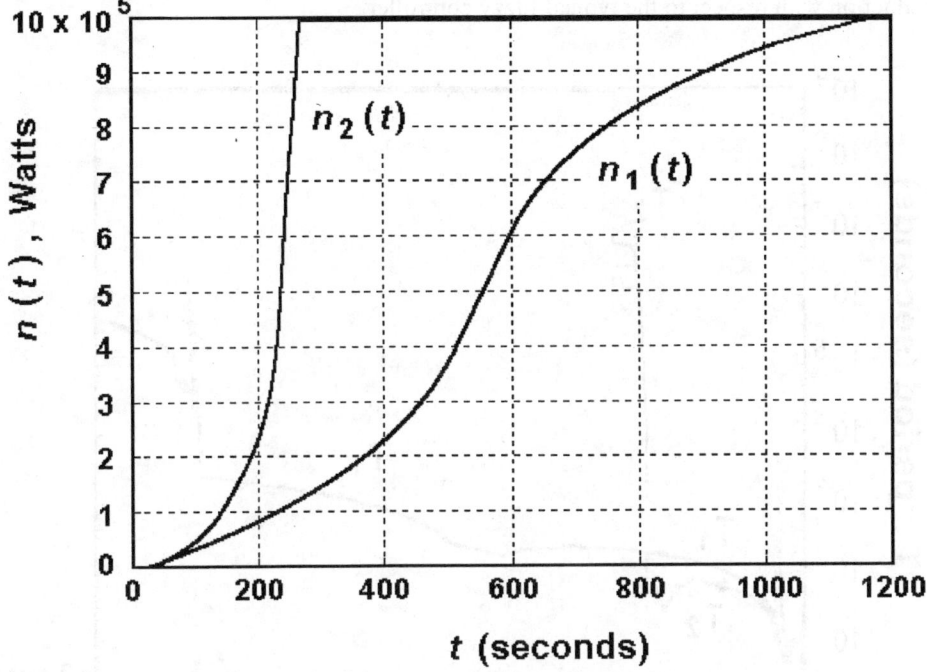

Fig. 4.2. Neutron power responses

Reactor period values throughout the control process are shown in Figure 4.3. Some interesting conclusions can be deduced from this figure in combination with Figures 4.1 and 4.2. First of all, both controllers drive the reactor period close to the lower period limit of 3 seconds, when the reactor operates at a low power level, during the first 50 seconds. Although the separation between $\rho_{ext1}(t)$ and $\rho_{ext2}(t)$, around the 50 seconds, is barely noticeable, the controller with SOMF adaptation starts producing greater values of the reactivity's rate of change, a fact that derives in lower period values around and after the 50 seconds in Figure 4.3. Still, the instantaneous period values are always maintained above the 3 seconds safety limit. The second conclusion is related to the possibility of power overshoots occurring around the desired power level (see curve $n_2(t)$ in Figure 4.2). In this regard, the combination of low %PD values and the SOMF fuzzy generation process, precisely at the time when the neutron power is about to attain the desired level, causes a sudden increase in the period values. In consequence, the controller reduces the rate of change of external reactivity inserted, thus arriving at the desired power level without any overshoot. As a third conclusion, the fuzzy controller with SOMF fuzzy generation needs fewer control cycles to attain a desired power level, which also means a faster response, approximately a 50% reduction with respect to the typical fuzzy controller.

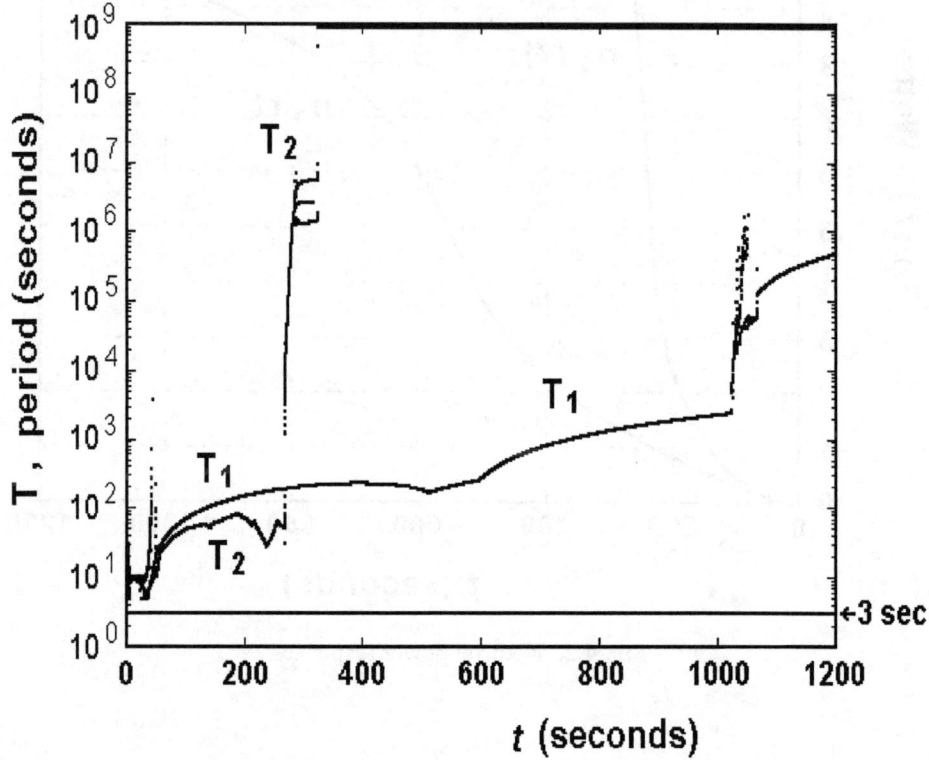

Fig. 4.3. Reactor period for a typical fuzzy controller, T_1, and a fuzzy controller with SOMF fuzzy generation, T_2

With respect to the implementation of the fuzzy algorithm on the reactor control console, long computing times per control cycle have been observed, qualitatively. The core of this console is a personal computer (PC), which performs a large amount of tasks such as supervision, interlocking, and man-machine interfacing. Some tests are being carried out to determine if this computational overload may be one of the causes of the slow execution of the algorithm. A possible solution to this potential problem could be the implementation of the algorithm on a microcontroller, which would receive the control commands from the PC. Likewise, the programming structure of the algorithm needs to be analysed in depth, and if required, to be optimised.

To conclude the chapter, the knowledge of the dynamic characteristics of the controlled system may be used to provide some kind of knowledge-based adjustment of the controller's parameters to improve the response of the closed-loop system.

References

Benítez-Read JS, Jamshidi M, Kisner R (1992), "Advanced Control Designs for Nuclear Reactors," *Control – Theory and Advanced Technology, C –TAT*, vol 8, (3), pp 447-464, September.

Benítez-Read JS, Jamshidi M (1992), "Adaptive Input-ouput Linearizing Control of Nuclear Reactors," Control –Theory and Advanced Technology, C –TAT, vol 8, (3), pp 525-546, September.

Benítez-Read JS, Abdallah C, Kumbla KK (1994), "Adaptive Nonlinear Control for Neutron Power Tracking in a Research Reactor," *Proc. of the European Simulation Symposium*, vol II, pp 195-199.

Benítez-Read JS, Vélez-Díaz D (1996), "Neutron Power Control in a Research Reactor Using a Fuzzy Rule Based System," *Soft Computing with Industrial Applications,* vol 5, pp 53-58, TSI Press Series.

Benítez-Read JS, Vélez-Díaz D (1998), "Comparative Study of Fuzzy Control Algorithms for a Nuclear Reactor," *Intelligent Automation and Soft Computing,* vol 1, pp ISORA-023.1-6, TSI Press CD ROMS Series.

Bereznai GT (1973), "Adaptive Nuclear Reactor Control Based on Optimal Low-Order Linear Models," *IEEE Transaction on Nuclear Science,* April, vol NS-20, 2, pp 72-79.

Bernard JA, Lanning DD, Ray A (1984), "Digital Control of Power Transients in a

Nuclear Reactor," *IEEE Transaction on Nuclear Science,* February, vol. NS-31, (1), pp 701-705.

Bernard JA (1986), "The Construction and Use of a Knowledge Base in the Real-Time Control of Research Reactor Power," *Proceedings of the Sixth Power Plant Dynamics, Control and Testing Symposium,* Knoxville, TN, April.
Bernard JA (1988), "Evaluation of 'Period-Generated' Control Laws for the Time-Optimal Control of Reactor Power," *IEEE Transaction on Nuclear Science,* February, vol NS-35, (1), pp 888-893.

Bubak M, Moscinski J (1983), "A Fuzzy-Logic Approach to HTR Nuclear Power Plant Model Control," *Annals of Nuclear Energy UK*, vol 10, (9), pp 467-471.

DeGroot MN (1968), "TRIGA Mark III Reactor: Instrumentation Maintenance Handbook," *Document No.GA-8585*, Gulf General Atomic, Inc.

Edwards RM, Garcia HE, Turso JA, Chavez CM, Abdennour AB, Weng CK, Ku CC, Ray A, Lee KY (1992), "Advanced Control Research at the Pennsylvania State University," *Advanced digital computers, controls, and automation technologies for power plants: Proceedings,* Electric Power Research Inst., Palo Alto, CA (United States), August, pp 26.1-26.10.

Hetrick DL (1971), *Dynamics of Nuclear Reactors*. The University of Chicago Press.

Jungin C, Yungjoon H, Unchul L (1993), "Automatic Reactor Power Control for a Pressurized Water Reactor," *Nuclear Technology,* May, vol 102, (2), pp 277-286.

Leopando LS (1998), "Incorporation of Personal Computers in a Research Reactor Instrumentation System for Data Monitoring and Analysis," Proc. of an IAEA Final Research Co-ordination Meeting, February, pp 97-115.

Lin C, Lee C, Raghavan R, Fahrner DM (1995), "Fuzzy Logic Control of Water Level in Advanced Boiling Water Reactor," *Proc. of the International Conference on Mathematics and Computations, Reactor Physics, and Environmental Analyses,* Portland, OR, USA, 30 Apr - 4 May 1995, vol 1-2, pp 32-38; La Grange Park, IL, USA, American Nuclear Society, Inc. 1995.

Lipinski WC, Vacroux AG (1970), "Optimal Digital Computer Control of Nuclear Reactor," *IEEE Transaction on Nuclear Science,* February, vol NS-17, (1), pp 510-516.

Moon BS (1993), "Fuzzy Logic Controllers for the Nuclear Power Plants: Simulation Experiences," *Hungarian Korean Symposium on Nuclear Energy*, Balatonfuered

113

(Hungary), 30 Mar - 2 Apr 1993, Uri,-G. (ed), *Nuclear Energy*, Budapest, Hungary, pp 237-250.

Nah MK (1995), "Application of Adaptive Control Theory to Nuclear Reactor Power Control," *Journal of the Korean Nuclear Society,* June, vol 27, (3), pp 336-343.

Nava W (1991), "Automatic Operation in Stationary Mode," (In Spanish), *Instruction: I.UR-10*, Rev 2, May 17, Instituto Nacional de Investigaciones Nucleares (ININ), México.

Pérez-Carbajal V (1994), *Input-Output Linearizing Control of a Nuclear Reactor*, (In Spanish), BSc Thesis, May 1994, Instituto Tecnológico de Toluca, México.

Raju GVS, Fadra UG (1973), "Design of an Optimal Noninteracting Control System," *IEEE Transaction on Nuclear Science,* February, vol NS-20, (1), pp 668-674.

Saif M (1989), "A Novel Approach for Optimal Control of a Pressurized Water Reactor," *IEEE Transaction on Nuclear Science,* February, vol NS-36, (1), pp 1317-1325.

Shibuya S, Iijima T (1996), "Application of Fuzzy Logic Control System for Reactor Feedwater Control," *Advances in the Operational Safety of Nuclear Power Plants, Proceedings of an International Symposium,* pp 601, International Atomic Energy Agency, Vienna (Austria).

Silvinsky CR, Schultz DG (1970), "State Variable Feedback and Series Compensator of Multivariable System," *Nuclear Science Engineering,* vol 41, (1), pp 125-129.

Sun BKH (1997), "Control and Automation Technology in United States Nuclear Power Plants," *Advanced control systems to improve nuclear power plant reliability and efficiency.international,* July, pp 177-184, Atomic Energy Agency, Vienna (Austria).

Takahashi Y, Takamatsu S (1965), "Digital Start-Up Control of a Research Reactor", *IEEE Transaction on Nuclear Science,* August, vol NS-12, (4), pp 355-366.

Tepper L (1975), "Suboptimal Control Study of a Nuclear Power Plant," *IEEE Transaction on Nuclear Science,* February, vol NS-22, (1), pp 812-819.

Vélez-Díaz D, Benítez-Read JS (1995), "Study of the Behavior of a TRIGA Reactor Point Kinetic Model, Based on Simulations," (In Spanish), *Technical report IT.ET.A-9514*, Instituto Nacional de Investigaciones Nucleares, México.

Vélez-Díaz D, Benítez-Read JS (1997), "Fuzzy System to Control the Neutron Power with Different Sets of Output Membership Functions," *Proc. 7th IFSA World Congress*, vol IV, pp 132-136, ISBN 80-200-0633-8, Prague, Czech Republic, June 25-29.

Weaver LE, Vanesse RE (1967), "State Variable Feedback Control of Multiregion Reactors," *Nuclear Science and Engineering,* vol 29, pp 264-271.

Weng CK, Edwards RM, Ray A (1994) "Robust Wide-Range Control of Nuclear Reactors by Using the Feedforward-Feedback Concept," *Nuclear Science and Engineering,* July, vol 117, (3), pp 177-185.

6 NPP Operator Support in Decision Making - Diagnostics of the Operation Failures Using Fuzzy Logic

Ivan Petruzela

I&C Energo, areal VU, budova c.2
190 11 Praha 9 - Bechovice, Czech republic
ipetruzela@ic-energo.cz

In large complex systems such as nuclear power plants (NPP) and chemical industry plants, various subsystems fulfil the needs of process control and safety. Continued operation of these systems has both economic and safety implications. Electric utilities seek continuously to improve the operation of power generating stations. The improvement in the NPP Dukovany (4x440 MW) is related to the introduction of the plant to frequency control, meaning that the plant is operated not only in the base load operational mode but in the load follow mode as well. To achieve the improvement of plant operation, it was necessary to provide modifications to the plant control system, plant information system and diagnostics. This chapter deals with utilisation of artificial intelligence (AI) methods in implementation of these plant modifications.

1 Use of Artificial Intelligence Methods in the Control of Complex Technological Systems

1.1 Control of Complex Technological Systems

In the operation of large and complicated systems, the ensuring a limitation of errors becomes more significant. Two factors influence the development and use of new procedures: safety and operation economy. Both factors induce the elimination of risk and optimal system performance.

The aim of company's management is to minimise losses and therefore minimise unscheduled outages of industrial plants. Analysis of recent catastrophic events in industrial areas and aviation has shown that the greatest factor in the reduction of losses is the improvement of the operator's ability to derive the correct conclusions from incomplete or fuzzy information. The problem areas include [1]:

- insufficient operator concentration;

- failure due to a breach of operational instructions;

- inability to anticipate equipment failure.

One of the possible solutions might be to develop maximal automated technological processes. Such systems should work correctly without frequent operator interventions. In case problems do occur. However, correct activity is expected from a stressed operator overloaded with information. Thus decision-making and its execution occurs in conditions of alarm and panic. Another solution is to respect and support the operator task in the control of such systems. Impetuous development of computer technology allows the use of artificial intelligence methods.

During the operation of a nuclear power plant (NPP), the operator controls simultaneously multiple technological subsystems, mutually linked through various physical processes. In each activity, there are different properties for a given control level. It is necessary to filter off information that is superfluous or even misguiding for the given conditions. The operator gains an overview of the overall condition from the information systems in the main control room. These information systems generate new information, differing qualitatively from the initial input data. The top of this information pyramid is formed with the "early failure detection system."

1.2 Artificial Intelligence Methods

Artificial intelligence (AI) in the past has been associated mainly with expert systems and artificial neural networks. Nevertheless, a number of new techniques have been developed, such as pattern recognition, modelling, decision-making theories, genetic algorithms, real time learning, and fuzzy systems. The aim of the implementation of such techniques is to obtain better control and prediction of highly non-linear systems behaviour. Their common characteristic is the non-algorithmic principle. AI methods from the viewpoint of their use in the NPP control will be summarised in the following sections.

1.2.1 Modelling

In relation to diagnostics in control systems, predictive models are referenced, i.e., models of a system that allow the prediction of the behaviour of the system in the near future with high probability level. These methods are not included in the artificial intelligence methods as they have a close relation to classical control and regulation methods.

The use of predictive methods is mostly based on an analytical description of a system. The foundation is a set of differential equations describing the system's properties. If we are able to create such a set of equations, nothing else is required in the creation of a mathematical model of the whole system on the computer by selecting an appropriate time step.

We can obtain important information by comparing the data from the technological process with the model outputs. It is also possible, by shorting the time step of the calculation, to anticipate the system under examination. The development of this process in the model then predicts the probable behaviour of the system in the near future. In this way, it is possible to detect some sorts of failures and the operator may correct them in time.

1.2.2 Expert Systems

Expert systems are computer programs simulating the decision-making process of an expert during the solution of complicated tasks by utilising his special knowledge took over from an expert. This knowledge is expressed in the way suitable for reaching the quality level of the expert in the decision-making within the selected problem area.

The term "expert system" first appeared at the end of the seventies and in the early eighties as a consequence of the finding that the quality of artificial intelligence depends on the quality of knowledge rather than on the quality of the mechanism for their use. Expert systems were construed from the early beginning as systems based on best expert knowledge .

In conventional programs, the knowledge of the experts is "spread" in the sequence of individual program instructions, given in advance. Expert systems, however, have expert knowledge expressed in a fully explicit way in the form of a "knowledge base" and only the strategy of its is given in advance.

Expert systems, nevertheless, did not approach the decision-making style and level of the experts. They hardly provide a piece of advice in case a part of required data is not available. Therefore greater efforts were directed to other AI areas. Artificial neural networks and fuzzy systems exhibited the greatest development.

1.2.3 Artificial Neural Networks (ANN)

Neural networks attracted considerable attention in many disciplines and their implementation has proven to be useful in certain areas of NPP. ANN solve problem situations by comparing them to information prepared in advance. ANN is being created using a special procedure, called training, where information about possible situations and the corresponding solutions are processed. Therefore, the knowledge of the specific dependencies between the input parameters and their solution is not needed for the ANN creation. ANN methods are particularly useful for the solution of situations, described by a great number of parameters with dependencies that are ambiguous or hardly definable.

ANN tested well in the subsystems diagnostics, but prefer their use for the control of the entire unit is controversial. It is difficult to create a complete kit of training sets and the generalising ability of ANN is limited.

Training sets are mostly obtained by capturing directly from the pertinent real process. We normally require that the function of neural network is to approximate the function of the process even in extreme situations occurring in real only very rarely or not at all. It is possible to obtain process functions representative enough for such situations only by simulating extreme situations using a mathematical model of the process.

1.2.4 Fuzzy Logic

Fuzzy logic is a mathematical discipline forming the basis for successful applications in the area of regulation and control. Fuzzy logic works with uncertainties and contradictory facts do not interfere with its work. Its structure corresponds to the natural human thinking and is even able to work with the normal usage of words.

These properties enable the use of ambiguously defined expert knowledge, to obtain a specific answer. Fuzzy logic often considered a trendy expression for the theory of probability. Both disciplines differ in both terminology and apparatus, however. The theory of probability serves to determine a degree of uncertainty, in which any of the presumed effects occurs. In the contrary fuzzy logic helps to determine which one of many effects defined in different way have occurred. There are also external, formally identical signs. Degree of appurtenance of the fuzzy sets is mostly defined in the interval [0,1] as in probability.

The main difference between the approaches of fuzzy logic and the classical technique is that fuzzy logic uses even uncertain information to describe the system, whereas the classical technique requires exact analytical relations. Therefore fuzzy logic offers the mathematical language appropriate for the description of the control of a system, the behaviour of which may be described in form of the 'if-then' rules.

119

2 A Proposal for a System to Support the Operator during NPP Operation in Frequency Control

2.1 NPP Control

Non-linear behaviour of the NPP systems make their control and optimisation very difficult. From this viewpoint it is necessary to understand the mental processes of the operators working in the NPP control room [2].

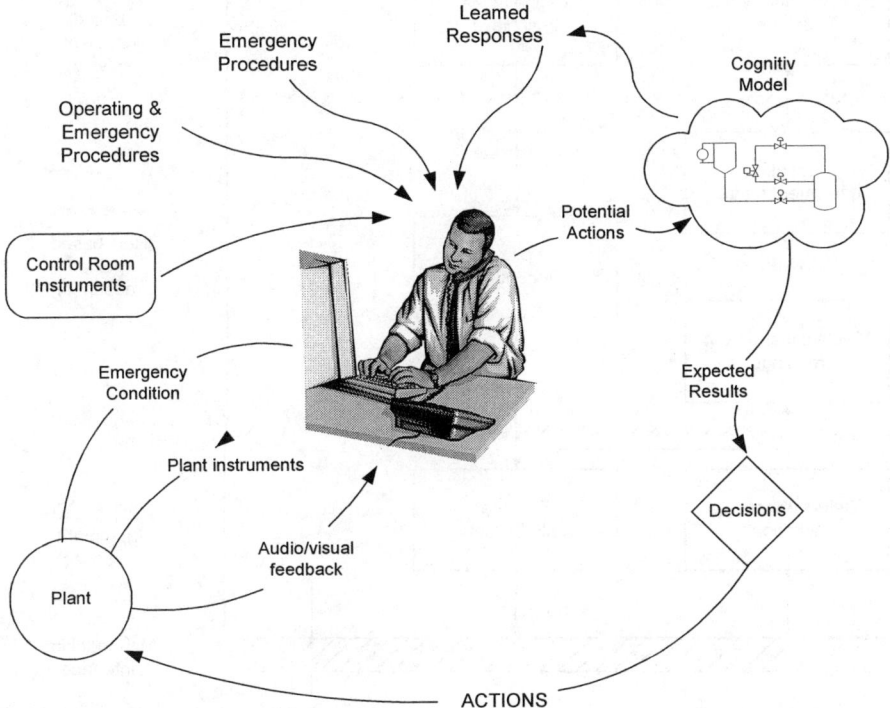

Fig.1. The role of the operator in the control room

Figure 1 shows the way the operator controls the power plant and the importance of using the proper predictive model for the plant. Each operator builds such models during time through their own experience, training, interaction with other operators and knowledge of operational procedures. The quality of the personal predictive model and the ability to use it for process control makes the difference between a good operator and an uninterested one.

The significance and quantity of information with which the operator is ready to occupy himself is variable. During nominal, failure-free, operation only a little basic information is needed for the control and there is a risk of underestimation of the need to follow the development of the situation. When an unexpected event occurs a great amount of information is needed for effective and safe decision-making.

The process of unit control may be explained using the information model developed by Rasmussen [3], depicted in Figure 2. In this model, the control process is divided from the operators' viewpoint into three categories:

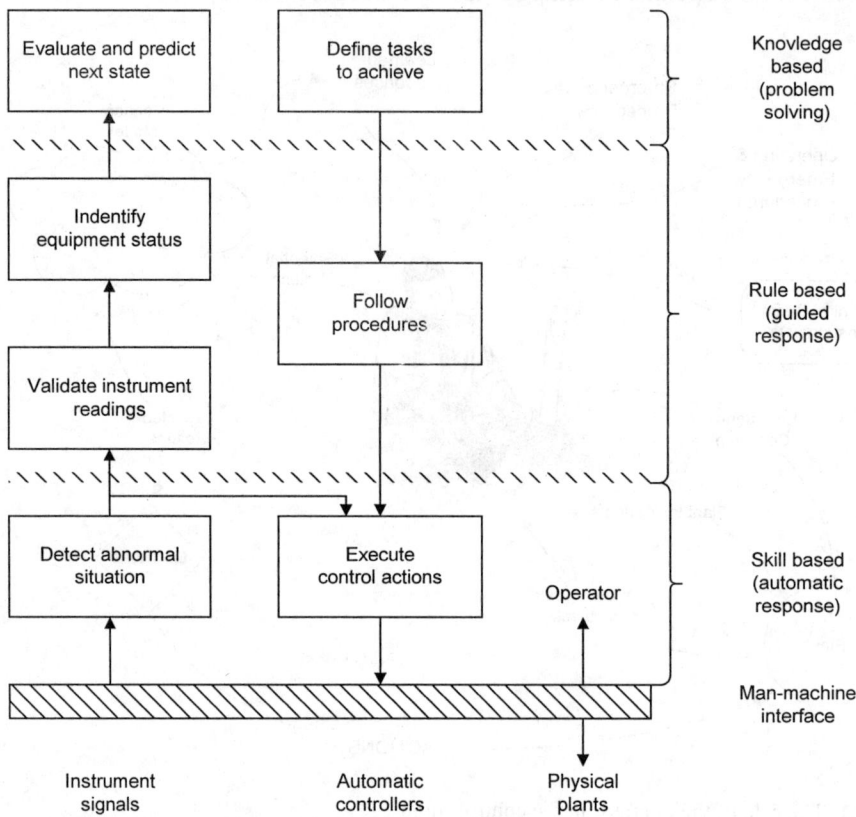

Fig. 2. Information model of the control room process

- Skill based, often practised tasks may be performed as more or less subconscious routines controlled by fixed models of behaviour;

- Rule based, while performing less known tasks the operator helps himself with remembered or written rules;

- Knowledge based, the performance of such tasks where typical models or rules may not be used directly. Activities follow only after processing information by diagnosis, planning and decision-making.

The ability to perform complicated real time calculations offers a new quality to the artificial intelligence systems: to maintain in each moment a graphic and dynamic model of the whole plant and the partial processes in their links to other plants. AI methods shorten the time necessary to elaborate more information and improve the response of both factors, the operator and autonomous regulators. As a consequence of this harmony of activities, the safety and reliability of the power plant is increased.

2.2 A Proposal for the Structure of an Early Failure Detection System

A substantial pre-requisite of a safe NPP operation is the correct operator response to system failures. The first method of failure detection in technological processes signalled exceeding the limit parameters. Today we try to detect a failure before dangerous areas are reached. Simultaneously, we try to locate it more precisely, including the initial cause of the failure condition.

In general, we may say that every time a problem is solved we are using knowledge about the given problem in form of measured data of correct or incorrect behaviour or in the form of an analytical description of the system. An effort is required to obtain a sufficient amount of knowledge about the problem. We may state the effort is approximately the same for all possible solutions. AI methods can be divided into the levels of the necessary amount of data and the necessary amount of analytical knowledge (as depicted in Figure 3), allowing selection of the most suitable method for the solution according to the specificity of the solved problem.

122

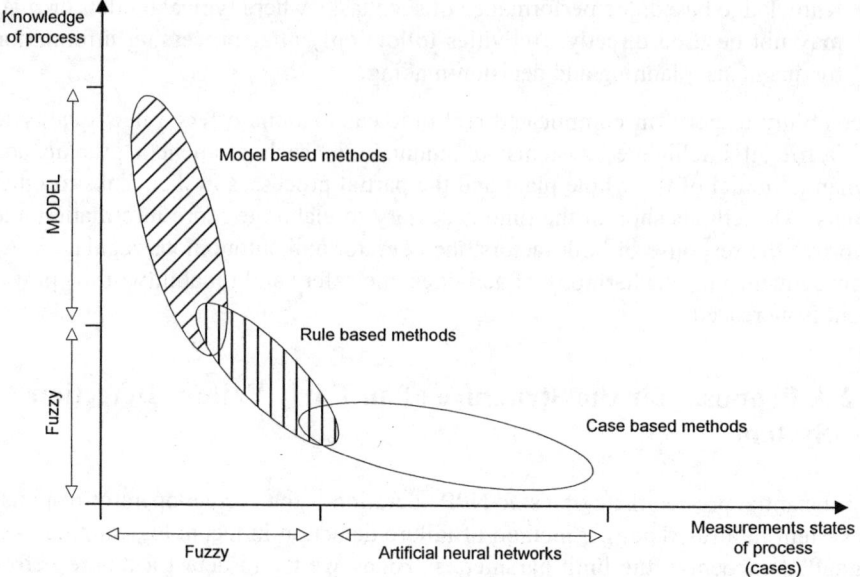

Fig. 3. The relation between AI methods

The design of the system of operator support for the NPP control in load follow mode is based on several facts. Many analytical findings were gained over several decades that were transformed into a number of models. The apparatus of fuzzy-logic allows us to describe the uncertainty and ambiguity of the operator behaviour based on the NPP state. Therefore the creation of the operator support [4] is based on modelling methods. Fuzzy logic can be used for the diagnostics of a variety of both measured and modelled data.

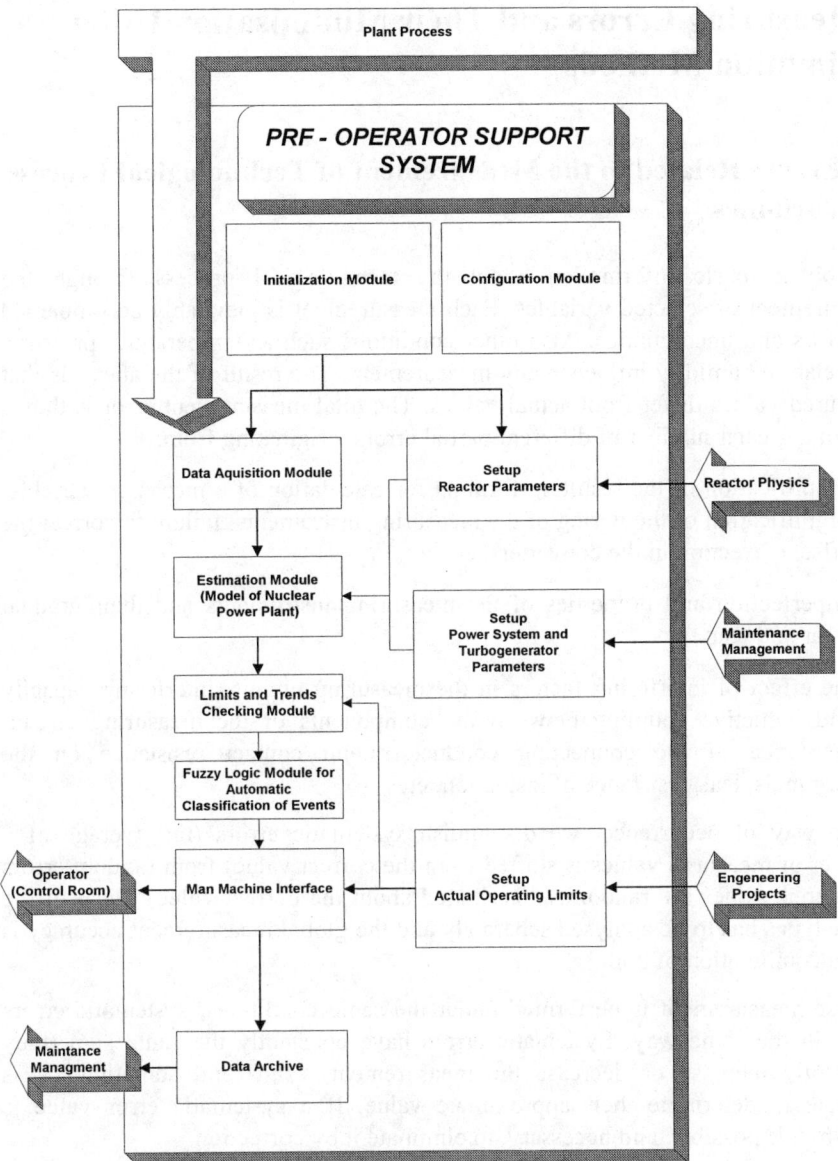

Fig. 4. Basic diagram of the PRF program

The basic diagram of the PRF program we propose is presented in Figure 4. In the next sections we will discuss in more detail two models, which perform the elimination of measuring errors by the estimation method and automated failure detection by fuzzy logic.

3 Measuring Errors and Their Minimisation Using Estimation Methods

3.1 Errors Related to the Measurement of Technological Process Variables

We obtain basic information about the technological process through the measurement of selected variables. Each measurement is inevitably accompanied by errors and uncertainties. Also other conditions such as temperature, pressure, and relative humidity influence any measurement. The result of the above is that measured values differ from actual values. The total measurement error is thus a sum of a greater number of different partial errors, originating from:

- simplification of the resulting formula for calculation of a measured variable, simplification of the wiring of the measuring instruments, failure to correct the offset correction in the converters, etc.

- imperfection and properties of the measuring instruments and their gradual wear and tear

- the effect of interfering factors in the measuring circuit (interfering capacity and inductive coupling between the components of the measuring circuit, resistance of the connecting conductors and contact resistance on the terminals, leak resistance of insulation, etc.)

As to way of occurrence, we distinguish systematic errors (the average of a number of measured values is shifted from the correct value) from random errors (measured values are randomly distributed about the correct value). Each of the above types has to be analysed separately and the global measurement accuracy is then a combination of both.

In case measurement is performed under the same conditions, systematic errors occur in the same way. Systematic errors have constantly the same sign (they constantly increase or decrease the measurement result) and sometimes it is possible to determine their approximate value. If a systematic error value is known, it is possible, and necessary, to eliminate it by correction.

Contrary to the above, random errors occur in a fully random way without any known response and often their origin is even unknown. Random error is an arbitrary deviation around a value, given by the systematic error. It is possible to reduce random error either by changing measuring equipment or by analytical methods. The relation between both types of errors is presented in Figure 5.

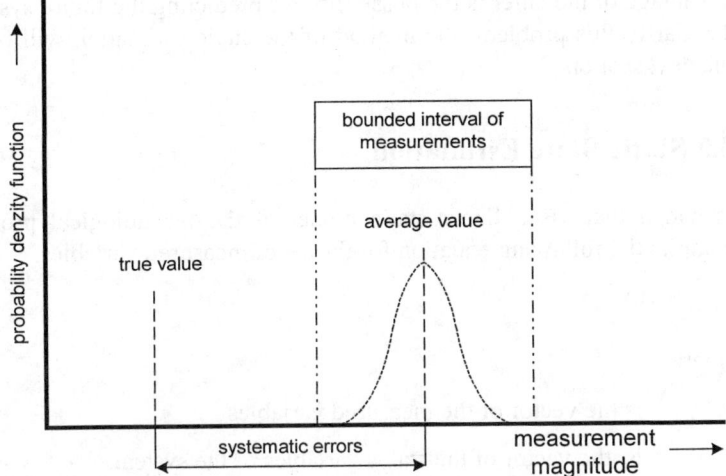

Fig. 5. Frequency distribution curve of measurement

3.2 Reduction of Errors Using Estimation

Due to limited accuracy and reliability of the measuring chains, all measured variables are inevitably affected with errors. Occurrence of errors results from the physical inconsistency of an obtained set of measured values. The set of "raw" measurements, related to a particular status of the system under control, is useless for linked analytical calculations and different corrections should be made.

Knowledge of the physical relationships between the measured variables (i.e., a knowledge of the model) allows us to determine analytically the system status. These mathematical models process the input and output variables based on a priori presumptions. The result is a set of new, estimated values of the measured variables, having minimum deviation between a priori presumptions and the process [5].

The estimator produces an optimal assessment of the system status, i.e., its most probable status at the moment of measurement sample capturing. The necessary presumption is a sufficient analytical and data redundancy. In case the set is affected with measurement errors, we may additionally eliminate rough measurement errors, which are being replaced by calculated values. Using estimation thus reduces the measurement errors.

Methods for plant status estimation are mostly broken down into the static and dynamic. Static estimation is sufficient for the calculation of steady state conditions of the unit. Dynamic estimation may be used for transient processes, based on time dependent modules, but its calculation is more complicated. The advantage of the latter is the possibility for predicting the future system condition. To clarify this problem, the method of the static estimation will be described in the next section.

3.3 Static State Estimation

Provided there is a linear static model of the technological process, we may compile the following equation for the set of measured variables:

$$z = H.x + v \qquad (3.1)$$

where:

z   is the vector of the measured variables

x   is the vector of the status variables of the system

v   is the vector of random measurement errors

H   is the matrix of the linear model of the measurement

The aim of this estimation is to minimise the vector of the deviations e, forming the difference between the measured variables z and the estimated variables, calculated using status variables x

$$e = z - H.x \qquad (3.2)$$

As it is a multiparameter task, the estimation uses the method of minimising the quadratic function J (weighted least squares method).

$$J(x) = e^T . R^{-1}.e = |e|^2 \qquad (3.3)$$

Where R^{-1} is the weight symmetrical matrix. Random vector v has usually the normal distribution with a mean value of zero and a covariance matrix R, being positively definite. In case individual components of the vector v are not correlated, the matrix R is a diagonal one and its diagonal is formed by relative errors δ^2 of the individual measurements. Their inverse values represent the measurement weights and the matrix R^{-1} is the so-called weight matrix.

By solving the criterion (3.3), i.e., by finding the extreme J(x) we obtain the calculation relation for the searched vector of the status variables

$$x = [H^T R^{-1} H]^{-1} H^T R^{-1} z \qquad (3.4)$$

The values of the relative errors of the estimated variables may be obtained for the covariant parameter matrix

$$Kov(x) = [H^T R^{-1} H]^{-1} \qquad (3.5)$$

In case no large measurement errors are present, the vector v has, according to the presumption, normal distribution with zero mean value and covariance matrix R.

The estimation criterion $J(x) = e^T R^{-1} e$ is scalar, relative to the vector e with normal distribution. The mentioned static properties are used for the search of large measurement errors. In the situation where no large measurement errors are present and the measurement noise is identical to the one of the model, the chi-squared distribution corresponds to the random variable $J(x)$. We may thus use, in accordance with the theory of the statistical testing of hypotheses, the statistical test

$$| J (x) | < a \qquad (3.6)$$

where a is the tabular value of the chi-squared distribution (m-n freedom degrees at a selected risk of acceptance of incorrect hypothesis). Thus, in the case for the obtained assessment x the inequality (3.6) will be fulfilled, we presume the measurement vector as being error-free. In the opposite case, we have to perform the identification of the particular erroneous measurements. Bad measurements are suspect and has to be replaced by estimated values. After that we will perform a new estimation. In this way we eliminate individual measurements. We will terminate the calculation only when $J(x)$ fulfils the condition (3.6).

3.4 Demonstration of the Use of Static Estimation Methods for the Calculation of Reactor Power

The reactor power N_R, is measured using ionisation chambers. Signal from the ionisation chambers has to be periodically calibrated and the measurement is affected with the error δN_R, which has variable value. Using the estimation method we may perform corrections of the measured reactor power to other measured variables, being in a certain physical relationship with N_R. In general, balance conservation equations are used (mass conservation, energy conservation laws, balance of moments, force, impulse, etc.). When analysing mutual dependencies of the measured variables, we are creating a mathematical/physical identification of the process, which may have different degree of detail [6].

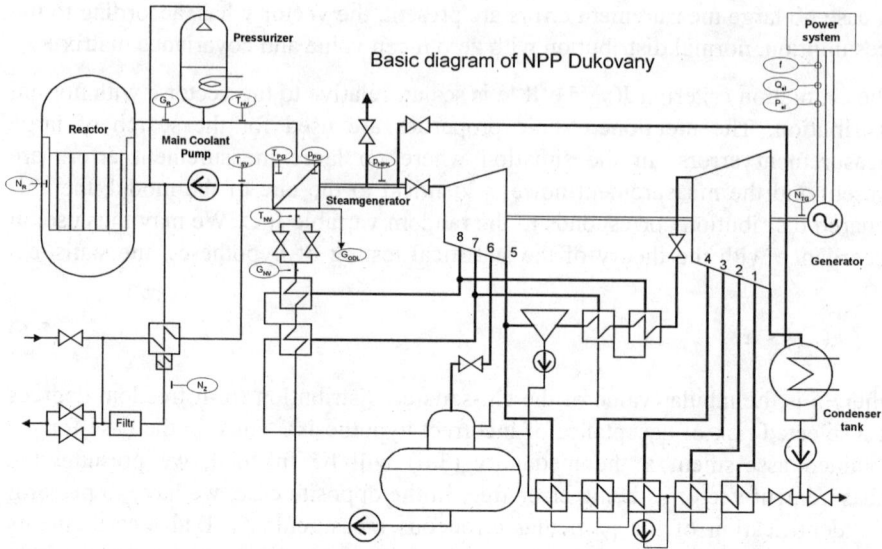

Fig. 6. Technological scheme of NPP Dukovany

The NPP unit (Figure 6) may then be broken down into main equipment, of which the parameters are measurable. From an electrical energy production view, the following are included

thermal reactor power	N_R
steam power of the steam generator	N_{PG}
turbine mechanical power	N_{TG}
alternator electrical power	N_{EL}

From the above values, only the electrical power of the alternator is being measured using a direct method. The remaining power values are related and can be calculated only when knowing the losses that occur throughout the production process. We are not able to measure the losses unambiguously and therefore we must calculate these power values from the measured variables and corrections. Data so obtained (mean values and their variances), nevertheless, do not fulfil the physical equations as the system is over-determined. Therefore, using the estimation method we calculate new values fulfilling the physical dependencies as well as differing from the measured ones as little as possible.

The basic step is the selection of a set of input and output variables, characterising with satisfactory accuracy the evaluated technological process. The principle for the selection is the registration of all statistically relevant links. The task will be simplified by the solution of a balanced condition. We will eliminate possible fluctuations by time smoothing of a certain number of samples. In case the time series is measured with a small variance, the measured status may be considered as a steady state.

A priori presumptions define for us equations for the calculation of reactor power (3.7) and steam generator power (3.8) from the status variables, mutual link (3.9), and the link of another measured variable to the status variable (3.10).

$$N_R = G_R * \left[\left(a_{hv} * T_{hv} + b_{hv} \right) - \left(a_{sv} * T_{sv} + b_{sv} \right) \right] \tag{3.7}$$

$$N_{PG} = G_{NV} * \left(a_{par} * p_{PG} + b_{par} \right) + G_{odl} * \left(a_{vs} * p_{PG} + b_{vs} \right)$$
$$- \left(G_{NV} - G_{odl} \right) * \left(a_{nv} * T_{NV} + b_{NV} \right) \tag{3.8}$$

$$N_R = N_{PG} + N_Z \tag{3.9}$$

$$T_{PG} = a_{ps} * p_{PG} + b_{ps} \tag{3.10}$$

Description of the individual constants is

a_i	slope of the linearized functions (water or steam enthalpy)
b_i	shift of the linearized functions (water or steam enthalpy)
G_{NV}	mass flow of the feedwater to SG
G_{odl}	blowdown mass flow from SG
G_R	mass flow through the reactor
N_Z	power of the losses
T_{HV}	temperature in the hot leg
T_{SV}	temperature in the cold leg
T_{NV}	temperature of the feedwater to SG
T_{PG}	temperature of the steam into SG
T_{SV}	temperature in the cold leg
p_{PG}	steam pressure into SG

The measured variables vector is formed by

N_R, N_{PG}, T_{PG}, T_{HV}, T_{SV}, p_{PG}, T_{NV}.

The status variables vector is formed by

T_{HV}, T_{SV}, p_{PG}, T_{NV}.

Estimated results are given in Table 1.

Table 1. Estimation results

measured variable					estimated variable		
name		value	absolute error	relative error	value	absolute error	relative error
N_R	MW	1372,029	27,50	2,00	1371,967	15,536	1,129
N_{PG}	MW	1374,095	27,50	2,00	1379,338	15,277	1,111
T_{PG}	°C	260,870	5,00	1,25	261,320	2,631	0,657
T_{HV}	°C	296,628	5,00	1,25	297,147	2,347	0,587
T_{SV}	°C	267,532	5,00	1,25	267,046	2,660	0,665
p_{PG}	MPa	4,807	0,08	1,25	4,793	0,020	0,330
T_{NV}	°C	223,559	5,00	1,25	226,892	1,574	0,394

As you can see from the results, estimation decreases the absolute error by approximately a half.

4 Fuzzy System of Automatic Failure Classification

4.1 General Description of a Fuzzy Logic Diagnostic System

The operator obtains information about the state of the technology using measured values. These values may be affected by different types of errors obligating him to use additional amount of knowledge, which he obtains from operational experience. During proper decision-making he takes into account links and divides potential situations into several categories:

- Normal operation

- Operation with a minor failure

- Failure condition

- Large failure.

The boundary between the individual conditions is ambiguous and may be, to a certain degree, partially subjective. Fuzzy logic is a very useful tool for the operator to solve such tasks. The operator recognises the failures based on detected symptoms. A man is limited in his capacity and he can recognise only dominant failures from a small number of symptoms. A system based on the fuzzy logic works in the same way and is able to encompass a much greater amount of information and links [7].

The most frequently used method is the so-called inversion method based on logical deduction rules modus ponens and modus tolens. Modus ponens says for two propositions P and Q, and the implication "P⇒Q" applies when the validity of P implies the validity of Q. On the other hand, modus tolens presumes the validity of the negation of the above implication, meaning the invalidity of the proposition Q implies the invalidity of the proposition Q. Using fuzzy application of the above rules we the apparatus obtained can be used in classifying the status of the diagnosed system. Based on the presence or absence of accompanying occurrences (signs) the potential for the failure being in the highest conformance with the occurrences can be derived.

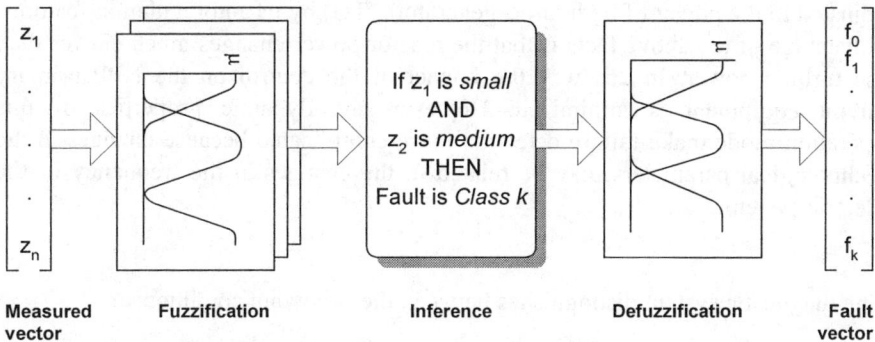

| Measured vector | Fuzzification | Inference | Defuzzification | Fault vector |

Fig. 7. Scheme of fuzzy diagnosis system

The task may so be converted to the solution of the relationship between the vector of failures that may occur and the vector of occurrences, which are indicators for failures. The proper procedure may be broken down into three parts, as presented in Figure 7.

During the first part called fuzzyfication fuzzy sets are created that determine the value of existence of the given occurrence. E.g. "steam pressure increase" is "small" or "large."

During the second part, a calculation of the appurtenance degree of all estimated failures is performed. The procedure is based on the fuzzy if - then rules. E.g. if the "steam pressure increase" is "small" then "the turbine controller failure" is "intermediate."

In the final part defuzzification is performed. It is useful to utilise prototypes, fuzzy sets corresponding to particular failure states, prepared in advance.

The above described simple scheme enables to postulate mutual relations of a number of technological variables. A great advantage is the fact that fuzzy if-then rules are described for a majority of failure situations in the operational instructions.

4.2 The Fuzzy Diagnostic System for the NPP Operation in the Primary Frequency Control

The design of the diagnostic system can be shown in more detail by using the NPP Dukovany operation in the primary frequency control as an example. The basic NPP mode is a steady state unit operation with constant (nominal) reactor power. The primary frequency control mode differs in that it is a dynamically stable operation of the unit. The unit power changes according to the needs of the power system within an allowed control range. At the same time, the reactor power is adjusted to the power of both turbo-generators (TG) by its autoregulation features. Advantage of the above facts is that the reactor power changes much slower than the turbine power. In this way the impact of the control on the NPP primary circuit equipment is minimized. However, the dynamic properties of this operation mode make failure detection more complicated because changes in the technological parameters may be related to the changes in the frequency of the electric system.

The diagnostic system distinguishes between the following conditions:

- correct NPP operation in the primary frequency control,

- operation close to limit conditions of the NPP operation in the primary frequency control,

- incorrect operation of the primary frequency control.

Correct NPP operation in the primary frequency control occurs when the turbine power and steam pressure in main steam collector is changing in the way required by the frequency changes.

One or a combination of following failures may be the cause of incorrect operation

- frequency corrector failure

- turbine controller failure

- failure of the proper PI controller

- reaching of the maximum opening of the turbine control valves

- reactor power decrease by burning out

The occurrences vector forms the following variables, going out of the estimation module

ΔN_{KF} output form the frequency corrector

N_Z turbine power setpoint

ΔN_{PI} input to the proper PI controller

$N - N_Z$ difference between the actual value and the setpoint of the turbine power

p_{22} turbine secondary oil pressure (control valves opening)

p_{HPK} main collector steam pressure

The existence of a failure sign is given by a specific condition, which may occur only up to a certain degree. This degree depends on the differences between the calculated and measured variables according to the given condition and, in some cases, on a time lag. Therefore we will evaluate it according to verbal description, using expressions such as "small, very small, intermediate, roughly large, large, etc." The fuzzy variable expresses the degree to which the ith condition is not fulfilled and is measured in an interval $[0, 100](\%)$. In this case 100% means the condition was completely infringed, and 0% that the condition is completely correct.

Relation between the infringement of the conditions and the failures is characterised in Table 2, having the following general form:

Table 2. Relation between the infringement and failures

Condition (sign) Failure	ΔN_{KF}	N_Z	ΔN_{PI}	$N - N_Z$	p_{22}	p_{HPK}
frequency corrector	r11	r12	r13	r14	r15	r16
turbine controller	r21	r22	r23	r24	r25	r26
PI controller	r31	r32	r33	r34	r35	r36
turbine control valves	r41	r42	r43	r44	r45	r46
Reactor burning out	r51	r52	r53	r54	r55	r56

The software LFLC1.5 (Linguistic Fuzzy Logic Controller - Dr. V. Novák, DSc) [7] is used as a fuzzy tool. The status of the whole system is given by the value of the TT operator "typical term" we obtain by selecting the maximum of the fuzzy variables of individual failure types and as a verbal description characterises the operation quality of the system.

5 Conclusion

The operator support system for NPP serves as a high-level decision support unit. It analyses the deviations between measurements and estimations and gives operator a decision.

The supporting system thus provides the operational staff with enough information to:

- Determine the condition of the controlled system.

- Assessment and control of the future condition of the controlled system.

- Diagnostics of the operational and failure conditions.

- Check of the efficiency of their own control interventions.

Connection of the real time modelling with the fuzzy logic systems creates a tool, which not only monitors and proposes suitable solutions, but also determines a diagnosis in an understandable way and predicts the future behaviour of the system.

It is useful when the situation is very complicated and the condition of the equipment is not clear from the monitoring systems, e.g. due to multiple failure or sensor failures. In this case the operators must evaluate the actual condition of the whole plant, predict the probable further development of the plant condition and to determine the further procedure to reach a safe condition.

References

[1] Computerization of operation and maintenance for nuclear power plants, IAEA-TECDOC-808, Viena, July 1995

[2] Development and implementation of computerized operator support systems in nuclear installations, IAEA-technical reports series No.372, Viena, 1994

[3] Rasmussen J., Skills, rules, knowledge: Signals, signs and symbols and other distinctions on human performance model, IEEE Trans. Syst. Man Cybern. SMC-13 3 (1983)

[4] Petruzela,I. Computer program for operator support in frequency control of NPP Dukovany, EGU Praha A-330-97-0016, Praha, 1997

[5] Nahi N.E., Estimation theory and applications, Wiley, New York, 1969

[6] Isermann R., Process fault detection based on modeling and estimation methods: A survey, Automatica, Vol.20, 1984

[7] Novak V., Fuzzy modelling, textbook, Ostrava, 1998

7 Optimal Operation Planning of Radioactive Waste Processing System by Fuzzy Theory

Jin Yeong Yang and Kun Jai Lee

Department of Nuclear Eng., Korea Advanced Institute of Science and Technology, Kusung-dong, Yusung-gu, Taejon, Korea, 305-701
yangjy@nucel0.kaist.ac.kr, kjlee@sorak.kaist.ac.kr

This study is concerned with the applications of linear goal programming and fuzzy theory to the analysis of management and operational problems in the radioactive processing system (RWPS). The developed model is validated and verified using actual data obtained from the RWPS at Kyoto University in Japan. The solution by goal programming and fuzzy theory would show the optimal operation point which is to maximize the total treatable radioactive waste volume and minimize the released radioactivity of liquid waste even under the restricted resources.

1 Introduction

The meaning of the optimization in the field of nuclear power was the economic evaluation based on the cost benefit analysis. The multi-objective optimization has not been applied to RWPS except Shimizu's (1981) work [1]. They had developed a kind of linear programming, RESTEM, to resolve multi-objective problem of RWPS but they did neither introduce the material balance on the system constraints, nor consider the priority order, and nor carry out the sensitivity analysis for the each goal in their work.

In this study Goal Programming (GP) based on modified modeling is used and to complement the demerit of GP, fuzzy theory is introduced. GP is a good decision method in modeling real world decision problems. GP extends linear programming to the problems which involve multiple objectives. It is necessary to specify aspiration levels for the objectives and aims to reduce the deviations from aspiration levels. In the case of a problem with non-equivalent goals the weight or priority of the goal is reflected through its deviation variables. Often, in real world problems the aspiration levels and priority factors of the decision maker, and sometimes even the weights to be assigned to the goals, are imprecise in nature. In such situations it is necessary to use of fuzzy set theory.

The main difference between GP-Fuzzy and GP is that the GP requires the decision maker to set definite aspiration values for each objective that we wish to achieve, whereas in fuzzy-goal programming these are specified in an imprecise manner. Throughout this chapter a fuzzy goal is considered as a goal with imprecise aspiration level. It was solved by introducing the fuzzy achievement function, $V(\mu)$. Unlike the conventional GP (minimizing the deviations) the fuzzy decision function consisting of μ_i is to be maximized in the GP-Fuzzy algorithm.

2 Fuzzy Algorithm

Fuzzy sets were introduced by Zadeh as a new way to represent vagueness in everyday life. They are a generalization of conventional set theory, one of the basic structures underlying computational mathematics and models. Computational pattern recognition has played a central role in the developments of fuzzy models because fuzzy interpretations of data structures are a very natural and intuitively plausible way to formulate and solve various problems.

Imprecision in data and information gathered from data about our environment is either statistical (e.g., the outcome of a coin toss is a matter of chance) or non-statistical (e.g., "apply the brakes pretty soon"). This latter type of uncertainty is called fuzziness. We all assimilate and use fuzzy data, vague rules, and imprecise information, just as we are able to make decisions about situations that seem to be governed by an element of chance. Accordingly, computational models of real systems should also be able to recognize, represent, manipulate, interpret, and use both fuzzy and statistical uncertainties. Statistical models deal with random events and outcomes; fuzzy models attempt to capture and quantify non-random imprecision.

Conventional (or crisp) sets contain objects that satisfy precise properties required for membership. The set of numbers H from 4 to 6 is crisp; we write $H = \{x \in R \mid 4 \leq x \leq 6\}$. Equivalently, H is described by its membership function,

$$\mu_H = 1; \; 4 \leq x \leq 6$$

$$0; \text{ otherwise}$$

The crisp set H and the graph of μ_H are shown in the Fig.1(a). Every real number x is either in H or is not in H. Because μ_H maps all real numbers $x \in R$ onto the two points $\{0,1\}$, crisp sets correspond to 2-valued logic (i.e., 1 or 0).

Fuzzy sets, on the other hand, contain objects that satisfy imprecise properties to varying degrees, for example, the "set" of numbers F that are "close to 5." The F, say μ_H, is called a membership function and maps numbers into the entire unit interval $[0,1]$. The value (x) is called the grade of membership of x in F. This construct corresponds to continuously valued logic; roughly speaking, all shades of gray between black (= 1) and white (= 0) can be described. Because the property "close to 7" is fuzzy, there is not a unique membership function for F.

Rather, it is left to the modeler to decide, based on the potential application and properties desired for *F*, what μ_H should be like. For example, the membership function (b) of Figure 1 might be a useful representation of *F*.

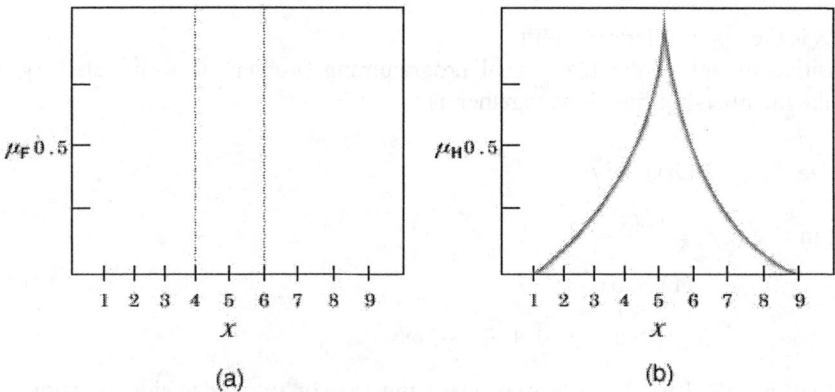

(a)   (b)

Fig 1. Fuzzy membership function for the word "close to 5"

Fuzzy-GP Alogrithm

Consider the fuzzy goal programming problem:
Find **X**
To satisfy $\mathbf{G_i(x)} \supseteq \mathbf{g_i}$, i = 1, 2, ..., m,
Subject to $AX \leq \mathbf{b}$,
$\qquad X \geq 0$,
Where **X** is an *n*-vector with components $x_1, x_2, ..., x_n$ and $AX \leq \mathbf{b}$ are system constraints in vector notation. The symbol '$\supseteq$' refers to the fuzzification of the aspiration level (i.e., approximately greater than or equal to). The *i*-th fuzzy goal $G_i(x) \supseteq g_i$ means that the decision maker is satisfied even if less than the g_i upto certain tolerance limit is attained. A linear membership function for the *i*-th fuzzy goal $G_i(x) \supseteq g_i$ can be expressed according to Zimmerman (1976) [9] as

$$\mu_i = \begin{cases} 1 & \text{if} \quad G_i(x) \geq g_i, \\ \dfrac{G_i(x) - L_i}{g_i - L_i} & \text{if} \quad L_i \leq G_i(x) \leq g_i, \\ 0 & \text{if} \quad G_i(x) \leq L_i \end{cases} \tag{1}$$

where L_i is the lower tolerance limit for the fuzzy goal $G_i(X)$. In case of the goal $G_i(x) \subseteq g_i$, the membership function is defined as

$$\mu_i = \begin{cases} 1 & \text{if} \quad G_i(x) \leq g_i, \\ \dfrac{U_i - G_i(x)}{U_i - g_i} & \text{if} \quad g_i \leq G_i(x) \leq U_i, \\ 0 & \text{if} \quad G_i(x) \geq U_i, \end{cases} \tag{2}$$

where U_i is the upper tolerance limit.

The additive model of the fuzzy goal programming problem is formulated by adding the membership functions together as

Maximize $\qquad V(\mu) = \sum_{i=1}^{m} \mu_i$

Subject to $\qquad \mu_i = \dfrac{G_i(x) - L_i}{g_i - L_i}$

$$AX \leq b, \tag{3}$$
$$\mu_i \leq 1,$$
$$X, \mu_i \geq 0, \quad i = 0, 1, 2, \cdots, m$$

where $V(\mu)$ is called the fuzzy achievement function or fuzzy decision function. This is a single objective optimization problem which can be solved by employing a suitable classical technique. Because the goals are fuzzy, unlike conventional goal programming (minimizing the deviation) the fuzzy decision function consisting of μ_i's is to be maximized here.

3 Goal Programming

Goal Programming method was proposed by Charnes and Cooper (1961) for a linear programming [2]. It has been further developed by Ijiri (1965), Lee (1972), and Ignizio et al. (1976) [3,4,5]. The method requires the decision maker to set goals for each objective that one wishes to attain. A preferred solution is defined as the one which minimizes the deviations from the set goals.

3.1 Terminology and Concepts

Objective: An objective is a relatively general statement (in narrative or quantitative terms) that reflects the desires of the decision maker. For example, one may wish to "maximize profit" or "minimize labor turnover" or "wipe out poverty."

Aspiration Level: An aspiration level is a specific value associated with a desired or acceptable level of achievement of an objective. Thus, an aspiration level is used to measure the achievement of an objective and generally serves to "anchor" the objective to reality.

Goal: An objective in conjunction with an aspiration level is termed a goal. For example, we may wish to "achieve at least X units of profit" or "reduce the rate of inflation by Y percent."

Goal deviation: Many of us aspire to be independently wealthy, but few will achieve this goal. The difference between what we accomplish and what we aspire to is the deviation from our goal. In all but trivial problems (or in cases where our aspiration levels are unrealistically low), we shall encounter deviations from our goals. That the deviation can be represented by over- as well as under-achievement of a goal.

3.2 Algorithm of Goal Programming

Goal programming algorithm is relatively simple and straightforward. It is the conversion of multiple objective problem into the single objective one's. It is based on its flexibility, efficiency, and ease of use and implementation. Goal Programming method has the most common form as following:

$$
\begin{aligned}
\text{Minimize} \quad & \left[P_1(d_1^+ + d_1^-), P_2(d_2^+ + d_2^-), \cdots, P_l(d_l^+ + d_l^-) \right] \\
\text{System Constraint} \quad & Z_1(x) + d_1^+ + d_1^- = b_1, \\
& Z_2(x) + d_2^+ + d_2^- = b_2, \\
& Z_3(x) + d_3^+ + d_3^- = b_3, \\
& \qquad \vdots \\
& Z_i(x) + d_i^+ + d_i^- = b_i, \\
\text{Absolute Goals} \quad & AX \leq B \\
& d_i^+, d_i^- \geq 0, \forall_i \\
& d_i^+ \times d_i^- = 0, \forall_i
\end{aligned}
\tag{4}
$$

where $b_j = 1, 2, ..., k$ are the goals set by the decision maker for the objectives, d_i^- and d_i^+ are respectively the under-achievement and over-achievement of the i-th goal. The P_i's are preemptive weights.

The solution algorithm for the above equation is that $P_1 (d^-, d^+)$ is minimized first. Next $P_2 (d^-, d^+)$ is minimized. If a lower ranking achievement function cannot be satisfied to the detriment of a higher ranking achievement function, this process continues until $P_l (d^-, d^+)$ is minimized. It is basically a modified simplex algorithm for linear problems.

Lexicographic Minimum:

Given an ordered array a of nonnegative elements a_k's, the solution given by $a^{(1)}$ is preferred to $a^{(2)}$ if

$$a_k^{(1)} < a_k^{(2)} \tag{5}$$

and all higher order elements (i.e., a_1 ,..., a_k) are equal. If no other solution is preferred to **a**, then **a** is the lexicographic minimum.

Thus, if we have two solutions, $a^{(r)}$ and $a^{(s)}$, where

$$a^{(r)} = (0, 17, 500, 77)$$
$$a^{(s)} = (0, 18, 2, 9)$$

$a^{(r)}$ is preferred to $a^{(s)}$.

Another term that is used to describe the lexicographic minimum notion is the concept of preemptive priorities. A solution that provides a lexicographic minimum to **a** also satisfies the concept of preemptive priorities. Any goal (or goals) at preemptive priority k (designated by P_k) will always be preferred to any at a lower priority $k+1$,..., K regardless of any scalar multiplier associated with these lower priorities.

Although the preemptive priority concept (or lexicographic minimum) may appear new to some readers, it was already used, implicitly, in single objective linear programming. Our first priority there was to find a solution that satisfied all constraints. Our next priority is to find a solution that minimizes or maximizes the single objective without violating the constraints.

Not only is the preemptive priority concept implicit in the simplex solution technique, it is also evident in real world decision making and has been documented in numerous studies. Consider, for example, the decision procedure that is often evident in the purchase of a home. The buyer's first priority may be to consider only a home that is within 10 mile radious form his or her place of work. All other homes are excluded from consideration. Next, the buyer may desire to limit the purchase price to under $200,000. Thus, even though homes outside the 10 mile radius may be below $200,000, they are not considered because they would conflict with the preemptive priority associated with this goal. After narrowing down the homes to those within the 10 mile radius and under $200,000, the next priority may be that the house has four bedrooms and a double car garage, and so on. Thus, the preemptive priority concept is used in the decision analysis as an iterative screening process.

Some of the criticisms leveled at goal programming center about the use of the lexicographic minimum (or preemptive priorities). This is unfortunate because, even if the lexicographic minimum does not supply the most desirable measure of achievement for a particular problem, it is generally extremely efficient in finding a good starting solution that may be improved upon by a straight forward relaxation of the strict interpretation of the lexicographic minimum. Further, it has never been intended that it be the rigid, inflexible concept that its critics describe. Consequently, we employ the lexicographic minimum or preemptive priority measure primarily because of its flexibility. That is, the notion may be relaxed or extended so as to encompass other measures of achievement.

Steps in Model Construction

The initial phase in the construction of the goal programming model is, or should be, the development of the baseline model. Once the baseline model has been constructed, we enter the next phase: the conversion of the baseline model into our specific form of the linear goal programming model (i.e., a multiple objective linear programming model). The assumptions necessary in this conversion are:

Aspiration levels may be associated with all objectives so as to transform them into goals.

Any rigid constraints (i.e., absolute goals) are ranked at the first priority. All remaining goals may be ranked according to importance.

With the exception of the first priority, all goals within a given priority must either be commensurable or, by means of weights, be made commensurable.

These are not particularly restrict assumptions and, in fact, are generally less severe than those employed to develop the single objective linear programming model. The steps in the formulation process may then be summarized as follows:
Step 1. Develop the baseline model.
Step 2. Specify aspiration levels for each and every objective.
Step 3. Include negative and positive deviation variables for each and every
goal and constraint.
Step 4. Rank the goals in terms of importance. The first priority is always
reserved for the rigid constraints.
Step 5. Establish the achievement function.

As mentioned earlier, there is more that must be considered in formulating the multiple objective model, but the reward is that the resultant model should better reflect the actual problem. Since any solution obtained is only as good as the model it was derived from, the solution to the multiple objective model should, in general, reflect an improvement over single objective models.

Table 1. Goal formulation

Goal Type	Goal Programming Form	Deviation Variables to be Minimized
$f_i(x) \le b_i$	$f_i(x) + d_i^- - d_i^+ = b_i$	d_i^+
$f_i(x) \ge b_i$	$f_i(x) + d_i^- - d_i^+ = b_i$	d_i^-
$f_i(x) = b_i$	$f_i(x) + d_i^- - d_i^+ = b_i$	$d_i^+ + d_i^-$

4 System Modeling

4.1 System Description

The radioactive waste treatment system consists of liquid, solid and gas processing system. The radioactive liquid wastes processing system collects and processes potentially radioactive wastes for recycling or for release to the environment. In the radioactive liquid waste treatment system, waste was segregated by its radioactivity. Low level radioactive waste was collected at IT-1, 2 (inlet tank) and treated by P-1 (chemical precipitation and filtration) or P-2 (chemical precipitation, filtration and ion exchange). Thereafter, it was diluted at DT-1, 2 (discharge tank) to meet the release limit of radioactivity and discharged to the environment. At P-4, the sludge from P-1 and P-2 was dehydrated and transferred to the S-1 for settling. Medium level radioactive waste was collected at IT-3 (inlet tank), processed by P-3 (evaporation) and transferred to the DT-3 (discharge tank) for dilution. Thereafter, it was released to the environment.

In the radioactive solid treatment system, solid waste was solidified with cement and drummed at P-6. Also, the sludge and slurry from S-1 and S-2 was solidified, drummed at P-5, and stored at the temporary storage area (TS) with solid waste.
In the radioactive gas treatment system, the off-gas from the liquid and solid waste was treated [8].

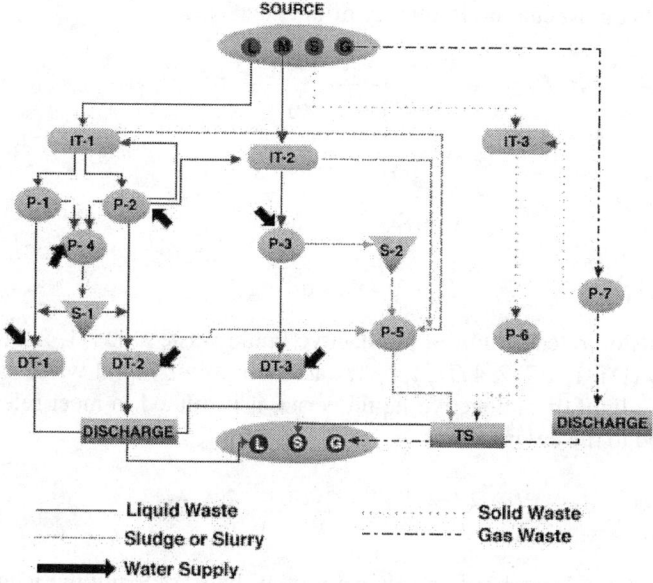

Fig 2. Reference System

4.2 Model Assumptions

It is assumed that the total treatable radioactive waste input volume (X_0: 532 m³/yr) and the design values are known: (1) the total cost limit, (2) the storage area limit, (3) the water consumption limit, (4) the total operation days limit, and (5) the surface dose limit in the storage area (P-5). These values were used in system criteria. The equipment capacities, engineering restrictions and material balances are listed as system constraints.

4.2.1 System Constraints

Equipment Capacities, Engineering Restrictions

Constraints on equipment capacity: it is assumed that each process operate from half of upper capacity to upper capacity.

$$0.5 f_{P-i} Y_i \le X_i \le f_{P-i} Y_i \ (i = 1, 2, 3, 4) \tag{6}$$

Constraints on sludge and slurry production rate: sludge production range between 2% ~ 5%, evaporator concentrates slurry range between 2% ~ 4%.

$$0.02 X_1 \le X_4 \le 0.05 X_1, \ 0.02 X_2 \le X_5 \le 0.05 X_2, \ 0.02 X_3 \le X_6 \le 0.04 X_3 \tag{7}$$

Constraints on evacuation frequency of inlet tank 1, 2:

$$\frac{X_L + ¥ã_{RLW} X_2}{Z_1} \le f'_{IT-1} \qquad , \qquad \frac{X_M + ¥ã_{RMW} X_2}{Z_2} \le f'_{IT-2} \tag{8}$$

$$Z_1 \le \frac{350}{t_{IT-1}} \qquad , \qquad Z_2 \le \frac{350}{t_{IT-2}} \tag{9}$$

$$Z_1 - 0.25 Y_3 \ge 0 \qquad , \qquad Z_2 - 0.4 Y_4 \le 0 \tag{10}$$

Constraints on release limit of radioactive liquid waste at DT-1, 2, 3: during these processes (DT-1, DT-2, DT-3) if the radioactivity of liquid waste is higher than the release limit of radioactive liquid waste, it is diluted to meet release limit and discharged to the environment.

$$\frac{A_{LLW,DT-1}}{X_7 + X_{10}} \ge RL, \quad \frac{A_{LLW,DT-2}}{X_8 + X_{11}} \ge RL, \quad \frac{A_{MLW,DT-2}}{X_9 + X_{12}} \ge RL \tag{11}$$

The RL is the release limit of radioactivity of the RWPS and the values of the RL is varied from 10^{-1} $\mu Ci/m^3$ to 20^{-1} $\mu Ci/m^3$.

Material Balance

At the IT-1 : $(1 - ¥â_{T-1}) X_L + \{(1 + w_{P-2})(1 - ¥â_{T-1}) ¥ã_{RLW} - 1\} X_2 - X_1 = 0$ (12)

At the IT-2 : $(1 - ¥â_{T-2}) X_M + \{(1 + w_{P-2})(1 - ¥â_{T-2}) ¥ã_{RMW}\} X_2 - X_3 = 0$ (13)

At the IT-3 : $\qquad\qquad\qquad\qquad\qquad\qquad\qquad X_S + X_{WF} - X_{P6} = 0$ (14)

Radioactive solid waste consists of compactible waste (COM), non-compactible (IMC) waste, and waste filter (WF) from the process, P-7.

$$X_S = X_{COM} + X_{IMC} + X_{WF}, \\ (X_{COM} = s_{COM} X_S, X_{IMC} = s_{IMC} X_S, X_{WF} = s_{WF} X_S) \tag{15}$$

At the P-1 : $\qquad\qquad\qquad\qquad\qquad\qquad\qquad X_1 - X_4 - X_{10} = 0$ (16)

At the P-2 : $\qquad\qquad\qquad\qquad\qquad\qquad X_2 + X_{P2W} - X_5 - X_{11} = 0$ (17)

At the P-3 : $\qquad\qquad\qquad\qquad\qquad\qquad X_3 + X_{P3W} - X_6 - X_{12} = 0$ (18)

At the P-4 : $\qquad\qquad X_4 + X_5 + X_{P4W} - 0.9 X_4 - 0.9 X_5 - X_{S1} = 0$ (19)

In this process, P-4, it is assumed that 90% water is dehydrated and transferred to the each system.

At the P-5 : $\qquad\qquad X_{S1'} + X_{S2} + \beta_{IT-1} X_0 + \beta_{IT-2} X_0 - X_{5TS} = 0$ (20)

At the P-6 : $\qquad\qquad\qquad\qquad\qquad\qquad\qquad\qquad X_{P6} - X_{6TS} = 0$ (21)

At the S-1 : $\qquad\qquad\qquad\qquad\qquad\qquad\qquad\qquad X_{S1} - X_{S1'} = 0$ (22)

At the S-2 : $\qquad\qquad\qquad\qquad\qquad\qquad\qquad\qquad X_6 - X_{S2} = 0$ (23)

At the DT-1 : $\qquad\qquad\qquad\qquad\qquad X_{10} + 0.9 X_4 + X_7 - X_{13} = 0$ (24)

At the DT-2 : $\qquad\qquad\qquad\qquad\qquad X_{11} + 0.9 X_5 + X_8 - X_{14} = 0$ (25)

At the DT-1, 2 the coefficients (0.9) of the X_4 and X_5 mean the water volume which is dehyrated from the process P-4.

At the DT-3 :
$$X_{12} - X_6 + X_9 - X_{15} = 0 \qquad (26)$$

When the wastes were solidified with cement in the P-5 and P-6, each volume is increased by 1.1 times.

Table 2. Parameter values estimated from empirical data [1]

Symbol	Estimated Value			Unit
g_i	LLW	MLW	SW	
	500/532	20/532	12/532	
α_i	LLW	MLW	SW	
	2.3	57.0	2.7	$\mu Ci/m^3$
s_i	COM	IMC	WF	
	2/7	1/7	4/7	
t_i	IT-1	IT-2	IT-3	
	7.0	23.3	10.0	Day
τ_i	IT-1(^{59}Fe)	IT-2(^{137}Cs)	IT-3(^{60}Co)	
	45.1d	30.0yr	5.24yr	

4.2.2 System Criteria

In the GP the deviation variables (d_i) are introduced and they are minimized in the objective function. In GP-Fuzzy the upper tolerance limit for the target values are set, by Eq. (2) system criteria are formed as following Table 3 and membership functions (μ_i) are maximized in the objective function.

Table 2. Parameter values estimated from empirical data (continued)

Symbol	Estimated Value					Unit
d_{0i}	P-1		P-2		P-3	
	0.480		0.087		5.80×10^{-3}	
d_{1i}	P-1		P-2		P-3	
	-0.317		-0.317		-4.70×10^{-2}	
d_{2i}	P-1		P-2		P-3	
	-6.85		-0.666		-7.14×10^{-3}	
q_i	SLG(^{59}Fe)		SLY(^{137}Cs)		SW(^{60}Co)	$\times 8.76$
	0.655		0.39		1.35	(mR•m^2/µCiyr)
w_i	P-2		P-3		P-4	
	0.28		2.0		150.0	
c_{0i}	P-1		P-2		P-3	$\times 10^4$¥/m^3
	0.04		0.32		0.93	
c_{1i}	P-1		P-2		P-3	$\times 10^4$¥/m^3
	1.4		1.4		-1.5	
c_{2i}	P-4			P-2 (REG)		$\times 10^4$¥/m^3
	0.5			0.029		
c_{3i}	SLG	SLY	COM	IMC	WF	$\times 10^4$¥/m^3
	60.0	60.0	15.0	46.0	23.3	
a_i	SLG	SLY	COM	IMC	WF	M^2
	0.084	0.084	0.325	0.130	0.36	
v_i	SLG	SLY	COM	IMC	WF	M^3
	0.02	0.02	0.2	0.05	0.54	
f_i'	IT-1			IT-2		M^3
	40.0			20.0		
r_i	RLW			RMW		
	0.2			0.08		
β_l	IT-1			IT-2		
	10^{-4}			10^{-3}		
f_i	P-1	P-2		P-3	P-4	m^3/d
	40.0	40.0		4.0	0.4	

Table 3. GP and fuzzy equations for each system criteria

System Criteria	GP	GP-Fuzzy
Cost (¥/yr)	$\left(C_T + C_D + C_W\right) + d_1^- \leq 10^7$	$\left(C_T + C_D + C_W\right) + 10^6\,\mu_1 \leq 1.1 \times 10^7$
Water (m³/yr)	$X_i + d_2^- \leq 5500$ $(i = P2W, P3W, P4W, 7,8,9)$	$X_i + 500\,\mu_2 \leq 5500$ $(i = P2W, P3W, P4W, 7,8,9)$
Operation Day (day/yr)	$Y_i + d_3^- \leq 200$ $(i = 1,2,3,4, MON, REG)$	$Y_i + 10\,\mu_3 \leq 200$ $(i = 1,2,3,4, MON, REG)$
Surface Dose (mR/yr)	$\dfrac{1}{2}\left\{\left(\dfrac{q_i}{50^2}\right)A\left(\dfrac{i}{TS}\right)\right\} + d_4^- \leq 5$ $(i = SLG, SLY, WF)$	$\dfrac{1}{2}\left\{\left(\dfrac{q_i}{50^2}\right)A\left(\dfrac{i}{TS}\right)\right\} + \mu_4 \leq 5$ $(i = SLG, SLY, WF)$
Storage Area (m²/yr)	$\left\{\dfrac{1}{2}\sum_i\left(\dfrac{a_i}{v_i}\right) \times X_{TS,i}\right\} + d_5^- \leq 90,$ $(i = (SLG, SLY, COM, IMC, WF))$	$\left\{\dfrac{1}{2}\sum_i\left(\dfrac{a_i}{v_i}\right) \times X_{TS,i}\right\} + \mu_5 \leq 90,$ $(i = (SLG, SLY, COM, IMC, WF))$
Objective Function	$Minimize \displaystyle\sum_{i=1}^{5} P_i d_i^-$	$Maximize \displaystyle\sum_{i=1}^{5} \mu_i$

5 Results and Discussion

The results obtained by GP, and RESTEM based on Shimizu's modeling are considerably different from each other (listed in the row SGP and RESTEM of each of the Tables 4~8). After adding material balances at each equipment and system constraints to Shimizu's modeling, the result by GP and Fuzzy theory was remarkably similar to that of RESTEM (listed in the row GP and GP-Fuzzy of each of Tables 4~8).

Table 4. Objectives of management I

	Treatable Amount (m³/yr)	Management Cost (10⁴¥/yr)	Discharged Radioactivity (μCi/yr)	Water Consumption (m³/yr)
Actual	532	600.0	126.7	2,484.7
RESTEM	1.5×532	900.0	190.0	3,727.0
SGP	2.0×532	1,014.1	84.82	5,000.0
GP	**1.5×532**	**999.9**	**156.4**	**4,999.9**
GP-FUZZY	**1.5×532**	**999.9**	**154.6**	**4,999.9**

Table 5. Objectives of management II

	Storage Area (m²)	Surface Dose (mR)	Water Cost (10⁴¥/yr)	Treatment Cost (10⁴¥/yr)	Disposal Cost (10⁴¥/yr)
RESTEM	33.21	4.51	18.635	224.3	657.02
SGP	22.85	5.18	24.941	155.2	840.00
GP	**20.65**	**4.91**	**25.000**	**248.9**	**726.01**
GP-FUZZY	**20.11**	**4.81**	**25.000**	**252.7**	**722.23**

Table 6. Specification of operation days and water consumption

	Operation Days (d)				Water Consumption (m³/yr)		
	Opera-tion	Moni-toring	Regen-Eration	Total	Regen-eration	Dilution	Cooling
RESTEM	100.5	91.7	7.8	200.0	109.33	1,013.3	2,604.3
SGP	94.7	59.4	4.7	158.8	66.19	1,184.7	3,737.2
GP	**131.2**	**58.9**	**7.6**	**197.7**	**105.85**	**2,219.6**	**2,674.4**
GP-FUZZY	**109.4**	**73.1**	**7.5**	**190.0**	**110.10**	**2,230.7**	**2,659.2**

Table 7. Radioactive liquid waste management plan

Process		Volume (m³/yr)	Operating Days (d/yr)	Volume per day (m³/d)	Sludge & Slurry rate (%)
P-1	RESTEM	437.61	21.88	20.00	2.00
	SGP	259.10	12.90	20.00	5.00
	GP	**468.06**	**23.40**	**20.00**	**2.00**
	GP-FUZZY	**438.73**	**21.93**	**20.00**	**2.00**
P-2	RESTEM	390.48	19.52	20.00	2.00
	SGP	236.40	11.80	20.00	5.00
	GP	**378.04**	**18.90**	**20.00**	**2.00**
	GP-FUZZY	**417.45**	**20.87**	**20.00**	**2.00**
P-3	RESTEM	63.08	17.71	3.56	2.00
	SGP	16.10	8.10	2.00	2.00
	GP	**68.08**	**33.26**	**2.04**	**2.00**
	GP-FUZZY	**72.10**	**33.78**	**2.13**	**2.00**
P-4	RESTEM	16.56	41.41	0.2	
	SGP	24.70	61.90	0.4	
	GP	**16.80**	**55.62**	**0.2**	
	GP-FUZZY	**16.70**	**42.80**	**0.2**	

Table 8. Discharged radioactive liquid waste

Process		Discharged Volume (m³/yr)	Radioactivity Limit (μCi/m³)	Discharged Amount (μCi/yr)
DT-1	RESTEM	1,303.30	0.1	130.33
	SGP	1,305.00	0.05	65.25
	GP	**2,355.17**	**0.05**	**117.75**
	GP-FUZZY	**2,207.63**	**0.05**	**110.38**
DT-2	RESTEM	499.59	0.1	49.96
	SGP	367.60	0.05	18.38
	GP	**708.90**	**0.05**	**35.44**
	GP-FUZZY	**782.80**	**0.05**	**39.14**
DT-3	RESTEM	97.27	01	9.73
	SGP	23.70	0.05	1.19
	GP	**64.32**	**0.05**	**3.21**
	GP-FUZZY	**103.25**	**0.05**	**5.10**

In order to carry out the validation and verification of model, the optimal point was compared with Shimizu's one and actual value was shown in Tables 4~8. In Tables 4 and 5, management cost, water consumption, surface dose, and operation days are closed to the design value 10^7 ¥/yr, 5000 m^3/yr, 5 mR/yr, and 200 days/yr, respectively. This means that the optimal operation point makes the best use of the restricted resources. When the solution was compared with actual value in Table 4, total treatable radioactive waste volume was increased to 1.5 times of actual value, and the discharged radioactivity was decreased for the treatable volume, i.e., reduced to 153 μCi/m^3. In Table 6, dilution water volume was more than the Shimizu's one, since the release limit was lowered to $20^{-1}\mu$Ci/m^3. Therefore, much water was required to dilute radioactive liquid waste in order to meet the release limit. In Table 7, slurry and sludge generation rate are all the same, while daily capacities of the P-1, P-2, P-3, and P-4 in column 4 show the lower bound of the design value. With lower bound, the operating efficiency can be reduced a little bit but it was one of the best way when the capacity for input volume is needed to be expanded in future and if the life time of the equipment is considered.

GP-Fuzzy

The GP-Fuzzy algorithm is applied to the radwaste treatment system based on the modified modeling which was added material balance at each component to the Shimizu's modeling. The objective function of the algorithm was composed of membership function, μ_i, and maximized. The results are presented in the GP-Fuzzy row of Tables 4~8.
The values of fuzzy decision function could be obtained as following:
$\mu_1 = 1.0000$, $\mu_2 = 1.0000$, $\mu_3 = 1.0000$, $\mu_4 = 1.0000$, $\mu_5 = 0.39829$
and by introducing these values to the system criteria the achieved goals are as following:

Cost $= 10^7$ (¥/yr),   Water consumption $= 5000$ (m^3/yr),
Operation days $= 190$ (day/yr),   Surface dose limit $= 4.81$ (mR/yr),
Storage area $= 20.11$ (m^2/yr).

These results mean that the limit values of each system criteria were appropriate except the storage area. The limit value of storage area is 50 (m^2/yr) originally but by the result we know that it was over-estimated.

5.1 Sensitivity Analysis

The target values of goals were changed, the priority order was rearranged and the goal constraints were added in or deleted from the priority structure for the sensitivity analysis.

The analysis, changing the target values of goal constraints, was conducted to analyze the resource requirements to achieve stated target values of goals or level of goal achievements for the capacity or level of resources available. Total treatable volume was varied from 532 m^3/yr to 1064 m^3/yr (= 2×532m^3/yr) increased by 0.5 times. Its optimal value was 798 m^3/yr (= 1.5×532m^3/yr). The goal level of cost limit was decreased to 9×10^6 ¥/yr, and the release limit of radioactivity from 10^{-1} μCi/m^3 to 20^{-1} μCi/m^3. And also other goal values (operation days, surface dose limit, and etc.) were varied. The target values of goals must be kept as original design value except the release limit. The design value of release limit was 10^{-1} μCi/m^3 originally but the optimal point was obtained at 20^{-1} μCi/m^3. It implies that the design value for the release limit can be lowered.

For the other sensitivity analysis the priority order was rearranged and the goal constraints were added in or deleted from the fundamental objective function which was composed of cost, operation days, and water consumption. Three types of result were obtained and were compared at row RESTEM, GP, and GP-Fuzzy in Tables 4~8. It is considered the various objective function which was added or deleted goal constraints of the storage area and surface dose limit in the fundamental objective function. Despite of adding or deleting the goal constraints of surface dose and storage area and the rearrangement of the priority order the result was not affected at all.

6 Conclusion

The linear fuzzy goal programming model of the RWPS presented in this chapter illustrates that how multiple and conflicting goals of the radioactive treatment system can be analyzed in such a way that total system's performance can be optimized under a desired goal-preference mechanisms. The linear goal programming technique is quite general even though it was presented with an actual case study in this chapter. All the RWPS have similar characteristics and can be studied with the linear goal programming model developed in this chapter.

In GP-Fuzzy algorithm fuzzy theory was introduced to solve the uncertainty of aspiration value of the goal programming by introducing the fuzzy decision function. But GP-Fuzzy had a problem that the result was very sensitive to the upper and low bound.

References

1. Yoshiaki, Shimizu. (1981), *J. Nucl. Sci. Technology*, 18, 773
2. A. Charnes and W. W. Cooper, (1961) *Management Models and Industrial Applications of Linear Programming.*, Vol. I. Wiley New York
3. Y. Ijiri, (1965) *Management Goals and Accounting for Control.*, North-Holland, Amsterdam
4. S. M. Lee, (1972) *Goal Programming for Decision Analysis.*, Auerbach Publishers, Philadelphia, Pensylvania
5. J. P. Ignizio, (1976) *Goal Programming and Extensions.*, Lexington Books, Massachusetts
6. A. Ravindran and, J. L. Arthur (1980) *PAGP, a partitioning algorithm for (linear) goal programming problems*. ACM Trans. Math. Soft. 6
7. R. N. Tiwari, S. Dharmar and J. R. Rao, (1987), 24, 27-34
8. Alan Moghissi, Herschel W. Godbee and Sue A. Hobart, (1986) *Radioactive Waste Technology*, The American Society of Mechanical Engineers, New York
9. H. J. Zimmerman, (1976) Descriptions and optimization of fuzzy systems, International Journal General Systems 2, 209-215

Nomenclature

a_i : Occupied area per drum (m^2) $\{ i = SLG, SLY, COM, IMC, WF \}$

c_i : Cost $\{ i = P-1, P-2, P-3, P-4, SLG, SLY, COM, IMC, WF \}$

d_i : Decontamination factor $\{ i = P-1, P-2, P-3 \}$

f_i : Upper limit per day capacity $(m^3 / day)\{ i = P-1, P-2, P-3, P-4 \}$

f_i' : Volume of inlet tank (m^3) $\{ i = IT-1, IT-2 \}$

g_i : Waste fraction $\{ i = LLW, MLW, SW \}$

q_i : Dose conversion factor $(mR \cdot m^2 / \mu Ci \cdot yr)$ $\{ i = SLG, SLY, SW \}$

s_i : Fraction of solid waste $\{ i = COM, IMC, WF \}$

t_i : Operation days (day) $\{ i = IT-1, IT-2, IT-3 \}$

v_i : Occupied volume per drum $(m^3/drum)\{ i = SLG, SLY, COM, IMC, WF \}$

w_i : Water fraction $\{ i = P-2, P-3, P-4 \}$

α_i : Specific radioactivity $(\mu Ci/m^3)$ $\{ i = SLG, SLY, COM, IMC, WF \}$

β_i : Sludge fraction $\{ i = IT-1, IT-2 \}$

γ_i : Recycled fraction $\{ i = RLW, RMW \}$

τ_i : Half life of representative nuclides $(i = {}^{59}Fe, {}^{137}Cs, {}^{60}Co)$

X_L : Low level radioactive liquid waste volume $(X_L = g_{LLW}X_0)$

X_M : Medium level radioactive liquid waste volume $(X_M = g_{MLW}X_0)$

X_S : Radioactive solid waste volume $(X_S = g_{SW}X_0)$

X_i : Radioactive waste volume at the each component $P-i (i=1, 2, 3, \cdots, 15)$

X_{P6} : Raioactive solid waste volume from the IT-3 to the P-6

$X_{COM, IMC, WF}$: Volume of radioactive solid waste

(COM : Compactible, IMC : Non-compactibe, WF : Waste Filter)

8 A Fuzzy Inference System for the Economic Calculus in Radioactive Waste Management

Pierre Kunsch[1], Antonio Fiordaliso[2], and Philippe Fortemps[2]

[1] ONDRAF-NIRAS, Place Madou 1, P.O. Box 25, 1210 Bruxelles, Belgique
[2] Faculté Polytechnique de Mons, Département de Mathématique et Recherche Opérationnelle, Rue de Houdain 9, 7000 Mons, Belgique
p.kunsch@nirond.be

This chapter illustrates a fuzzy inference system (FIS) developed to assist the economic calculus in radioactive waste management (RWM). The extended time horizons and, in addition, the first-of-a-kind nature of many RWM systems induce large cost uncertainties in project funding. The traditional approach in economic calculus is to include contingency factors in basic cost estimates. A distinction is made between T-factors, used for technological uncertainties, and P-factors, used for project contingencies. In the particular case of nuclear projects, the Electric Power Research Institute (EPRI) has developed specific recommendations for defining both contingency factors. The approach is based on the statistical interpretation of past experience data in the field. As a generalisation of the EPRI results, a new methodology using fuzzy inference rules is proposed. The inputs to the FIS are derived from the answers of experts regarding both the degrees of technological maturity and project advancement. Inferred T- and P-factors proposed by the FIS are given either as single estimates as possibility intervals. The latter are shown to possess all suitable dynamic properties for cost estimates in RWM projects, compatible with the EPRI recommendations.

1 Introduction

In Belgium, the management of all radioactive waste on the national territory has been entrusted to a state-owned organisation, ONDRAF/NIRAS. Like in most other European countries, waste management activities and projects are financed on the basis of the triple P's, i.e., the Polluter Pays Principle. It implies that all charges and costs are charged to the radioactive waste producer. The practical use of PPP is delicate for projects still in a planning stage, such as final disposal for high level waste for which the concept has not yet been fully decided. For that reason, cost estimates are only ex-ante best estimates.

On the other hand, PPP requires immediate down payment from the producer at waste delivery. This payment is equal to the net present value of the future cost stream. For that reason, sufficiently accurate assessment of costs is crucial in the case of future projects. Should the producer's payment be insufficient, the ensuing funding shortfall will place the agency in a difficult financial position. Should the payment be, on the contrary, largely overestimated, it might become an unacceptable or unjustified burden to the waste producer.

In all circumstances however, a clear bias will exist in favour of avoiding any funding shortfall. To be more precise, the board of ONDRAF/NIRAS requires that the risk of underfunding must remain below 10% for all projects. In order to respect this constraint, it is necessary to have a well-designed methodology for assessing uncertainties. Section 2 describes the approach currently applied in the Belgian agency. Basically, uncertainties are represented by suitable contingency factors. Section 3 gives information on how they are derived today. Section 4 elaborates on the fuzzy inference system proposed as an advanced approach for obtaining these factors. Some results from actual calculations are shown in section 5. Finally, Section 6 provides useful conclusions for the practical implementation in the current ONDRAF/NIRAS' practice.

2   The Conventional Approach of Economic Calculus

Risk analysis of the costs of radioactive waste management projects is traditionally based on well-known statistical techniques. The final "best-cost estimate" is obtained by applying suitable contingency factors to the initial "basic-cost estimate," originating from the project engineers. The resulting cost figure must provide a sufficient protection against blatant underestimates. This implies that, in the course of the actual implementation of the project, the probability of exceeding the best-cost estimate is kept below the imposed 10% risk limit.

According to modern financial practice in portfolio management, the best-cost estimate represents the "Value at Risk," in short VaR (see for example [1]). It should be noted that, for an activity like final disposal of radioactive waste, several VaR's are considered. Individual sub-projects are separately assessed, each corresponding to a partial implementation stage in the extended life cycle of the repository. Each stage is characterised by specific cost uncertainties. The latter generally increase as the time horizon recedes.

As an example, it is clear that the initial front-end investment cost necessary to open a repository site is much more accurately known than the closure costs to be incurred in more than 50 years from now. Post-monitoring costs, which arise after the repository closure, are still completely preliminary with respect to their exact nature or duration. In the following, these elementary project stages, or sub-projects, are simply called "projects."

Assuming that the engineers' basic-cost estimate is available, contingency factors are determined for each project. This is done according to the recommendations of the Electric Power Research Institute (EPRI) in USA. The "Technical Assessment Guide" of EPRI ([2]) gives recommendations for the economic calculus of projects related to nuclear power generation. Details on the uncertainty typology and the decommissioning cost calculations are given in [3].

The Belgian agency has adopted the same approach for its own projects, assuming it is also applicable to radioactive waste management. The steps in the economic calculus are thus the following:

- In $t = t_0$, the basic-cost estimate is provided by the project engineers in charge of the field-implementation, according to the "overnight cost" principle. This means that the prevailing safety rules and legislation, as well as the present state-of-the-art technology define the general framework for the idealised overnight implementation of the project. Assume C_0 to be this basic-cost estimate.
- The economists in charge of the cost assessment extrapolate the overnight cost C_0, with the help of technical and scientific experts. An estimate of the uncertainties is prepared, considering a more realistic time horizon $[t_0, t^*]$ specific to the project. This is equivalent to weighting C_0 with a contingency factor $(1 + \alpha)$ larger than one, giving the best-cost estimate C available in $[t_0, t^*]$ as:

$$C = (1 + \alpha) \quad C_0 \tag{1}$$

where $\alpha > 0$. In principle α is unbounded from above. As an example, a value larger than one is not uncommon in decommissioning cost assessments.

- The economist confirms by scenario analysis that the probability of having the actual cost exceeding C descends below the 10% VaR limit. An example of this analysis is available in the report [4].

3 The Nature of Contingency Factors in Radioactive Waste Management

The EPRI methodology is based on a detailed typology of uncertainties to be considered for long-term nuclear projects. Two contingency factors are being considered. In the formula above, the total weight factor $(1 + \alpha)$ needed for obtaining the best-cost estimate C from C_0, is given by the following equation:

$$1 + \alpha = (1 + Y) \quad (1 + Z). \tag{2}$$

The two independently determined contingency factors $(1 + Y)$ and $(1 + Z)$ are called respectively project (P) and technology (T) factors further on in this chapter. The P-factor (Project), with value $(1 + Y)$, expresses the uncertainties related to the advancement status of the project on the drafting boards. Four levels are being considered, ranging from "simplified" to "finalised" estimates. The corresponding values applicable for Y are given as ranges in Table 1.

The T-factor (Technology), with value $(1 + Z)$, expresses the uncertainties related to the available status of scientific and technological maturity to implement the project. The level of maturity depends on the availability of data from anterior projects comparable to the current one. Five levels are being considered, ranging from "no or limited data" to "operational data." The corresponding values applicable to Z are given as ranges in Table 2.

Table 1. The EPRI recommendations for Y

	Levels P	Ranges for Y	
(P_1)	Simplified	[0.30, 0.50]	(Y_1)
(P_2)	Preliminary	[0.15, 0.30]	(Y_2)
(P_3)	Detailed	[0.10, 0.20]	(Y_3)
(P_4)	Finalised	[0.05, 0.10]	(Y_4)

Table 2. The EPRI recommendations for Z

	Levels T	Ranges for Z	
(T_1)	No or limited data	[0.40, z_1 (to specify)]	(Z_1)
(T_2)	Bench scale data	[0.30, 0.70]	(Z_2)
(T_3)	Small pilot plant data	[0.20, 0.35]	(Z_3)
(T_4)	Full scale module data	[0.05, 0.20]	(Z_4)
(T_5)	Operational data	[0.00 0.10]	(Z_5)

Experts are asked to determine the adequate levels for both P- and T-factors. Experts also have to select the adequate crisp values within the provided two ranges respectively given in Tables 1 and 2. Finally, opinions resulting from peer review interrogations are aggregated, using suitable weighting techniques. In the remainder of this chapter, it is assumed that the results of this aggregation are available as inputs to the fuzzy inference system.

The judgements of the experts are revised as the project comes closer to its actual completion, i.e., as both aspects of project advancement and technological maturity change. Also, some external shock may impose a revision, e.g., a change in regulatory impositions quite common in the nuclear industries. From the simple description just given, it is immediately clear that casting the EPRI recommendations into fuzzy rules has a big potential in easing their practical use:

- Judgements of experts and peer reviews conveniently provide the inputs to the rules.
- Rules can be applied separately for each uncertainty component, i.e., project (P) and technology (T).
- Evolution due to the advancement and maturing of the project is easily handled by updating the inputs.

The basic ingredients of the fuzzy inference system described in the remaining sections of this chapter are herewith set in place.

4 A Fuzzy Approach for Radioactive Waste Management Funding

Tables 1 and 2 represent a set of rules that could be written as

$$\textbf{If } P \textbf{ is } P_i \quad \textbf{then } Y \textbf{ is } Y_i \qquad (i = 1, \ldots, 4) \tag{3}$$

or

$$\textbf{If } T \textbf{ is } T_j \quad \textbf{then } Z \textbf{ is } Z_j \qquad (j = 1, \ldots, 5) \tag{4}$$

where the states P_i, Y_i, T_j are defined on the above-mentioned Tables. P-states (e.g., "Preliminary estimate") or T-states (e.g., "Operational data") are vague and not precisely defined. The same goes for Y and Z. Indeed, one may question the validity of the bounds proposed by EPRI to define Y_i and Z_j. The idea exploited here is to represent such imprecise notions with the help of fuzzy sets (P_i, Y_i, T_j, Z_j) and to extent the inference process given in Tables 1 and 2 to the fuzzy case.

4.1 Methodology

The input variable (P-factor) results from the opinions of experts on the advancement of the project. Each expert is requested to give a score on a scale between 0 and 1. Opinions are aggregated using for example simple weighting of each expert, according to his degree of credibility or knowledge in relation with the project. The result $x \in [0, 1]$ of the experts' interrogation is used to determine the position of the project within four regions of advancement of the draft. The same approach is applicable to the T-factor. For each aspect (P and T), a bank of rules enables to separately determine the related contingency factors (Y and Z). The estimated cost C is then computed as

$$C = (1 + Y)(1 + Z)C_0. \tag{5}$$

4.2 Membership Functions

In the remainder of this chapter, we will denote by μ_A the membership function (m.f.) relative to fuzzy set A. When many data points are available, it is possible to automatically adjust the number and the shape of the membership functions. Some possible techniques are based on the use of artificial neural networks, genetic algorithms or constraint nonlinear optimisation.

The above-mentioned techniques are, however, not adequate in our case because of the lack of historical data. Hence, we will develop a systematic approach based on the information contained in Tables 1 and 2. The methodology - explained here for the P-rules of type (3) - is of course applicable to the T-rules.

Let us consider first the m.f. μ_{P_i} relative to the "**if**"-parts of the rules. We will make the following assumptions.

- μ_{P_i} are triangular m.f..
- At most two states (P_i) can be active at the same time. This assumption is natural in the sense that it is highly improbable that one expert hesitates between more than two states.
- The sum of the membership values computed at $x \in [0, 1]$ is equal to 1. This assumption is often used in fuzzy clustering or in control.
- $\mu_{P_1}(0) = 1$ and $\mu_{P_4}(1) = 1$.

With this in mind, we only need to fix two parameters, namely a and b, to determine the four states P_i (see Figure 1). In the same way, the determination of the five m.f. μ_{T_i} requires 3 parameters c, d and e (see Figure 2).

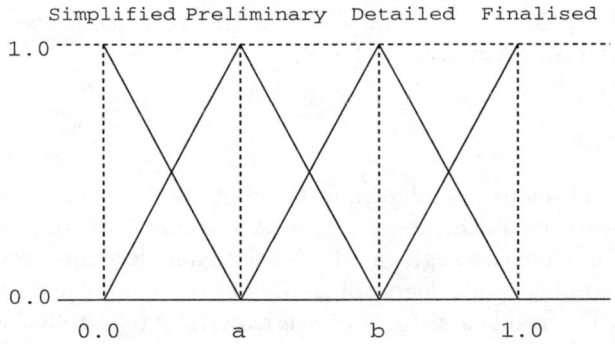

Fig. 1. Membership functions μ_{P_i}

Concerning the μ_{Y_i} m.f., we can deform the crisp (rectangular) functions proposed by EPRI to allow for gradual membership. One way to achieve

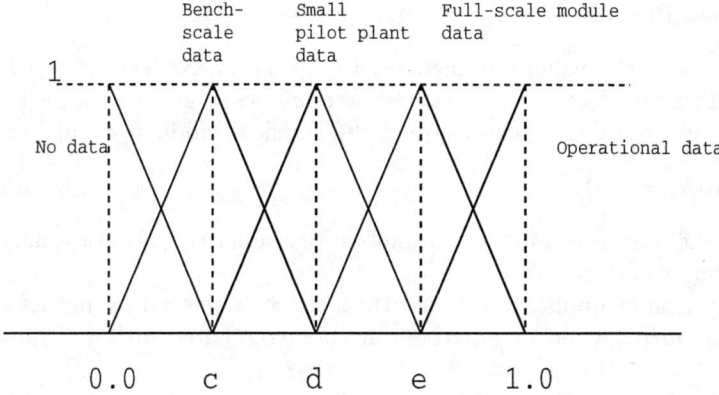

Fig. 2. Membership functions μ_{T_i}

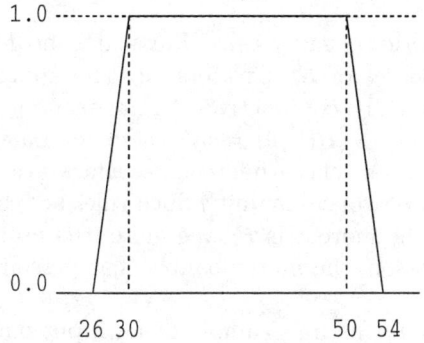

Fig. 3. An example of trapezoidal m.f. ($\lambda = 0.4$)

this is to enlarge the support of the proposed functions by multiplying the length of the rectangles' bases by a factor λ in such a way that the resulting trapezium is isosceles (see Figure 3). The value of the λ parameter will be used for the μ_{Y_i} and the μ_{Z_i} m.f. (by default, $\lambda = 0.4$).

In summary, building the input and output m.f. can be achieved by setting the values of seven parameters. By default, we set $a = 1/3, b = 2/3, c = 1/4, d = 1/2, e = 3/4, \lambda = 0.4$ and $z_1 = 0.8$. Note that the values chosen for a, b ensure μ_{P_i} to be similar triangles and the overlapping between two adjacent m.f. to be constant. The same is true for μ_{T_i} because of the values chosen for c, d, e. The z_1 value required to specify Z_1 (see Table 2) is set to 0.8 [1].

---

[1] Note that the z_1 value may be greater than 1.

4.3 Inference Process

Let us denote by x, the value obtained via the group of experts for the P-factor. The ith rule of the fuzzy inference system allows to generate a partial conclusion $\mu_{\tilde{Y}_i}(y)$ obtained by extension of the classical Modus Ponens [5]:

$$\mu_{\tilde{Y}_i}(y) = R(u_i, \mu_{Y_i}(y)) \tag{6}$$

where $u_i = \mu_{P_i}(x)$ and R is a fuzzy conjunction (conjunctive rules) or a fuzzy implication (implicative rules).

The aggregation of implication-based rules can be achieved by means of any t-norm (e.g. min), while conjunctive rules are aggregated with a disjunction using any t-conorm (e.g. max) [2]:

$$\mu_{\tilde{Y}}(y) = \min_i \mu_{\tilde{Y}_i}(y) \qquad \text{for implication-based rules}$$

$$\mu_{\tilde{Y}}(y) = \max_i \mu_{\tilde{Y}_i}(y) \qquad \text{for conjunctive rules.}$$

We know there are several categories of fuzzy rules. Particular meanings can be obtained by using special classes of R operators [5]. The approach followed by control engineers is typically conjunctive. As an example, we can cite the Mamdani's inference process ($R(u, v) = min(u, v)$) or Larsen's inference ($R(u, v) = uv$) (see Figure 4). These inference schemes are well known and widely used in the fuzzy control community. Such rules are called *possibilistic* since they express that the more x is P_i, the more it is possible that y lies in Y_i (e.g. "the older a person, the more possible that person has been at the university").

Indeed, viewing the pattern (P_i, Y_i) as an example of an input-output pair, a conjunctive rule means: "if P_i is observed, Y_i is possible." Hence, possibilistic rule-based systems are well adapted for describing a set of data points or for example-based reasoning. As a matter of fact, they are widely used in interpolation related fields such as functions approximation, control or forecasting.

As a matter of fact, conjunctive inference has been adopted in a previous paper dealing with the problem of determining the funding requirements in radioactive waste management [6]. However, the approach developed in this chapter will be implication-based. The motivation for choosing an implication-driven inference is that it appears that when a rule-based system is used to encode some expertise or knowledge, the conjunctive approach has some drawbacks compared to the implication-based inference [5,7].

For example, adding more and more possibilistic rules in a decision system will result in a conclusion $\mu_{\tilde{Y}}$ that is less and less precise (the support

---

[2] Note that this is coherent with the fact that conjunctive rules generate possible conclusions, while the conclusions obtained with implication-based rules can be seen as constrains.

of $\mu_{\tilde{Y}}$ is larger and larger) since the partial conclusions are aggregated with a t-conorm. This is counterintuitive because a decision rule is supposed to encode some expertise or information. Moreover, the fuzzy conclusion generated by a possibilistic rule is generally subnormalised. This is because the obtained output is a lower bound of the actual possibility distribution [7]. In the context of possibility theory, this means that for a given rule, no output y is fully possible, which seems a bit strange for the less. More details about the comparison of these two kinds of inference can be found in [5,7].

The preceding remarks have motivated our choice of using an implication-based inference scheme. In the following, we explain our approach for choosing an adequate implication. Theoretically, the semantic interpretation of fuzzy rules enables an implication function to be selected in accordance with the meaning of the rule. Basically, two types of meaning for fuzzy implication-based rules emerge.

- Certainty rules are obtained with S-implications such as the Kleene-Dienes implication:

$$R(u, v) = \max(1 - u, v). \tag{7}$$

The meaning of the rule is: the more x is P_i, the more it is *certain* y lies in Y_i (e.g. "the younger a person, the more certain that person has not yet been at school"). Another example of such rules is obtained with the Lukasiewicz implication:

$$R(u, v) = \min(1 - u + v, 1). \tag{8}$$

The drawback of this type of inference is that a level of uncertainty appears in the computation of the partial conclusion $\mu_{\tilde{Y}_i}$ as well as in the conclusion $\mu_{\tilde{Y}}$. This means that all the y values are at least $1 - u$ possible (see Figure 4).

- Gradual rules are obtained with R-implications such as the Gödel implication:

$$R(u, v) = 1 \quad \text{if } u \le v$$
$$= v \quad \text{else}$$

or the Goguen implication:

$$R(u, v) = 1 \quad \text{if } u = 0$$
$$= \min(1, \frac{v}{u}) \quad \text{else.}$$

The meaning of the rule is: the more x is P_i and the more y is the image of x, the more y is Y_i. An example of gradual rule is: "the better a student, the higher his marks." Actually, such kind of rules produces an interpolation reasoning. The reason for this is described hereafter.

Consider the two rules "if P is P_1 then Y is Y_1" and "if P is P_2 then Y is Y_2." Let us denote by x the observed value for P. It is clear that the core of the conclusion $\mu_{\tilde{Y}}$ is the intersection of the cores relative to the partial conclusions $\mu_{\tilde{Y}_1}$ and $\mu_{\tilde{Y}_2}$. In other words, a precise input x between P_1 and P_2 will produce a conclusion lying between Y_1 and Y_2. This phenomenon is illustrated with two Gödel-type rules in Figures. 10, 11, 12 and 13. This interpolation reasoning property explains why Dubois and Prade [7] suggest this kind of fuzzy inference should be used in fuzzy control.

In our context, one may think the semantic of fuzzy gradual rules corresponds to the subjective approach followed by engineers and experts since these rules translate knowledge like: "the more preliminary the project, the more the initial cost estimate must be weighted."

Note that as opposed to the Goguen implication, Gödel implication can generate discontinuous partial conclusions $\mu_{\tilde{Y}_i}$ (see Figure 4). This aspect may be counterintuitive for a decision-making problem unless there is strong evidence that the partial conclusion should be discontinuous.

Considering the preceding remarks, the choice of the Goguen implication now appears to be natural. The last stage of the fuzzy inference system is to defuzzify the aggregated output $\mu_{\tilde{Y}}(y)$. The system output $\hat{y}$ is computed as the mode of the possibility distribution $\mu_{\tilde{Y}}(y)$. This value corresponds to the most possible contingency factor. Note that when several modes exist, $\hat{y}$ is chosen as the median mode. The inference process for the P-factor rules bank is illustrated in Figure 5.

5 Results

5.1 Generating Single Estimates

Figure 6a (resp. Figure 6b) shows the contingency factor Y (resp. Z) versus the P-levels (resp. T-levels). Figure 7 shows factor $(1+Y)(1+Z)$ as a function of (P, T).

We observe several plateaus in the inference surface depicted in Figure 7. This can be counterintuitive because it expresses that the decision value $(1 + Y)(1 + Z)$ may change sharply in relation to a small variation affecting the P or T factors. The plateaus of Figure 7 appear because of the way rules have been built. In fact, P-rules and T-rules are activated in pairs. As a result, a plateau occurs when four rules are activated. A sudden fall is bound to happen when one of the four rules is replaced by another one.

One way to avoid the presence of these brutal transitions is to replace the input m.f. by their smooth counterparts, that is by gaussian m.f.. Indeed, gaussian m.f. are such that *all* the rules will be activated at the same time. Practically, the input m.f. are now given by

$$\mu(x; m, \sigma) = e^{-\frac{1}{2}(\frac{x-m}{\sigma})^2} \tag{9}$$

where m is the location of the center and σ controls the spread of the gaussian.

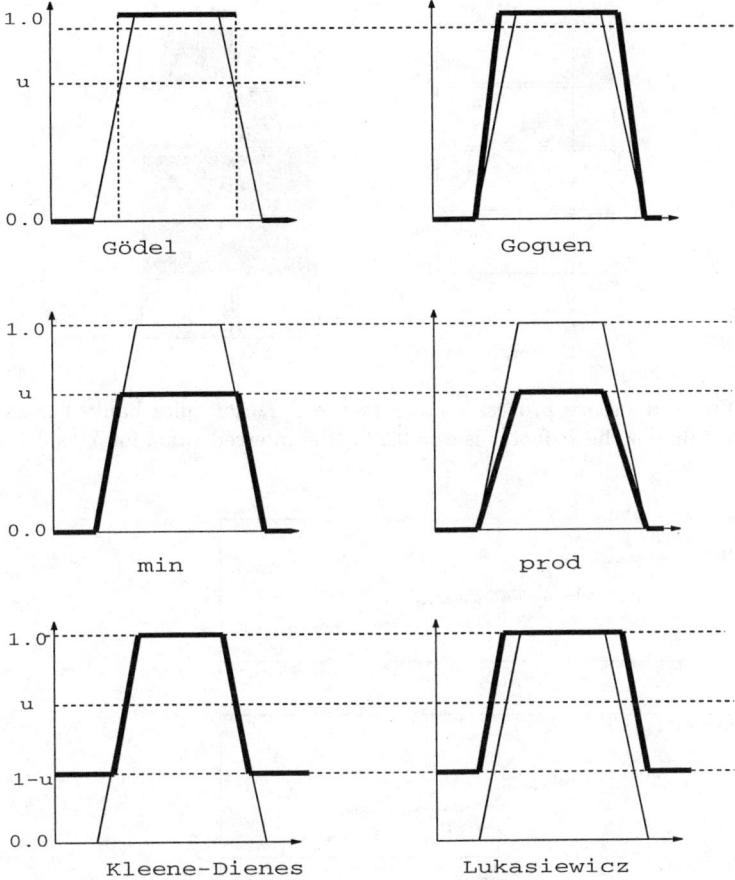

Fig. 4. Response to several fuzzy rules of the type "if P is P_i then Y is Y_i" with one precise input x ($\mu_{P_i}(x) = u$). Illustration with the Gödel, Goguen, Kleene-Dienes and Lukasiewicz implications and with the minimum and product conjunctions

However, when triangular m.f. are replaced with gaussian m.f., we observe output $\mu_{\tilde{Y}}$ results in the null m.f. expressing that no value is possible for a given x (total contradiction). Of course, in this case, no output value can be inferred. This is illustrated in Figure 8 (note the output value 0.3 is the default value corresponding to the center of the output range).

The explanation of what happens in Figure 8 is given below. Since Gaussian m.f are never equal to 0, the activations $u_i = \mu_{P_i}(x)$ are always distinct from 0. Since the trapeziums have bounded supports and $u_i \neq 0$, the Goguen inference will generate partial conclusions $\mu_{\tilde{Y}_i}$ having the same support as μ_{Y_i}. Finally, the partial conclusions are aggregated with the min, and the

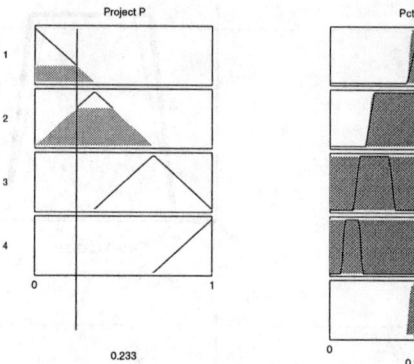

Fig. 5. Fuzzy inference process related to the P-factor rules bank. For example, when the value of the P-factor is $x = 0.233$, the inferred value for Y is 0.291

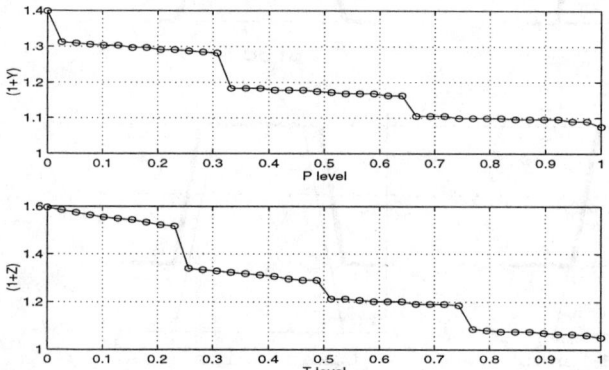

Fig. 6. (a) Upper graph. Factor $(1+Y)$ as a function of P. **(b)** Lower graph. Factor $(1 + Z)$ versus T

final output $\mu_{\tilde{Y}}$ is the null m.f.. One way to avoid this drawback is to use fuzzy conclusions with unbounded support. In our case, a natural choice is to consider generalised bell curves m.f. (gbell) defined by

$$\mu(y; a, b, c) = \frac{1}{1 + |\frac{y-c}{a}|^{2b}} \qquad (10)$$

where $b > 0$ and c locates the center of the curve. Practically, we can adjust c and a to vary the center and the width of the m.f. and then use b to control the slope at the crossover point. Figure 9 represents the smooth version of the trapezium shown on Figure 3.

It is also interesting to note that a system composed of Gödel-type rules requires sufficient overlapping between the rules conclusions in order to achieve gradual transitions between them. For illustrating purposes, let us consider

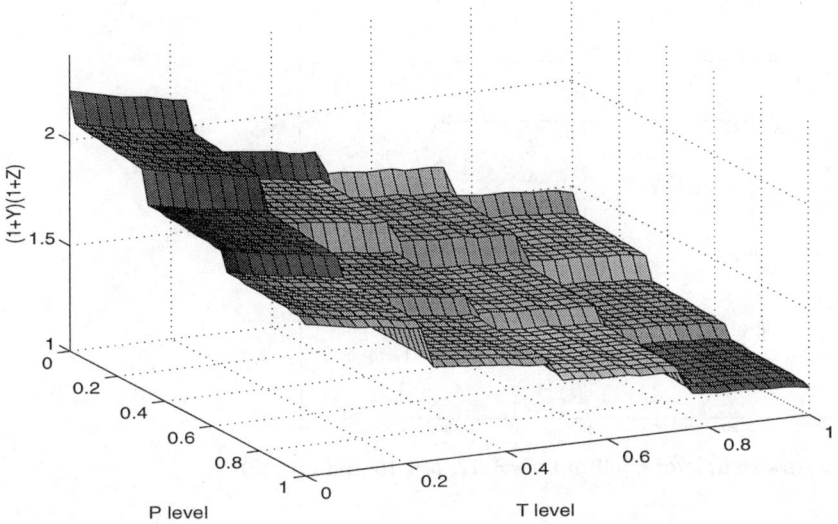

Fig. 7. Weighting factor $(1 + Y)(1 + Z)$ as a function of (P, T)

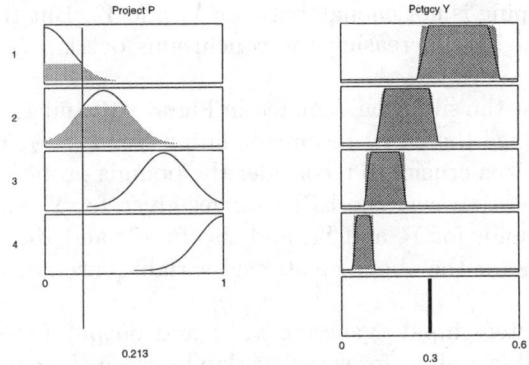

Fig. 8. P-factor rules bank: aggregation of partial conclusions with the *min* operator may give no output

the two following rules "if P is P_1 then Y is Y_1" and "if P is P_2 then Y is Y_2." Figures 10, 11, 12 and 13 show the Gödel inference process for two cases: the first one corresponds to a small overlapping between Y_1 and Y_2 (Figures 10 and 11) and the second deals with a larger overlap (Figures 12 and 13). For each case, two precise input values have been considered: $x = 0.05$ and $x = 0.25$. When $x = 0.05$ (resp. $x = 0.25$), the first (resp. second) rule is dominant.

In both cases, we observe that the *individual* rules preserve the gradual aspect: "the more x is P_1 (resp. P_2), the more y is Y_1 (resp. Y_2)." But what about the system or *overall* conclusion? In the first case, as opposed to what

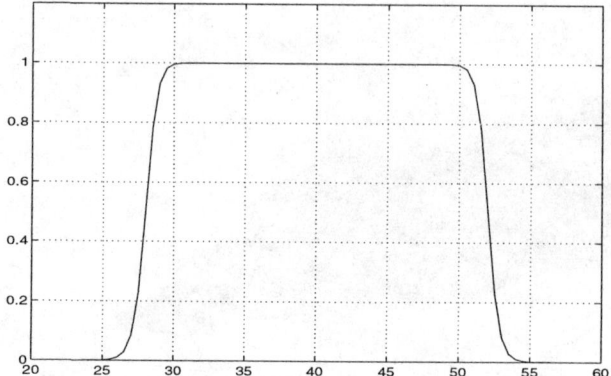

Fig. 9. An example of gbell m.f. ($a = 12$, $b = 15$ and $c = 40$)

we would expect, the aggregation phase does not generate a fuzzy set that resembles neither Y_1 (see Figure 10) nor Y_2 (see Figure 11). This results from the fact that the overlapping is not enough between Y_1 and Y_2. But the expected effect can be obtained by increasing the conclusions overlap (see Figures 12 and 13).

It should be noted here that the situation depicted in Figures 10 and 11 is the one obtained by "translating" the EPRI recommendations for P_1, P_2, Y_1 and Y_2 as explained early. So, it is crucial to reconsider the bounds proposed by EPRI. Indeed, one may question why the EPRI ranges given for Y_1 and Y_2 do not overlap. The same holds for Y_3 and Y_4, and also for Z_3 and Z_4. In a sense, the proposed intervals for the above-mentioned variables are not in the spirit of gradual rules.

The parameters of the system input (gaussian m.f.) and output (gbell m.f.) fuzzy sets reported in Table 3 allow for some overlap between the cores related to the conclusion variables. The results obtained with this second set of rules is shown in Figures 14 and 15.

5.2 Generating Possibility Intervals

The defuzzification stage destroys some information that could be helpful for the decision-maker. More specifically, the possibility distributions $\mu_{\tilde{Y}}$ and $\mu_{\tilde{Z}}$ could be used to generate possibility intervals for the Y and Z contingency factors. This can immediately be achieved by taking the α-cut of the conclusion $\mu_{\tilde{Y}}$ (and $\mu_{\tilde{Z}}$). Actually, the α-cut related to $\mu_{\tilde{Y}}$ returns the interval $I(\alpha)$ defined by:

$$I(\alpha) = \{y \in \Re \mid \mu_{\tilde{Y}}(y) \geq \alpha\}. \tag{11}$$

The length of $I(\alpha)$ informs about the doubt affecting the decision to be taken at level α (in the following, we will set $\alpha = 0.5$). The advantage of such an approach is that it can deal with the fact that the decision-makers'

Table 3. Final parameters for the input and output m.f.

States	Parameters	States	Parameters
P_1	$m = 0.00, \sigma = 0.1100$	Y_1	$a = 0.19, b = 5, c = 0.35$
P_2	$m = 0.33, \sigma = 0.1100$	Y_2	$a = 0.19, b = 5, c = 0.26$
P_3	$m = 0.66, \sigma = 0.1100$	Y_3	$a = 0.12, b = 3, c = 0.13$
P_4	$m = 1.00, \sigma = 0.1100$	Y_4	$a = 0.10, b = 3, c = 0.04$
T_1	$m = 0.00, \sigma = 0.1189$	Z_1	$a = 0.30, b = 5, c = 0.60$
T_2	$m = 0.25, \sigma = 0.1189$	Z_2	$a = 0.30, b = 5, c = 0.50$
T_3	$m = 0.50, \sigma = 0.1189$	Z_3	$a = 0.18, b = 2, c = 0.30$
T_4	$m = 0.75, \sigma = 0.1189$	Z_4	$a = 0.18, b = 2, c = 0.16$
T_5	$m = 1.00, \sigma = 0.1189$	Z_5	$a = 0.12, b = 3, c = 0.05$

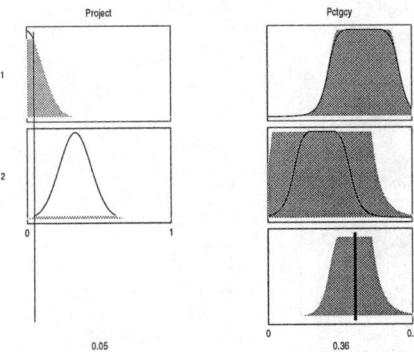

Fig. 10. Illustration of the Gödel inference process. Small overlap between the rules conclusions. Evaluation is made for $x = 0.05$

attitude may change as measure as the project approaches its maturity or the technology becomes more and more operational. For example, if the project is very simplified, he may wish to adopt a more conservative attitude, opting for the upper bound of $I(\alpha)$, while the lower bound may be preferred if the stage of development of the project is more detailed.

By considering the bounds of the 0.5-cuts of $\mu_{\tilde{Y}}$ (and $\mu_{\tilde{Z}}$), one may generate two curves expressing conservative (upper bound) and optimistic (lower bound) attitudes relatively to the decision problem. These curves are denoted by Y_M (resp. Y_m) and Z_M (resp. Z_m) when taking the upper (resp. lower) bounds of intervals $I(\alpha)$ (see Figure 14). The zone lying between Y_M and Y_m may be interpreted as the decision zone for computing the Y and Z contingency factors. The decision zone for the weighting factor $(1 + Y)(1 + Z)$ can

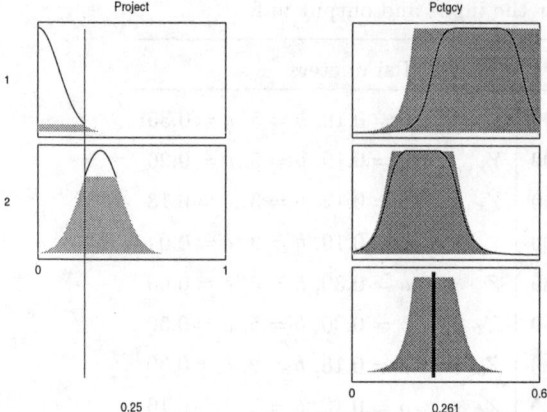

Fig. 11. Illustration of the Gödel inference process. Small overlap between the rules conclusions. Evaluation is made for $x = 0.25$

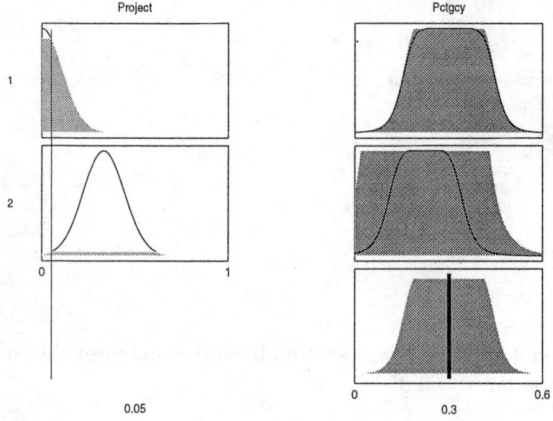

Fig. 12. Illustration of the Gödel inference process. The overlap between the rules conclusions has been increased. Evaluation is made for $x = 0.05$

be obtained by reporting the surfaces $(1 + Y_M)(1 + Z_M)$ and $(1 + Y_m)(1 + Z_m)$ versus (P, T) (see Figure 15).

As expected, the distance between these curves (that is $Y_M - Y_m$ and $Z_M - Z_m$) shows a decreasing tendency. Actually, the narrowing of the Y (resp. Z) possibility intervals is due to the fact that the widths of the partial conclusions $\mu_{\tilde{Y}_i}$ (resp. $\mu_{\tilde{Z}_j}$) - related to the Y (resp. Z) contingency factor - gets smaller and smaller as measure as the Project (resp. Technology) factor strives against 1. Note that the decreasing of the decision intervals lengths

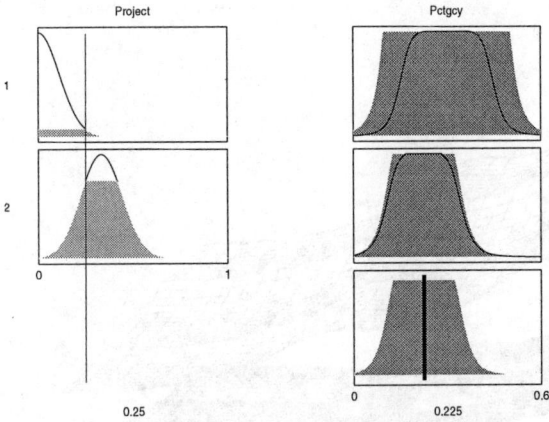

Fig. 13. Illustration of the Gödel inference process. The overlap between the rules conclusions has been increased. Evaluation is made for $x = 0.25$

can be gradually achieved owing to the existence of some overlapping between the fuzzy rules conditions.

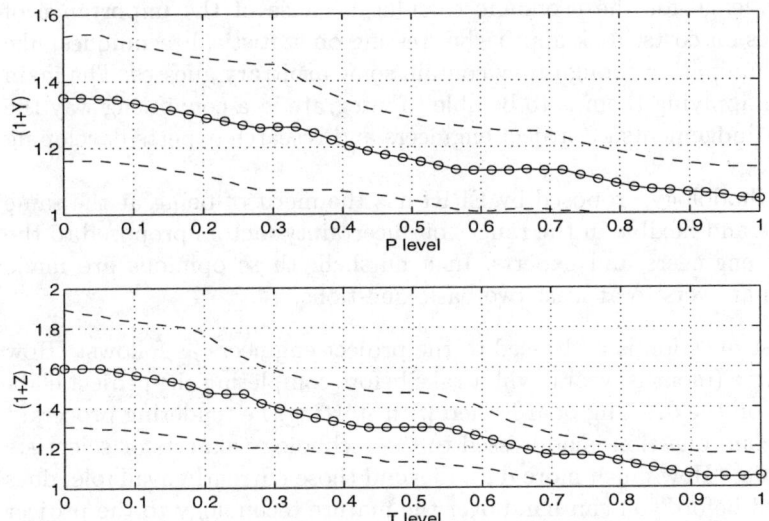

Fig. 14. (a) Upper graph. Factors $(1 + Y)$ (*solid line*), $(1 + Y_M)$ and $(1 + Y_m)$ (*dashed line*) versus P. **(b)** Lower graph. Factors $(1 + Z)$ (*solid line*), $(1 + Z_M)$ and $(1 + Z_m)$ (*dashed line*) versus T

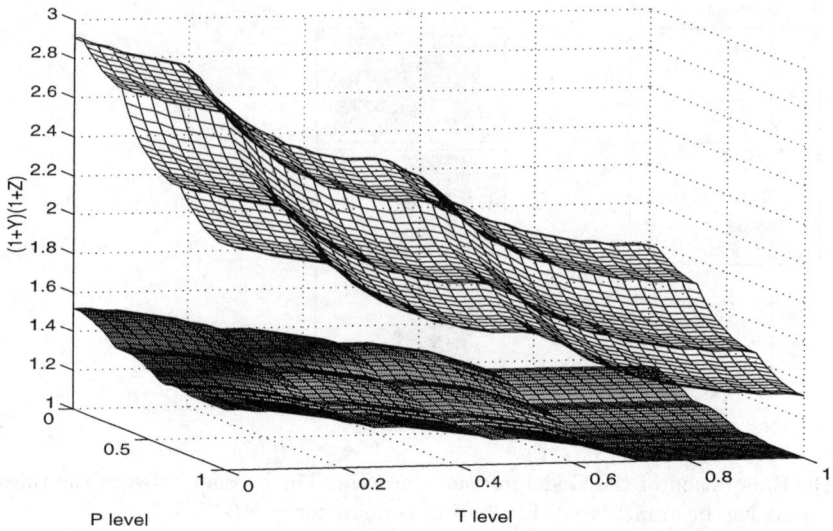

Fig. 15. Weighting factors $(1 + Y_M)(1 + Z_M)$ and $(1 + Y_m)(1 + Z_m)$ versus (P, T)

6 Conclusions

The funding requirement of radioactive waste management projects represents a challenge for the economic calculus because of the importance of uncertainties on costs. Risk approaches resting on statistical techniques, like VaR, are illuminating though they contain some arbitrary guesses. The main difficulty in applying them is to be able to integrate in a convincing way the uncertainty judgements of project engineers and research experts developing the technology.

The methodology proposed by EPRI has the merit of being at the same time precise and flexible in the ranges of uncertainty factors proposed to the opinions of engineers and experts. In a nutshell, these opinions are made available as answers to at least two basic questions:

- The first question is addressed to the project engineers as follows: "How much time (months, years) will it take before completing the project elaboration on the drafting boards, herewith starting the tendering process?"
- The second question is addressed to the technology experts or scientists as follows: "How much more data, beyond those currently available, does one need before you can hand over the mature technology to the project engineers? In other words, what is the amplitude of the outstanding R&D work?"

The semantic form of the basic questions confirms the adequacy of the fuzzy inference system (FIS) for the further treatment of the answers, as proposed by the authors in this chapter. In addition, the evolving character of the projects makes gradual rules a sensible approach. The Goguen implication herewith allows to elaborate possibility intervals of contingency factors with acceptable characteristics for engineers and experts. In particular, there is a need for less and less conservatism, as the project becomes more and more real.

In its present status described in this chapter, the FIS is not fully operational yet. Several topics are worth developing before the FIS will be applicable in practice. One pre-eminent topic regards the design of a man-machine interface for conveniently calibrating the membership functions and inference rules by using the experience of engineers and experts. Other important topics are, on the one hand, collecting answers to the mentioned basic questions and, on the other hand, aggregating by means of suitable rules the answers of the engineers and experts participating in peer reviews.

In summary, the interpretation of the EPRI methodology as an FIS for the economic calculus in radioactive waste management seems to be value adding to economists in the field. At the same time, the proposed FIS is not betraying the genuine EPRI philosophy of uncertainties. The semantic language is, beyond doubt, closer to the reality of the cost estimates than conventional risk analysis. That is why the authors think that this approach is rather generic. It will be useful to economists facing uncertainties or ignorance in actual costs of technological projects covering long-term horizons.

References

1. Esch L., Kieffer R. et al. (1997) Value at Risk. De Boeck, Brussels
2. Electric Power Research Institute (1986) Technical Assessment Guide, vol. 1, P-4463s-SR. CA EPRI, Palo Alto
3. Biewald, B., Bernow, S. (1991), Confronting uncertainty: contingency planning for decommissioning. The Energy Journal 12, 233–245
4. ONDRAF/NIRAS (1997) Report NIROND 97-04 on the disposal options for short-lived and low-level waste, Brussels
5. Dubois, D., Prade, H. (1996), What are fuzzy rules and how to use them. Fuzzy Sets and Systems 84, 169–185
6. Kunsch, P.L., Ajdler, A. (1998) Determination of funding requirements in radioactive waste management projects using fuzzy reasoning. Proc. FLINS'98, 353–358
7. Dubois, D., Prade, H. (1996), Logique floue, interpolation et commande. RAIRO 30, 607–644

9 Neuro-Fuzzy Control Applications in Pressurized Water Reactors

Man Gyun Na

Nuclear Engineering Department
Chosun University
375 Seosuk-dong Dong-gu Kwangju 501-759
Republic of Korea
magyna@mail.chosun.ac.kr

In large-scale systems like nuclear systems, automation frees operators from vigilance over routine and tedious tasks by emulating the human expertise in a faster and reliable fashion. The nuclear power plant operational data indicate that the conventional control system may fail when plant nonlinearities and their parameter changes become significant. Typical examples in pressurized water reactors (PWRs) are the power oscillations due to nonlinear xenon behavior, and large level swings of steam generators due to the swell and shrink effects during startup. Since the conventional automation technologies are not completely suitable, their operations are primarily dependent on plant operators. Since the power distribution and steam generator level controls have been the most challenging control problems in the nuclear field, there have been a number of research activities in these areas. Among many controllers proposed to replace the manual operations, the neuro-fuzzy control method is generally regarded as a suitable control method due to its human-like characteristics.

In this chapter we review the general neuro-fuzzy control methods and introduce two neuro-fuzzy control applications in PWRs such as the power distribution and steam generator water level controls, which are representative control problems in nuclear field.

1 Introduction

In spite of various positive aspects of using advanced digital controllers, for many reasons modern control systems have not been incorporated extensively in nuclear power plants. However, problems created by growing obsolescence of existing technology have stimulated interest in upgrading these systems. Recently, there have been a number of researches on applications of advanced control methods to nuclear power plants. Modern technology can improve plant availability and reduce the instrumentation and control contribution to escalating operation and maintenance costs [Wilkinson et al. 1992].

The complex nuclear power plant control systems such as the power distribution control system and steam generator level control system during startup are controlled with better performance by human operators than by conventional automatic controllers. The control strategies employed by experienced operators can often be formulated as a number of rules that are simple to carry out manually but difficult to implement by using conventional control algorithms. This difficulty is due to fact that human beings use qualitative fuzzy terms like *small*, *big*, *medium*, and so on rather than quantitative terms when describing various decisions to be taken as a function of different states of the process. It is this qualitative or fuzzy nature of man's way of making decisions that has encouraged control engineers to try to apply fuzzy logic to process control [Ostergaard 1977].

The fuzzy controllers incorporate the experience of human operators into their design. Conventional ones, however, cannot incorporate the fuzzy linguistic. A fuzzy control algorithm is designed from a set of linguistic rules, which describe the operator's control strategy. The main advantages of the fuzzy controller are the possibility of implementing *rule of thumb* experience, intuition, heuristics and the fact that it does not need a mathematical model of the process [Kickert and Mamdani 1978].

A neuro-fuzzy system is a fuzzy logic system equipped with a training algorithm. The fuzzy logic system is constructed from a collection of fuzzy *if-then* rules and the training algorithm adjusts the parameters of the fuzzy logic system based on numerical information (mainly input-output pairs). Neuro-fuzzy systems combine linguistic and numerical information. Because fuzzy logic systems are constructed from fuzzy *if-then* rules, linguistic information can be directly incorporated. On the other hand, numerical information is incorporated by training the fuzzy logic system to match the input-output pairs.

Since the power distribution control due to nonlinear xenon behavior and steam generator level control due to the swell and shrink effects during startup have been the most challenging control problems in PWRs, there have been extensive research activities in these areas. Since their conventional automation technologies are not completely suitable, their operations are primarily dependent on plant operators. Therefore, among many controllers proposed to replace the manual operations, the neuro-fuzzy control method is generally regarded as a suitable control method due to its human-like characteristics.

The objective of this chapter is to investigate the effectiveness of the neuro-fuzzy control method in application to PWR operations. First, we review the general neuro-fuzzy control methods (in Section 2). Then we introduce neuro-fuzzy control applications to the steam generator water level (in Section 3) and power distribution (in Section 4) in PWRs.

2 Neuro-Fuzzy Control Methods

2.1 Fuzzy Control System

2.1.1 Introduction

A fuzzy control action consists of situation and action pairs where conditional rules in *if-then* statements are generally used. A fundamental issue to be considered in the implementation of a fuzzy control system is to establish the required conditional rules. There are two approaches for building rule-bases. One has a heuristic and qualitative nature in which the control experience of an expert is extracted and expressed typically by a set of *if-then* linguistic statements. The other has some connection with machine learning in which the suitable rule-base is created by automatically learning from input and output data of the process, and has been used extensively in neural network systems. Adapting fuzzy systems for on-line application would be the desirable objective. Such neuronal improvements of fuzzy systems as well as the fuzzification of neural network systems aim at exploiting the complementary nature of the two approaches: the fuzzy and neural network approaches. Their composite is usually called a neuro-fuzzy system or an adaptive fuzzy system.

The basic configuration of a neuro-fuzzy control system is shown in Figure 1 where the rule base consists of a collection of fuzzy if-then rules. The fuzzy inference engine that is a component of the fuzzy controller as shown in Figure 2, uses the fuzzy *if-then* rules to determine a mapping from fuzzy sets in input universe of discourse $V \subset R^L$ to fuzzy sets in the output universe of discourse $W \subset R$ based on fuzzy logic principle.

In order to use the fuzzy controller, whose inputs are real-valued variables, we have to add a fuzzifier to the input. The fuzzifier maps crisp points in V to fuzzy sets in V. For example, if V is a collection of objects denoted by x, then a fuzzy set A in V is defined as a set of ordered pairs:

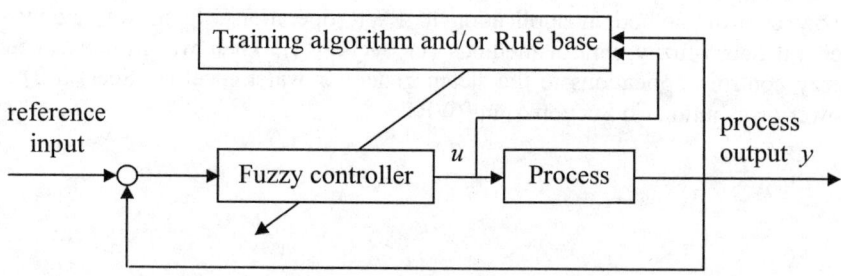

Fig. 1. Neuro-fuzzy control system

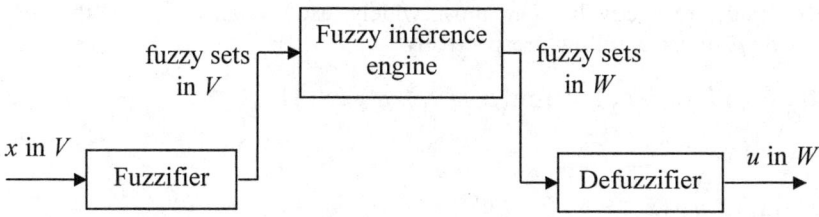

Fig. 2. General configuration of a fuzzy controller

$$A = \left\{ (x, \mu_A(x)) \big| x \in V \right\}, \tag{1}$$

where $\mu_A(x)$ is called the membership function of x in A. The membership function maps each element of V to a continuous membership value between zero and one. There is no restriction on the shape of a membership function. The Gaussian, triangular, trapezoid and bell-shaped functions are the most commonly used types in the formula of the membership function. Also, since the process takes only real values as inputs, the controller output should be real values. Therefore, we have to add a defuzzifier to the controller output.

2.1.2 Fuzzy Sets

Generally, the arbitrary i-th *if-then* rule is of the following form:

$$R_i : \text{If } x_1 \text{ is } A_{i1} \ AND \cdots AND \ x_L \text{ is } A_{iL}, \text{ then } u_i \text{ is } B_i, \tag{2}$$

where $\mathbf{x} = (x_1, \cdots, x_L)^T \in V (= V_1 \times \cdots \times V_L)$ and $u \in W$ are input and output linguistic variables, respectively, A_{ij} and B_i are fuzzy sets, and $i = 1, 2, ..., n$. And n is the number of fuzzy *if-then* rules and L the number of input variables. The fuzzy rule like Eq. (2) is called the Mamdani fuzzy model (Mamdani and Assilian 1975). These fuzzy *if-then* rules provide a convenient framework to incorporate human experts' knowledge. Without loss of generality, we consider multi-input single-output fuzzy logic system, since a multi-output system can always be decomposed into a group of single-output systems.

In case there exist multiple rules, the interpretation of multiple rules is usually taken as the union (fuzzy *OR* operator) of the fuzzy relations corresponding to the fuzzy rules. Since there exist the fuzzy *AND* and/or *OR* operator in the rule base, we need to know the mathematical interpretation of the fuzzy *AND* and *OR* operators. Usually, we adopt *min* for the fuzzy *AND* operator, and *max* for the fuzzy *OR* one. However, these are only one possible choice of operators for the fuzzy *AND* and *OR*, respectively. Other definitions for the fuzzy *AND* and *OR* have been proposed under the names *T-norm* and *T-conorm* operators [Dubois and

Prade 1980], respectively. The most widely used *T-norm* (T) is the *min* or *algebraic product* as follows, respectively:

$$\mu_{A_1}(x_1) T \mu_{A_2}(x_2) = \min\left(\mu_{A_1}(x_1), \mu_{A_2}(x_2)\right), \tag{3}$$

or

$$\mu_{A_1}(x_1) T \mu_{A_2}(x_2) = \mu_{A_1}(x_1) \cdot \mu_{A_2}(x_2). \tag{4}$$

A very common *T-conorm* (S) is the *max* or *logical sum* as follows, respectively:

$$\mu_{A_1}(x_1) S \mu_{A_2}(x_2) = \max\left(\mu_{A_1}(x_1), \mu_{A_2}(x_2)\right), \tag{5}$$

or

$$\mu_{A_1}(x_1) S \mu_{A_2}(x_2) = \mu_{A_1}(x_1) + \mu_{A_2}(x_2) - \mu_{A_1}(x_1) \cdot \mu_{A_2}(x_2). \tag{6}$$

2.1.3 Fuzzy Inference System

In a fuzzy inference engine as shown in Figure 2, fuzzy logic principles are used to combine the fuzzy *if-then* rules in the fuzzy rule base into a mapping from fuzzy sets in V to fuzzy sets in W.

If the Mamdani fuzzy model, Eq. (2), is assumed to have two antecedents and two rules, and triangular membership functions are used, Figure 3 is an illustration of how a fuzzy inference system derives the overall output u when subjected to two crisp inputs x_1 and x_2. First, we find the degree of matching w_1 and w_2 as, respectively.

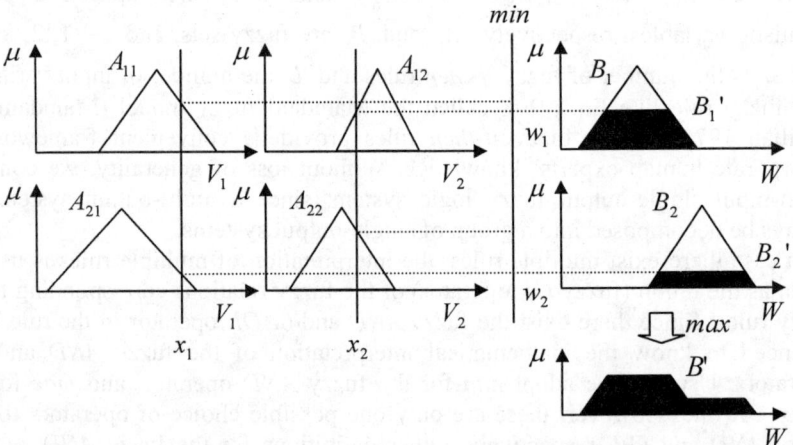

Fig. 3. Mamdani fuzzy inference system using *min* and *max* for fuzzy *AND* and *OR* operators, respectively

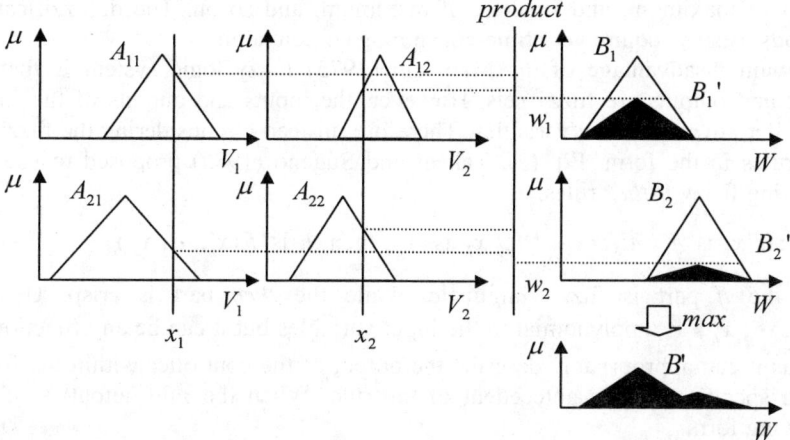

Fig. 4. Mamdani fuzzy inference system using *algebraic product* and *max* for fuzzy *AND* and *OR* operators, respectively

$$w_1 = \min\left(\mu_{A_{11}}(x_1), \mu_{A_{12}}(x_2)\right),$$

$$w_2 = \min\left(\mu_{A_{21}}(x_1), \mu_{A_{22}}(x_2)\right).$$

Then the membership function of the resulting B_1', $\mu_{B_1'}(u)$, is equal to the membership function of B_1 clipped by w_1, and that of the resulting B_2', $\mu_{B_2'}(u)$, is that of B_2 clipped by w_2. Last, the aggregated output membership function, $\mu_{B'}(u)$, is $\max\left(\mu_{B_1'}(u), \mu_{B_2'}(u)\right)$. This is called a *max-min* composition.

If we adopt *algebraic product* and *max* as our choice for the fuzzy *AND* and *OR* operators, respectively, then the resulting fuzzy reasoning is shown in Figure 4 where the inferred output of each rule is a fuzzy set scaled down by its firing strength via the *algebraic product*. This is called a *max-product* composition.

Since the process takes only crisp values as inputs, we have to use a defuzzifier to convert fuzzy sets to crisp points. The defuzzifier performs a mapping from a fuzzy set in W to a crisp point $u \in W$. The most commonly used defuzzification method is the centroid of area, which is defined as

$$u = \frac{\int_U \mu_{B'}(u)u\,du}{\int_U \mu_{B'}(u)\,du}, \tag{7}$$

where $\mu_{B'}(u)$ is the aggregated output membership function. We can use other defuzzification methods, which include bisector of area, mean of maximum, largest of maximum, and smallest of maximum, and so on. The defuzzification methods usually require very time-consuming computation.

A main disadvantage of the Mamdani (1975) fuzzy logic system is that its inputs and outputs are fuzzy sets. However, the inputs and outputs of the fuzzy controller are real-valued variables. Therefore, instead of considering the fuzzy *if-then* rules in the form, Eq. (2), Takagi and Sugeno (1985) proposed to use the following fuzzy *if-then* rules:

$$R_i : \text{If } x_1 \text{ is } A_{i1} \ AND \cdots AND \ x_L \text{ is } A_{iL}, \text{ then } u_i \text{ is } f_i(x_1, \cdots, x_L), \tag{8}$$

Here the *if* part is fuzzy linguistic, while the *then* part is crisp. Usually $f_i(x_1, \cdots, x_L)$ is a polynomial in the input variables but it can be any function as long as it can appropriately describe the output of the controller within the fuzzy region specified by the antecedent of the rule. When the rule output is of the following form:

$$f_i(x_1, \cdots, x_L) = \sum_{j=1}^{L} q_{ij} x_j + q_{i0}, \tag{9}$$

the resulting fuzzy inference system is called a first-order Takagi-Sugeno fuzzy model. If $f_i(x_1, \cdots, x_L)$ is a constant, it is called a zero-order Takagi-Sugeno fuzzy model.

Figure 5 shows the fuzzy reasoning procedure for a first-order Takagi-Sugeno fuzzy model in case there exist two antecedents and two rules. Since each rule has a crisp output, the time-consuming defuzzification procedure is avoided and the overall output is obtained via the weighted average operator. Sometimes the

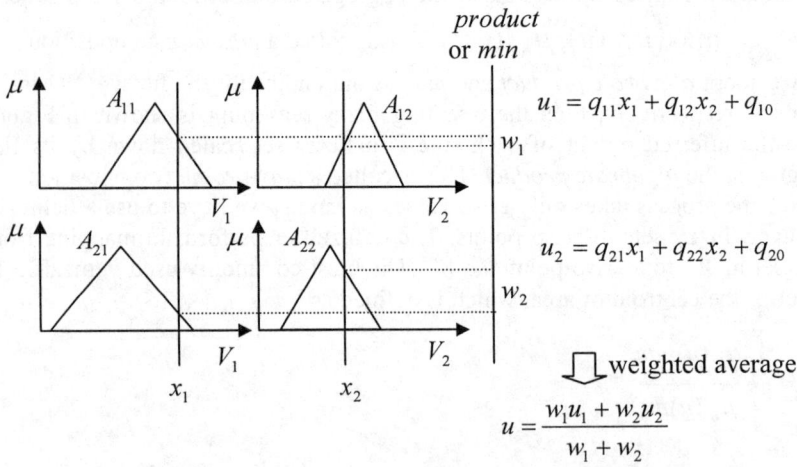

Fig. 5. Takagi-Sugeno fuzzy inference system

weighted average operator is replaced with the weighted sum operator (that is, $u = w_1 u_1 + w_2 u_2$). However, this simplification may lead to the loss of membership function linguistic meanings unless the sum of firing strengths is close to unity.

According to different choices of fuzzy *AND* (*T-norm*) and *OR* (*T-conorm*) operators, defuzzification method, membership function, and the Mamdani or Takagi-Sugeno fuzzy model, we obtain other fuzzy inference systems. In both the applications in Sections 3 and 4, we will use a Gaussian membership function, a first-order Takagi-Sugeno fuzzy model, and *algebraic product* for fuzzy *AND* operator. In addition, we will use the weighted average method to obtain the overall controller output. From now on, we will consider the fuzzy inference system that will be used commonly in both the applications of the fuzzy inference system in Sections 3 and 4.

The symmetric Gaussian membership function is of the following form:

$$\mu_{ij}(x_j) = e^{-\frac{(x_j - c_{ij})^2}{2\sigma_{ij}^2}}, \tag{10}$$

where c_{ij} is the center position of a peak of a membership function for the i-th rule and the j-th input, and σ_{ij} is its sharpness.

The output of an arbitrary i-th rule, f_i, is composed of the first-order polynomial of inputs. The output of the fuzzy inference with n rules is obtained by weighting the real values of consequent part for all rules with the corresponding membership grade. The controller output is obtained as follows:

$$\begin{aligned}
u &= \sum_{i=1}^{n} \overline{w}_i f_i \\
&= \overline{w}_1 (x_1 q_{11} + \cdots + x_L q_{1L} + q_{10}) + \cdots + \overline{w}_n (x_1 q_{n1} + \cdots + x_L q_{nL} + q_{n0}),
\end{aligned} \tag{11}$$

where

$$\overline{w}_i = \frac{w_i}{\sum_{i=1}^{n} w_i}, \tag{12}$$

$$w_i = \prod_{j=1}^{L} \mu_{ij}(x_j). \tag{13}$$

The output can be rewritten as the following vector form:

$$u = \mathbf{w}^T \mathbf{q}, \tag{14}$$

where the vectors $\mathbf{w}$ and $\mathbf{q}$ are defined as

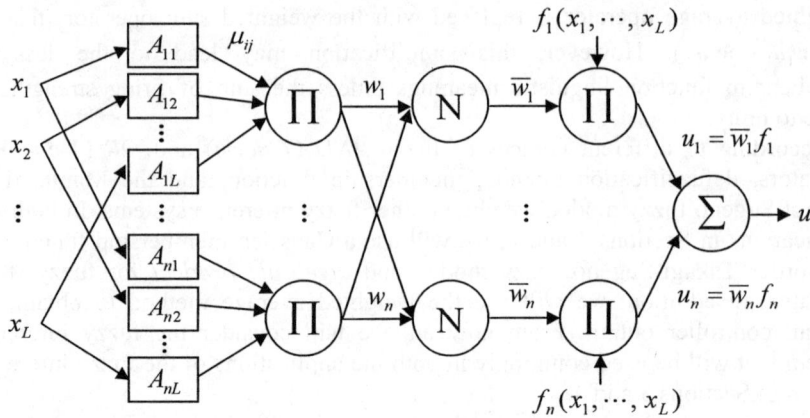

Fig. 6. Fuzzy inference system

$$\mathbf{w} = \left[\overline{w}_1 x_1 \cdots \overline{w}_n x_1 \cdots\cdots \overline{w}_1 x_L \cdots \overline{w}_n x_L \ \overline{w}_1 \cdots \overline{w}_n\right]^T, \tag{15}$$

$$\mathbf{q} = \left[q_{11} \cdots q_{n1} \cdots\cdots q_{1L} \cdots q_{nL} \ q_{10} \cdots q_{n0}\right]^T. \tag{16}$$

The membership value for rule i, w_i, means a compatibility grade between antecedent parts of a rule. The multiplicative weight in Eq. (13) is preferred over the minimum weight because of its smoothness properties. The fuzzy system described above is shown in Figure 6. In Figure 6, x_1, x_2 and x_L are the input values to the fuzzy inference system. μ_{ij} means the membership function of the j-th input for the i-th rule. When the figure is explained from left to right, the signs Π and N are expressed as Eqs. (13) and (12), respectively. The sign Σ means the summation of the input values. The second sign Π and the sign Σ are expressed as Eq. (11).

2.2 Training Methods

The antecedent membership function parameters (that is, c_{ij} and σ_{ij}) and the consequent parameters (that is, q_{ij} and q_{i0}) of the fuzzy inference system must be optimized for the good performance of a fuzzy controller.

2.2.1 Back-propagation Algorithm

Most neuro-fuzzy system adaptations rely on back-propagation. The back-propagation algorithm is a general method for recursively solving for parameter optimization. It uses a gradient descent method. The gradient descent method tunes the antecedent and consequent parameters so that the objective function, E, predefined by control designers is minimized. In order to train the arbitrary parameter a_{ij} such as c_{ij}, σ_{ij}, q_{ij}, and q_{i0}, we use the following iterative calculation:

$$a_{ij}(t+1) = a_{ij}(t) - \eta_a \frac{\partial E}{\partial a_{ij}}\Big|_t , \tag{17}$$

where $i = 1, 2, ..., n$, $j = 1, 2, ..., L$, $t = 0, 1, 2, \cdots$, and η_a is a learning rate for a parameter a. The gradient descent method is very stable when the learning rate is small, but has slow convergence property. Several techniques for speeding up the gradient descent optimization have been used including momentum and a variable learning rate [Moller 1993].

2.2.2 Least-Squares Method

The back-propagation algorithm was developed to train the fuzzy logic system so as to match desired output-actual output pairs. Because the fuzzy logic system is nonlinear in its adjustable parameters, the back-propagation algorithm implements a nonlinear gradient optimization procedure and can be trapped at a local minimum and converges slowly. If we fix some parameters of the fuzzy logic system, the resulting fuzzy system is equivalent to a series expansion of some basis functions. This basis function expansion is linear in its adjustable parameters. Therefore, we can use the least-squares method to determine the remaining parameters. For example, if we fix the membership function parameters in the first-order Takagi-Sugeno (1985) fuzzy model, the inference system output is written by Eq (14). When some input-output pattern data for training are given, the consequent parameters are chosen such that the data satisfy the following equation:

$$\mathbf{u} = \mathbf{W}^T \mathbf{q} , \tag{18}$$

where $\mathbf{u}$ is the output data and the matrix $\mathbf{W}$ as shown in Eq. (15) includes the input data defined as, respectively

$$\mathbf{u} = \begin{bmatrix} u^1 & u^2 & \cdots & u^N \end{bmatrix}^T ,$$

$$\mathbf{W} = \begin{bmatrix} (\mathbf{w}^1)^T & (\mathbf{w}^2)^T & \cdots & (\mathbf{w}^N)^T \end{bmatrix}.$$

N is the number of the input-output data pairs. The controller outputs are represented by $N \times 3n$-dimensional matrix with N rows equal to the number of data pairs and $3n$ columns equal to 3 times the number of rules. In order to solve the parameter vector $\mathbf{q}$ in Eq. (18), the matrix $\mathbf{W}$ should be invertible but is not usually a square matrix. Therefore, we solve the vector using the pseudoinverse as follows:

$$\mathbf{q} = \left(\mathbf{W}\mathbf{W}^T \right)^{-1} \mathbf{W}\mathbf{u} . \tag{19}$$

The least-squares method is a one-pass regression procedure and is therefore much faster than the back-propagation algorithm.

2.2.3 Genetic Algorithm

The term *genetic algorithm* is derived from the fact that its operations are based on the mechanics of genetic adaptation in biological systems. Genetic algorithms are search algorithms based on the mechanics of natural selection and natural genetics. Genetic algorithms for optimization were formally introduced in the 1970s by Holland (1975). More details about genetic algorithms can be found in Goldberg (1989) and Mitchell (1996). Genetic algorithms have been proven to be efficient in many different areas, such as nuclear fuel loading optimization [DeChaine and Feltus 1996; Parks 1996] and fuzzy logic controller design [Kim et al. 1995; Na 1998; Park et al. 1995].

The genetic algorithm is a method for moving from one population of chromosomes to a new population by using a kind of natural selection together with the genetics-inspired operators of selection, crossover, and mutation. Many optimization methods move from a single point in the decision space to the next using some transition rule to determine the next point. This point-to-point method is dangerous because it is a perfect prescription for locating false peaks in many peaked search spaces. By contrast, genetic algorithms work from many points simultaneously climbing many peaks in parallel. Thus, the probability of finding a false peak is reduced over methods that go point to point. Therefore, genetic algorithms are less susceptible to being stuck at local optima than conventional search methods.

Many search techniques require much auxiliary information in order to work properly. For example, gradient techniques need derivatives in order to climb the current peak. By contrast, genetic algorithms do not need all this auxiliary information. Also, genetic algorithms use probabilistic transition rules to guide their search but the use of probability does not suggest that the method is some simple random search. Genetic algorithms use random choice as a tool to guide a search toward regions of the search space with likely improvement [Goldberg 1989]. Despite of these advantages, however, genetic algorithms tend to be computationally expensive.

In genetic algorithms, the term *chromosome* typically refers to a candidate solution to a problem, generally encoded as a bit string. Each chromosome can be thought of as a point in the search space of candidate solutions. The genetic algorithms process populations of chromosomes, successively replacing one such population with another. The genetic algorithms require a fitness function that assigns a score to each chromosome in the current population. The fitness of a chromosome depends on how well that chromosome solves the problem at hand

[Mitchell 1996]. We need to properly define the fitness function according to the given problem.

Initially, after an initial population of chromosomes is randomly generated, then the typical genetic algorithm evolves the population through the following three operators.

1) Selection Operator: This operator selects individuals (chromosomes) in the population for reproduction. The goodness of each individual depends on its fitness. Fitness may be determined by an objective function. The fitter the chromosome, the more times it is likely to be selected to be reproduced.

2) Crossover Operator: After two individuals are chosen from the population using the selection operator, the crossover operator randomly chooses a crossover site along the bit strings and exchanges the subsequences before and after that crossover site between the two individuals to create two offspring. For example, the strings *000000* and *111111* could be crossed over after the second locus in each to produce the two offspring *110000* and *001111*. The two new offspring created from this mating are put into the next generation of the population. By recombining portions of good individuals, this process is likely to create even better individuals.

3) Mutation Operator: With some low probability, a portion of the new individuals will have some of their bits flipped. Mutation can occur at each bit position in a string with some small probability. Its purpose is to maintain diversity within the population and inhibit premature convergence.

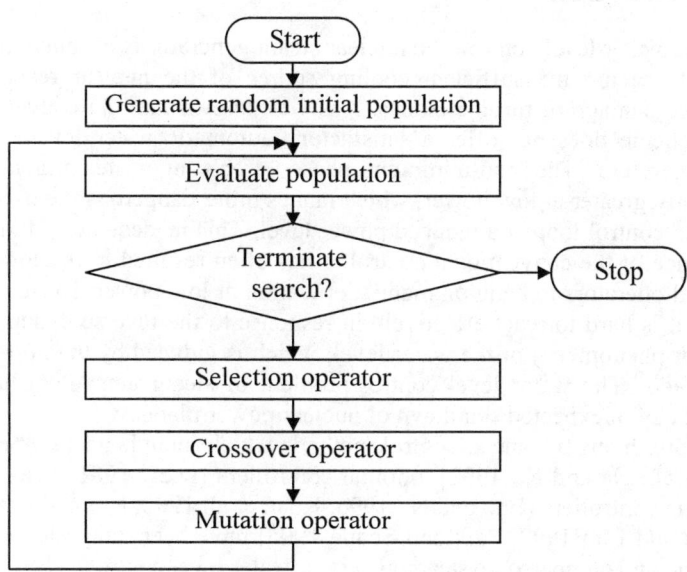

Fig. 7. Diagram of genetic algorithms

Most genetic algorithms follow the procedures in Figure 7 as explained above.

To use a genetic algorithm, we must represent a solution to a given problem as a chromosome. The genetic algorithm then creates a population of solutions (chromosomes) and applies genetic operators such as selection, crossover and mutation to evolve the solutions in order to find the best one. The three most important aspects of using genetic algorithms are (1) definition of the objective function, (2) definition and implementation of the genetic representation, and (3) definition and implementation of the genetic operators.

We can use two or more learning algorithm (hybrid learning algorithm) to train the parameters of neuro-fuzzy inference systems. For example, we can train the antecedent parameters by using a back-propagation method and the consequent parameters using a least-squares method. The consequent parameters are updated first using a least-squares method and then the antecedent parameters are updated by back-propagating the errors that still exist [Jang 1993]. Also, we can train the antecedent parameters by using a genetic algorithm and the consequent parameters using a least-squares method.

3 Application to the Steam Generator Level Control

3.1 Introduction

The proper water level control of a nuclear steam generator is of much importance in order to secure the sufficient cooling source of the nuclear reactor and to prevent the damage of turbine blades. It is well known that the conventional P-I control scheme does not offer a satisfactory automatic water level control for steam generators. The non-minimum phase effects in a steam generator are significantly greater at low power, which makes more dangerous the use of a high gain of the control loop at a reduced power level. This inadequate and insufficient performance of the conventional controller has often resulted in reactor shutdown and forced operators to hang on manual operation at low power. Even to a skilled operator, it is hard to react effectively in response to the reverse dynamics (swell and shrink phenomena) of the water level, which is induced by the non-minimum phase effects. The water level control problem of steam generators has been a main cause of unexpected shutdown of nuclear power plants.

Therefore, many advanced control methods which include adaptive controllers [Irving et al.; Na and No 1992], optimal controllers [Feely 1981; Lee 1994], and fuzzy logic controllers [Cho and No 1996; Kuan et al. 1992; Lee and No 1993; Na 1998; Na and Lim 1997; Park and Seong 1995] have been suggested to improve the chronic and cumbersome steam generator water level control.

The fuzzy control method has some limitations from the fact that its performance largely depends on initial membership function parameters and how to settle the rule base. Especially, several computational approaches for training a fuzzy controller of non-minimum phase systems have not been successful [Park et al. 1994]. To train the fuzzy controller with the conventional error backpropagation methods, system output errors must be backpropagated through the system because the errors at the output of the fuzzy controller are required. Since the inverse system of the non-minimum phase system has unstable dynamics, however, it is difficult to train the fuzzy-controller via the backpropagation of the system output error [Park et al.1995]. Moreover, non-minimum phase systems show an undershoot phenomenon for a step reference input. Therefore, to reduce the undershoot effect and shorten the settling time of non-minimum phase systems, we apply a genetic algorithm to train the fuzzy controller for the nuclear steam generator water level which has non-minimum phase characteristics.

3.2 Controller Design

3.2.1 Fuzzy Rule Base

Since the fuzzy control method has a difficulty that the number of rules becomes large by adding input variables, the number of input variables is normally limited by two, or separated rules are developed for the other variables. Thus, a simplified and compact-type rule base is required. In this section, we assume that a fuzzy controller has two inputs (water level error and flowrate error). In addition, we assume that the number of membership functions for each input variable is the same. Figure 8 shows the fuzzy rule table.

We tune the neuro-fuzzy controller off-line by a genetic algorithm. We configure the fuzzy inference system so that its parameters have reasonable values and it has small number of parameters, which are tuned by the genetic algorithm. Also, for simplification of the genetic coding we normalize the input signals x_1 and x_2 as follows.

$$\bar{x}_1 = \frac{x_1}{a_1}, \quad \bar{x}_2 = \frac{x_2}{a_2}, \tag{20}$$

where a_j is a normalizing parameter for each input signal.

The general fuzzy inference system, which is given by Eq. (8), has too many parameters. Therefore, we assume that the membership functions have the same sharpness for each input signal. s_1 and s_2 are the sharpness for the input signals $\bar{x}_1$ and $\bar{x}_2$, respectively. Because the a_j becomes a normalizing parameter for each input, the center values of membership functions for *NB*, *ZO* and *PB* are -1, 0 and 1, respectively. In addition, we assume that if the center value for PS_k is c_{kj} [$k = 1, 2, \cdots, (m-3)/2$, $j = 1, 2$], one for NS_k is $-c_{kj}$, where m is the number of membership functions for each input. Therefore, all information about the

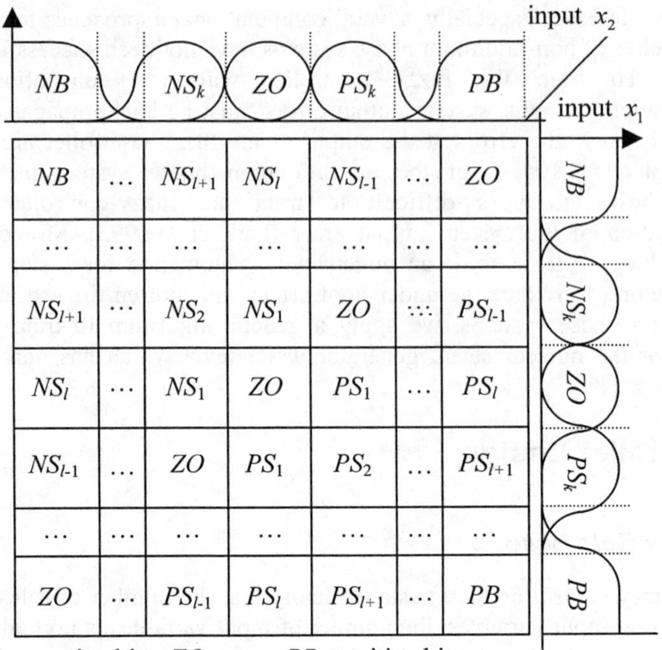

input x_2

input x_1

NB	...	NS_{l+1}	NS_l	NS_{l-1}	...	ZO
...	...	...	...	...	...	...
NS_{l+1}	...	NS_2	NS_1	ZO	::::	PS_{l-1}
NS_l	...	NS_1	ZO	PS_1	...	PS_l
NS_{l-1}	...	ZO	PS_1	PS_2	...	PS_{l+1}
...	...	...	...	...	...	...
ZO	...	PS_{l-1}	PS_l	PS_{l+1}	...	PB

NB: negative big; *ZO*: zero; *PB*: positive big
NS_k: negative small of size k, [k = 1, 2, $(m-3)/2$]
PS_k: positive small of size k, [k = 1, 2, $(m-3)/2$]
Control actions: NB, NS_1, NS_2, ..., NS_{m-2}, ZO, PS_1, PS_2, ..., PS_{m-2}, PB
m: the number of membership functions for each input

Fig. 8. Generalized fuzzy rule table

membership functions of the fuzzy inference system can be derived from the parameters a_j and s_j and the center values c_{kj} of the PS_k membership functions. The number of the antecedent parameters, which will be optimized, is $m + 1$.

The real values of the consequent part for rule i and input j, q_{ij} and q_{i0}, are defined for each control action of the fuzzy rule table in Figure 8 as follows.

$$q_{ij} = q_{i0} = 0 \text{ for } ZO, \tag{21}$$

$$q_{ij} \text{ for } PS_l \text{ or } PB = q_{lj} \, [\, l = 1, \cdots, (m-2) \text{ for } PS_l, \, l = m-1 \text{ for } PB, \, j = 1, 2\,], \tag{22}$$

$$q_{i0} \text{ for } PS_l \text{ or } PB = q_{l0} \, [\, l = 1, \cdots, (m-2) \text{ for } PS_l, \, l = m-1 \text{ for } PB\,]. \tag{23}$$

As shown in Figure 8, the real values on the same diagonal line are the same. Also, we choose the real values for NS_l and NB as the negative of the values for

PS_l and PB, respectively. Therefore, the consequent parameters of only the light ash-colored boxes in Figure 8 are needed and those of other boxes are unnecessary. The number of the consequent parameters, q_{ij} and q_{i0}, which will be optimized, becomes $3(m-1)$ instead of $3n$. The total number of the antecedent and consequent parameters, which will be optimized, is $4m-2$, and the parameters are a_j, s_j, c_{kj}, q_{lj} and q_{l0}.

3.2.2 Genetic Algorithm for Parameter Optimization

We will evaluate the performance of the water level controller by how well it tracks the setpoint and how effectively it deals with the steam flowrate disturbance. Therefore, we require that the controller have desirable response to the step changes of setpoint (water level) and external disturbance (steam flowrate). As it were, the fuzzy membership functions and rule base have to be optimized so that the controller has optimal response to the setpoint and external disturbance changes. A nuclear steam generator is a non-minimum phase system that shows the undershoot and overshoot phenomena for step increase and decrease of the input (feedwater flowrate), respectively. Therefore, we propose a cost function that evaluates the extent to which each individual (chromosome) is suitable for the given objectives such as small undershoot and overshoot together with small overall error. The fitness of an individual is calculated by means of the energy (cost function) of the individual. The chromosome that has lower energy has higher fitness. Each chromosome contains the antecedent parameters c_{kj}, a_j and s_j, and the consequent parameters q_{lj} and q_{l0} which describe the fuzzy membership functions and rule base. In order to generate the reference response specification, we assume that the setpoint increases from 0 mm to 100 mm at 1 min and the steam flowrate increases by 50% of the steady-state value at 12 min.

To accomplish the aforementioned objectives, we define the three energy functions expressed by

$$E_1 = \int_0^T \left| y_d(t) - y(t) \right| dt , \tag{24}$$

$$E_2 = \int_0^T \left| \min(y(t), 0) \right| dt , \tag{25}$$

$$E_3 = \int_0^T \left| \max(y(t) - y_d(t), 0) \right| dt , \tag{26}$$

where $y_d(t)$ and $y(t)$ denote the step setpoint change and the actual output response, respectively, and T is a training time interval. E_1, E_2 and E_3 are overall sums of absolute errors, absolute value of undershoot, and absolute value of overshoot, respectively. We define the fitness function as follows.

$$F = \exp(\frac{\delta - E_t}{\rho}) , \tag{27}$$

where $E_t = \alpha E_1 + \beta E_2 + \gamma E_3$ which is called as a total weighted error from now, and α, β and γ are the weighting coefficients. We introduce the parameters δ and ρ in Eq. (27) so as to prevent outreaching the calculation range of a computer.

To increase the efficiency of the conventional genetic algorithm (Na 1998), the proposed genetic algorithm has initial coarse tuning characteristics by initially representing each parameter in a chromosome by a small bit number. If we represent the parameters in a chromosome by big bit numbers, the genetic algorithm can find the accurate optimal points in a limit of resolution but needs much more time to reach a convergence point. However, because the genetic algorithm is a time-consuming algorithm, it is unnecessary to represent by a big bit number from the beginning the parameters in a chromosomes which are distant from the optimal values. However, it is necessary to represent it by a big bit number as many chromosomes (solutions) gradually approach the optimal points. Therefore, when the simulation generation reaches one third of the maximum generation, the bit number is increased by one third of its initial bit number. And then, when the simulation generation reaches two thirds of the maximum generation, the bit number is increased by two thirds of its initial bit number. By this method, the genetic algorithm has initial coarse tuning and final fine tuning characteristics.

The crossover site is selected by two ways when a chromosome contains many parameters. The first is that the crossover site is randomly selected anywhere in a chromosome. The second is that the crossover site is randomly selected between only parameters in a chromosome. This method slows an initial premature convergence without reaching optimal solutions and speeds up a final convergence. The first way increases diversity since the crossover site is randomly selected anywhere in a chromosome. Therefore, this way prevents a premature convergence without reaching optimal solutions. However, this way slows the final convergence due to excessive diversity in a final stage. Therefore, in a final stage, it is necessary to reduce the application probability of this way, which means that some diversity decreases. The application probability of the two ways depends on the number of the tuned parameters. In this work, two ways go fifty-fifty.

Last, some chromosomes with higher fitness in a priori generation are added to a new generation. In that case, the new population size of the total new generation increases. Then the increased portion of chromosomes with lower fitness in the total new generation are removed. This is to inhibit final drifting without convergence.

3.3 Numerical Simulations

Let us perform numerical simulations to study the performance of the proposed algorithm. The dynamics of a steam generator is described in terms of input (feedwater flowrate; u), output (water level; y) and measurable disturbance (steam flowrate; v). Irving (1980) derived the following fourth-order Laplace transfer function for steam generators:

$$y(s) = \frac{G_1}{s}[u(s) - v(s)] - \frac{G_2}{1 + \tau_2 s}[u(s) - v(s)]$$

$$+ \frac{G_3 s}{\tau_1^{-2} + 4\pi^2 T^{-2} + 2\tau_1^{-1}s + s^2} u(s), \tag{28}$$

where s is a Laplace variable. The parameter values of a steam generator for 5% powers are given in Table 1. The sampling time is chosen to be 2 sec.

Table 1. Parameters of a steam generator model at 5% power level

Power level (%)	G_1	G_2	G_3	τ_1 (sec)	τ_2 (sec)	T (sec)	V_0 (kg/sec)
5	0.058	9.630	0.181	41.9	48.4	119.6	57.4

Before we apply the proposed controller to the steam generator water level, we have to optimize the control parameters off-line by the genetic algorithm using the two signals; the water level error and the flow error (difference between steam and feedwater flowrates).

Since it is assumed that each of input variables has three membership functions, the total number of parameters that have to be optimized is 10. The parameters are $a_1, a_2, s_1, s_2, q_{11}, q_{21}, q_{12}, q_{22}, q_{10}$ and q_{20}. Therefore, each chromosome consists of a binary string, which include the 10 parameters. Initially, each parameter consists of 10 bits. Therefore, a chromosome has 100 bits. An initial population is randomly chosen. The initial values for this genetic algorithm are as follows:

Population size = 30,

Crossover probability = 100%,

Mutation probability = 3%,

Maximum generation = 100,

Constants for the fitness function; $\alpha = 1$, $\beta = 2$, $\gamma = 2$, $\delta = 200$, $\rho = 10$.

The weighting parameters β and γ are two times more heavily weighted than the parameter α so that the proposed controller has small shrink and swell phenomena.

Table 2. Finally tuned parameter values of the neuro-fuzzy and P-I controllers at 5% power level

	a_1	a_2	s_1	s_2	q_{11}
Neuro-fuzzy Controller	204.2645	8.8381	0.3883	0.8544	7.2308
	q_{21}	q_{12}	q_{22}	q_{10}	q_{20}
	16.9341	7.2634	14.9221	17.2074	12.6157
P-I Controller	K_{p1}	K_{i1}	K_{p2}		K_{i2}
	232.0360	7.5751	299.5620		100.0000

Under the setpoint and steam flowrate changes for the aforementioned reference response specification, the membership functions and rule base that have the maximum fitness, are selected by undergoing 100 evolutions of the genetic algorithm. We choose very severe convergence criteria to observe the full simulation results of the genetic algorithm for the neuro-fuzzy controller.

To verify the performance of the proposed controller, it is necessary to compare this controller with a conventional proportional-integral (P-I) controller. The simple diagram of the P-I controller is given in Figure 9. The same genetic algorithm that is used to optimize the proposed neuro-fuzzy controller optimizes the control gains of the P-I controller. The tuning gains of the P-I controller are K_{p1}, K_{i1}, K_{p2} and K_{i2}. Table 2 shows the final optimized parameters of the neuro-fuzzy controller and the optimized P-I gains. Figure 10 shows the fitness functions versus generation. The fitness function initially increases fast and later increases gradually. From Figure 10, we can predict that the performance of the neuro-fuzzy controller will be better than that of the P-I controller.

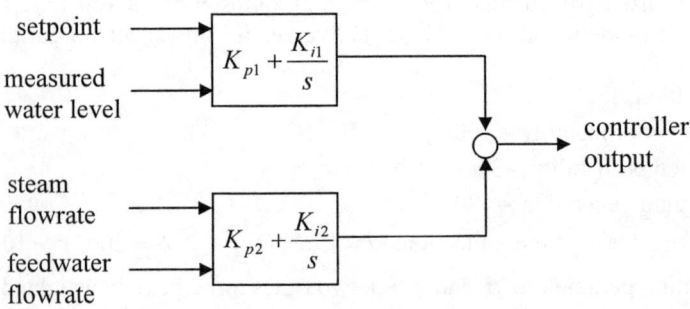

Fig. 9. Schematic diagram of a P-I controller

During the real transient simulations, a steam generator has been operated at a steady state. At 5 min, the setpoint of the water level increases from 0 mm to 100 mm. At 20 min, the 50% step increase of the steady-state steam flowrate takes place. Since the nuclear steam generator level control has troubles at below 20

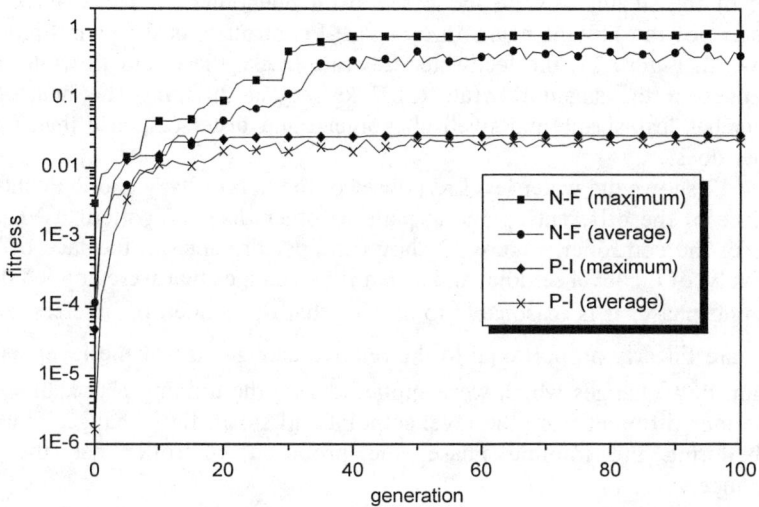

Fig. 10. Fitness at the training stage of the neuro-fuzzy (N-F) and P-I controllers

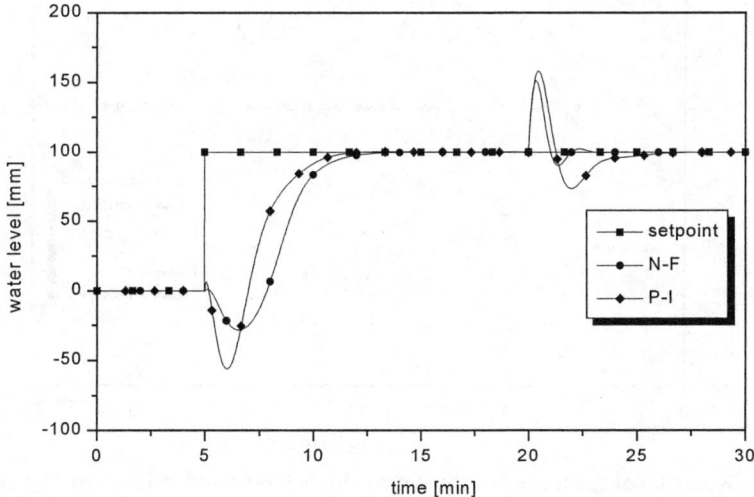

Fig. 11. Water level responses of the neuro-fuzzy (N-F) and P-I controllers

percent power, we simulate the neuro-fuzzy and P-I controllers at 5% power level. The ranges of the feedwater and steam flowrates and the water level are normalized around the operating points.

Figure 11 shows the water level responses of the neuro-fuzzy and P-I controllers at 5% power level. At 5 min, such a sudden increase of the setpoint requires more feedwater, which brings a shrink phenomenon. At 20 min, the step increase of the steam flowrate induces a swell phenomenon. The control input (feedwater flowrate) by the neuro-fuzzy and P-I controllers is shown in Figure 11. As shown in Figure 11, the feedwater flowrate tracks the steam flowrate in the disturbances of the steam flowrate (28.7 kg/sec) at 20 min. The neuro-fuzzy controller has less shrink and swell phenomena and faster response than the P-I controller does.

Figure 12 shows the water level responses of the neuro-fuzzy and P-I controller in the face of the different operating patterns other than the pattern used in the training of the controller. Figure 12 shows the performance in the face of 50%, respectively, of the level setpoint and steam flow changes that were applied during the training phase. It is reasonable to assume that the trained parameters a_l, q_{lj} and q_{l0} are linearly proportional to the relative change size of the level setpoint and steam flow changes which were applied during the training phase. In spite of the situations different from the level setpoint and steam flow changes that were applied during the training phase, the proposed controller has the good performance.

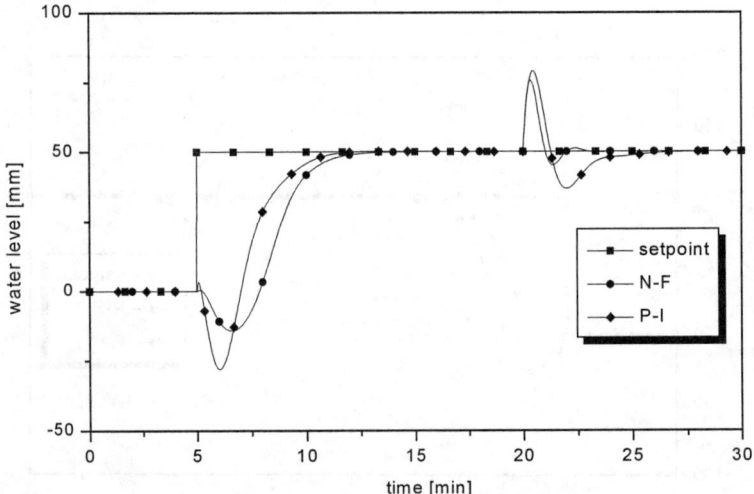

Fig. 12. Water level responses of the neuro-fuzzy (N-F) and P-I controllers (in the face of 50% of the level setpoint and steam flow changes that were applied during the training phase)

4 Application to the Power Distribution Control

4.1 Introduction

Changes in xenon concentration, power level and burnup have an influence on the spatial core power distribution. Axially nonuniform buildup and removal of xenon cause the core power distribution to oscillate between the core top and its bottom with a period of 20 to 30 h. Xenon oscillation is particularly important because of the large thermal absorption cross section of xenon. Its effects in the reactor are delayed because only a small fraction of xenon is produced directly by fission but the major portion is formed by the decay of the iodine precursor. The axial xenon oscillation in nuclear reactors is a highly nonlinear phenomenon that is a function of several time-variant parameters such as boron level, rod position and power level.

Maintaining the local core power within acceptable limits is a common objective for control problems. The control of xenon induced power oscillations in conventional reactors is mainly regulated with the control rods by plant operators and the control rods are inserted in radially symmetric groups. The radial power shape can be changed by moving independent rod groups. Control of the core power distribution is mostly concentrated in the axial direction. Axial power shaping in pressurized water reactors is achieved by insertion or withdrawal of groups of full-length and part-length control rods and changes in boron concentration in the coolant [Frogner 1978].

The fact that there is no direct way of measuring the xenon concentration often causes operators a great deal of difficulty in anticipating the amplitude, direction, and the rate of change of the xenon imbalance that is closely related with axial power shape. Since the power distribution control has been one of the most challenging control problems in the nuclear field, there has been extensive research in this area, especially using optimal control methods.

The design of an optimal controller is, in general, based on an assumed linear model that is an approximate representation of a nonlinear plant. Moreover, the controller needs precise measurements or estimations of plant variables. On the contrary to this model-based controller, the neuro-fuzzy controllers do not rely on an accurate description of the plant but are generally based on an expert's knowledge of the underlying process. Also, the controller can be designed to be automatically fine-tuned or calibrated using the process data to obtain the desired performance.

For the power control of a nuclear reactor, Akin and Altin (1991), Hah and Lee (1994), Heger et al. (1995), Park and Cho (1995) and Ramaswamy et al. (1993) applied fuzzy logic and neural network methods. Na and Upadhyaya (1998) applied a fuzzy logic and neural network method for the core power distribution control. The fuzzy logic and neural network methods have been used a lot for the power control and have been used only a little for the core power distribution control.

4.2 Controller Design

4.2.1 Introduction

While the conventional neuro-fuzzy control schemes may be capable of dealing with single-variable linear or nonlinear systems, considerable difficulty is encountered when applied to multivariable nonlinear systems like the core power distribution control. The difficulty stems not from the development of control algorithm but from the construction of the rule base, which is important to implement the neuro-fuzzy controller. The reactor core is axially divided into several regions and each region has one input and one output. Therefore, we can apply a conventional neuro-fuzzy controller for each region. However, since each region is coupled with other regions, we should take into account the interaction among these regions to obtain good performance. However, it is difficult to construct a rule-base due to the presence of interactions between control channels. Here, it is accomplished through an approximate decoupling scheme.

First, we describe a neuro-fuzzy controller for a SISO (single input and single output) process and then derive an approximate decoupling scheme to consider some decoupling effect between channels.

In a fuzzy control system of a channel, we express a fuzzy rule like Eq. (8) and use the same controller output formula as Eq. (11).

4.2.2 Back-propagation Method for Parameter Optimization

We use the gradient descent method to tune the parameters of the membership functions by minimizing the objective function defined as follows:

$$E = \frac{1}{2}\left\{\sum_{p=1}^{t}\lambda^{t-p}\left[\alpha u_p^2 + \left(u_p - u_p^r\right)^2\right]\right\},$$ (29)

where λ is a forgetting factor, α a weighting factor, u_p the output of the neuro-fuzzy controller at time step p, u_p^r its reference output, and t the current time step. A forgetting factor is introduced to take into account for an exponential decay of the past data so that the control rules are modified fast according to the change of process dynamics. If the process dynamics is changed, the practical implementation has necessitated the employment of the forgetting factor with a value between zero and unity. The forgetting factor is, in general, chosen within the range $0.9 < \lambda \leq 1$. The results of previous time steps are weighted less than those of near present ones. And the membership function parameters are tuned so that excessive control effort is prevented by containing an input-squared term in addition to an error-squared term in the objective error function.

The parameters which minimize the above objective function, can be obtained by the following iterative calculation derived from Eq. (17):

$$a_{ij}(t+1) = a_{ij}(t) - \eta_a \sum_{p=1}^{t} \lambda^{t-p} \left[(\alpha+1)u_p - u_p^r \right] \frac{\partial u_p}{\partial a_{ij}}, \tag{30}$$

where a_{ij} denotes the antecedent and consequent parameters such as c_{ij}, σ_{ij}, q_{ij} and q_{i0}. To update the membership function parameters (antecedent parameters) and consequent parameters, we need to evaluate $\dfrac{\partial u_p}{\partial a_{ij}}$. We use the following relationship to optimize the antecedent parameters:

$$\begin{aligned}
\frac{\partial u_p}{\partial a_{ij}} &= \frac{\partial u_p}{\partial w_i} \frac{\partial w_i}{\partial w_{ij}} \frac{\partial w_{ij}}{\partial a_{ij}} \\
&= \frac{f_i - u_p}{\sum_{i=1}^{n} w_i} \frac{w_i}{w_{ij}(x_j)} \frac{\partial w_{ij}}{\partial a_{ij}}.
\end{aligned} \tag{31}$$

In Eq. (31), $\dfrac{\partial w_{ij}}{\partial a_{ij}}$ depends on the input membership function type. Since we use the symmetric Gaussian membership function, we need the following derivative to update the parameter c_{ij} :

$$\frac{\partial w_{ij}}{\partial c_{ij}} = w_{ij} \frac{x_j - c_{ij}}{\sigma_{ij}^2}. \tag{32}$$

In addition, to update the parameter σ_{ij}, we need the following derivative:

$$\frac{\partial w_{ij}}{\partial \sigma_{ij}} = w_{ij} \frac{(x_j - c_{ij})^2}{\sigma_{ij}^3}. \tag{33}$$

Similarly, we use the following relationships to optimize the consequent parameters q_{ij} and q_{i0}:

$$\frac{\partial u_p}{\partial q_{ij}} = \frac{\partial u_p}{\partial f_i} \frac{\partial f_i}{\partial q_{ij}} = \overline{w}_i x_j, \tag{34}$$

$$\frac{\partial u_p}{\partial q_{i0}} = \frac{\partial u_p}{\partial f_i} \frac{\partial f_i}{\partial q_{i0}} = \overline{w}_i. \tag{35}$$

Consequently, we adapt the parameters by the following equations:

$$c_{ij}(t+1) = c_{ij}(t) - \eta_c \sum_{p=1}^{t} \lambda^{t-p} \left[(\alpha+1)u_p - u_p^r \right] \left(f_i - u_p \right) \overline{w}_i \frac{x_j - c_{ij}}{\sigma_{ij}^2}, \tag{36}$$

$$\sigma_{ij}(t+1) = \sigma_{ij}(t) - \eta_\sigma \sum_{p=1}^{t} \lambda^{t-p} \left[(\alpha+1)u_p - u_p^r \right] \left(f_i - u_p \right) \overline{w}_i \frac{\left(x_j - c_{ij} \right)^2}{\sigma_{ij}^3}, \tag{37}$$

$$q_{ij}(t+1) = q_{ij}(t) - \eta_q \sum_{p=1}^{t} \lambda^{t-p} \left[(\alpha+1)u_p - u_p^r \right] \overline{w}_i x_j, \tag{38}$$

$$q_{i0}(t+1) = q_{i0}(t) - \eta_{q0} \sum_{p=1}^{t} \lambda^{t-p} \left[(\alpha+1)u_p - u_p^r \right] \overline{w}_i. \tag{39}$$

4.2.3 Decoupling Scheme

The procedure of designing a SISO neuro-fuzzy controller described above indicates that only little qualitative knowledge about the process is required for deriving the rule-base. Such a controller will not function well when applied to multivariable systems in the presence of interactions among channels. To achieve better performance, it is necessary to consider coupling effects.

We assume that we are able to suitably identify dominant interactive sources. Then, we use an approximate adaptive technique [Nie 1997] to counter interactive effects. Let the control input to the process consist of the outputs from the SISO neuro-fuzzy controller and the decoupling unit as follows:

$$u = u_f + u_d, \tag{40}$$

where u_f denotes the output of the SISO neuro-fuzzy controller, and u_d the output of the decoupling unit.

In case that there is no interaction among channels, the closed system of the arbitrary k-th channel employing the neuro-fuzzy controller may be represented by

$$g^k \left(x_1^k, x_2^k, \cdots, x_L^k \right) = u_f^{ki}, \tag{41}$$

where the superscript 'k' denotes the k-th channel, g^k denotes all dynamic terms for the channel, and u_f^{ki} is the corresponding neuro-fuzzy controller output. From now on, all of the superscripts 'k' denote the k-th channel. In case that a dominant interactive term d^k is added to the channel but there does not exist the introduction of any decoupling mechanism, the closed system would be as follows:

$$g^k\left(x_1^k, x_2^k, \cdots, x_L^k\right) + d^k = u_f^k, \tag{42}$$

where u_f^k is the fuzzy control effort when d^k exists, and is assumed as follows:

$$u_f^k = u_f^{ki} + \delta u_f^k \cong \left(1 + \zeta^k\right) u_f^{ki}. \tag{43}$$

Now we want to use an additional control effort u_d^k to reduce the interactive effect produced by d^k, represented as

$$g^k\left(x_1^k, x_2^k, \cdots, x_L^k\right) + d^k = \left(1 + \zeta^k\right) u_f^{ki} + u_d^k. \tag{44}$$

Then the interactive term becomes

$$d^k = \mu^k u_f^k + u_d^k, \tag{45}$$

where $\mu^k = \dfrac{\zeta^k}{1 + \zeta^k}$.

We assume that we can approximately express the interaction component d^k of the k-th channel and its predicted effort u_d^k as a linear combination of some dominant neuro-fuzzy inputs as follows:

$$\hat{d}^k \cong l_{d1}^k x_{d1}^k + l_{d2}^k x_{d2}^k + \cdots, \tag{46}$$

$$\hat{u}_d^k \cong \hat{l}_{a1}^k x_{d1}^k + \hat{l}_{a2}^k x_{d2}^k + \cdots, \tag{47}$$

where the hat sign denotes approximate values, x_{d1}^k and x_{d2}^k denote some inputs which have a strong interaction with the k-th channel, and l_{a1}^k and l_{d1}^k are unknown. We define the following cost function to solve the coefficients $\hat{l}_{a1}^k$, $\hat{l}_{a2}^k$, and so forth:

$$\Delta J \equiv \frac{1}{2}\left(\mu^k u_f^k\right)^2 = \frac{1}{2}\left[\left(l_{d1}^k x_{d1}^k + l_{d2}^k x_{d2}^k + \cdots\right) - \left(\hat{l}_{a1}^k x_{d1}^k + \hat{l}_{a2}^k x_{d2}^k + \cdots\right)\right]^2. \tag{48}$$

Then we can adapt the coefficients $\hat{l}_{a1}^k$ and $\hat{l}_{a2}^k$ by forming the gradients of the function ΔJ with respect to $\hat{l}_{a1}^k$ and $\hat{l}_{a2}^k$. The adaptation law is given by

$$\hat{l}_{a1}^k(t+1) = \hat{l}_{a1}^k(t) - \rho_1 \frac{\partial \Delta J}{\partial \hat{l}_{a1}^k} = \hat{l}_{a1}^k(t) + \eta_{l1} u_f^k x_{d1}^k, \tag{49}$$

$$\hat{l}_{a2}^k(t+1) = \hat{l}_{a2}^k(t) - \rho_2 \frac{\partial \Delta J}{\partial \hat{l}_{a2}^k} = \hat{l}_{a2}^k(t) + \eta_{l2} u_f^k x_{d2}^k, \tag{50}$$

where $\eta_{11} = \rho_1 \mu^k$ and $\eta_{12} = \rho_2 \mu^k$.

The architecture of the proposed neuro-fuzzy control system consists several SISO neuro-fuzzy controllers in Figure 6 in parallel whose the number is the same as that of channels, and the SISO neuro-fuzzy controller works independently.

Then the predicted additional control effort $\hat{u}_d^k$ from each decoupling unit is added to the corresponding SISO controller output.

4.3 Numerical Simulations

4.3.1 Xenon Spatial Oscillation Model

We will use the axial xenon oscillation model [Onega and Kisner 1978; Na and Upadhyaya 1998] developed previously to demonstrate the neuro-fuzzy control algorithm. The model employs the nonlinear xenon and iodine balance equations and a one-group, one-dimensional, neutron diffusion equation having nonlinear power reactivity feedback. The flux, xenon, and iodine distributions are assumed to consist of two-term spatial, harmonic series. The reactor model is made as nearly critical as possible using a variational estimate of the eigenvalue of the one-dimensional diffusion equation. The total power of the reactor core is held constant although the power density varies as a function of both time and position. The one-dimensional reactor is reduced to a two-point representation by dividing the reactor into two equal halves and integrating over each half to find the average values for the flux, xenon and iodine spatial distributions. The neutron flux and the xenon and iodine concentrations averaged over the lower half of the core are given by the following equations:

$$\overline{\varphi}_1(t) = \frac{\int_{-H/2}^{0} \varphi(z,t)dz}{\int_{-H/2}^{0} dz} = \frac{2}{\pi}\left[1 - A(t)\right], \tag{51}$$

$$\overline{x}_1(t) = \frac{2}{\pi}\left[1 - B(t)\right], \tag{52}$$

$$\overline{y}_1(t) = \frac{2}{\pi}\left[1 - C(t)\right], \tag{53}$$

where $A(t)$, $B(t)$ and $C(t)$ are the amplitude functions of the flux, xenon and iodine concentrations. The equations for the upper half of the core are

$$\overline{\varphi}_2(t) = \frac{2}{\pi}\left[1 + A(t)\right], \tag{54}$$

$$\overline{x}_2(t) = \frac{2}{\pi}\big[1 + B(t)\big],$$
(55)

$$\overline{y}_2(t) = \frac{2}{\pi}\big[1 + C(t)\big].$$
(56)

The dynamics of the amplitude functions are given as (Na and Upadhyaya 1998):

$$-\beta_2 A^2 + 2(\beta_1 - \beta_3)A + \beta_2 = 0,$$
(57)

$$\frac{dB(t)}{dt} = \left(\gamma_X \Sigma_f \frac{\phi_0}{X_0}\right)A(t) + \left(\lambda_I \frac{I_0}{X_0}\right)C(t) - \lambda_X B(t) - \frac{2}{3}\sigma_X \phi_0 \big[A(t) + B(t)\big],$$
(58)

$$\frac{dC(t)}{dt} = \left(\gamma_I \Sigma_f \frac{\phi_0}{I_0}\right)A(t) - \lambda_I C(t),$$
(59)

where many parameters have their usual meanings and some other parameters are denoted by

$$\beta_1 = \frac{1}{\Sigma_f}\left[4D\left(\frac{\pi}{H}\right)^2 + \frac{1}{2}(\Sigma_{a1} + \Sigma_{a2}) + \frac{32}{15\pi}(\sigma_X X_0 + 3\alpha_F \phi_0 \overline{\Sigma}_a)\right],$$
(60)

$$\beta_2 = \frac{1}{\Sigma_f}\left[\frac{8}{3\pi}(-\Sigma_{a1} + \Sigma_{a2}) + \frac{64}{15\pi}\sigma_X X_0 B(t)\right],$$
(61)

$$\beta_3 = \frac{1}{\Sigma_f}\left[D\left(\frac{\pi}{H}\right)^2 + \frac{1}{2}(\Sigma_{a1} + \Sigma_{a2}) + \frac{8}{3\pi}(\sigma_X X_0 + \alpha_F \phi_0 \overline{\Sigma}_a)\right],$$
(62)

$$I_0 = \frac{\gamma_I \Sigma_f \phi_0}{\lambda_I},$$
(63)

$$X_0 = \frac{(\gamma_I + \gamma_X)\Sigma_f \phi_0}{\lambda_X + \frac{\pi}{4}\sigma_X \phi_0}.$$
(64)

Σ_a is expressed as the combination of absorption cross section of the fuel, moderator, structure and control poison. Σ_{a1} and Σ_{a2} control the lower and upper halves of the core, respectively. Table 3 lists the one-group diffusion parameters of the foregoing dynamic equations.

Table 3. One-group diffusion parameters of the axial xenon oscillation model

Parameters	ϕ_0 [cm^{-2}·sec^{-1}]	σ_X [cm^2]	α_F [cm^2·sec]	γ_I	γ_X	λ_I [sec^{-1}]
values	2.1×10^{13}	2.6×10^{-18}	3.6×10^{-16}	0.061	0.003	2.87×10^{-5}
Parameters	λ_X [sec^{-1}]	D [cm]	H [cm]	Σ_f [cm^{-1}]	$\nu\Sigma_f$ [cm^{-1}]	$\overline{\Sigma}_a$ [cm^{-1}]
values	2.09×10^{-5}	0.375	365.8	0.65	1.56	1.523

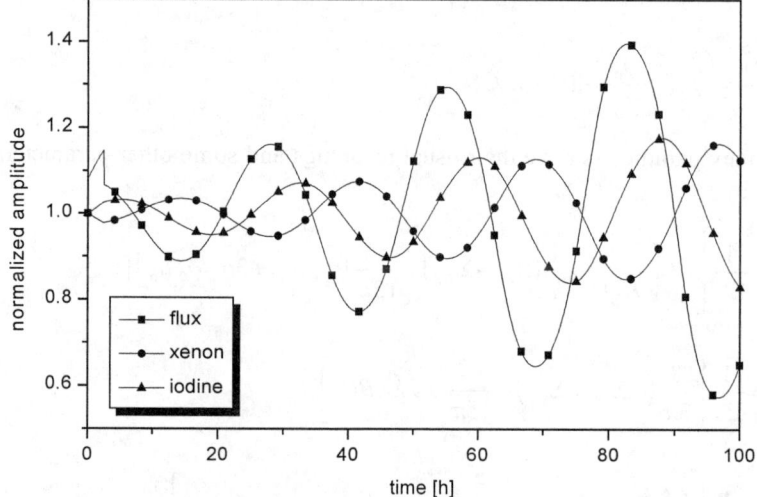

Figure 13. Normalized flux, xenon and iodine oscillations in the lower half of reactor core due to a step reactivity change

Now let's simulate the model for the case where the reactor has been at 100 percent power level with steady-state xenon concentration before a perturbation is applied. The perturbation is the change in Σ_{a1} which corresponds to a 0.1% change of absorber from its equilibrium position, and lasts 2.5 h. The normalized flux, xenon and iodine oscillations in the lower half of the core are initiated at $t = 0$ by a step decrease in Σ_{a1} and are shown in Figure 13. The initial sudden insertion of the positive reactivity induces the prompt jump like the initial peak in Fig. 13. The magnitude of the oscillations increases versus time.

4.3.2 Implementation

Since the axial xenon oscillation model is divided into the lower and upper regions, the proposed control system consists of two channels. We choose the first channel inputs to the neuro-fuzzy controller as follows:

$x_1^1(t) = w_1(t) - y_1(t)$: the difference between the normalized target flux and the normalized neutron flux in the lower half of the reactor,

$x_2^1(t) = \Sigma_{a1}(t) - \Sigma_{a1}(t-1)$: the difference in absorber cross section between two neighboring time steps in the lower half of the reactor,

where w and y denote the normalized target neutron flux and the normalized neutron flux, respectively. Then we choose the second channel inputs to the neuro-fuzzy controller as follows:

$$x_1^2(t) = w_2(t) - y_2(t),$$
$$x_2^2(t) = \Sigma_{a2}(t) - \Sigma_{a2}(t-1).$$

We choose the dominant interaction input for the first channel as x_1^2 and that for the second one as x_1^1. Therefore, the additional inputs for decoupling of the first and second channels consist of only one term as follows, respectively:

$$\hat{u}_d^1 = \hat{l}_a^1(w_2 - y_2), \tag{65}$$

$$\hat{u}_d^2 = \hat{l}_a^2(w_1 - y_1). \tag{66}$$

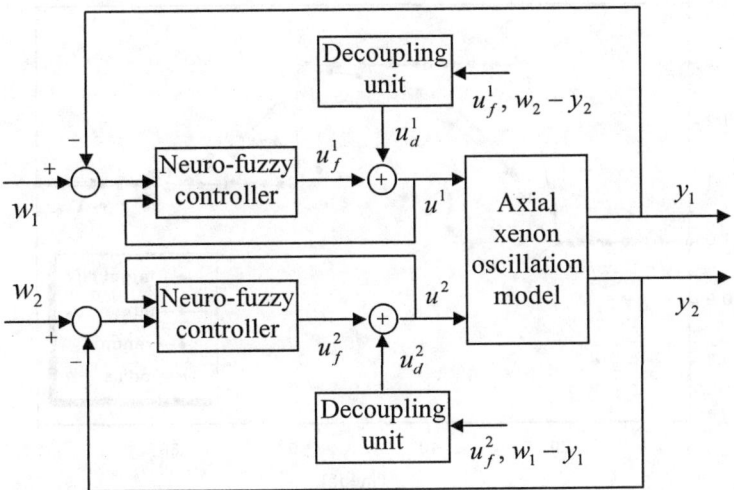

Fig. 14. Architecture of the neuro-fuzzy controller for neutron flux shape control

The adaptive law is

$$\hat{l}_a^1(t+1) = \hat{l}_a^1(t) + \eta_l^1 u_f^1 (w_2 - y_2),$$ (67)

$$\hat{l}_a^2(t+1) = \hat{l}_a^2(t) + \eta_l^2 u_f^2 (w_1 - y_1).$$ (68)

Figure 14 shows the overall control architecture for the axial power distribution. In the figure, the "neuro-fuzzy controller" blocks generate the control inputs in case of the SISO without coupling between the bottom and top regions. The "decoupling unit" blocks generate the additional control inputs to remove the coupling effect because the coupling actually exists.

We use the control input from the adaptive controller [Na et al. 1998] developed previously as the desired output of the neuro-fuzzy controller to train the controller parameters on-line, since the adaptive controller gives good performance. On the other hand, we can train the neuro-fuzzy controller off-line so that the neutron flux traces the desired target output like Section 3. The variation between two neighboring time steps is used for control input. We add to the normalized flux a measurement noise signal that has a Gaussian distribution with mean 0 and variance 0.0001, to simulate more realistic plant environment. The proposed controller has 9 rules and two inputs for each channel, and the number of the channels is two.

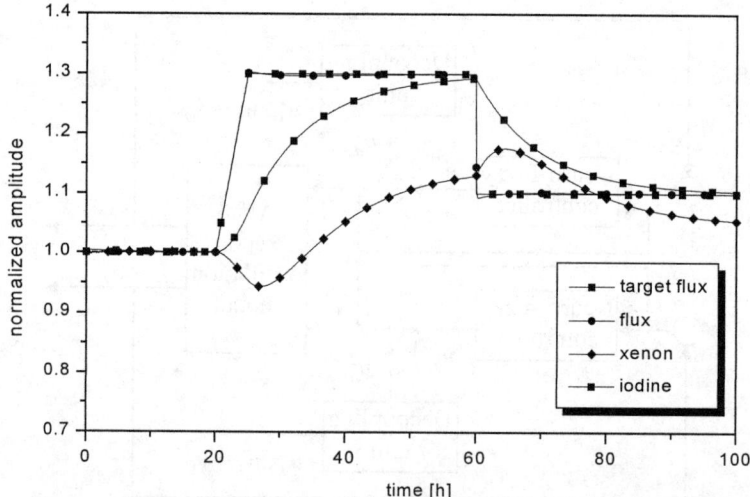

Fig. 15. Normalized flux, xenon and iodine responses in the lower half of reactor core induced by the ramp and step changes of target neutron flux (the neuro-fuzzy controller with the decoupling unit)

4.3.3 Simulation Results

We perform three different simulations to demonstrate the proposed controller for three cases; 1) tracking of the target axial shape, 2) the effect of the decoupling unit, and 3) dampening of the oscillations induced by a perturbation. In all simulations, we assume that the reactor has been in steady state at 100 percent power level with steady-state xenon concentration before this controller is applied.

Figure 15 shows the performances of the controller due to the ramp and step changes of the target axial shape. We want the neutron flux distribution at the bottom and top regions to be symmetric, which means that their normalized neutron flux is one. However, to investigate the performance of the controller, we assume that the normalized target neutron flux changes from one. When the normalized target flux increases by a ramp at $t = 20$ h and decreases by a step at $t = 60$ h, the proposed controller traces the target flux without overshoot.

To investigate the effect of the decoupling unit, Figure 16 simulates the circumstances similar to the prior simulation with the decoupling unit removed. We can find out that the decoupling unit provides good characteristics against the interactions between the channels by comparing Figure 16 with Figure 15. Therefore, when there exist interactions among channels, it is necessary to consider coupling effects to achieve better performance.

Figure 17 shows how well this controller damps the oscillations when we induce some oscillations on purpose. During the first 20 h, all circumstances are the same as the first simulation. A perturbation is suddenly initiated at 20 h and

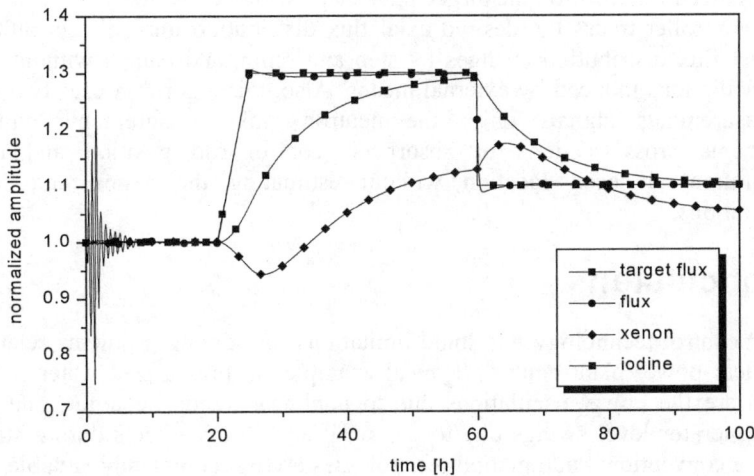

Fig. 16. Normalized flux, xenon and iodine responses in the lower half of reactor core induced by the ramp and step changes of target neutron flux (the neuro-fuzzy controller without the decoupling unit)

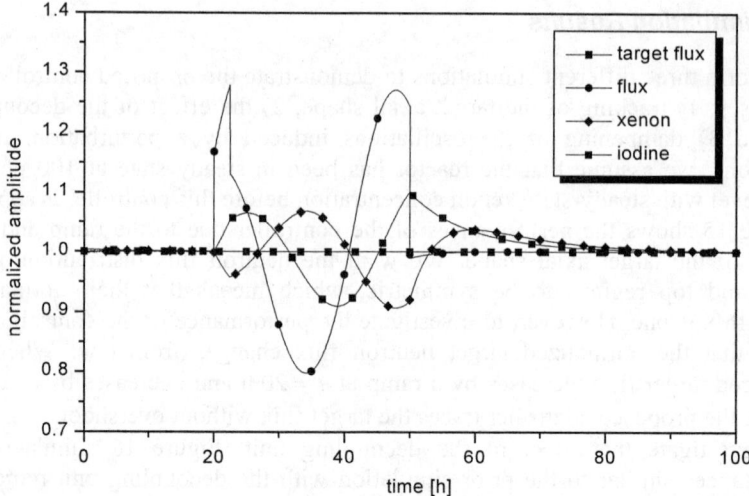

Fig. 17. Normalized flux, xenon and iodine responses in the lower half of reactor core induced by the added free oscillations (the neuro-fuzzy controller with the decoupling unit)

lasts 2.5 h. Its amount is a 0.2% change of absorber in the lower region at that time. Some free oscillations of the flux, xenon and iodine take place without any controller action for 27.5 h after the initiation of the perturbation. We apply the neuro-fuzzy controller at 50 h. After that time, the oscillations stop promptly and the normalized flux follows the target axial shape without any delay.

This controller traces the desired axial flux distribution immediately although the target flux distribution changes by step and ramp, and damps without delay some oscillations induced by external means. Also, this algorithm uses two kinds of measurements signals only: the neutron flux measurements and the macroscopic cross sections of absorbers (control rod position and boron concentration) at each location without estimating the xenon and iodine concentrations.

5 Conclusions

The old control technology has some limitations for solving problems related to the nuclear power plant control. Typical examples in pressurized water reactors (PWRs) are the power oscillations due to nonlinear xenon behavior, and large steam generator level swings due to the swell and shrink effects during startup. Since the conventional automation technologies are not completely suitable, their operations are primarily dependent on plant operators.

The control technology has improved rapidly in recent years and offered new possibilities in the area of control. Traditional analog systems are being replaced by their digital counterparts in many recent industrial applications. Because on-line computation of the new technology is possible with cheap and efficient microprocessor, a number of advanced control algorithms has found feasible grounds for implementation. However, in spite of various positive aspects of using advanced digital controllers, for many reasons modern control methods have not been incorporated extensively in nuclear power plants, especially because of their conservative characteristics.

The main objective of this chapter was to investigate the effectiveness of the neuro-fuzzy control method in application to PWR operations. Therefore, in this chapter we dealt with the applications of the neuro-fuzzy controller, which has great popularity in process industry, to pressurized water reactors. The application results of this chapter showed that the neuro-fuzzy control algorithm could be applied successfully to the troublesome control problems in PWRs.

References

Akin HL, Altin V (1991) Rule-based fuzzy logic controller for a PWR-type nuclear power plant. IEEE Trans Nucl Sci 38: 883-890

Cho BH, No HC (1996) Design of stability-guaranteed fuzzy logic controller for nuclear steam generators. IEEE Trans Nucl Sci 43: 716-730

DeChaine MD and Feltus MA (1996) Fuel management optimization using genetic algorithms and expert knowledge. Nucl Sci Eng 124: 188-196

Dubois D, Prade H (1980) Fuzzy sets and systems: theory and applications. Academic, New York

Feely JJ (1981) Optimal digital estimation and control of a natural circulation steam generator. EG&G Idaho, Idaho Falls, ID

Frogner B, Rao HS (1978) Control of nuclear power plants. IEEE Trans Auto Cont 23: 405-417

Goldberg DE (1989) Genetic algorithms in search, optimization, and machine learning. Addison-Wesley, Reading, MA

Hah YJ, Lee BW (1994) Fuzzy power control algorithm for a pressurized water reactor. Nucl Tech 106: 242-253

Heger AS et al. (1995) Application of fuzzy logic in nuclear reactor control Part 1: An assessment of state-of-the-art. Nucl Safety 36: 109-121

Holland JH (1975) Adaptation in natural and artificial systems. Univ Michigan Press, Ann Arbor

Irving E et al. (1980) Toward efficient full automatic operation of the PWR steam generator with water level adaptive control. In: Harding BJ (ed) Boiler dynamic and control in nuclear power stations 2. BNES, London, pp 309-329

Jang JSR (1993) ANFIS: Adaptive network-based fuzzy inference system. IEEE Trans Syst Man Cyber 23: 665-685

Jang JSR, Sun CT (1995) Neuro-fuzzy modeling and control. Proc IEEE 83: 378-406

Kickert WJM, Mamdani EH (1978) Analysis of a fuzzy logic controller. Fuzzy Set Syst 1: 29-44

Kim J et al. (1995) Designing fuzzy net controllers using genetic algorithms. IEEE Cont Syst 15: 66-72

Kuan CC et al. (1992) Fuzzy logic control of steam generator water level in pressurized water reactors. Nucl Tech 100: 125-134

Lee YJ (1994) Optimal design of the nuclear S/G digital water level control system. J Kor Nucl Soc 26: 32-40.

Lee JY, No HC (1993) A 9-rule fuzzy logic controller of the nuclear steam generator. J Kor Nucl Soc 25: 371-380

Mamdani EH, Assilian S (1975) An experiment in linguistic synthesis with a fuzzy logic controller. Int J Man-Machine Studies 7: 1-13

Mitchell M (1996) An introduction to genetic algorithms. MIT Press, Cambridge Massachusetts

Moller MF (1993) A scaled conjugate algorithm for fast supervised learning. IEEE Trans Neural Networks 6: 525-533

Na MG (1998) Design of a genetic fuzzy controller for the nuclear steam generator water level control. IEEE Trans Nucl Sci 45: 2261-2271

Na MG et al. (1998) Adaptive control for axial power distribution in nuclear reactors. Nucl Sci Eng 129: 283-293

Na MG, Lim JH (1997) A fuzzy controller based on self-tuning rules for the nuclear steam generators. KSME Int J 11: 485-493

Na MG, No HC (1992) Design of an adaptive observer-based controller for the water level of steam generators. Nucl Eng Des 135: 379-394

Na MG, Upadhyaya BR (1998) A neuro-fuzzy controller for axial power distribution in nuclear reactors. IEEE Trans Nucl Sci 45: 59-67

Nie J (1997) Fuzzy control of multivariable nonlinear servomechanisms with explicit decoupling scheme. IEEE Trans Fuzzy Syst 5: 304-311

Onega RJ, Kisner RA (1978) An axial xenon oscillation model. Annals Nucl Energy 5: 13-19

Ostergaard JJ (1977) Fuzzy logic control of a heat exchanger process. In: Gupta MM et al. (eds) Fuzzy automata and decision processes. Amsterdam New York, pp 285-320

Park MG, Cho NZ (1995) Self-tuning control of a nuclear reactor using a Gaussian function neural network. Nucl Tech 110: 285-293

Park YM et al. (1994) An inverse dynamics controller for power system stabilizing using artificial neural networks. In: Proc Int Conf Power Syst Tech 2: 1326-1329, Beijing, China

Park S et al. (1995) A neuro-genetic controller for non-minimum phase systems. IEEE Trans Neural Networks 6: 1297-1300

Park GY, Seong PH (1995) Application of a fuzzy learning algorithm to nuclear steam generator level control. Annals Nucl Energy 22: 135-146

Parks GT (1996) Multiobjective pressurized water reactor reload core design by nondominated genetic algorithm search. Nucl Sci Eng 124: 178-187

Ramaswamy P et al. (1993) An automatic tuning method of a fuzzy logic controller for nuclear reactors. IEEE Trans Nucl Sci 40: 1253-1262

Takagi T, Sugeno M (1985) Fuzzy identification of systems and its applications to modeling and control. IEEE Trans Syst Man Cyber 15: 116-132

Wang LX (1994) Adaptive fuzzy systems and control. Prentice-Hall, Engelwood Cliffs

10 Neural and Fuzzy Transient Classification Systems: General Techniques and Applications in Nuclear Power Plants

Davide Roverso

Institutt for energiteknikk, OECD Halden Reactor Project,
POBox 173, N-1751 Halden, Norway
roverso@computer.org

In this chapter the problem of identifying events in dynamic processes (e.g., faults, anomalous behaviours, etc.) is tackled with soft computing techniques aimed at the classification of the process transients generated by such events. A review of previous work is followed by a discussion of several alternative designs and models which employ both fuzzy and neural systems. These have been developed during an ongoing reserch program which was initiated by the need of finding new principled methods to perform alarm structuring/suppression in a nuclear power plant alarm system. This initial goal was soon expanded beyond alarm handling, to include diagnostic tasks in general. The application of these systems to domains other than NPPs was also taken into special consideration. A systematic study was carried out with the aim of comparing alternative neural network designs and models. Four main approaches have been investigated: radial basis function (RBF) neural networks and cascade-RBF neural networks combined with fuzzy clustering, self-organizing map neural networks, and recurrent neural networks. The main evaluation criteria adopted were: identification accuracy, reliability (i.e., correct recognition of an unknown event as such), robustness (to noise and to changing initial conditions), and real time performance. A series of initial tests on a small set of BWR transients was recently followed by more advanced tests on PWR transients corresponding to various occurrences of rapid load rejection events (plant islanding). The chapter is closed by a discussion of open issues and future directions for research and applications.

1 Introduction

In this chapter we will present and discuss various techniques for transient classification which have been developed and tested within the framework of the prototype system ALADDIN. The main focus of this chapter will be on techniques which combine fuzzy clustering and artificial neural networks (ANNs) to approach the problem of identifying events in dynamic processes. The main motivation for the development of such a system derived from the need of finding new principled methods to perform alarm structuring/suppression in a nuclear power plant (NPP) alarm system. One such method consists in basing the alarm structuring/suppression on a fast recognition of the event generating the alarms, so that a subset of alarms sufficient to efficiently handle the current fault can be selected for the operator, minimizing the operator's workload in a potentially stressful situation. The scope of application of a system like ALADDIN goes however beyond alarm handling, to include diagnostic tasks in general. The possible application of the system to domains other than NPPs was also taken into special consideration during the design phase.

In this chapter we report on the first phase of the ALADDIN project which consisted mainly in the design and comparative study of a series of ANN-based approaches to transient classification, and the development of a system prototype which, eventually, will be integrated in existing alarm, diagnosis, accident management, and condition monitoring systems such as CASH [1], IDS [2], CAMS [3], and PEANO [6].

2 Related Work

In recent years the range of experimental applications of ANNs to nuclear power plants (NPPs) has been growing steadily and with encouraging results. Diagnostic and monitoring tasks such as fault identification, event classification, and signal validation are among the most popular. In particular signal validation has been at the center of ongoing research here in Halden [4,5,6]. It is on this body of work and accumulated experience that the ALADDIN project was based.

2.1 NPP Transient Classification

Focusing more specifically on transient classification in NPPs, we note that among the first to demonstrate the feasibility of using ANNs were Bartlett and Uhrig [7]. That work was developed further and enhanced in [8] where a modular architecture of two ANNs is shown to identify 27 transients of a BWR by using 97 plant variables. The first ANN is used to determine whether the plant is in a normal condition or not, while the second ANN performs the actual classification. A limitation of this model is that it bases the classification on the values assumed by the chosen variables at a single instant of time, i.e., it does not take into consideration the evolution in time of the observed variables. This choice was made to keep the solution simple, but it also meant an increase in the number of variables which was needed to reach a satisfactory classification (i.e., 97 variables).

An important contribution was made by Bartal, Lin, and Uhrig [9], where they recognized the necessity for a classifier of being able to provide a "don't-know" answer when presented with a transient of a kind not contained in its accumulated knowledge base. They developed a classifier based on probabilistic neural networks which was shown to classify 72 scenarios of 13 different types of transients, given in input the observed values of 76 variables. They also introduced a mechanism of evidence accumulation by which classification results obtained at previous time steps are used as supporting evidence toward the final classification. The neural classifier is still working on independent instants of time (i.e., temporal information is not used), but the final classification is computed using a majority vote of the values obtained at each time step.

An interesting method for event classification was proposed by Furukawa, Ueda, and Kitamura in [10]. They construct an independent classifier for each observed variable which receives in input the whole time-series of the selected variable, and produces in output the best possible *partial* classification in class sets (i.e., an event is assigned to one of the *super-classes*[1] which are separable given the information conveyed by a single variable). The intersection of the class sets generated by all the classifiers produces the final classification. This approach has the advantage of being very robust, since it constructs a classification based on multiple independent classifiers, but it would most probably not suffice when the needed discrimination function becomes more complex and requires an analysis of the interaction between two or more variables.

An alternative way of dealing with temporal data using an *implicit time measure* was proposed by Jeong, Furuta, and Kondo [11]. A discrimination function which uses a series of past values of the input time series has the problem of identifying the initiation time of a transient so that the classifier can "synchronize" with the new event. Their proposed solution to this problem comes from measuring implicitly time by integrating a selected parameter which decreases (or increases) monotonically with time for a given set of events (say the primary pressure in Steam Generator Tube Leak SGTL or Rod Cluster Control Assembly Ejection RCCAE events). Transients are then synchronized by the integration of the variation of the selected parameter, with the normal value at steady state used as a starting point.

The same authors recently proposed [12] the *adaptive template matching algorithm* which allows to describe transients in a two-dimensional continuum of time and severity level.

Since the motivation behind ALADDIN derived from the problem of alarm structuring/suppression in an NPP alarm system, we need to mention here the work of Ohga et al. [13] which integrated a simple ANN based event identification module into an alarm handling system. Even if the neural classification model was tested on only three types of events out of the nine for which it was designed, nonetheless the results showed the feasibility of this approach to alarm handling.

---

[1] Which might contain more than a single event class.

As a final note on related work we should say that neural networks are by no means the only models which are being used to tackle event classification in dynamic processes. One alternative method which needs special mention is the use of Hidden Markov Models [14,15]. Other related methods not covered by this review are for example statistical and bayesian methods for time series classification[2].

In the following we will present a series of neural models for event classification in dynamic processes, with a special emphasis on fast transients and on the ability to produce a "don't-know" classification whenever presented with a previously unseen transient.

3 Neural Models for Transient Classification in ALADDIN

Artificial Neural Networks (ANNs) are particularly suited to deal with the problem of event classification in dynamic processes for several reasons (for a general reference on neural networks see [16]). First of all ANNs can approximate any well-behaved function with an arbitrary accuracy, which is an essential advantage on methods based on regression when the problem at hand presents essential nonlinearities. One should stress that, in some applications, ANNs do not outperform other system detection methods. The biggest advantage of ANNs manifests itself when dealing with hard problems, e.g., in the case of significantly overlapping patterns, high background noise, and dynamically changing environments.

The ANNs characteristics of adaptive learning, generalization ability, fault tolerance, robustness to noisy data, and parallel processing make them a very interesting candidate for approaching the classification of dynamic events. In the following we will present four ANN-based architectures which were devised and used to tackle this problem.

3.1 Fuzzy Clustering and RBF Neural Classifier

The first approach which was taken consists in a two-step process involving a fuzzy and possibilistic fuzzy clustering phase followed by a classification phase. The objective of the clustering is to transform an event description so as to make the classification at the same time simpler and robust (both to noise and to changing initial conditions).

In this context, fuzzy clustering is used to partition the N-dimensional space of the observed variables into regions through which event trajectories pass. The idea is that a record of the regions through which a new event passes could provide

---

[2] These methods are generally based on the extraction of features from a time series (e.g., FFT, autocorrelation coefficients, etc.) which are then used as the basis for the classification.

enough information to properly classify the event, and still be robust to noise or changing initial conditions. A clustering algorithms obtained combining ISODATA [17] and *Fuzzy C-Means* [18] was used for this purpose.

One drawback of fuzzy clustering is that even to points which are relatively far from the cluster centers (i.e., far from the regions 'populated' by the trajectories of the events under consideration) are assigned membership values which have to add to 1. This implies that such a clustering will not be able to produce a "don't know" answer in case it is presented with an input vector very different from those it has been designed to recognize, and will assign a high membership to the cluster to which the vector is closest. To overcome this problem a *possibilistic* [19] fuzzy clustering algorithm was employed. Simply stated, *possibilistic* fuzzy clustering relaxes the constraint that the sum of the membership values be always equal to one. The result is that the memberships of points falling in areas not 'covered' by the clusters will all be zero. This property is particularly important if the clustering has to be used for the classification of events of safety-critical systems like NPPs, for which the cost of a misclassification can be very high.

The algorithm which generates the possibilistic fuzzy clustering tends to move the clusters towards the most populated regions of the input space. Relatively to an event classifier (which generates a classification from the memberships), if on one hand this property increases the confidence in the classification (i.e., it reduces the chances of misclassifying an 'unknown' event), on the other it can deteriorate its accuracy (i.e., it increases the chances of misclassifying a 'known' event). A combination of a fuzzy and a possibilistic fuzzy clustering was therefore chosen, where a fuzzy clustering generates a membership signature for the event classifier, while a possibilistic fuzzy clustering generates a 'confidence signature,' i.e., an estimate of how 'common' each vector is for the set of 'known' events.

The actual event classifier is based on radial basis function (RBF) ANNs [20, 21]. An RBF neural network is a supervised[3] feedforward network based on a set of localized receptive field neurons which are centered on the regions populated by the training vectors. Regions of the input space close to the training vectors are mapped to high output values by the neurons' local receptive fields, while regions which are distant from the training data produce low output values. This property contrasts with the behaviour of the most common supervised neural network, the standard feedforward multilayer neural network with sigmoidal hidden neurons, which can produce a high output even in regions very far away from the training vectors. Since one of our main concerns is to be able to identify events which are different from the ones included in the classifier design, and therefore avoid a misclassification, the choice of an RBF architecture was a natural one.

---

[3] 'Supervised' because it is trained using a set of examples of the input-output mapping which the network is supposed to reproduce.

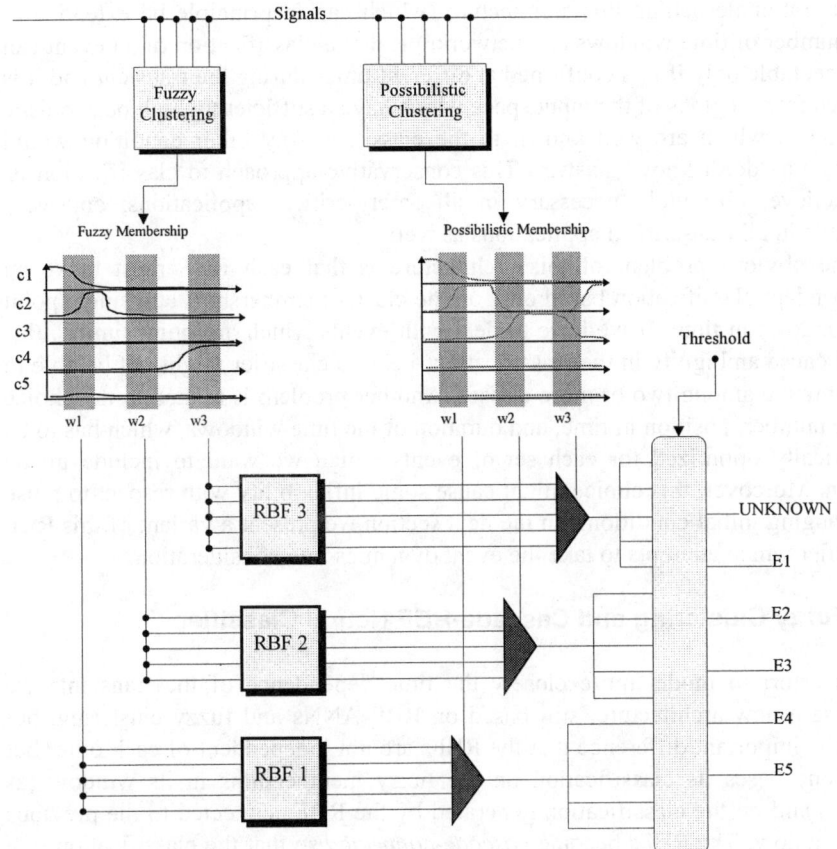

Fig. 1. The architecture of the RBF classifier

One problem of both kinds of ANNs is that they are not able to directly model time-dependent data, i.e., they can only map a static input vector to an output value. Our problem is that we want to classify time-dependent fuzzy signatures.

The solution adopted here is that of sampling the input data by averaging the fuzzy memberships in three time windows, and using those values as inputs to three independent RBF classifiers. Each RBF network will then have one input for each cluster and one output for each event class. The RBFs are trained to recognise a prototypical event for each event class. During operation, the classifications generated by the RBFs are weighted according to a confidence value that corresponds to the maximum possibilistic membership in the respective time windows, and combined to produce a final classification of the event. A final threshold gate will decide whether the classification is acceptable or if we are instead in a "don't know" situation. This architecture is shown in Figure 1.

The rationale behind this architecture (which can in principle be extended to any number of time windows and networks) is that a classification of an event can be acceptable only if it is confirmed at different times during the transient and it is derived from regions of the input space which have a sufficiently high possibilistic value, i.e., which are well known to the classifier. Any other condition would generate a "don't know" answer. This conservative approach to classification is, we believe, absolutely necessary in all safety-critical applications, and very valuable in all cost-critical applications as well.

One obvious problem of this architecture is that each RBF must make an independent classification based only on the cluster memberships at a single point (or window) in time. If we have to deal with events which are quite similar, this could cause ambiguity in the classification, i.e., the classifier might not be able to discriminate among two or more classes. Another problem is related to the choice of the number, position in time, and duration of the time windows, which has to be empirically optimized for each set of events which we want to include in the design. Moreover, this choice might cause some inflexibility with respect to noise or changing initial conditions. In the next section we present a variant of this RBF classifier which attempts to take the event dynamics into consideration.

3.2 Fuzzy Clustering and Cascade-RBF Neural Classifier

In an effort to model more closely the time dependence of the transients we propose a new architecture, still based on RBF ANNs and fuzzy clustering, but with the important difference that the RBFs are not independent of each other but each one bases its classification on the fuzzy memberships in its window (as before) and on the classification generated by the RBF connected to the previous time window. The RBFs become *cascade-connected* so that the classification of a network gets directly influenced by the classification given by the previous network, and indirectly influenced by all the previous classifications.

The first RBF (i.e., the one connected to the first time window) will be the same as in the previous model, thus having one input for each cluster and one output for each class. Each subsequent network will have one input for each cluster plus one input for each class, and one output for each class. As in the previous model, each classification is weighted by the corresponding possibilistic value, while the combination of the classifications is automatically achieved by the connection in cascade. The final classification will be the weighted output of the last RBF network which is then passed through the same threshold gate to screen out the "don't know" cases. The whole architecture is shown in Figure 2.

This new cascade architecture should be able to 'follow' a transient in time as it moves through the fuzzy clusters membership landscape, allowing the model to build up in time its confidence in the classification, and possibly eliminate ambiguities which the previous RBF model could not solve. The parameter optimization problems are also less critical here since only the window width needs to be decided.

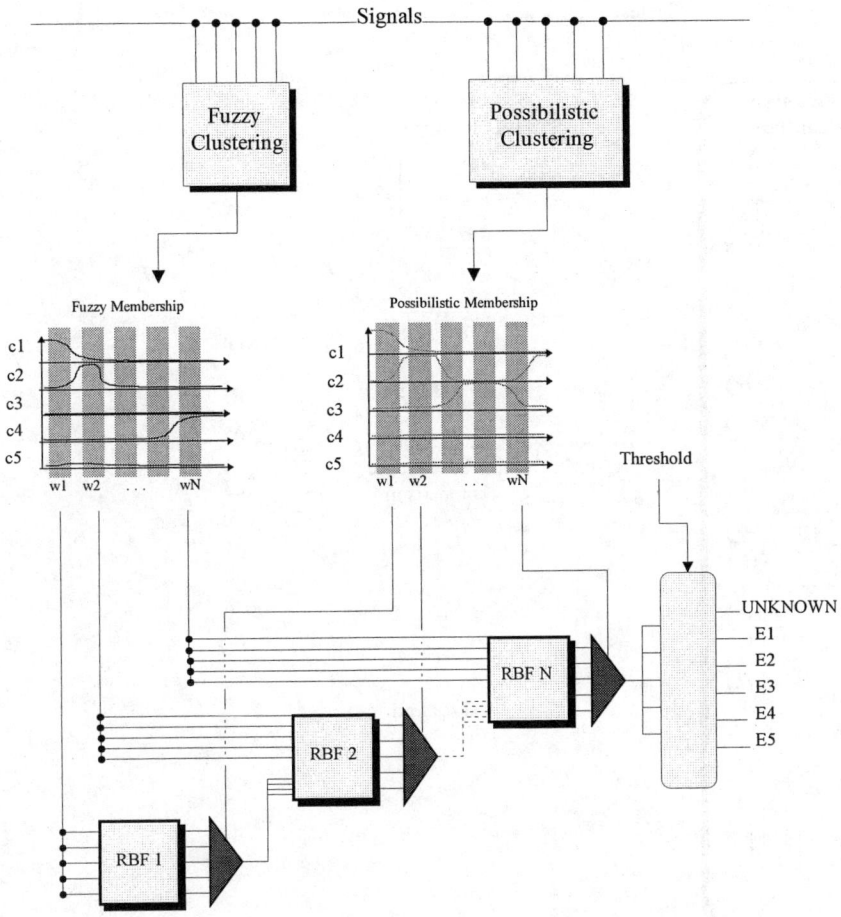

Fig. 2. The architecture of the Cascade-RBF classifier

3.3 SOM Neural Classifier

A different approach was devised which is based on the Self Organizing Map (SOM) ANNs [22], also known as Kohonen Maps. An SOM is a neural network which can be trained to construct a mapping between an N-dimensional input vector space and a (typically) 2-dimensional matrix of neurons, such that the topological relationships among the input vectors are preserved, and are represented in the neuron matrix in terms of a spatial distribution of neural activity. To each neuron is associated an N-dimensional weight vector and, given an N-dimensional input vector, each neuron receives an activation proportional to the distance of its weight vector to the input.

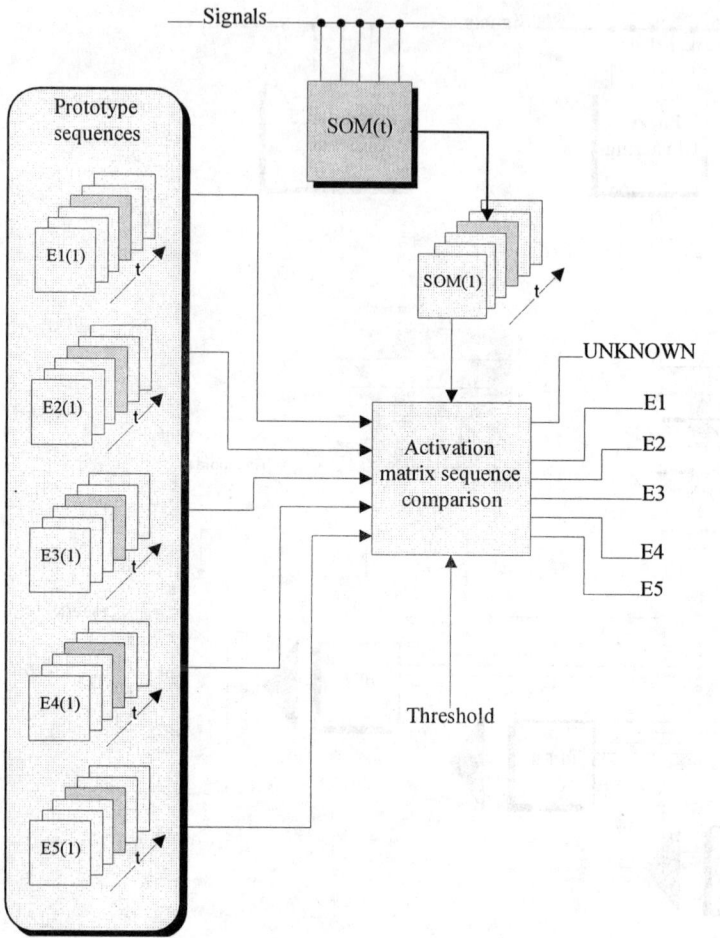

Fig. 3. The architecture of the SOM classifier

The network training iteratively adjusts the neurons weight vectors so as to generate a topographic map of the input space in which similar input vectors activate nearby units. In other words the SOM generates a mapping between the N-dimensional input space and the 2-dimensional map of neurons such that the topology of the input space is preserved, i.e., the mapping preserves neighbourhood relations (but not necessarily Euclidean distances). It has to be noted that the SOM training in unsupervised[4]. Another feature of the SOM

---

[4] 'Unsupervised' because during training it is presented only with input vectors. The network itself discovers the structure of the data, i.e., the common features of the input

network is that it also captures the probability distribution of the input data, in the sense that denser regions in the input space attract more neurons than less dense regions, and are therefore modelled more closely. Hence the SOM performs also an indirect form of clustering, where a cluster in the input space is represented by a group of neurons on the map whose weight vectors are comparatively close together.

As in the case of the RBF networks, also SOMs are not directly designed to model time-dependent data[5]. Furthermore, SOMs are not classifiers *per se* so that their output has to be interpreted in order to come to a classification.

In our case we are dealing with an input consisting of a sequence of N-dimensional vectors, each generating an activation pattern on the SOM. Hence, the problem of classifying the transients' trajectories becomes here the problem of classifying a sequence of activation patterns, i.e., a kind of 'activation trajectory,' in the 2-dimensional map space.

To solve this problem we propose a system in which, after the training of an SOM on all the available transient data, the activation pattern sequences of prototypical transients for each event class are stored. During operation, all stored activation sequences are dynamically compared with the activation sequence generated by the current transient, while this evolves in time, to eventually form a classification. The architecture of the system is shown in Figure 3.

Comparing this approach to the RBF models proposed previously in this chapter, one notices that the mapping performed by the SOM is a transformation of the complex N-dimensional input space which, in a way, corresponds to the transformation performed by the fuzzy clustering. The weight vector of each neuron of the SOM can be interpreted as a cluster center[6], and the activation of the neurons as a membership value. Both methods attempt to reduce the complexity of the problem before performing a classification, but both methods also have problems in dealing with time-dependent inputs.

One possible additional problem of the SOM model might be its complexity, i.e., the space required to store the activation sequences of the event class prototypes, and the time required for the comparison of the prototype sequences with the current sequence. In fact, one activation sequence consists of an activation matrix (the size of the SOM) for each time step of the sequence, and the matrix comparison (performed in this case by combining row- and column-wise correlation measures) has to be performed for each prototype and for each time

---

examples, and 'reorganizes' its internal parameters to reflect these features (from which the name 'Self Organizing').

[5] Modifications to the original SOM in the direction of temporal sequence processing have been attempted by various authors, for example with the Temporal Kohonen Map [23] and the Recurrent Self-Organizing Map [24].

[6] Note however that a cluster in the input space is generally represented by a group of neurons in the SOM.

step. One could think of ways to optimize this process by, for example, storing only the portions of the matrices which change and adopting a simpler matrix similarity measure. At this stage, however, our objective is primarily in comparing the classification performance of the various models, while computational performance optimization is left for a later stage.

3.4 Elman Recurrent Neural Classifier

A substantially different approach was attempted which makes use of a special kind of *recurrent* ANN: the Elman network [25]. Recurrent neural networks are a class of neural networks which are able to deal with temporal input signals thanks to an internal architecture which is *recurrent*, i.e., it makes use of feedback connections among the neurons. The Elman recurrent network is a supervised neural network with a partially recurrent architecture (i.e., not all neurons have feedback connections) whose main characteristic is an array of neurons which record the internal status of the network. At each time step this internal status is fed-back as an additional input to the network which then computes the new status and the output vector.

This property allows us to feed directly the N-dimensional event trajectories to the ANN and train it to produce in output the required classification. This simple architecture is shown in Figure 4.

This simplicity, however, comes at a price. First of all recurrent networks are inherently hard to train. This difficulty comes from the fact that the network needs to construct a classification from a sequence of inputs, learning how to change its internal state in time and how to use the feedback connections. All these factors generally lead to a long training phase.

Another problem is the absence of a direct way of influencing how much the network generalizes from the training examples. Note that the amount of generalization needs to be controlled in order to balance the requirements of class discrimination, on one side, and robustness to noise and different initial conditions on the other. In the RBF network models presented above, for example, a parameter influences directly the size of the receptive fields of the RBF neurons. Bigger receptive fields mean that inputs more distant from the prototypes still activate the neurons and classify the input[7]. In the SOM model a similar control is easily achieved in the matrix comparison phase, allowing more 'different' matrices to match if a higher generalization is required. In the case of the Elman model, the only way to control generalization is through the examples which are used for training. If a higher generalization is required, new examples can be generated by adding increasing amounts of noise to the prototype cases and

---

[7] However, one should note that bigger receptive fields can also mean higher ambiguity in the classification, since they augment the eventual overlap among the classes.

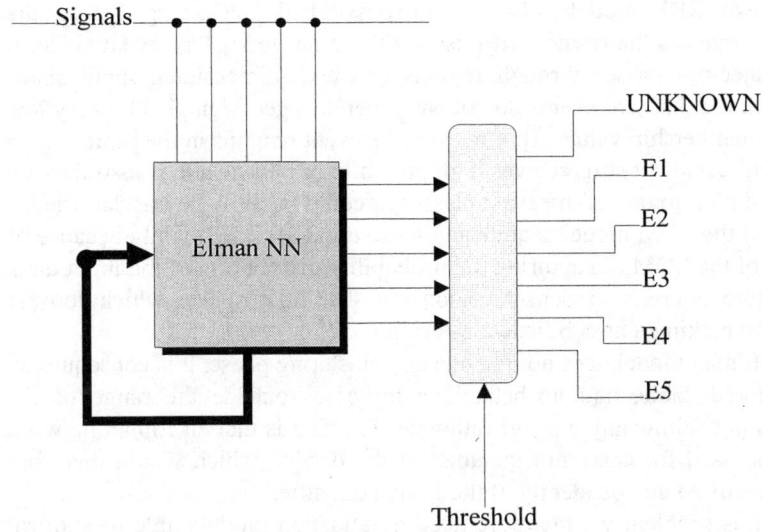

Fig. 4. The architecture of the Elman classifier

'forcing' the network to recognize the noisy examples as still belonging to the same class. This method works very well but has the disadvantage of further increasing the number of training examples and therefore the training time and cost.

An advantage of the recurrent model is that it can discriminate event classes which differ even slightly from each other. This is due to the fact that the Elman model is fed directly with the event signals, which are not 'transformed' by a clustering phase as in the RBFs and SOM models. This property, however, might make the Elman model more susceptible to noise. Another advantage of the recurrent model is that it can recognize event transients in slightly different time scales, i.e., it can compensate for faster or slower instances of the same transient. A last advantage, which comes with the architectural simplicity of the Elman model, is its speed and consequent suitability to work in a real-time environment.

3.5 Validation of Classification

A major problem with most classifiers based on generalization from examples (or prototypes) is that their response is not always predictable when they are presented with a new, previously unseen type of input. In the case of event classification, the occurrence of an event which was not included in the design of the classifier can lead to an incorrect classification. This phenomenon is true for all the neural models described in the previous sections, even though at different degrees of seriousness.

For the two RBF models, the use of possibilistic clustering before the classification, reduces the chances of misclassifying an 'unknown' event since, if the event trajectory passes through regions of the N-dimensional input space through which the 'known' events do not pass, then this gets signalled by very low possibilistic membership values. If, however, the event remains in the same region of the 'known' events, it still receives high possibilistic values and, if its trajectory is somehow similar to one of the event classes, it could possibly be misclassified.

The case of the SOM model is quite similar to the RBF case, mainly because of the property of the SOM of capturing the probability distribution of the input data, allocating more neurons to denser regions of the input space, which loosely corresponds to making a possibilistic clustering.

Since the Elman model does not include any clustering phase, it is consequently the most unpredictable, and its behaviour for cases outside the range of the training does not follow any a priori rationale. The fact is that an Elman network should not be used for cases not included in the design, which would therefore need to be identified independently of the Elman classifier.

To solve this problem we therefore need a validation module able to confirm that the answer given by one of the above classification models is a meaningful one and not simply coincidental. In the following we propose a solution which makes use of a technique which we have already applied to the classification problem, namely possibilistic fuzzy clustering.

3.5.1 Possibilistic Fuzzy Clustering for Validation

Both RBF models described previously already use possibilistic fuzzy clustering to identify inputs which are outside the regions covered by the clusters. However, in both cases the possibilistic clustering is performed on the single N-dimensional input vectors which are part of an event trajectory, and not on the trajectories as a whole (i.e., the sequences of vectors). This means that the clustering can only tell us if an event trajectory remains within the regions covered by the clusters, but not if a trajectory which remains within the clusters is of a class which was included in the design.

In order to use possibilistic clustering for validation, we need to make the clustering in the space of the trajectories, not the space of the N-dimensional input vectors. If a trajectory is composed of say m time steps, then the clustering will need to be performed in $m*N$ dimensions. Given a possibilistic fuzzy clustering in $m*N$ dimensions, a new event trajectory will receive a sufficiently high membership value only in the case that it is sufficiently close to the events used to form the clusters, i.e., it is of an event class which was included in the design. To tune the sensitivity of this validator, and its robustness to noise and changing initial conditions, one can include in the clustering phase new examples generated by adding increasing amounts of noise to the prototype cases of each class, and define an appropriate membership threshold below which an event will be classified as "unknown."

One problem with this validation method is that the $m*N$ clustering space has to be fixed in advance, i.e., all the events need to be synchronized; the same event trajectory shifted in time will in general not be recognizable. This limits the applicability of this method only to domains in which a trigger signal is available which signals the beginning of an event, and which can be used as a synchronization point.

4 Case Study 1: Simulated BWR Events

For a comparative evaluation of these models, an NPP simulator was used to generate data relative to a small set of events. The simulated plant is the *Forsmark 2* BWR, which is a 969-MWe ABB reactor in Sweden, and it was simulated using the APROS simulation environment, developed by IVO and VTT, Finland. In order to test the classification accuracy of the introduced methods, we required a set of events which showed a range of degrees of similarity among them. The simulated events were:

- Turbine trip with bypass valve operational (TTWBP)
- Turbine trip without bypass (TTWOBP)
- Main steam isolation valve closure (MSIV)
- Feedwater heating loss (FWL)
- Feedwater controller failure (FWC)

Process variables were sampled at 8Hz and the following were recorded:

- Reactor water level (RWL)
- Feedwater flow (FWF)
- Steam flow (STF)
- Core pressure (CP)
- Power (P)

All these five events generate a *scram*, and this signal was used to synchronize the events for the classifiers and the validator. All the events were simulated at 100%, 80%, 62%, and 49% power, except for the FWL event which was simulated only at 100% power (since it didn't lead to a *scram* at lower power values). Given only these variables, it is clear that the transients relative to the TTWOBP and MSIV events will be the most similar (see Figure 5). In order to study the robustness to noise of the various models, a series of tests were performed by generating noisy transients from the original obtained from the simulator by adding gaussian noise ranging from 5% to 15%. In total 510 transients were in this way generated. As an example, in Figure 5 are shown the first 100 samples (i.e., 12.5 sec.) of all the recorded signals for the five events at 100% power with an added noise component of 5%. All signals were then passed through a signal processing phase consisting of four steps:

222

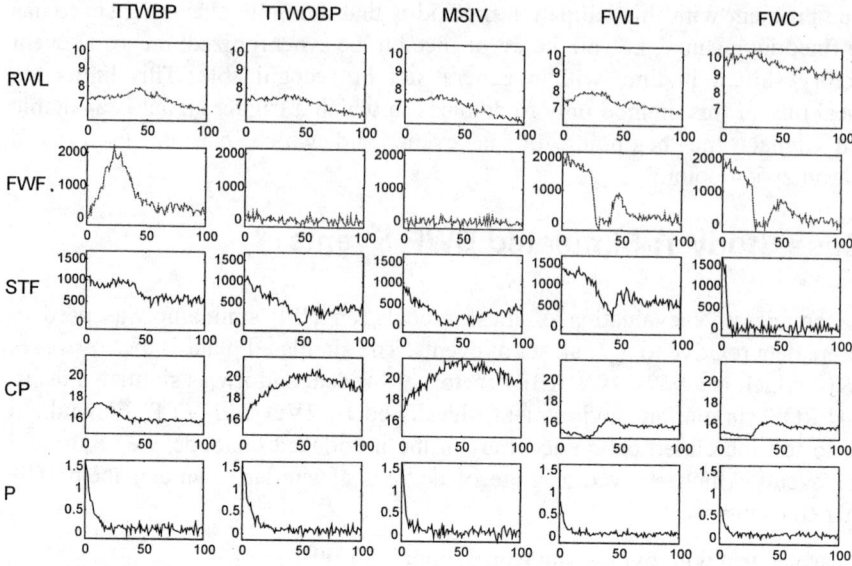

Fig. 5. Simulated events at 100% power with added noise (5%)

1. signals are independently normalized with respect to their steady state value, i.e., $s'(t) = \left(s(t) - s(t_0)\right)/s(t_0)$
2. signals are passed through a zero-phase digital filter
3. signals are normalized with respect to their global range
4. signals are squashed in the range [-1,1] by the bipolar *tansigmoid* function $y = 2/\left(1 + e^{-3x}\right) - 1$

The purpose of this signal processing is to both filter part of the noise and prepare the data for the clustering and neural network models.

4.1 RBF Classifier

The first step in the development of the RBF classification model consisted in the fuzzy and possibilistic clustering phase performed on the data corresponding to the five event transients at 100% power, which were used here as prototypes of the five event classes. In addition, another two sets of five events each were added to these by including a 1% noise component to the original events. The ISODATA clustering procedure produced 10 cluster centers which were then passed to the fuzzy clustering which iteratively updated the centers and constructed the fuzzy memberships. The result of the fuzzy clustering was then passed on to the possibilistic clustering algorithm which produced a possibilistic membership landscape. The next step in the development of the RBF classification model was

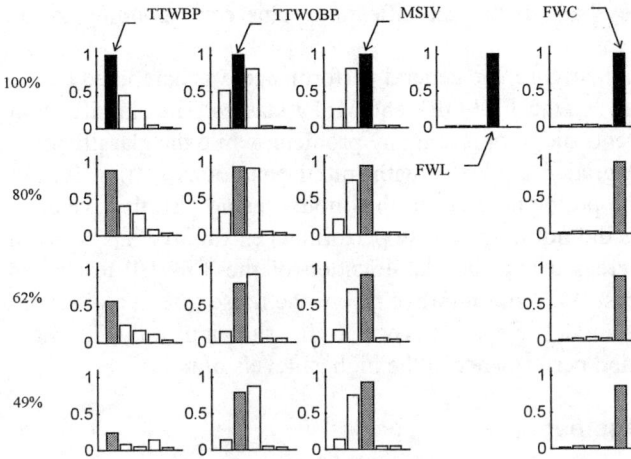

Fig. 6. Classification performance of the RBF model

the empirical optimization of the parameters which define the number, position in time, and duration of the time windows to each of which an actual RBF neural network is to be associated.

Since one of the main points in favour of this design is its simplicity (compared with the Cascaded-RBF design), it was decided to limit the number of time windows to three. The best combination was found to be windows of 0.75 seconds (i.e., 6 samples), separated by 0.5 seconds (i.e., 4 samples), and starting at time t=0, or more specifically:

- Window 1, from t=0 to t=0.75 (samples 1 to 6)
- Window 2, from t=1.25 to t=2 (samples 11 to 16)
- Window 3, from t=2.5 to t=3.25 (samples 21 to 26)

The final architecture has thus three RBF neural networks, each with 10 inputs (one for each cluster) to which the average membership values in the corresponding window are passed, and 5 outputs, one for each event class. The outputs of the networks assume values between 0 and 1, representing degrees of recognition. The RBF were then trained to recognize the 5 events at 100% power (with no noise) by centering the receptive fields of their respective neurons on the membership vectors generated by these class prototypes. A parameter which controls the size of the neuron's receptive fields was then adjusted for an optimal recognition of the remaining events (i.e., the ones at 80%, 62%, and 49% power).

Figure 6 shows the results of the classification tests on all the 17 events. On the first row are the 5 events at 100% power which were used as class prototypes. Each bar represents the final recognition value (i.e., the combination of the outputs of the three RBFs, weighted by the respective possibilistic evaluation) for the corresponding event class, as pointed by the arrows. The black bars are the class prototypes, while the grey bars are the classification value corresponding to the 'correct' class of the test cases.

As it can be seen from the figure, the general performance was hampered by the anticipated ambiguity between the TTWOBP and MSIV classes. The classification of the FWL and FWC events did not present any problem, while the classification of the TTWBP events degraded gradually with initial conditions getting further and further away from the prototype case. In this model, using also the events at 49% power as prototypes did not improve the performance. The overlap between TTWOBP and MSIV increased and the classification of the TTWBP test cases (i.e., the 80% and 62% cases) did not improve. From the noise tests we observed that, overall, the RBF model responds very well, showing only a slight degradation of classification performance at the highest levels of noise.

4.2 Cascade-RBF Classifier

The Cascade-RBF classifier used the same clustering which was performed for the RBF classifier. The next step was again the empirical optimization of the parameters which define the number, position in time, and duration of the time windows to each of which an actual RBF neural network is to be associated and connected in cascade with the adjacent networks.

The best combination was found to be 10 windows of 0.25 seconds (i.e., 2 samples) each adjacent in time, and starting at time t=0, or more specifically:

- Window 1, from t=0 to t=0.25 (samples 1 and 2)
- Window 2, from t=.25 to t=.5 (samples 3 and 4)
- . . .
- Window 10, from t=2.25 to t=2.5 (samples 19 and 20)

The final architecture has thus 10 RBF neural networks, the first with 10 inputs (one for each cluster) to which the average membership values in the corresponding window are passed, and 5 outputs, one for each event class, while the rest have 15 inputs (one for each cluster plus one for each output of the previous RBF) and 5 outputs, one for each event class. The outputs of the networks assume values between 0 and 1, representing degrees of recognition. The output of the last RBF (weighted by the possibilistic score) is taken as the final output. The networks were trained similarly to the previous model.

The cascade-RBF model general performance was quite similar to that of the simpler RBF model. The more complex cascade-RBF architecture was still unable to properly discriminate between the TTWOBP and MSIV classes, however the classification was sharper, i.e., the separation among the classes was more

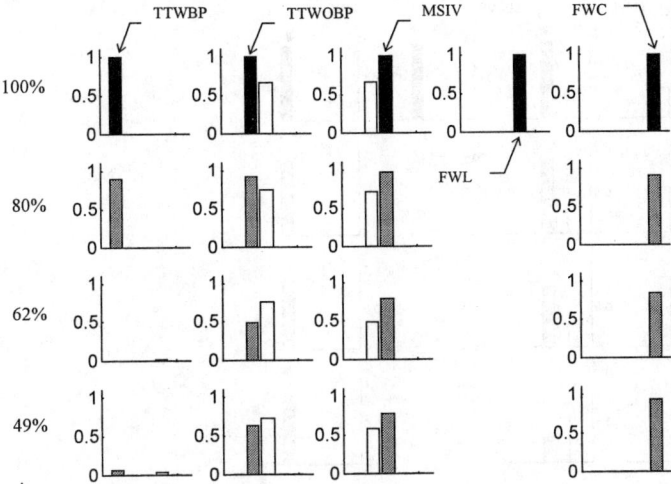

Fig. 7. Classification performance of the Cascade-RBF model

marked[8]. This is of course highly desirable and is the main advantage of this approach. Figure 7 shows the results of the classification tests on all the 17 events.

As for the robustness to noise, the Cascade-RBF models is slightly more sensitive when compared to the simple RBF classifier. This can be easily explained by observing that the architecture of the Cascade-RBF model is such that the effects of noise accumulate from one neural network to the next in the cascade; this is because the 'partial' classifications, perturbed by noise, are forwarded from one network to the next.

As a last observation we have to say that also in this model, the use of the events at 49% power as additional prototypes (together with the 100% cases) did not improve the performance.

4.3 SOM Classifier

The SOM classifier constructed for this test case is based on a 6x8 matrix of neurons which was trained to map the 5-dimensional input data on the SOM using Kohonen's competitive learning rule. The training data consisted of the 5-dimensional vectors constituting the first 50 samples of 30 events, namely the 5 simulated events at 100% power, plus 5 events with 5% added noise for each of

---

[8] This comes naturally from the fact that whether in the RBF case two events had to match in only 3 windows (and independently of each order) to receive a similar classification, in the cascade-RBF case two events have to match throughout the development of the transient, i.e., in all the 10 windows and in the correct order.

226

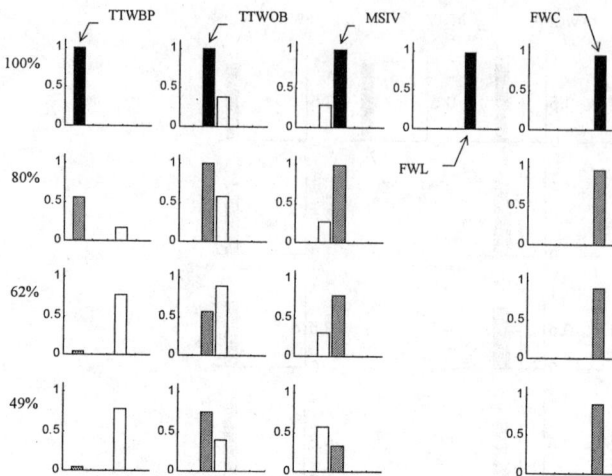

Fig. 8. Classification performance of the SOM model

these simulated events. A SOM classifier was then constructed based on the 5 events at 100% power (with no noise) as class prototypes. Figure 8 shows the results of the classification tests on all the 17 simulated events.

With the SOM model, the general performance was similar to the Cascade-RBF model. The SOM architecture was better at discriminating between the ambiguous TTWOBP and MSIV classes (at the 100% and 80% power levels), but was less good at generalizing from the prototypes. This led for example to a gross misclassification of the TTWBP events at 62% and 49% power.

The current architecture of the SOM classifier allows for only one prototype per class. Therefore it was not possible in this case to use of the events at 49% power as additional prototypes (together with the 100% cases) as it was attempted in both RBF cases. As for the robustness to noise, overall the results showed that the SOM classifier is very robust. In conclusion, the SOM model makes for a very robust, stable, and sharp classifier which however does not generalize too well.

4.4 Elman Classifier

The Elman classifier implemented for this test case is composed of a network of eight recurrent neurons which was trained to classify directly the sequences of 5-dimensional vectors derived from the first 20 samples of the simulated events. The training data consisted of the 5 simulated events at 100% power, the 4 simulated events at 49% power, plus 2 events with 3.5% added noise, 2 events with 7% added noise, and 2 events with 10.5% added noise for each of these 9 simulated events. The Elman classifier was the only model which could satisfactorily discriminate between the ambiguous TTWOBP and MSIV classes. This result is

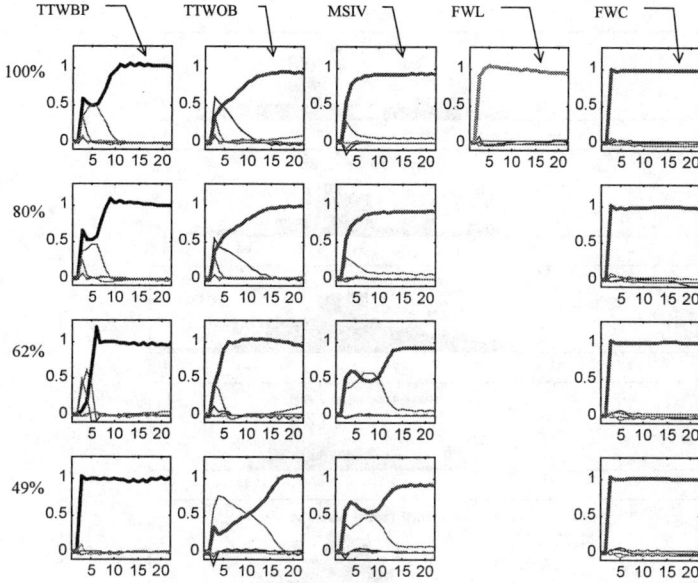

Fig. 9. Performance of the Elman classifier

shown in Figure 9. Since the Elman classifier produces a continuous classification in time, the figure shows the development in time of the classification values of each class, where the thick lines correspond to the classification value of the correct class. The classification was performed within the first 2.5 seconds, making it a good candidate for alarm filtering applications. The good results of the Elman model are only partly due to the ease with which the Elman ANN is able to use for training several prototypes for each class[9]. The main advantage of this recurrent model still lies in its being designed to deal directly with sequences of input vectors.

4.5 Possibilistic Validation

A similar series of tests was conducted with the aim of evaluating the performance of the possibilistic validation technique. A total of five sets of tests were conducted, each set consisting of a training phase conducted on only four 'known' event classes, leaving one event class out to be used as 'unknown.' Each test set was composed of about 150 'known' events (i.e., events whose class was included in the training) and 150 'unknown' events (i.e., events of the left-out class), each

---

[9] This did not help in the RBF and cascade-RBF cases and was not possible in the SOM case.

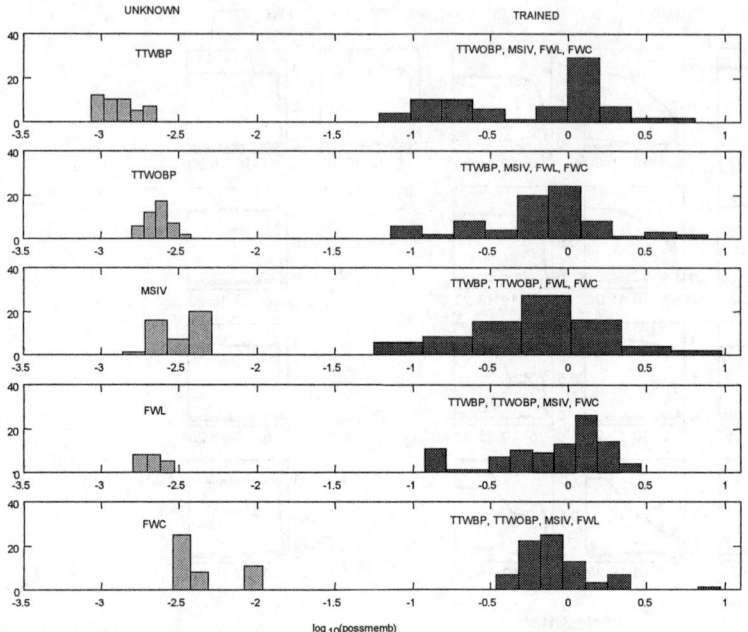

Fig. 10. Performance of the Possibilistic Validation

obtained by picking one of the simulated events and adding a 5% noise component. It was decided to use an event length of 20 samples[10], i.e., the first 20 samples of each transient were used as input to the validator, defining in this way a clustering space of 20*5=100 dimensions. The clustering was of course performed only on the 'known' events. The results are summarized in Figure 10, where each graph corresponds to one test set and shows the distribution of the possibilistic scores (on the horizontal axis) in logarithmic scale. In lighter color are the 'unknown' cases, while in the darker color are the 'known' cases. As it can be seen, the 'unknown' cases are very well separated from the 'known' (or 'trained') cases.

Overall we can say that this approach is very promising and is a very good candidate for inclusion in a dynamic event classification system, coupled with an accurate classifier such as the Elman model described in Section 3.4 and tested in Section 4.4. This is the configuration that we are currently proposing in the ALADDIN prototype, which is schematically shown in Figure 11.

In this architecture the decision on whether the classification output of the Elman network is accepted, or an UNKNOWN classification is to be produced,

---

[10] To match the event length used in the Elman classifier.

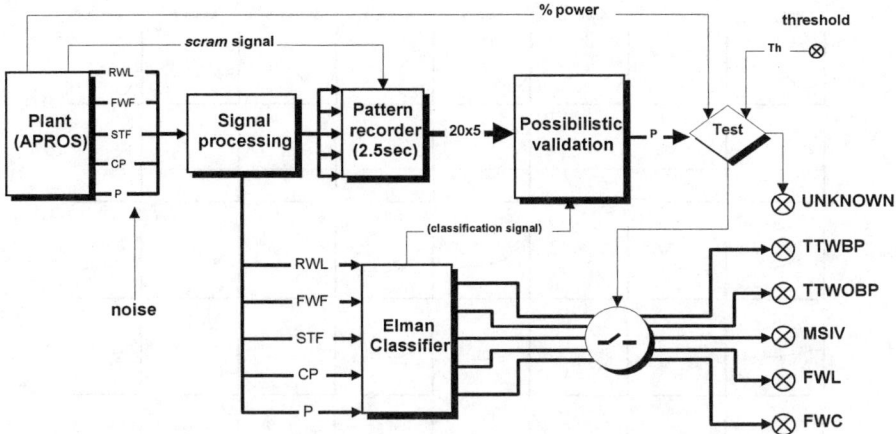

Fig. 11. The architecture of the ALADDIN prototype

depends on the possibilistic membership value P produced by the validation module, and on whether the plant operating power is within the design range of the classifier. The *scram* signal is used here to trigger the pattern recorder and, after 2.5 seconds (or 20 samples), the possibilistic validation of the classifier's output. As opposed to the other classification models described in Section 3, the Elman classifier does not need a trigger signal. This is because it does not use a fixed window on the input signals to make a classification, but rather it produces a (continuous) classification from continuous inputs.

This feature of the Elman model could be used in the future to extend the system to deal also with events which do not generate a *scram*. The problem of how to identify a window where to apply the possibilistic validation could be solved by using a 'classification signal,' coming from the Elman model when it reaches a sufficiently high classification, to trigger the validation on the buffered patterns.

5 Case Study 2: PWR Plant Islanding Events

Tests of the current prototype system have been recently carried out on simulated data of PWR transients, corresponding to various occurrences of plant islanding, i.e. various occurrences of rapid load rejection events, inluding particular malfunctions on plant I&C systems (instrumentation, actuators or closed-loop control systems). CEA (Commissariat à l'énergie atomique, France) in connection with EDF (Electricité de France) provided the data. These transients are initiated by a rapid closure of the main steam admission valves to the turbine, following an

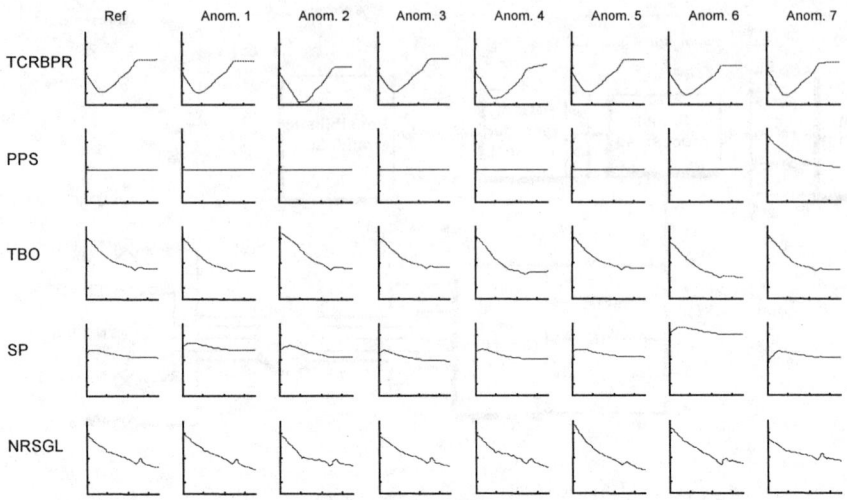

Fig. 12. Simulated Plant Islanding transients from t_0+50sec to t_0+400sec

electrical accident on the grid, and the final stable state obtained is highly sensitive to the relative dynamic behaviour of the primary and secondary power controls.

The data provided consisted of 7 anomalous transients plus a normal reference case. Additionally, 4 blind test transients were provided (where the amplitude of the failures differ from the transients used for training) to assess the robustness of the system and its sensitivity to scaling effects. The 7 anomalies are respectively:

1. Failure to open one group of condenser steam dump valves
2. Failure to extract half of the R control rod group (temperature closed-loop control in the G mode primary power control of the French plants)
3. Failure to close of two condenser steam dump valves (half a group) in the stabilization phase of the transient
4. Failure of the derivative branch of the temperature closed-loop control function (delay in the control rod insertion)
5. Failure of the derivative branch of the steam generator closed-loop control function
6. Shift in the reference value of the secondary steam pressure control (delay in the steam dump valves opening/closing)
7. Secondary power measurement failure due to pressure sensor response time increase

Each transient was described by the recorded values of 13 process variables (600 samples with a one second time step). In these tests it was decided to use only the stabilisation phase of the transients, and more specifically the data corresponding to the time interval t_0+50 seconds to t_0+400 seconds, where t_0 is the time of initiation of the transients (i.e., the time corresponding to the sharp decrease in the power setpoint). This decision was taken even if it was clear that Anomaly 6 would be best discriminated in the initial phase of the transient (i.e., the first 50 seconds).

A closer look at the data showed that a subset of the 13 recorded variables might suffice to the purpose of transient discrimination, and the decision was taken to use only the following 5 variables:

- Temperature control rod bank position R (TCRBPR)
- Plant power setpoint (PPS)
- Turbine bypass opening (TBO)
- Steam pressure (SP)
- Narrow range steam generator level (NRSGL)

The evolution in time, through the stabilisation phase of the reference and the 7 anomaluos transients, of these 5 variables is shown in Figure 12. As it can be seen, the transients are in general very similar to each other. Due to the fact that the Elman recurrent neural networks used by the ALADDIN prototype are best trained with relatively short sequences [26], it was decided to compress the transients using a wavelet transformation[11], and use for training the difference between the reference and the anomaluos transients. This compression reduced the length of the transients from 350 data points to only 10, without loosing the main features of the transients.

Since only a single transient for each anomaly was available for training, it was decided to increase the generalisation ability of the system (i.e., decrease its sensitivity to scaling effects) by generating additional training transients for each anomaly through the introduction of random scale distortions. This forced the system to extrapolate from the given transients but had the consequence of making the use of the validation module of the system virtually impossible[12]. A more recommendable approach would be to train the system on several transients for each anomaly, covering the expected range of actual occurrence.

In order to increase the robustness of the trained system, the introduction of randomly generated, scale-distorted transients, enabled us to use the so called *boosting* [27, 28] technique, which consists in training a number of classifiers (in this case 10 Elman networks) on different sets of training data (i.e., transients) and then combine the classification results by averaging. The results obtained are presented in Figure 13, which shows the output of the "boosted" neural classifier. The system was trained to respond with a value of 1 for the correct class and -1 for all the other classes. Each line corresponds to one anomaly class, and in each graph the thick line correspont to the correct class. Each unit in the horizontal axis

---

[11] This transformation would have also had the effect of removing noise in the signals, should there had been any.

[12] The "unknown" transients could in principle be overlapping with the distorted versions of the prototype transients used for training, making the validation useless.

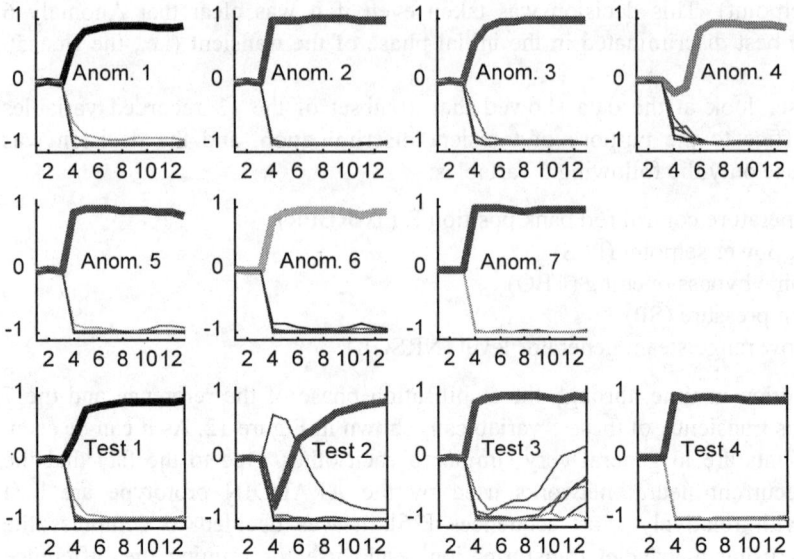

Fig. 13. Performance of the Boosted Elman model on the Plant Islanding transients

corresponds to roughly 35 seconds of the process transient (due to the wavelet compression).

As it can be seen the system behaves very well, even in cases like Test 2, in which, as we were told later, the anomaly size is much smaller than in the training case.

Similar tests were also performed with the other three models presented in this chapter, i.e., the RBF classifier, the Cascade-RBF classifier, and the SOM classifier. The results obtained are in line with the ones obtained in the first case study of BWR transients, with the Cascade-RBF model performing slightly better than the plain RBF model, and with the SOM classifier again showing generalization problems.

6 Future Developments

Even though the results obtained so far are very encouraging, further work still needs to be done in this area. One of the most immediate tasks is the investigation of how these techniques scale up to a larger set of event classes and signals. With a large number of classes and signals, training problems might emerge and new solutions might become necessary (for example the modularization of the neural classifiers). Another primary task is the investigation and development of techniques for the detection and classification of transients which are not associated with a trigger signal (i.e., the *scram* of our first case study, or the plan power setpoint drop of the second case study). A possible solution to this problem was hinted on at the end of Section 4.5, but no practical tests have been conducted so far. A more direct and systematic comparison with competing techniques (e.g., time series classification techniques) is also a highly desirable endeavour.

Looking more closely at the ALADDIN prototype, once the practical viability of the system has been confirmed, the next step would be its integration in an actual alarm system such as CASH [1]. An integration of ALADDIN in the computerized accident management support system CAMS [3] is among the planned activities, and so is its integration in the integrated diagnosis system IDS [2] and the process and condition monitoring system PEANO [6].

References

1. N.T. Førdestrømmen, B.R. Moum, B. Torralba, C. Decurnex, "CASH: An Advanced Computerized Alarm System", in *Proceedings of the ANS-meeting*, Philadelphia, Pennsylvania, USA, June 1995.
2. T.S. Brendeford "General Knowledge Structure for Diagnosis", in *Proceedings of the IAEA's specialist's Meeting on Monitoring and Diagnosis Systems to Improve Nuclear Power Plant Reliability and Safety*, Barnwood, UK, 14-17 May 1996.
3. P. Fantoni, A. Sørenssen, and G. Mayer "CAMS: A Computerised Accident Management System for Operator Support During Normal and Abnormal Conditions in Nuclear Power Plants", in *Proceedings of the Second OECD Specialist Meeting on Operator Aids for Severe Accident Management (SAMOA-2)*, September 1997.
4. P.F. Fantoni and A. Mazzola, "Multiple-Failure Signal Validation in Nuclear Power Plants using Artificial Neural Networks", *Nuclear Technology*, March 1996.
5. P.F. Fantoni, "Neuro-Fuzzy Models Applied to Full Range Signal Validation in Nuclear Power Plants", *ANS International Topical Meeting on Nuclear Power Plant Instrumentation, control and Human Machine Interface*, The Pennsylvania State University, 1996.
6. P.F. Fantoni, S. Figedy, A. Racz, B. Papin, "A Neuro-Fuzzy Model Applied to Full Range Signal Validation of PWR Nuclear Power Plant Data", in *Proceedings of FLINS'98, the 3rd International FLINS Workshop on Fuzzy Logic and Intelligent Technologies for Nuclear Science and Industry*, Antwerp, Belgium, September 1998.
7. E.B. Bartlett and R.E. Uhrig, "Nuclear Power Plant Status Diagnostics Using an Artificial Neural Network", *Nuclear Technology*, Vol. 97, 1992, pp. 272-281.
8. A. Basu and E.B. Bartlett, "Detecting Faults in a Nuclear Power Plant by Using a Dynamic Node Architecture Artificial Neural Network", *Nuclear Science and Engineering*, Vol. 116, 1995.
9. Y. Bartal, J. Lin, and R.E. Uhrig, "Nuclear Power Plant Transient Diagnostics Using Artificial Neural Networks that Allow "Don't-Know" Classifications", *Nuclear Technology*, Vol. 110, 1995.
10. H. Furukawa, T. Ueda, and M. Kitamura, "Use of Self-Organizing Neural Networks for Rational Definition of Plant Diagnostic Symptoms", Proceedings of the International Topical Meeting on Computer-Based Human Support Systems: Technology, Methods, and Future, The American Nuclear Society Inc., La Grange Park, IL, 1995, pp. 441-448.
11. E. Jeong, K. Furuta, S. Kondo, "Identification of Transient in Nuclear Power Plant Using Neural Network with Implicit Time Measure", *Proceedings of the International Topical Meeting on Computer-Based Human Support Systems: Technology, Methods, and Future*, The American Nuclear Society Inc., La Grange Park, IL, 1995, pp. 467-474.

12. E. Jeong, K. Furuta, S. Kondo, "Identification of Transient in Nuclear Power Plant Using Adaptive Template Matching with Neural Network", *Proceedings of the International Topical Meeting on Nuclear Plant Instrumentation, Control, and Human-Machine Interface Technologies, NPIC&HMIT'96*, The American Nuclear Society Inc., La Grange Park, IL, 1996, pp. 243-250.

13. Y. Ohga, S. Arita, T. Fukuzaki, N. Takinawa, Y. Takano, S. Shiratory, and T. Wada, "Evaluation Test of Event Identification Method Using Neural Network at Kashiwazaki Kariwa Nuclear Power Station Unit No.4", *Journal of Nuclear Science and Technology*, Vol. 33, No. 5, 1996, pp. 439-447.

14. K.C. Kwon, C.S. Ham, and J.H. Kim, "An Application of Hidden Markov Model to Transient Identification in Nuclear Power Plants", *Proceedings of the International Topical Meeting on Computer-Based Human Support Systems: Technology, Methods, and Future*, The American Nuclear Society Inc., La Grange Park, IL, 1995, pp. 275-280.

15. K.C. Kwon and C.S. Ham, "A Stochastic Approach with the Hidden Markov Model for Accident Diagnosis in Nuclear Power Plants", *HRP Workshop on Intelligent Decision Support Systems for Emergency Management*, Halden, Norway, 1997.

16. M.H. Hassoun, *Fundamentals of Artificial Neural Networks*, The MIT Press, Cambridge, 1995.

17. J.T. Tou and R.C. Gonzalez, *Pattern Recognition Principles*, p.97. Addison-Wesley, Reading (MA), 1974.

18. J.C. Bezdek, *Pattern Recognition with Fuzzy Objective Function Algorithms*, Plenum Press, 1981.

19. R. Krishnapuram and J. Keller, "A possibilistic approach to clustering", *IEEE Transactions on Fuzzy Systems*, Vol. 1, No. 2, 1993.

20. D.S. Broomhead and D. Lowe, "Multivariable functional interpolation and adaptive networks", *Complex Systems*, Vol. 2, pp. 321-355, 1988.

21. S. Chen, C.F.N. Cowan, and P.M. Grant, "Orthogonal Least Squares Learning Algorithm for Radial Basis Function Networks", *IEEE Transactions on Neural Networks*, Vol. 2, No. 2, pp. 302-309, 1991.

22. T. Kohonen, *Self-Organization and Associative Memory*, 3rd edn., Springer-Verlag, Berlin, 1989.

23. G.J. Chappel & J.G. Taylor, "The Temporal Kohonen Map", *Neural Networks*, Vol. 6, pp. 441-445, 1993.

24. T. Koskela, M. Varsta, J. Heikkonen, and K. Kaski, "Time Series Prediction using Recurrent SOM with Local Linear Models", *International Journal of Knowledge Based Intelligent Engineering Systems*, 1997.

25. J.L. Elman, "Finding structure in time", *Cognitive Science*, Vol. 14, pp. 179-211, 1990.

26. Y. Bengio, P. Simard, and P. Frasconi, "Learning Long-Term Dependencies with Gradient Descent is Difficult", *IEEE Transactions on Neural Networks*, Vol. 5(2), pp. 157-166, 1994.

27. Y. Freund, "Boosting a weak learning algorithm by majority", *Information and Computation*, Vol. 121(2), pp. 256-285, 1995.

28. Y. Freund & R.E. Shapire, "Experiments with a New Boosting Algorithm", in *Machine Learning, Proceeding of the 13th International Conference (ICML '96)*, pp. 148-156, Morgan Kaufmann, Bari (Italy), 1996.

11 Neural Networks in Signal Processing

Rekha Govil

Department of Computer Science & Electronics
Banasthali Vidyapith - 304 022, India
rekha@bv.ernet.in

Nuclear Engineering has matured during the last decade. In research & design, control, supervision, maintenance and production, mathematical models and theories are used extensively. In all such applications signal processing is embedded in the process. Artificial Neural Networks (ANN), because of their nonlinear, adaptive nature are well suited to such applications where the classical assumptions of linearity and second order Gaussian noise statistics cannot be made. ANN's can be treated as nonparametric techniques, which can model an underlying process from example data. They can also adopt their model parameters to statistical change with time. Algorithms in the framework of Neural Networks in Signal processing have found new applications potentials in the field of Nuclear Engineering. This paper reviews the fundamentals of Neural Networks in signal processing and their applications in tasks such as recognition/identification and control. The topics covered include dynamic modeling, model based ANN's, statistical learning, eigen structure based processing and generalization structures.

1 Introduction

During the infancy of the development of Neural Networks technology, one thing that excited people's interest was its analogy to biological systems. Even though not all has been understood about the learning processes of human neural systems, Artificial Neural Networks (ANN) have, without a doubt, provided solutions to many problems in different application areas.

The brain is a highly complex, nonlinear, and parallel information processing system. It consists of about one hundred billion (10^{11}) neural cells, each connected to about 10,000 neighbouring neurons and receiving signals from there. The brain routinely accomplishes perceptual recognition tasks (e.g., recognizing a familar face in a scene) in about 100-200 msec. The neuron, the basic information processing element (PE) in the central nervous system plays a very important and diverse role in human sensory processing, control and cognition. The brain is able to do complex tasks by its ability to learn from experience. An Artificial Neural Network is designed to model the working of human brain.

The ANN is usually implemented using electronic components (digital or analog) and/or simulated on a digital computer. It employs massive interconnection of simple computing cells called "neurons" or "processing elements (PE)," It resembles the brain in two ways:

- knowledge is aquired by the network through learning process
- inter neuron connection strengths (synaptic weights) are responsible for storing the knowledge.

The way the synaptic weights change is what makes the design of ANNs. Such an approach is close to linear adaptive filter theory, which is well established and is used in many diverse fields such as communication, control, sonar, radar, and biomedical engineering.

An ANN works as follows:

A neuron receives inputs from a large number of other neurons or from an external stimulus. A weighted sum of these inputs is fed into a nonlinear activation function. The output of this function is fanned out (distributed) to connections to other neurons. The topology of neurons connections defines the flow of information in the network. The way the weights are adjusted in the network constitutes the learning process. Thus the three essential components of an ANN computational system are - activation function, architecture, and, the learning law. Due to the differences in these three components, different ANN structures are explored for various applications and these structures differ in their computational complexities and requirements. The taxonomy along with the interrelationship of neural structures is shown in Figure 1.

The main attributes of neural processing are its nonlinear and adaptive learning capability, which enables machines to recognize possible variations of a same object or pattern and/or to identify unknown functions and mappings based on a finite set of training data, which can be noisy with missing information. Based on this 'Training by example' property with strong support of statistical and optimization theories, neural networks are becoming one of the most powerful and appealing nonlinear and adaptive data analysis tools for a variety of signal processing applications. In the present article we review the recent developments in the field of Neural Networks applications in signal processing which are as applicable in Nuclear Engineering as in any other engineering discipline. The topics covered include pattern recognition and synthesis, control, image analysis and several others. Essentially, neural networks have become a very effective tool in signal processing, particularly in various recognition and/or identification tasks.

1.1 Learning in Neural Networks

Learning is the most important aspect of an ANN. All the knowledge in the ANN is due to interconnection weights between different neurons and it is the learning process that determines the weights. On the basis of learning, ANN's can be classified into two broad classes; supervised and unsupervised learning models. The supervised learning models are trained by exposing them to

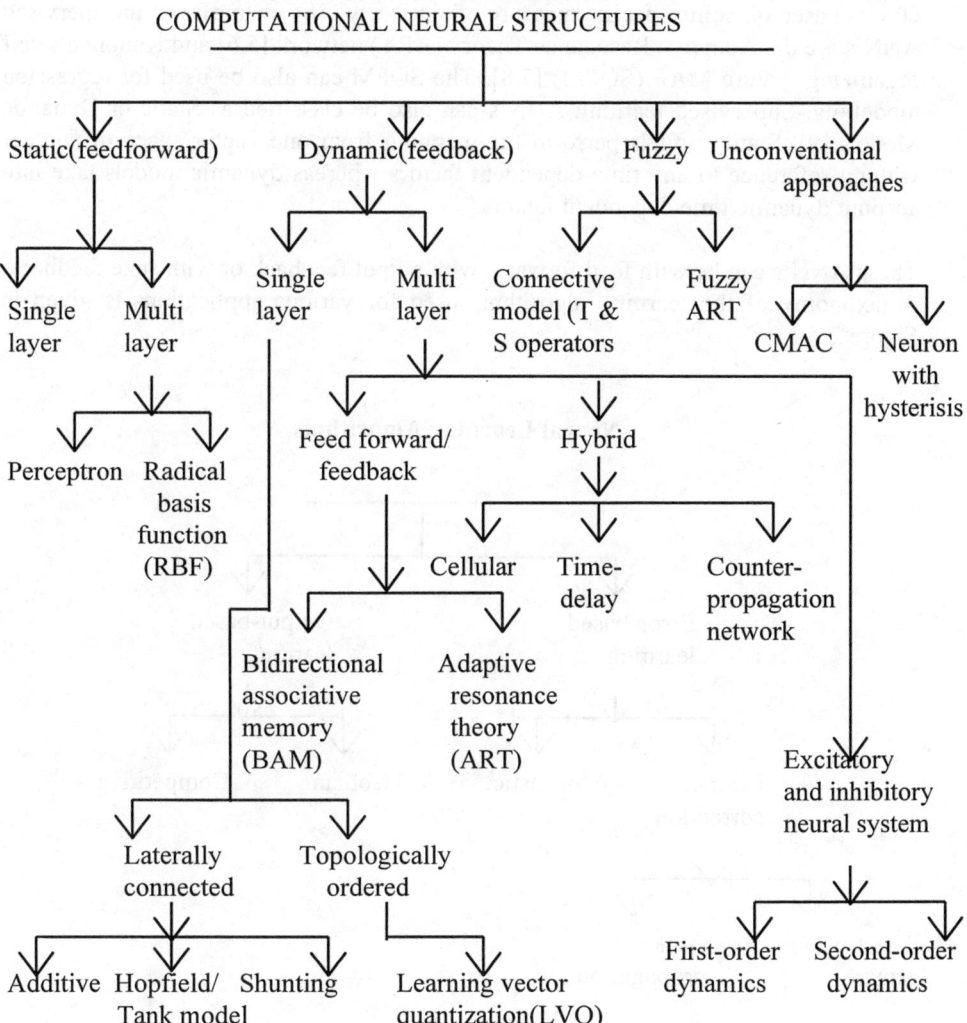

CLASSIFICATION OF VARIOUS
COMPUTATIONAL NEURAL STRUCTURES

Fig. 1. The Taxonomy of ANN structures

example input/output vector pairs to implement mappings that match the examples or underlying equations as closely as possible. The popular examples of supervised learning models are Multi-layer Perceptron neural network (MLP) [1,2] and Radial Basis Function neural network [3,4]. MLP is the most popular although network design and learning and adaptation complexities are faced while implementing it. RBF is relatively faster and is gaining popularity due to its closer relationship to Bayesian theory.

The unsupervised or self organizing models group the input sample into self similar classes based on some specific measure of similarity. The examples of unsupervised ANN's are the Adaptive Resonance Theory (ART) network [5,6] and Kohonen's self organizing feature MAP (SOFM) [7,8]. The SOFM can also be used for regression modelling. Supervised learning ANN's can also be classified as Static or Dynamic Models [9]. Static models perform the mapping from one vector space to another without reference to any time dependent factors whereas dynamic models take into account dynamic/time dependent factors.

These ANN's can be with feed forward, with output feedback or with state feedback. A taxonomy of the learning algorithms used for various applications is given in Figure 2.

Neural Learning Algorithms

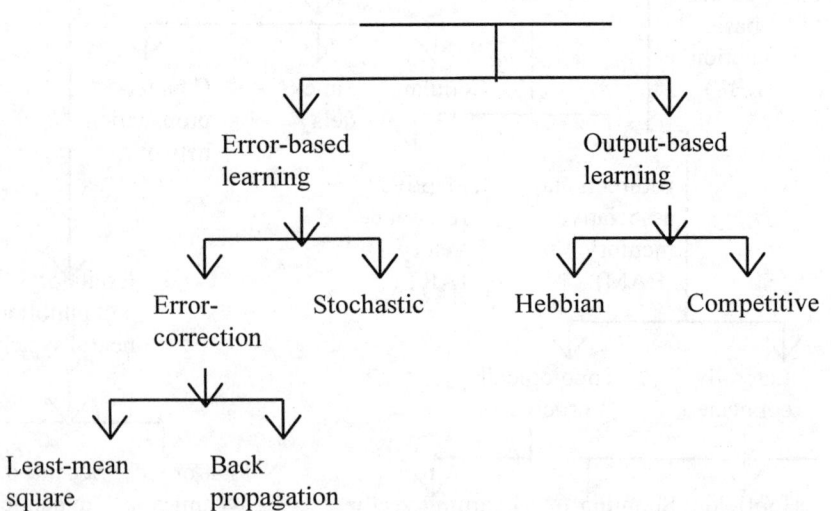

Fig. 2. The Taxonomy of ANN algorithms

1.2 Benefits of ANN

An ANN derives its power through its massive parallel distributed structure, and, its ability to learn and therefore generalize. Generalization refers to getting outputs for inputs not encountered during training. Some of the typical characteristics of the use of ANN are:

- Nonlinearity
- Input-output Mapping
- Adoptability
- Fault Tolerance
- VLSI implementability.

All these characteristics make the ANN an ideal tool for use in adaptive pattern classification, signal processing, and control. The VLSI implementation provides a means for capturing truly complex behaviour in hierarchical fashion and thus suitable for real-time applications.

2 The Use of ANN for Signal Processing

ANN's, because of their inherently non-linear nature are well suited to signal processing applications where the classical assumptions of linearity and second order Gaussian noise statistics cannot be made [10]. Real signals are usually generated by dynamic processes that are non-linear and non-Gaussian, in which case, the classical approach may not always produce an optimal solution [11]. There are two main approaches to engineering problems - parametric and non-parametric. The parametric approach is based on apriori models derived from the scientific knowledge about the problem. The nonparametric approach is based on the use of more general models trained to replicate desired behaviours using sufficiently representative data sets. In practice, the best solution is often desired through a mixture of parametric and nonparametric techniques. In this context, ANN's are useful because they can be treated as a nonparametric technique, which can model an underlying process from example data. They can also adapt their model parameters to statistical change with time.

The general goal for ANN signal processing algorithms in practical application is, given noisy and imprecise non-linear data, to enhance desired responses and reduce irrelevant and unwanted responses. The two extremes of this general goal are:

- Pattern Recognition, and
- Signal Conditioning problems.

In pattern recognition, a discrete decision is made regarding the presence or absence of a desired pattern, whereas in signal conditioning, the pattern is to be recovered. Most signal processing problems are usually related to a time or spatial data series. For example

- pattern recognition and signal detection;
- signal and system modeling and inverse modeling;
- signal filtering and smoothing;
- signal and system prediction.

The application of ANN's to nonlinear signal processing requires considerable judgement to design and signal processing issues. An ANN design involves

- raw data preprocessing,
- feature extraction from the preprocessed data,
- selection of network model and type, and
- network testing and evaluation.

The design process is a complex iterative and interactive problem specific task. The choice of the network model and type is based on the precise requirements of the problem. Different ANN models and types have different features that may be typically suitable for particular application. In fact the arithmetic calculating power of a computer has far prevailed over the human capability. But, for complicated signal processing problems, where large quantity of data and sophisticated inference are involved in real time, the performance of electronic hardware is still inferior to human. To perform associatively is essential for many intellectual activities. Therefore, one has to take advantage of parallel processing capability of neural networks in developing intelligent machines, which can have both superior computing speed and profound associative and inference abilities. A configuration of a generic neural network based intelligent system is shown in Figure 3.

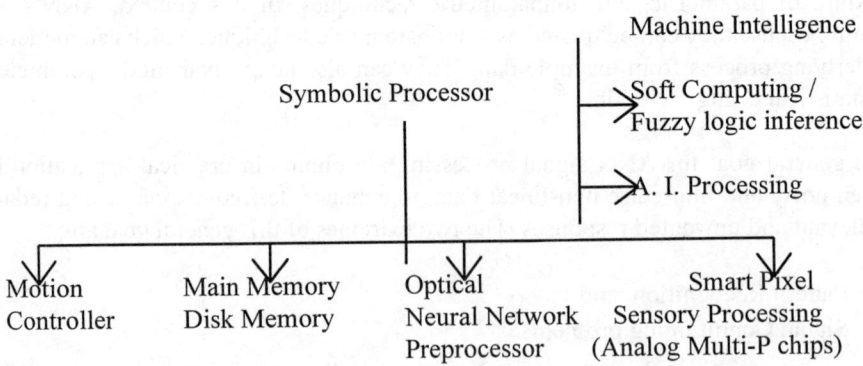

Fig. 3. A hybrid scheme of neural network based intelligent system

A high-level symbolic processor can access data from different sources, such as digital data from its cache memory, the main memory and hard disks. This data could also be from the electronic/optical smart sensors with local processing. Inference and decision made by the neural network system can then be used to activate the controller.

In what follows we review the fundamentals of Neural Networks in Signal processing and their applications. The topics covered include dynamic modeling, model-based ANN's, statistical learning, eigen structure based processing, and, optical and superconducting signal processing.

3 Dynamic Modeling

For models to explain data, linear models have been extensively used for static data in regression and for time series in optimum filtering. Dynamic modeling is a concept devoted to non-linear regression and its time counterpart. Its goal is to create a non-linear model for the incoming time series, using only available samples of the time series. In fact this formulation is an extension of the prediction formulation for non-linear systems.

A first step in dynamic modeling is to create an embedding of the time series, to create the *reconstruction space*. It is a space where the dynamics of the autonomous non-linear system that created the time series can be reconstructed.

It has been shown that an MLP with time delayed inputs can predict a chaotic time series [12]. Broomhead [13] showed that instead of MLP's, RBF networks can be used to implement the reconstruction space prediction, but then one has to go from prediction to dynamic modeling. Prediction only models the short term behaviour while dynamic modeling must also model the long term behaviour of the dynamics. ANN's are suitable for dynamic modeling because they learn from the time series directly and then map powerfully [14]. In dynamic modeling, one seeds the obtained model with a point in the trajectory, and then the output is fed back to its input to reconstruct the original time series as close as possible.

From a methodology perspective, dynamic modeling is able to model highly complex phenomena beyond the capabilities of the linear model. This methodology is able to model sea-clutter in radar, improving drastically the signal-to-noise ratio for faint targets [15]. Similar applications may exist in oher engineering problems where one can model the background activity and then subtract it from the incoming signal to improve detectability of faint transients. Dynamic modeling can also be used to capture the variability of real-world-data and encapsulate it in synthetic models.

An important practical problem in dynamic modeling is the noise omnipresent in real time series. Noise destroys the dynamics and hence it affects the accuracy of dynamic models. One solution to deal with noise is to seek local linear models instead of global models. Another alternative could be to explore filter embeddings instead of delay embeddings.

Dynamic modeling offers great potential for engineering. The accepted methodology to deal with weak signals in background noise has always been to concentrate on the signal characteristics because nothing can be done with the noise. But dynamic modeling changes this perspective because it can model the background noise. This may affect the design of detectors. Another aspect is signal generation, as very complex time wave forms may be synthesized with very little effort. The research in this area is in its infancy and many problems exist concerning the limit of dynamic modeling and noise. But now we are ready to start synthesizing very complex time waveforms, eventually mimicking real-word signals when needed. This may in future impact communication systems.

4 Model Based ANN's for Image Processing

The biological and computational facts lead us to believe that designing appropriate network architecture for a particular task is as important as adjusting the connection weights through training in a fully connected network. Biologically, natural neural networks have a hierarchically clustered architecture that is locally dense but globally sparse. Computationally, fully connected ANN's may not generalize well and may be difficult in hardware implementation. Hence, in recent years, Model based Neural Networks (MNN's) are playing important roles in architecture design for image processing applications. The major trend is to use some a priori knowledge to put together a complex network using simple (and fully connected) building blocks [16]. In applications like filtering and segmentation in image processing feature extraction, and classification in pattern recognition, these blocks are arranged in a serial order, each committing itself to a processing stage. In other applications, the blocks may have parallel structure, each committing itself to a particular processing task. The final results are the synthesized results from these individual processing blocks. More complex models may have both serial and parallel structures. Learning is performed after the model is structured. Each of the blocks may be trained separately.

A more sophisticated approach is to integrate the architecture and perform learning in one step, which is simulating the biological neural networks more closely. Some of the typical networks in this category are a network of networks [17], a hierarchical associative memory model and the self generating map [18].

MNN's are finding large applications in image processing/analysis/coding and pattern recognition [19,20]. Processing methods based on MNN's have also been used in a broad range of application oriented image and pattern analysis problems such as face detection and recognition [21], 3-D object classification, character and handwritten neural recognition [22] and signature verification [23], etc.

There is a great potential in MNN, since its true power lies in application understanding and system engineering, instead of individual algorithm design. But no doubt MNN's are a convenient tool for coherently integrating visual information processing modules and may lead to development of truly intelligent signal/image processing and computer vision systems.

5 Statistical Learning Networks

A learning network estimates an unknown function from representative observations of the relevant variables. A multivariate model free regression problem can be stated as follows:

Let there be n pair of vectors (y_i, x_i), i=1,2,...,n which have been generated from q unknown models

$$y_{ik} = g_k(x_i) + \in_{ik}, \quad i=1,2,...,n; \ k=1,2,...,q.$$

Here $\{y_i\}$ are called the multivariate "response" vectors and the $\{x_i\}$ are called the "independent variables." The $\{g_k\}$ are unknown smooth nonparametric (model free) functions. The only assumption made about $\{g_k\}$ is about the smoothness of the function from p-dimensional Euclidean space to the real time:

$$g_k : R^p \to R, \ k = 1,2,...,q.$$

The $\{\in_k\}$ are random variables with zero mean, $[\in_k] = 0$, and are independent of $\{xi\}$. The goal of the regression is to construct estimation, $\hat{g}_1, \hat{g}_2, ..., \hat{g}_q$, which are functions of the data (y_i, x_i), i = 1,2,...,n ,to best approximate the unknown functions, $g_1, g_2,...,g_q$, and then use these estimates to predict a new y, given a new x. Then

$$\hat{y}_j = \hat{g}_j(x), \quad\quad\quad j = 1,2,...,q$$

A similar formulation can be made for *model-free classification*, which is another major signal processing application. In classification formulation, the response vectors y, are set to be binary vectors for each independent variable, x, i.e., y=[0,0,0----,0,1,0----,0], where the selection of "1" indicates which class the x belongs to.

For model free statistical regression the following approaches are being followed by the statistical community.

- Additive models [24,25];
- Projection Pursuit Regression (PPR) [26];
- Classification and regression trees [27];
- Multivariate Adaptive Regression Splines (MARS) [28].

Additive models though useful in multivariate regression, can solve only a thin subclass of model free problems. Regression trees, MARS and projection-pursuit regression can, in principle, model very general classes of smooth $\{g_i\}$ with arbitrary precision.

An MLP provides a very powerful model-free approximation capability. But it uses back propagation learning (BPL), which has very inefficient and slow training since only the first derivative (or gradient) information about the training error is utilized. The training process could be speed up by employing second order optimization algorithms [29]. Several attempts have been proposed to improve the choice of nonlinear activations, from Sigmoid to Gaussian or bell shaped.

The Projection Pursuit Learning (PPL) is a statistical procedure proposed for multivariate data analysis using a two-layer network. It interprets high dimensional data through well-chosen lower dimensional projections. The "Pursuit" part of the name refers to optimization with respect to the projection directions. Three sets of PPL parameters: projection directions, projection "strengths", and the unknown smooth activation functions are estimated via the least squares criteria of minimizing the squared error loss function. Each response variable is modeled as a different linear combination of the activations. Each activation is taken as a nonlinear smooth function of a different linear combination of the independent variables. Similar to a BPL perceptron, a PPL network forms projections of the data in directions determined from the interconnection weights. But unlike BPL, which uses a fixed set of nonlinear activations (sigmoid), a PPL estimates the nonlinear activations based on an optimization approach that involves use of a one dimensional nonparametric (model free) data smoother.

A PPL learns neuron-by-neuron and layer-by-layer cyclically after all the training patterns are presented (unlike BPL, where the weights of all layers are estimated simultaneously). Specifically, in PPL the output layer weights are estimated using linear least square, the nonlinear activation of each hidden neuron is estimated by a one dimensional data smoother and the input layer weights are estimated by Gauss - Newton nonlinear least squares[30].

6 ANN for Eigen-Structure-Based Signal Processing

Eigen decomposition (ED) and singular-value-decomposition (SVD) methods have found very wide use in many fields of signal processing such as signal compression, signal coding, adaptive filtering, system identification, signal separation, signal prediction, noise rejection, feature extraction, and signal classification [31]. In these applications it is a primary requirement to adaptively perform ED or SVD from the sampled data.

The adaptive algorithm demands good performance and efficient implementation. ANN approaches, specifically Principal Component Analysis (PCA) networks, Minor-Component-Analysis (MCA) networks and Independent-Component-Analysis (ICA) networks are recently receiving great interest in this context. The key features of these ANN's are unsupervised learning (self-organization) and parallel processing capabilities. In these algorithms, the time series is assumed to have been generated by mutually exclusive, piecewise, stationary dynamical systems with switching between these dynamical systems occurring rapidly. The average switching rate is much slower. The dynamical systems that produce the time series are unknown, as are the switching times. Dividing a signal according to stationary intervals is called *segmentation* and modeling each one of the stationary intervals is called *identification*. The goal is to segment the time series into stationary region, model the resulting segments, and identify them. The great appeal of ED and SVD framework is its ability to project a collection of random vectors to a linear subspace with the best preservation of their variance. The axes of the projection space are the eigenvectors of the autocorrelation function. The axes are thus signal dependent, which normally implies more computation when compared with a priori determined bases such as Fourier analysis. However, the compactness of the representation is very appealing for data compression [32], and seeking fast, online methods to compute subspace decompositions is an ongoing research topic in communications and signal processing. The eigen compositions of continuous time stationary random process have been formulated by Karhunen and Loeve [33]. For stationary discrete time signals, when PCA is used, it is called temporal PCA. A temporal PCA architecture is shown in Figure 4. To incorporate the feature of time series to be piecewise stationary, with rapid transition between the stationary region, prediction is used as the competitive mechanism, where several adaptable "experts" compete to explain the same data [34].

The primary applications of competitive temporal PCA could be in:
- signal segmentation;
- noise reduction;
- adaptive encoding;
- sinusoidal frequency estimation.

It is best suited to signal dominated by a few strong linear modes. Common examples of such signals are earthquake measurement, speech, and music. More importantly, these algorithms correspond to continuous - time differential equations, which can be immediately used to construct analog neural networks to perform the desired ED and SVD on the order of hundreds of nanoseconds independent of the dimension of the matrix. This can provide very high speed signal processing in building coprocessors.

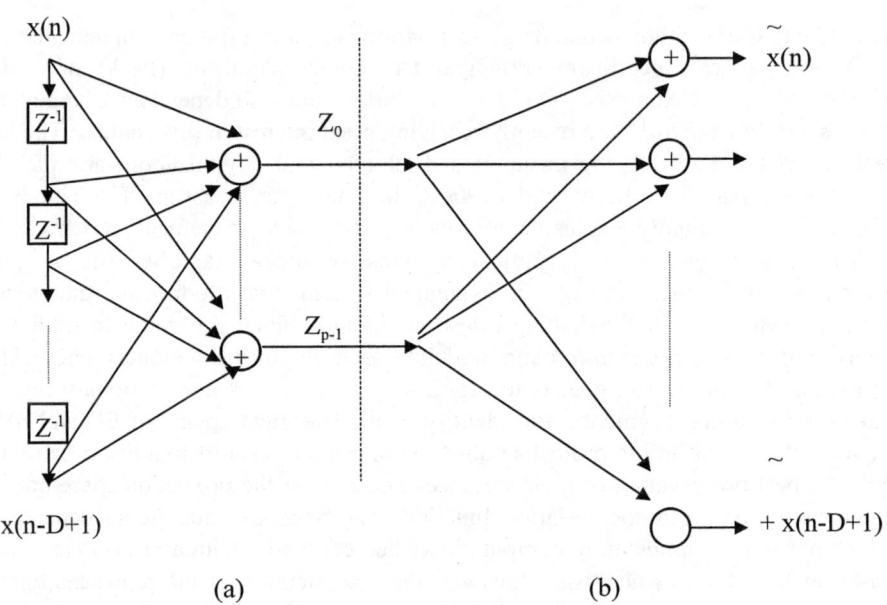

(a)   (b)

Fig. 4. (a) PCA network (b) Reconstruction network

7 The Functional Link ANN

Though capable of universal approximation, an MLP suffers from slow convergence rate and high computational complexity. An alternate structure, functional link ANN (FLANN) provides large reduction in computational requirement, with capability of formation of complex decision boundaries. In the MLP successive layers carry out a sequence of mappings until a final representation. In the FLANN, the initial representation of a pattern is described in a space of increasing dimensions. The concept of functional link is described in Figure 5(a).

In this model, each component of the input vector is acted on by the functional link to yield the enhanced representation of the original pattern. The functions used may be a subset of orthonormal basis functions spanning over an n-dimensional representation space. The net affect is the mapping of input pattern into a large pattern space. With careful selection of functional expansions, this single layer structure is capable of universal approximation.

In many practical signal processing applications, the Polynomial Perceptron Network (PPN) structure offers satisfactory results, for example, in channel equalization. But in many cases PPN does not converge and shows high instability. The other functions which have been successfully used for the purpose of multidimensional functional approximations are orthogonal functions such as Legendre, Chebyshev and Trigonometric polynomials. The advantage of using such polynomials is that after training the FLANN weights represent a multidimensional Fourier series decomposition of a periodic version of the desired response function.

Enhanced input pattern

$\varphi_1(x_1) ... \varphi_N(x_1)$ $\varphi_1(x_2) ... \varphi_N(x_2) ...\varphi_1(x)_n .. \varphi_N(x_n)$

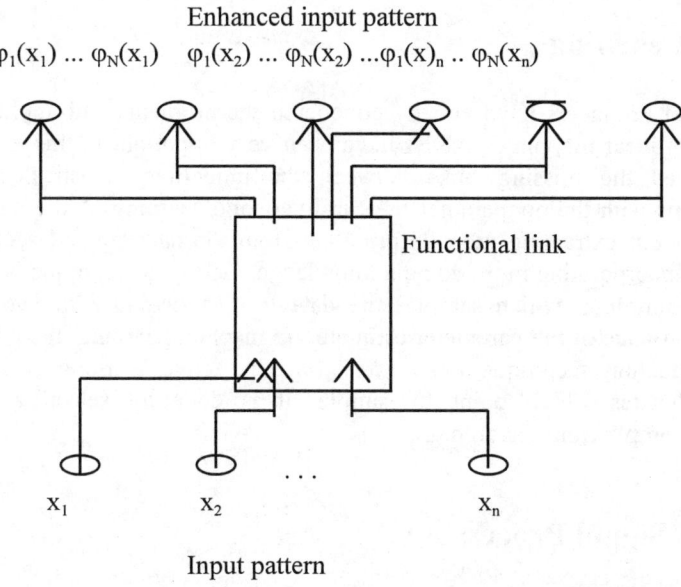

Functional link

x_1 $\qquad$ x_2 $\qquad$ ... $\qquad$ x_n

Input pattern

Fig. 5(a). The scheme of functional expansion model

A schematic of a generalised FLANN structure is described in Figure 5(b). An input pattern $[x_1, x_2,...,x_n]^T$ is enhanced by the functional expansion. The enhanced input pattern with proper weights is then presented to the single layer network.

Trigonometric functions have been successfully used in controlling a simulated two joint-planar robot arm and control of a robot manipulator. Use of trigonometric polynomials for functional expansions has also given better results in learning of a two-variable function.

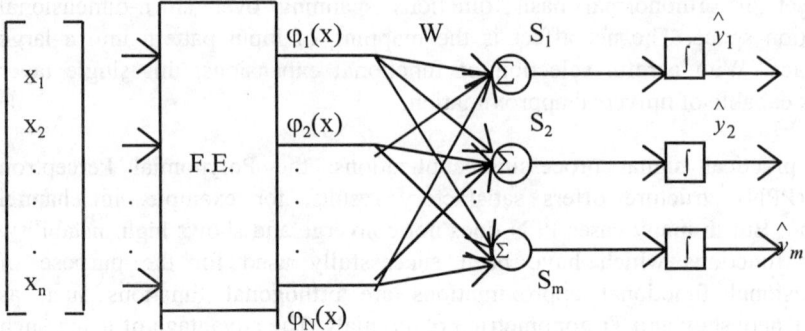

Fig. 5(b). The structure of a generalized FLANN

8 Active Learning

Scientists and Engineers have already conquered the problems and applications of statistical nonlinear mappings. ANNs have been very important in this arena. They have provided the missing links between the modelfree statistical regression /classifications with the nonparametric neural networks learning. A big challenge is how to go about extracting more information from the data beyond second order statistics by incorporating more domain knowledge. One such technique is known as importance sampling, which samples the data from a *twisted distribution* which reduce the variance of the parameter estimate. In machine-learning, there is another variance reduction technique *active learning*. In active learning, the learning algorithm dictates which point to sample. It is done by selecting the most informative sample from the domain.

9 Optical Signal Processing

The prospects of optical computing are well reported by Kaminuma, et al. [35]. Coherent laser light can be guided to pass through two convex lenses as a broad parallel beam. If an object of varying light intensity is placed in front of the first lens, the output of the second lens will be a copy of the original image with a reversed orientation. In such an arrangement, the image goes through a two-dimensional Fourier transformation by the first lens and brought to its focal point and this transformed result undergoes a reverse Fourier transformation by the second lens.

Optical Signal Processing can easily perform *correlational computation, convolution, and spatial filtering* on a two-dimensional visual image. It can be applied to tasks of object recognition and automated defect detection with operational speeds many order of magnitude faster as compared with digital computer approach.

But current digital computers will not be entirely replaced by optical components. Optical processing modules could serve as peripheral sensors and interface units for the digital system. Large scale optoelectronic neurocomputing machines could be designed with photonics. Optics offers the advantages in realizing parallelism, massive interconnectivity and plasticity required in construction of such machines. Analog optoelectronic hardware implementation of neural nets can combine the best of the two worlds: the massive interconnectivity and parallelism of optics, and the flexibility, high gain and decision making capability through inexpensive nonlinearity by microelectronics.

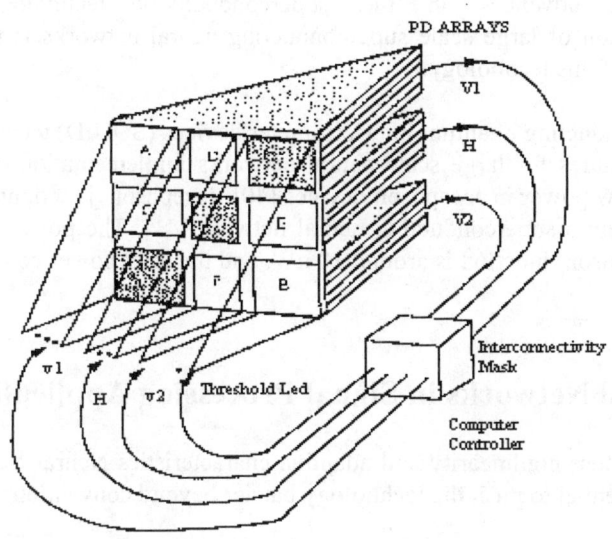

Fig. 6. Optoelectronic-neurocomputer architecture for multilayer nets

Two approaches to large scale optoelectronic neurocomputing machines have been proposed [36]: one using interegated optoelectronic neural chips with interchip optical interconnects that enable their clustering into large neural networks; and another based on nodes with 2-D arrangement of neurons, 4-D connective matrices for increased packaging density and compatibility with two-dimensional data. A simple architecture for optoelectronic neurocomputer could be full connectivity, an optical crossbar switch to perform the vector matrix multiplication, the state vector of the neural network represented by the linear light emitting array (LEA), the synaptic weight matrix W implemented in a photographic transparency mask, or a 2-

dimensional spatial light modulator (SLM) (for adaptive learning). The activation potential to be sensed with a photodiode array (PDA), which is shown in Figure 6.

In essence, attempts are being made to implement optoelectronic neural machines to realize fast parallel computing for various signal processing applications such as automatic target recognition and tracking [37]. ART networks also have been implemented with optoelectronics which can process input fields in excess of 10^7 nodes [38]. Practical applications of ART networks include intelligent design, laser radar processing and pattern recognition of occluded objects [39].

10 Superconducting Neural Networks

With rapid advances in the superconductivity technology, monolithic implementation of large scale superconducting neural networks is the recent most application of this technology.

The superconducting quantum interferrometer devices (SQUID) technology provides desirable features for large scale neural networks implementation due to ultrahigh speed and low power in Josephson devices [40]. Josephson junctions have also been designed using a superconducting neural network [41]. The power consumption at such a Josephson junction is around 10 nW, and the total power consumption below 100 nW.

11 Neural Networks in Signal Processing Applications

Because of their nonlinearity and adaptive characteristics Neural Network Systems have the potential to push the technology barrier beyond conventional approaches.

Whether it is control, supervision, maintenance, production research or design, the application of classical mathematical models does not completely solve the problem. Nuclear Engineering is different from other disciplines in the following way:

- experimentation is possible only to a certain degree,
- the consequence of misjudgment could be horrendous,
- waste disposal is the most difficult problem yet unsolved ,
- the abuse of nuclear material for criminal purposes can have unprecedented consequences,
- The public opinion is controversial and influential.

In areas such as Control in and of Nuclear Power Plants safety management and environmental protection, accounting of Nuclear waste and design of Nuclear safeguards, model based ANN's or dynamic modeling techniques could be useful.

The potential of ANN's in signal processing applications in Nuclear Engineering has just begun to be explored and much literature does not exist. Some results have been reported by D. Ruan et al. [42] and Uhrig [43]. Some recent general Engineering examples are described in the following. These give a fair idea of the ANN application potential in signal processing which can be used in process monitoring, control, radiation protection, data analysis and various other Nuclear Engineering applications.

(i)   Optical Character Recognition:
        These NN topologies are specialized MLPs with convolution layers and shared weights. The best results are obtained by LeNet5 developed by AT&T where they achieved 99.1% accuracy in automatically reading bank cheques [44].

(ii)   Image Restoration:
        The problem of color image restoration presents a unique difficulty in that the cross channel correlations need to be exploited. ANN's are especially well suited to this task because they can effectively adapt to the local nature of the problem Neural processing techniques have recently led to efficient VLSI architectures for image restoration due to their highly parallel nature [45].

(iii)   Automobile Control:
        Researchers believe that recurrent networks are the most promising technology to help maintain the emission levels of the clear air in automobiles. Ford Corporation has used an adaptation of the Decoupled extended Kalman filter training to control the engine idle and misfiring [46].

(iv)   Self Organizing Feature Maps (SOFM):
        SOFM converts complex, nonlinear, statistical relationship between high dimensional data into simple, geometrical relationship on a low dimension. It provides compression of information while preserving the relationships of the primary data elements. SOFM is therefore applicable to a large number of practical systems [47].

(v)   Natural Speech Synthesis:
        Text to speech synthesizer using multiple neural networks, each specializing in a particular area of human natural language ability represents better naturalness and adaptivity over conventional rule based approaches [48].

(vi)   Neural Vision System:
        A neural vision system can be used for a number of complex real-world image-processing applications such as computer-assisted diagnosis of

various biomedical systems, identification of underwater mines through sonar-image processing etc. Model based image processing is now a well-known research field.

(vii)   Cocktail Party Problem:

Separation of blindly mixed signals (such as voices in a party - the cocktail party effect), where neither the knowledge of the signals nor of the mixing process is available, is an important problem in inverse modeling and communications (channel equalization, teleconferencing). ANN have been shown to be naturally applicable to such problems due to their nonlinear activation functions and learning by example capabilities [49].

(viii)   Chaotic Dynamic Representations:

Model complex natural phenomena generated by deterministic nonlinear phenomena such as sea clutter can be dynamically reconstructed by ANN. The universal approximation properties of ANN (either MLP or radial-basis-functions) coupled with regularization have been shown to identify the nonlinear system that produced the time series [50].

(ix)   3-D Object Representation:

ANN have also been used in representing 3-D objects e.g. an ANN architecture is used to represent 3-D endocardial (inner) and epicardial (outer) heart contours and quantitatively estimate the motion of left ventricles of human hearts from ultrasound images [51].

(x)   Wavelength-Division Multiplexing (WDM):

Optical beams with different wavelengths propagate without interferring one another. The WDM scheme effectively increases the information carrying capacity of a waveguide or a fibber. Wavelength can be used as another dimension for parallel computing. With the 2-D chip and wave length as third dimension, the 3-D parallel computing can be achieved. The multiwave optoelectronic ICs can provide this multiplexing with a performance of delays below 100 picosecond [52].

12 Future Challenges in Neural Networks Signal Processing

The great advantage of MLP is that they are universal approximator and are very efficient for function approximation in high dimensional spaces. Unlike polynomial approximation, where the rate of the convergence of the error decreases with the dimension of the input space, the rate of convergence of error in MLP is independent of the dimensionally of input space. This is why MLPs perform better than other statistical procedures for larger dimensional spaces.

Doing well on unseen data is one of the most striking properties of neural systems. However, the training with MSE and the MLP topology do not control the generalization capability and hence *Time Processing* is probably the greatest challenge with neural networks. There has not been a genuine attempt to solve time varying problems with nonlinear structure. Time is considered as an extra dimension for the representation of information instead of being utilized directly for information processing.

MLPs are purely static and are incapable of processing time information. One way to extend MLPs to time processing is by creating a time window over the data to serve as memory of the past as was done in the Time - Delay Neural Network (TDNN) [53]. Alternately, dynamic neural networks bring the memory inside the neural network topology. This can be done in two ways: either the PEs are extended with local memory structures or the network topology becomes recurrent. Memory structures are generally linear, which means that the first type can be used as a simple and stable method. The concepts of filtering and linear modeling can be used here.

Recurrent connections across the topology do not show stability and they cannot be trained with standard back propagation. Temporal sequence data is dealt with the partially recurrent network, also called Simple Recurrent Networks (SRN). An SRN is a feed forward network but includes a carefully chosen set of fixed feedback connections. But these networks need to be studied for model stability.

In terms of learning algorithms, there are two basic approaches to train dynamic networks: real time recurrent learning (RTRL) and back propagation through time (BPTT). RTRL is global in the topology and local in time, while BPTT is local in the topology and global in time (anticipatory). Hence, for on-line learning RTRL is to be used and for forecasting BPTT is chosen.

In speech recognition, Hidden Markov Models (HMM) have been successfully employed. An HMM is a doubly stochastic process with an underlying stochastic process that is not observable (i.e., hidden), but can only be observed through another set of stochastic processes that produce the sequence of observed symbols [54]. The learning technique used in HMMs has a close algorithmic analogy with that used in the BPTT networks. Though ANN's have not surpassed the power of HMM's for speech recognition, but it is felt that integrating NN techniques into HMMs may provide new ways of looking at time processing.

The performance of ANN in the task of time series prediction has opened up a lot of applications in engineering - in control, production and management. Many technical challenges such as the principal methods of dealing with missing data, fusion of information from different sources, dealing with the large variability in experts behavior in treating information are the near future applications of ANN.

Multimedia technologies represent a new opportunity for research interactions among a variety of media such as speech, audio, image, video, text and graphics. Future multimedia technologies will need to handle information with an increasing level of intelligence, for example, automatic recognition and interpretation of multimodal signals. The key attribute of neural processing essential to intelligent multimedia processing is their adaptive learning capability. Now Neural Networks serve as a core technology for several vital multimedia functionalities, such as

- efficient representations for audio/visual information,
- detection and classification techniques,
- fusion of multimodal signals,
- multimodal conversion and synchronization.

A strong early motivation for interest in neural networks was our interest in understanding computations in the biological systems. But, interesting applications resulting from statistical signal processing, exploiting the nonlinar function approximation capabilities of neural networks have come up. Neural networks are now playing an important role in solving the problems of classification, regression, generalization and approximations. In fact they have injected a new blood in Applied Statistics and there is a lot more in store for the future.

References

[1]   D.E. Rumelhart and D.E.   Hinton, and R.J. Williams, "Learning representation by backpropagating errors," *Nature*, vol. 323, no. 9, pp. 533-536, 1986.

[2]   D.E. Rumelhart and J.L. McClelland, "Parallel distributed processing, explorations in the microstructure of cognition," in *Foundations*, vol 1. Cambridge, MA:MIT Press, 1986.

[3]   M.J.D. Powell, "Radial basis functions for multivariate interpolation: A review," Tech. Rep. DAMPT 1985/NA12, Dept. Appl. Math. Theoretical Phys., Cabridge Univ., Cabridge, U.K., 1985.

[4]   D.S. Broomhead and D.Lowe, "Radial basis-function, multi-variable functional interpolation and adaptive networks," Royal Signals Radar Est. Memo. 4148, Mar. 28, 1988.

[5]   G.A. Carpenter and S.A. Grossberg, "ART2: Self-organization of stable category recognition codes for analog input patterns," *Appl. Opt.,* vol. 26, no. 3, pp. 4919-4930, 1987.

[6]   _____, "The ART of adaptive pattern recognizing neural network," *IEEE Comput. Mag.,* pp. 77-88, Mar. 1988.

[7].   T. Kohonen, "Self-organized formation of topologically correct feature maps," *Biolog. Cybern.*, vol. 43, pp. 59-69, 1982.

[8]   _____, "Self-organizing maps: Optimization approaches," in *Proc. Int. Conf. Artif. Neural Networks*, Espoo, Finland, June 1991, pp. 981-990.

[9]   D.R. Hush and B.G. Horne, "Progress in supervised neural networks," *IEEE Signal Processing Mag.*, vol. 10, pp 8-39, Jan. 1993.

[10]   A. Lapedes and R. Farber, "Nonlinear signal processing using neural networks: Prediction and system modeling," Tech. Rep. LA-UR87-2662, Los Alamos Nat. Lab., Los Alamos, NM, 1987.

[11]   S. Haykin, "Neural networks expand SP's horizons: Advanced algorithms for signal processing simultaneously account for nonlinearity, nonstationarity, and non-Gaussianity," *IEEE Signal Processing Mag.*, vol. 13, pp 24-49, Mar. 1996.

[12]   A. Lapedes and R. Farber, "Nonlinear Signal Processing using Neural Networks: Prediction and System Modeling," LA-VR-87-2662, Los Alamos National Laboratory, New Mexico, 1987.

[13]   D. Broomhead, D. Lowe, "Multivariable Functional Interpolation and Adaptive Networks," *Complex Systems* 2, pp. 321-355, 1988.

[14]   I.W. Sandberg and L. Xu. "Uniform Approximation and Gamma Networks," *Neural Networks*, 10:781-784, 1997.

[15]   S. Haykin and X. Li, "Detection of Signals in Noise," *Proc. of IEEE,* 13. S. Roberts and L. Tarassenko, "Automated Sleep EEg Analysis Using an RBF Network," in *Applications of Neural Networks,* (ed. Murray), Kluwer, 1995.

[16]   T.M. Caelli, D.McG. Squire and T. P. J. Wild, "Model-Based Neural Networks" *Neural Networks,* vol. 6, pp. 613-625, 1993.

[17]   J.A. Anderson and J. P. Sutton, "A Network of Networks : Computation and Neurobiology," *World Congress of Neural Networks*, vol. 1, pp. 561-568, 1995.

[18]   W.X. Wen, H. Liu and A. Jennings, "Self-Generating Neural Networks," *Proc. IJCNN'92*, Baltimore, June 1992.

[19]   L. Guan, J.A. Anderson and J.P. Sutton, "A Network of Networks Processing Model for Image Regularization," *IEEE Trans. Neural Networks,* vol. 8, no. 1, pp. 169-174, 1997.

[20]   N. Srinvasa and R. Sharma, "SOIM: A Self - Organizing Invertible Map With Applications in Active Vision," *IEEE Trans. Neural Networks*, vol. 8, no.´3, pp. 758-773, 1997.

[21]   S.H. Lin, S.Y. Kung, and L.J. Lin, "Face Recognition/Detection   by Probabilistic Decision-based Neural Networks," *IEEE Transactions on Neural Networks*, vol. 8, no. 1, pp. 114-132, 1997.

[22]   J. Cao, M. Ahmad and M. Shridhar, "A Hierarchical Neural Network Architecture for Handwritten Numeral Recognition," *Pattern Recognition*, vol. 30, no. 2, pp. 289-294, 1997.

[23]   R. Bajaj and S. Chaudhury, "Signature Verification Using Multiple Neural Networks," *Pattern Recognition*, vol. 30, no. 1, pp. 1-7, 1997.

[24]   L. Breiman and J.H. Friedman, "Estimating Optimal Transformations for Multiple Regression and Correlation," *Journal of the American Statistical Association*, 80:580-619, 1985.

[25]   R. Tibshirani, "Estimating Transformations for Regression via Additive And Variance Stabilizing," *Journal of the American Statistical Association*, Vol. 83, pp. 394-405, 1988.

[26]   J.H. Friedman and W. Stuetzle, "Projection Pursuit Regression," *Journal of the American Statistical Association*, Vol.76, No. 376, pp. 817-823, December 1981.

[27]   L. Breiman, J.H. Friedman, R. Olshen and C.J. Stone, Classification and Regression Trees, *Wadsworth,* Belmont, California, 1984.

[28]   J.H. Friedman, "Multivariate Adaptive Regression Splines (MARS), with Discussion," Annals of Statistics, Vol. 19, No. 1, pp. 1-141, March 1991.

[29]   J.N. Hwang and P.S. Lewis, "From Nonlinear Optimization to Neural Network Learning," In *Proc. 24th Asilomar Conf. on Signals, Systems, Computers,* pp. 985-989, Pacific Grove, CA, November 1990.

[30]   J.N. Hwang, S.R. Lay, M. Maechler, D. Martin, J. Schimert, "Regression Modeling in Back-Propagation and Projection Pursuit Learning," *IEEE Trans. on Neural Networks*, 5(3): 342-353, May, 1994.

[31]   Fa-Long Luo and Rolf Unbehaueu, " Neural Networks for Eigen-Structure Based Signal Processing," *IEEE Signal Processing Magazine*, pp 28-40, Nov. 1997.

[32]   S.Haykin, Neural Networks; A compreheusive foundation, New York MacMillan, 1994.

[33]   N. Van Trees, Detection, Estimation and Modulation Theory: Part I. New York: Willy, 1968.

[34]   M.I. Jordon and R.A. Jacobs, "Hierarchical mixtures of experts and the EM algorithm," Neural Compute.,Vol 6,pp181-214 1994.

[35]   T. Kaminuma G. Mastsumoto, "Biocomputers," Chapman and Hall publishers, New York, 1991.

[36]   N.H. Farhat, "Optoelectronic Neural Networks and Learning Machines," *IEEE Circuits and Devices magzine*, pp 32-41, Sept. 1989

[37]   T.H. Chao, "An Integrated Optoelectronic ATR processor," *Proc. of JPL workshop on Neural Network Practical Application and Prospects*, pp. 193-208, Pasadena, CA, May 1994..

[38]   T.P. Caudell, S.P. Smith, G.O. Johnson, D.C. Wansch, "An Application of Neural Networks to Group Technology," *Proc. at SPE,* Application of Neural Networks II, Vol. 1469,pp. 612-621.

[39]   P. Kolodzy, "Multidimensional Machine Vision using Neural Networks," *IEEE Int'l Conf. on Neural Networks* , Vol., II pp 747-758, 1987.

[40]   Y. Harada, E. Goto, "Artificial Neural Networks with Josephson Devices," *IEEE Transaction on Magnetics*, Vol. 27, pp. 2863-2866, Mar. 1991.

[41]   Y. Mizugaki, K. Nakajima, Y. Sawada, T. Yamashita, "Super Conducting Implementation of Neural Networks using fluxon pluses," *IEEE Trans. on Applied Superconductivity*, Vol. 3, No. 1, pp. 2765-2768, Mar. 1993.

[42]   D. Ruan, P. D'hondt, P. Govaerts and E. Kerre (eds.), Fuzzy Logic and Intelligent Technologies in Nuclear Science. Singapore (1994).

[43] R.E.Uhrig, Artificial Neural Networks and Potential Applications to Nuclear Power Plants. In: A.S. Jovanovic, A.C. Lucia and S. Fukuda (eds.) Knowledge-Based System Applications in Power Plant and Structural Engineering. EUR 15408 EN (1984), pp. 23-42.

[44] L. Yaeger, R. Lyon, B. Webb, "Effective Training of A Neural Network Character Classifier for Word Recognition," *Proc. Advances in Neural Information Processing Systems* 9, pp. 807-813, MIT Press, 1997.

[45] A. Katsaggelos and S. P. R. Kumar, "Signal and Multistep Iterative Image Restoration and VLSI implementation," *Signal Processing*, Vol. 16, pp. 29-40, Jan. 1989.

[46] G. Puskurius, L. Feldkamp, "Neurocontrol of Nonlinear Dynamical Systems with Kalman Filter Trained Recurrent Networks," *IEEE Trans. Neural Networks*, 5(2): 279-297, 1994.

[47] T. Kohonen, E. Oja, O.Simula, A Visa, J. Kangas, "Engineering Applications of the Self-Organizing Map," *Proceedings of IEEE*, 84(10): 1358-1384, October 1996.

[48] O. Karaali, G. Corrigan, I. Gerson and N. Massey: "Text-to-Speech Conversion with Neural Networks: A Recurrent TDNN Approach," *Eurospeech* 1997.

[49] T. Bell, T. Sejnowski, "An Information-Maximization Approach to Blind Source Deconvolution," *Neural Computation*, 7:1129-1159, 1995.

[50] S. Haykin and J. Principe,"Dynamic Modelling of Chaotic Time Series with Neural Networks," IEEE Signal Processing Magazine, 13(2):24-49,1996.

[51] Y.H. Tseng, J.N. Hwang, and Florence Sheehan, "Three-Dimensional Object Representation and Invariant Recognition Using Continuous Distance Transform Neural Networks," *IEEE Trans. on Neural Networks*, special issue on Pattern Recognition, Vol. 8, No. 1, pp. 141-147, January 1997.

[52] S. Maeda, T. Aoki, and T. Higuchi, "Toward multiwave optoelectronics for 3-D parallel computing," *IEEE Int'l Solid-state Circuits Conference*, pp 132-133, San Francisco, CA, Feb. 1993.

[53] A. Waibel, T. Hanazawa, G. Hinton, K. Shikano, K.J. Lang, "Phoneme Recognition Using Time-Delay Neural Networks," *IEEE Trans. On Accoustics, Speech, and Signal Processing,* 37(3): 328-339, March 1989.

[54] L.R. Rabiner, B.H. Juang, "An Introduction to Hidden Markov Models," *IEEE ASSP Magazine*, pp. 4-16, Jan. 1986.

12 Application of Neural Networks in Reactor Diagnostics and Monitoring

I. Pázsit,[*] N. S. Garis[+], and P. Lindén[*]

[*]Department of Reactor Physics, Chalmers University of Technology,
SE-412 96 Göteborg, Sweden
[+]Swedish Nuclear Power Inspectorate
SE-106 58 Stockholm, Sweden
imre@nephy.chalmers.se

This chapter gives an account of the use of neural network techniques in reactor diagnostics and monitoring through a few concrete examples of successful practical applications. Diagnostic problems require the solution of a so-called inverse task, namely to determine the normal or abnormal values of some system parameters ("noise sources"), that cannot be directly measured, by observing their effect on other, measurable parameters ("reactor noise"). In the past, such inversion or unfolding techniques were possible to perform only if the direct task, i.e., calculation of the induced noise from the noise sources, could be made with a compact analytical solution. This condition hindered wide-spread practical applications. The use of artificial neural networks (ANN) presents a very powerful solution to the unfolding problem, since it does not require an analytical relationship between the cause and the reason. ANNs can be trained on simulated data. In the nuclear industry very powerful and accurate numerical methods exist to calculate process variables for operating plant, thus ANNs trained on simulated data can be used in real applications. This will be demonstrated through examples in this chapter.

1 Introduction

Problems in reactor diagnostics and monitoring have most often the character of an inverse task. One measures the effect or consequence of the change, or fluctuation, of some system parameter on other parameters or their spatial or frequency dependence. Usually, if the change or fluctuation of the source parameter (flow, temperature, vibrations etc.) is known, the induced changes in the measured parameters (i.e., neutron flux) can be calculated through the transfer properties of the system. This is also called the direct task. In diagnostics, however, the task is the reverse, namely one needs to unfold the change of source parameters from the measured consequences.

The possibilities of performing such inversions, and thus to perform effective diagnostics, have been improved vastly by the use of neural networks in diagnostic problems. In the past, successful inversion could only be performed when the direct task was possible to solve by analytical means. This reduced the applications to extremely simple cases or model calculations, since in core neutron physics and thermal hydraulics, realistic cases can only be calculated numerically. With the advent of the use of neural networks, diagnostics could be extended to cases which can only be treated numerically.

The burden of the inapplicability of analytical methods in nuclear and thermal hydraulic problems in the power industry is of course also a merit. Namely, the need of accurate prediction methods for safe and economic operation lead to the assessmental development of very powerful, albeit specialised, codes that can predict system behaviour with a very high degree of accuracy. Due to especially the safety aspect, core neutronic and thermal hydraulic calculations are among the most accurate and best tested methods in all categories.

This means that for the solution of an inverse task, a neural network can be trained by simulated data of high accuracy. The use of simulated data is inevitable since for obvious reasons, not the least reasons of economy, there is no possibility of obtaining a training set experimentally. Due to the high accuracy of simulated data, real-life diagnostic problems can be tackled by neural networks.

The fact that neural networks can be used for reactor diagnostics, monitoring and control has been realised quite some time ago (for a review in the field we refer to Ref. [1]). Neural networks have been used extensively in the nuclear engineering field for both parameter estimation and diagnostics. Successful applications or pilot studies include diagnostics of steam generators, vibration properties, sensor validation, valves, feedwater flow, estimation of moderator temperature coefficients, BWR stability margins, detection of anomalies, and noise spectra analysis. However, as explained in [1], for a long time most published work contained feasibility studies of hypothetical cases, with no or little practical applications. It is mostly the use of advanced calculation methods to generate high-quality training data that makes practical applications possible.

In this chapter several such applications are described. The first application regards the determination of the axial position of a control rod from the distortion of the axial neutron flux shape around the rod. The second case concerns the localisation of a vibrating control rod in a reactor core by measurement of the space dependent neutron noise induced by the vibrations. The last application treats the analysis of pulsed neutron activation data of water flow in a pipe in order to determine mass flow with high precision. All applications were tested in real measurements and will be applied in practice at operating plant routinely.

2 Determination of Axial Control Rod Elevation from the Measured Flux Shape with Neural Networks

This work was initiated by the fact that according to operational experience, the electromechanical position indicators, giving the axial elevation of a control rod in a pressurized water reactor (PWR), can get de-calibrated during operation. In certain cases the error may be in the range of 20 cm. The control rod position is an important operational variable and thus its accurate value, both for a bank of control rods and for the individual rods separately, must be known accurately, with a significantly better precision than the above value. How such errors occur and on other technical details we refer to other publications Refs [2,3]. Here we only conclude that there is a need for an alternative method to calibrate the standard rod positioning instrumentation.

This alternative method is based on the observation that a partially inserted rod causes a distortion in the axial flux shape in its vicinity and that the distortion is different for different rod elevations. In other words, there is an information in the flux shape regarding the rod elevation. Since the axial flux shape can be measured within a fuel assembly containing a control rod assembly (or in its neighbour) with a movable neutron detector, there is in principle a possibility to unfold rod elevation from the measured flux shape.

The principle is illustrated in Fig. 1 that shows the axial flux shape in the upper part of the core as measured within the assembly, in a neighbouring assembly, and in the case of a fully withdrawn control rod. It is seen that at least when measured in the control assembly, there is a distortion of the flux shape close to the rod which is clearly related to the rod elevation. The problem is that this relationship is very implicit, and cannot be unfolded by some simple algorithm.

While the above inverse problem of unfolding rod elevation from the flux shape is difficult, the direct task of determining the flux shape corresponding to a certain rod insertion is conceptually simple. It requires the solution of neutron diffusion or transport equations with given parameters in a reactor core. Although in a concrete core such a calculation is quite a formidable task despite its conceptual simplicity, there exist advanced in-core fuel management (ICFM) codes which solve such a problem with good precision. An example is shown in Fig. 2 in which measured and calculated values of the axial flux are compared. The measurement was taken in the Swedish PWR Ringhals-4 at the beginning of cycle 13, and the calculations were made by the ICFM code SIMULATE ([4,5]).

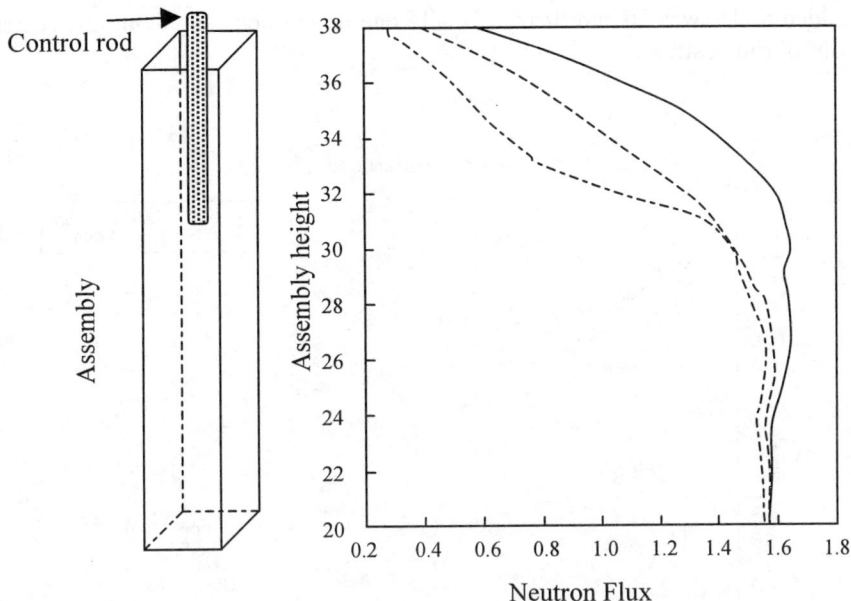

Fig. 1. The neutron flux profile for a partially inserted control rod in the upper quarter part of the core both within the assembly (point-dashed line) and from a neighbouring one (dashed line). The flux profile for a withdrawn control rod is also included (solid line)

At this point it is obvious that this inversion problem can ideally be solved by artificial neural networks. A training set can be generated by calculating the flux shapes in an actual core loading, corresponding to various rod elevations. Then, the trained network can take a measured neutron flux shape as input and determine the rod elevation as output. Such a method was elaborated and tested on measured data from the Ringhals-4 PWR. This method and the results from its test will be briefly described here.

Performing the neutron flux calculation requires the handling of certain subtle questions such as using non-equilibrium Xenon distribution as well as the detector count option (as opposed to flux calculation) in the calculations. With these details we however refer to [2] and here we concentrate on the description of the neural algorithm instead.

The neural network used in this work, as in most of our work, is a simple three-layered feed forward network with backward error propagation. For more details on neural networks [6,7] are referred to. The structure of the network is seen in Fig. 3. The number of input nodes is equal to the number of points in which the flux was calculated, with some comments that follow below. The number of the hidden nodes was 30 and there was only one output node supplying the searched control rod position.

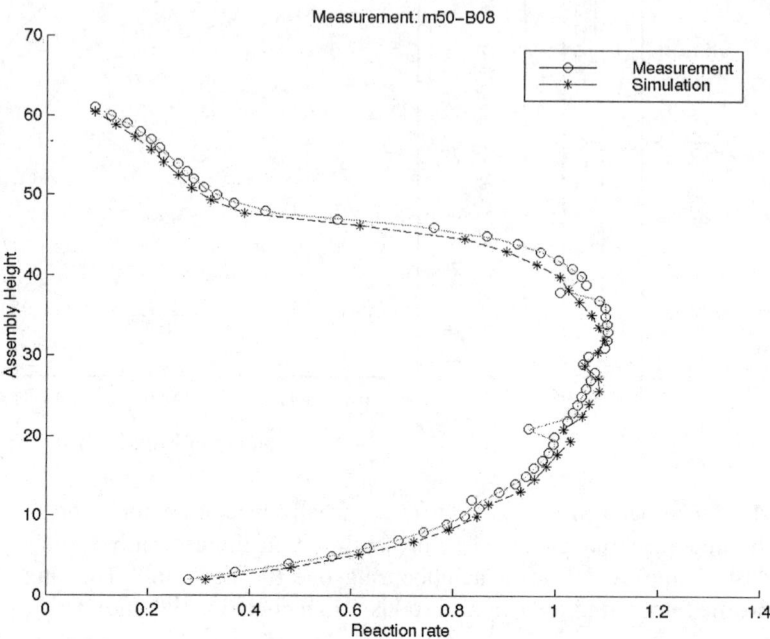

Fig. 2. Comparison between measured and calculated axial flux shape for assembly B08 with a partially inserted rod

For maximum performance of the network, a certain reduction of all available training data was made. Although the flux shape can be calculated for any control rod elevation from top to bottom, in an experiment only less than halfway inserted rods are used. This is because one constraint is that the measurement should make as little as possible perturbation of the reactor operation. For this reason, in the training only flux patterns with at most halfway inserted rods were used. Another point is that the flux distortion dies away with increasing distance from the rod tip, thus in one input pattern, it is not necessary to use the axial dependence of the flux shape in the whole core. Hence the number of input nodes (points in which the flux was calculated) in the present case was only 58, less than the number of points in which the flux is calculated. Altogether 65 training patterns were calculated; out of these 33 were used for the training of the network and the rest was used for its testing before it was used with real measurement data.

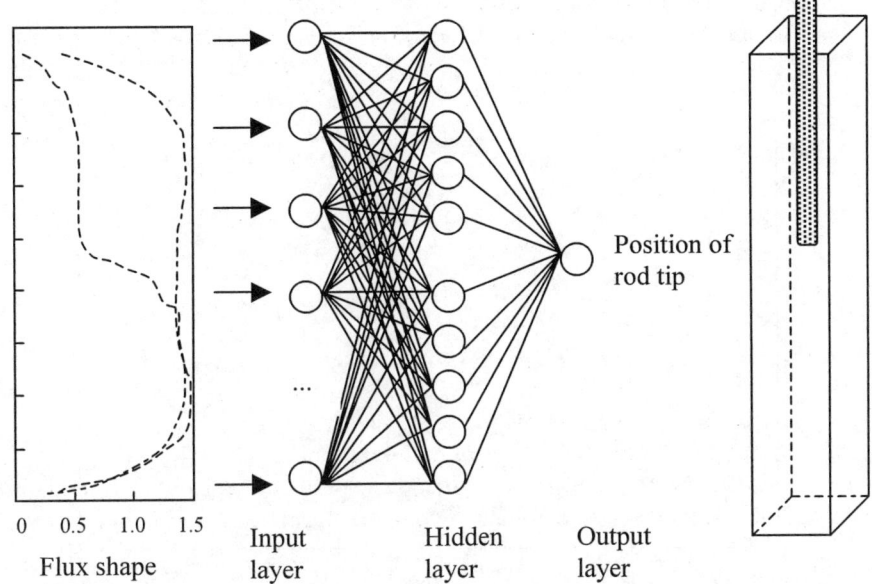

Fig. 3. The structure of a standard three-layered feed-forward neural network used in the rod positioning method

One particular result, including both testing with simulated data and one single test with measurement, is shown in Fig. 4. The symbols "*" indicate the errors of the tests that were obtained by using the calculated flux shapes as input (as mentioned earlier these flux shapes were not used during the training). It is seen that the deviations are in the range of 1-2 cm. The symbol "o" shows the error obtained by using the measured data as input. This gives only one single result since there was only one measurement made. The error is less than 2 cm, which is completely satisfactory for the method to be useful. On the whole, four more experimental tests were made with similar results.

It is thus demonstrated that neural network based identification is suitable to determine the axial elevation of a partially inserted rod in a PWR. A permanent installation of the method is underway and its routine use is planned at the Ringhals PWRs.

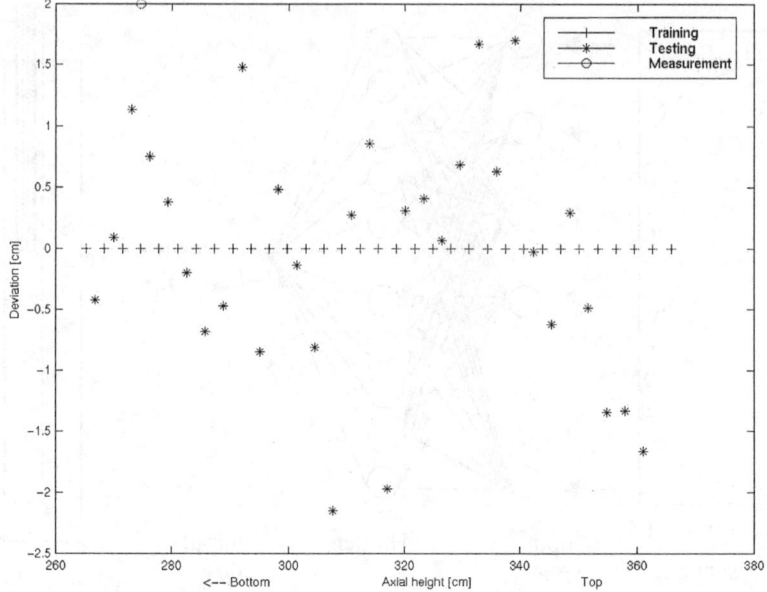

Fig. 4. Deviation (error) of network output and original values for the control rod inserted in position B08 and moving only in the upper quarter of the core. A ±5% Gaussian noise was added to the simulated flux shapes that were used in the training

3 Localisation of a Vibrating Control Rod from Neutron Noise Measurements

The possibility and occurrence of excessive vibrations of control rods in certain type of PWRs, driven by the coolant flow, has been known for a long time ([8,9]). Such an excessive vibration occurs as a consequence of some defect in the material properties of the rod that develop during operation. The only type of sensors that carry information on the vibrations in an operating plant are the in-core neutron detectors. The vibration of the faulty rod leads to space- and time-dependent fluctuations of the neutron flux, commonly called neutron noise, which can be measured by in-core-detectors. Thus the vibrations can be diagnosed by reactor noise analysis methods (for a description of noise analysis methods we refer to [10]). Diagnostics means partly detecting an anomalously vibrating rod as well as identifying the rod (also called rod localisation) as well as determining vibration properties (amplitude, anisotropy of vibration and preferred direction). Out of these, localisation, i.e., identifying the vibrating rod, is the far more important, but also the most complicated task.

The localisation method is based on an expression for the measured neutron noise which consists of a convolution over the noise source, represented by the vibrating rod, and the neutron physical transfer function of the system as

$$\delta\phi(\mathbf{r},\omega) = \int_{V_R} G(\mathbf{r},\mathbf{r}',\omega)\, S(\mathbf{r}',\omega)d\mathbf{r}'. \tag{1}$$

Equation (1) is common to all neutron diagnostic problems. Here, $S(\mathbf{r},\omega)$ is the perturbation represented by the fluctuations of the nuclear cross sections in the core, given rise by e.g., control rod vibrations. $G(\mathbf{r},\mathbf{r}',\omega)$ is the system transfer function which is only dependent on the parameters of the unperturbed system. This is a consequence of using linear theory, based on the smallness of the perturbation and the induced noise. The transfer function $G(\mathbf{r},\mathbf{r}',\omega)$ can thus be calculated for a reactor independently of the type of perturbation. It is actually equal to the Green's function of the Fourier transformed time-dependent diffusion equations with a time-dependent source. Since it is a result of a Fourier transform, the transfer function is complex and it contains information on both the spatial attenuation of the noise away from the source as well as its phase delay.

The task of diagnostics is to quantify the noise source $S(\mathbf{r},\omega)$ from the measured neutron noise $\delta\phi(\mathbf{r},\omega)$, using the knowledge of the transfer function. Determining $S(\mathbf{r},\omega)$ from $\delta\phi(\mathbf{r},\omega)$ through (1) is an inverse task. This inverse task is complicated by the fact that $\delta\phi(\mathbf{r},\omega)$ is only known in a few discrete points, i.e., in the detector positions. Thus usually some assumption is made on the functional form of the noise source, corresponding to a given perturbation, which only contains a few parameters. These few parameters may then be unfolded from (1). For instance, the spatial dependence of the absorption cross section of a control rod is represented by a spatial Dirac delta function at a point $\mathbf{r}_p$ in the 2-dimensional horizontal cross section of the core (where the localisation need to be performed), and thus a rod whose 2-D vibrations are described by the time-dependent vector $\vec{\varepsilon}(t)$ represent a perturbation in the form

$$S(\mathbf{r},t) \propto \delta\Sigma_a(\mathbf{r},t) = \gamma\left[\delta(\mathbf{r} - \mathbf{r}_p - \vec{\varepsilon}(t)) - \delta(\mathbf{r} - \mathbf{r}_p)\right]. \tag{2}$$

Using (2) in (1) and using a one-term Taylor expansion in the vibration components by utilising the smallness of the vibration amplitude $|\vec{\varepsilon}(t)|$, one obtains an explicit expression for the vibration induced neutron noise that contains the parameters $\mathbf{r}_p$ and $\vec{\varepsilon}(t)$. Switching to the auto- and cross-spectra of the neutron noise and the displacement components, the spectra of the detector signals at various positions $\mathbf{r}_i$, $i = 1,2,...$, will be related to the auto- and cross-spectra of the displacement components $\varepsilon_x(\omega)$ and $\varepsilon_y(\omega)$ as follows:

$$APSD_{\delta\phi_i}(\omega) = \frac{\gamma^2}{D^2}\left\{\left|G_{ix}(\mathbf{r}_p,\omega)\right|^2 S_{xx}(\omega) + \left|G_{iy}(\mathbf{r}_p,\omega)\right|^2 S_{yy}(\omega) \right.$$
$$\left. + 2\,\mathrm{Re}\left[G_{ix}(\mathbf{r}_p,\omega)\,G_{iy}(\mathbf{r}_p,\omega)\,S_{xy}(\omega)\right]\right\} \qquad (3)$$

and

$$CPSD_{\delta\phi_i\delta\phi_j}(\omega) =$$
$$\frac{\gamma^2}{D^2}\left\{G_{ix}(\mathbf{r}_p,\omega)\,G^*_{jx}(\mathbf{r}_p,\omega)\,S_{xx}(\omega) + G_{iy}(\mathbf{r}_p,\omega)\,G^*_{jy}(\mathbf{r}_p,\omega)\,S_{yy}(\omega) + \right. \qquad (4)$$
$$\left. G_{ix}(\mathbf{r}_p,\omega)\,G^*_{jy}(\mathbf{r}_p,\omega)\,S_{xy}(\omega) + G_{jx}(\mathbf{r}_p,\omega)\,G^*_{iy}(\mathbf{r}_p,\omega)\,S^*_{xy}(\omega)\right\}$$

Here, G_{ix} etc. is a shorthand notation for $\partial/\partial x_p\,G(\mathbf{r}_i,\mathbf{r}_p,\omega)$. Further, $S_{xx}(\omega)$, $S_{yy}(\omega)$ and $S_{xy}(\omega)$ are the auto- and cross-spectra of the vibration displacement components. As described in other publications, the possible variety of displacement component spectra are parametrized by two variables, an ellipticity (anisotropy) parameter $k \in [0,1]$ and the preferred direction of the vibration $\alpha \in [0,\pi]$ as

$$S_{xx} \propto 1 + k\cos 2\alpha \qquad (5)$$

$$S_{yy} \propto 1 - k\cos 2\alpha \qquad (6)$$

$$S_{xy} \propto \quad k\sin 2\alpha \qquad (7)$$

For an isotropic vibration $k = 0$, while vibration along a straight line has $k = 1$. Between these two extreme values the amplitude distribution of the vibration is an ellipse with the main axis lying in the direction α.

Based on the above, an algorithmic procedure may be elaborated by the unfolding of $\mathbf{r}_p$ from the measured neutron noise. Such a procedure was elaborated earlier [9]. The method was tested with both simulated and real measurement data.

As it was also discussed in earlier work, the inversion procedure (traditional algorithm) for control rod localisation has certain limitations and drawbacks [11]. It was therefore a logical idea to replace the inversion algorithm by artificial neural networks (ANNs) [11]. With the model described by (1)-(7), and using explicit expressions for the transfer function and the vibration parameters k and α, the neutron noise induced by the vibrations of all rods with various vibration properties can be calculated, and thus input patterns (neutron noise data) can be generated.

With these, a 3-layered backward propagation neural network can be trained to identify the vibrating rod. The results have shown that, similar to other applications, the use of a neural network is much faster, more effective and simpler than the traditional algorithm. The network identification was also tested in a real measurement with success. This identification was made on data taken in a Hungarian PWR when excessive vibrations of a control rod occurred. After training on simulated data the network identified the vibrating rod correctly.

At this point a general characteristic of neutron noise diagnostics need to be mentioned. The transfer function used so far, also in [9] and [11], was extremely simple. It has been based on the so-called power reactor approximation, which results in a simple and real solution in the form

$$G(r,\varphi,r_0,\varphi_0 = 0) = -\frac{1}{4\pi}\log\left[\frac{R^2 + (r\,r_0/R)^2 - 2\,r\,r_0\cos\varphi}{r^2 + r_0^2 - 2\,r\,r_0\cos\varphi}\right] \qquad (8)$$

where R is the radius of the bare cylindrical reactor. Physically, such a transfer function describes the spatial attenuation of the noise in a certain frequency range in reactors of a certain size, but it does not account for the phase delay of the signal propagation.

The use of simple transfer functions is generic in all core diagnostic methods where, traditionally, analytical solutions have been preferred. The main reason for this is that traditional unfolding is much easier and faster if an analytical solution of the direct task is available. In certain cases lack of such an explicit expression may completely prohibit the construction of an unfolding algorithm. A final reason is that there are no codes or methods available that could calculate the frequency-dependent complex transfer function for a realistic inhomogeneous core, partly because their usefulness in traditional unfolding methods is not clear.

One advantage of using neural networks in noise diagnostics, such as rod localisation is, however, that one is not constrained to use simplified core models for the transfer function. For the ANN-based method to work, the transfer function need not be either analytic or simple. Thus with the application of neural networks, the class of diagnostic problems that can be tackled becomes significantly larger than before. This also led to the development of codes that calculate the transfer function of realistic cores [12].

While such complex transfer functions for real cores are under development, it is worthwhile to study the performance of ANNs for somewhat more complicated transfer functions, which can be still calculated analytically. Results of such a study will be reported here. We have applied a complex transfer function in which neglection of the imaginary terms are not neglected. The transfer function used in the present study is given by the following expression:

$$G(r,\varphi,r',\varphi'=0,\omega) = \frac{1}{4}\Big[Y_0(B(\omega)|\mathbf{r}-\mathbf{r}'|)$$
$$-\sum \frac{\delta_n Y_n(B(\omega)\,R)\,J_n(B(\omega)\,r')}{J_n(B(\omega)\,R)} J_n(B(\omega)\,r)\cos(n\varphi)\Big] \tag{9}$$

where $\delta_n = 1$ for $n = 0$ and $\delta_n = 2$ for $n > 0$. Further, φ is the angle between $\mathbf{r}$ and $\mathbf{r}'$. $B^2(\omega)$ is the frequency dependent buckling which contains the static buckling and the zero reactor transfer function and is therefore complex ([1,13]), thus making also complex. Physically, this means that the phase delay effects in the signal propagation are also accounted for. The main purpose of this section is to report on the performance of the ANN-based localisation technique with this more realistic, complex transfer function. More details can be found in a current publication [13].

In the ANN based procedure, a large number of neutron noise spectra are generated by (3) and (4), corresponding to various rod positions and vibration parameters. These should be varied over all cases that may occur in the concrete reactor for which the network is trained. The noise spectra are given as input and the network should supply the rod position (or rather, the identity of the control rod out of a discrete set of rods in the core) as output.

In this study, similarly to the preceding Section, a standard three-layered feed-forward network with backward error propagation was used. The structure of the network used in the present paper is shown in Fig. 5, and the core geometry, number and position of the control rods and the detectors is shown in Fig. 6. The input nodes consist of the neutron detector auto-spectra and the real and complex parts, or absolute value and phase, of the cross-spectra. The input data are normalised such that they all lie within the same range, usually between 0.1 and 0.9. The number of the nodes is determined by the number of detectors used. In the present study 4 detectors have been used, because adding one more detector to the required minimum leads to a substantial increase in performance. Thus the number of input nodes with 4 detectors is 16 (4 auto and 2x6 cross-spectra). The number of output nodes is equal to the number of the control rods with binary outputs 0 and 1. Since in the previous work the method was tested in a core with a control rod bank consisting of 7 rods, the same structure was used here, giving 7 output nodes. All output nodes are assumed to yield a zero output except one, which indicates the identity of the vibrating rod.

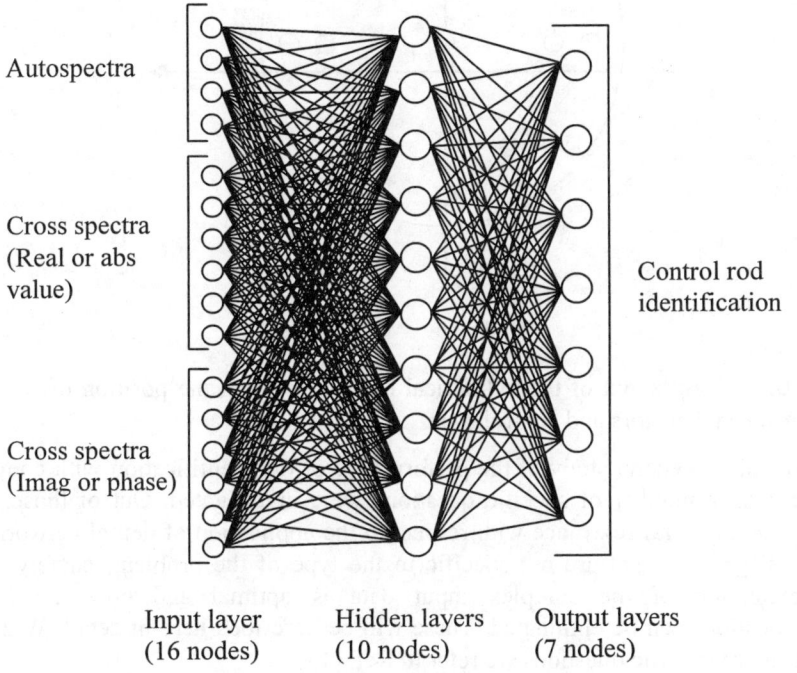

Autospectra

Cross spectra
(Real or abs
value)

Control rod
identification

Cross spectra
(Imag or phase)

| Input layer | Hidden layers | Output layers |
| (16 nodes) | (10 nodes) | (7 nodes) |

Fig. 5.   Structure of the implemented neural network

A performance test, similar to the previous work [11] was performed. Such an investigation consists of training of the network with simulated data (training patterns) and a subsequent test of it by further data in which the success and reliability ratios of its performance were investigated. In a general case, such an investigation includes even finding the optimum value of some freely adjustable parameters such as the learning and momentum rates and the number of hidden nodes. However, for the sake of comparison, the same values of these parameters were used as in earlier work. This also means that the network was not specifically optimised for the present investigation.

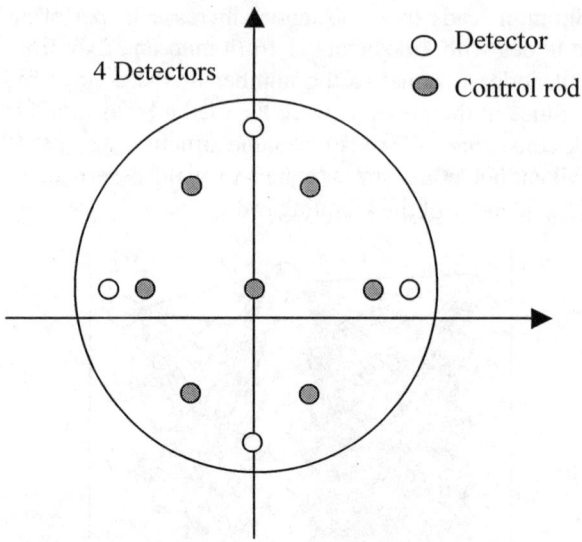

Fig. 6.   Core layout of the cylindrical reactor showing the position of neutron detectors and control rods

Apart from a general study of the performance of the identification with complex input data, a number of specific questions were investigated. Out of those, two have some general relevance with respect to the application of neural networks in plant diagnostics that are not specific to the type of the problem, namely what representation of the complex input data is optimal and how the faulty identifications can be eliminated. These will be described here in detail. With the other, more specific questions we refer to Ref. [13].

4 Choice of Input Data Representation

What regards complex input data, there are two choices, namely either absolute value and phase or real and imaginary parts can be used. It is not clear in advance which one is more effective. By performing calculations for both cases, it was found that if amplitude and phase values are normalised separately, such that the available range is better made use of for all signals, then the use of amplitude and phase is more advantageous than using real and imaginary parts. This is mostly because it requires fewer training cycles for a given user-defined value of the total r.m.s. output error. The result is demonstrated in Table 1 where both input data representations are compared quantitatively with respect to training cycles, success ratios, etc.

This conclusion may appear intuitively logical since the amplitude and phase have clear physical meaning, and they both have a monotonic dependence on the distance between the noise source and the detector. However, neural networks work purely algorithmically and from the algorithmic point of view the answer cannot be predicted by intuition. Nevertheless, it turned out that the amplitude and phase suit also the ANN-based localisation.

Table 1. Results of the efficiency of the trained network for several frequencies. The number of input pattern pairs used is 10 000 and the user-defined value of the total rms output error is 0.05. The success and reliability ratios were determined using $4x10^4$ new input data after the training was completed

| Frequency [rad/s] | Input data representation | | | | | |
| | Amplitude and phase of spectra | | | Real and imaginary parts of spectra | | |
	Training cycles	Success ratio [%]	Reliability ratio [%]	Training cycles	Success ratio [%]	Reliability ratio [%]
0.005	137	99.5	57.3	136	99.0	95.2
0.01	53	99.2	94.5	48	99.1	95.2
0.05	20	99.7	99.1	57	99.3	97.2
0.10	45	99.6	98.2	94	99.9	99.1
0.50	190	99.5	94.9	397	99.6	98.8
2.0	184	99.6	82.3	456	99.7	99.1
5.0	188	99.6	93.8	460	99.7	99.1
20.0	178	99.3	96.6	467	99.6	98.7
50.0	91	99.6	95.7	405	99.9	99.4
100.0	43	99.6	96.8	130	99.9	97.9
200.0	14	99.1	93.5	280	99.9	84.7

5 Investigation of the Erroneous Identifications of the Trained Network

In order to investigate the distribution of faulty (erroneous) identifications it was suggested in an earlier paper, see [11], that one way to improve the efficiency of the neural network would be to check whether the faulty identifications constitute a special subset in the values of anisotropy (k) and preferred direction of the vibration (α). The idea is that if the k and α parameters, belonging to the cases when the localisation failed, are clustered around a few centres, then this information can be used to increase the confidence of the localisation procedure. Namely, the identifications with parameters lying around those centres can be discarded as "unreliable," and thus the ratio of successful identifications is increased. This increase of the confidence is important, since earlier work showed that a trained ANN has a certain, even if low, failure rate (erroneous identification of the vibrating rod). Since power reactor diagnostics deals with individual events, it is vital to decrease the failure rate to, say, below 10^{-6}. This is possible with the methods described below.

In Fig. 7, the distribution of faulty identifications is shown w.r.t. control rod number (1 to 7), k and α for the first three cases in Table 1. It can be seen that the faulty identifications for all three cases are indeed clustered around a few control rods and direction values. Another interesting result is that faulty identifications occur only for very high k-values. Since the cases with high k-values correspond to extreme anisotropy which are unlikely to occur in reality, this means that the efficiency of the network can be improved by just neglecting these extreme values in both the training and the identification.

In addition to the above, a number of other questions were investigated. These were the following: a) dependence of the performance on the frequency; b) effects of detector geometry on the performance; b) use of various approximations of the transfer function; c) inclusion of the parameters k and α directly into the network as parameters to be identified. These are relatively specific questions that fall outside the scope of this Chapter. They are described in Ref. [13].

The main conclusion of this study is that for the ANN based localisation technique a more complex transfer function does not constitute a difficulty. The possibility of using neural networks in localising a vibrating rod in a real case has been already proven earlier [11]. Thus application of ANN based neutron noise diagnostics is expected to have more application with the development of more realistic transfer functions. Such a work is already under way [12].

6 Determination of Mass Flow of Water with Pulsed Neutron Activation and Neural Networks

Determination of mass transport or volumetric flow of water is important in all current energy producing systems for determination of produced energy. This is because the most accurate and reliable way of energy or power measurement is the calorimetric one. The energy is determined from measurements of temperature difference and mass flow of water.

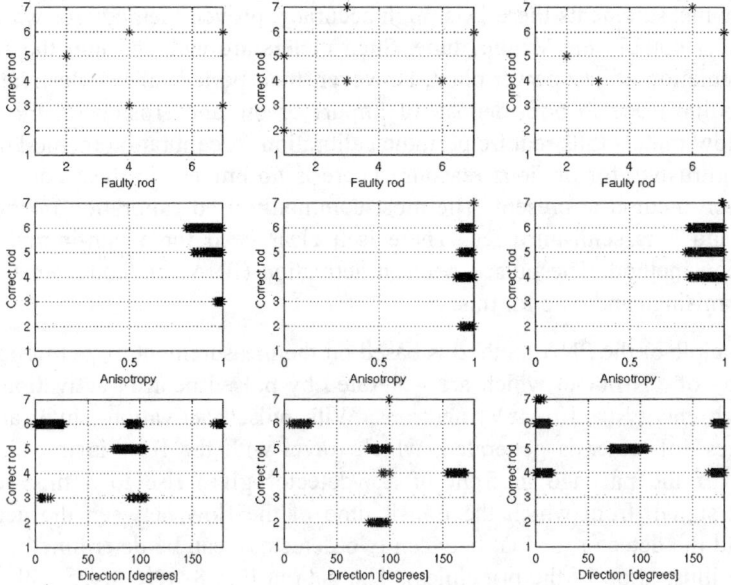

Fig. 7. The distribution of faulty identifications w.r.t. anisotropy (ellipticity) and preferred direction of the vibration for the first three cases in Table 1

This is the case even with nuclear reactors, whose instantaneous power level needs to be continuously monitored for safety and economical reasons. The motivation for the present work arose actually for the need of accurate measurements of reactor power through measuring the feed-water flow (and of course inlet and outlet temperature) in nuclear reactors. Precise knowledge of produced power and instant power level is necessary not only for operational and economical reasons but even to normalise nuclear calculations for burnup, transuranium production etc.

One would think that the power level of a nuclear plant could be accurately measured by pure nuclear methods, i.e., by in-core measurements of the neutron flux. However, there are several reasons why this is not the case, and nuclear measurements could easily result in 5 to 10% error in total power determination. On the other hand, a calorimetric power determination is in general accurate to better than one percent, basically because both mass flow and temperature are integral measurements (in contrast to the local signal of a neutron detector) that are performed outside the core.

For flow measurements there exist high accuracy, proven methods for water mass flow, most notably the Venturi tube. Such meters are included into the standard instrumentation of any power plant. However, their performance deteriorates with time (fouling) due to both deposit of impurities on, and erosion of, the orifice. These flow-meters thus require periodic calibration. A calibration method needs to be non-intrusive, for obvious reasons. There is no universally best non-intrusive calibration method at present. The most commonly used calibration method uses tracers and it is semi-intrusive. There is a clear need for a non-intrusive flow calibration method. The pulsed neutron activation (PNA) method seems to be a very promising candidate for this.

The principle of the PNA method is based on the measurement of γ-emission from the decay of ^{16}N nuclei which are generated by pulsed neutron activation of ^{16}O with high energy (> 10 MeV) neutrons. With pulsed activation, small activated "packages" of water is generated, which travel with the flow in the pipe. The passage of the package in front of a γ-detector gives rise to a time-resolved detector signal, from which the transit time of the flow between the activation point and the detector, or that between two detectors, can be determined (see Fig. 8 for an illustration of the principles). The data in Fig. 8 were taken with a pulse length of 100 ms, and application of a few hundred pulses. The measurement was performed at Chalmers in the project FlowAct and is described in more detail in [14,15].

The possibility of extracting mass flow data from PNA measurements has been investigated in a number of works, both theoretically and experimentally. The experimental proof-of-principle is relatively straightforward. However, the extraction of the mass flow from the detector signal with high precision is not a simple task. The difficulty is, as seen from Fig. 8, that the measurement serves a function, the time resolved detector signal(s), whereas one is only interested in one single parameter, i.e., a functional of the measured function. This parameter is the average time delay of flow between the pulse and the detector or between two detectors, which, together with the distance between the corresponding points, gives an area-averaged velocity from which the mass flow can be determined by multiplying with the pipe cross section.

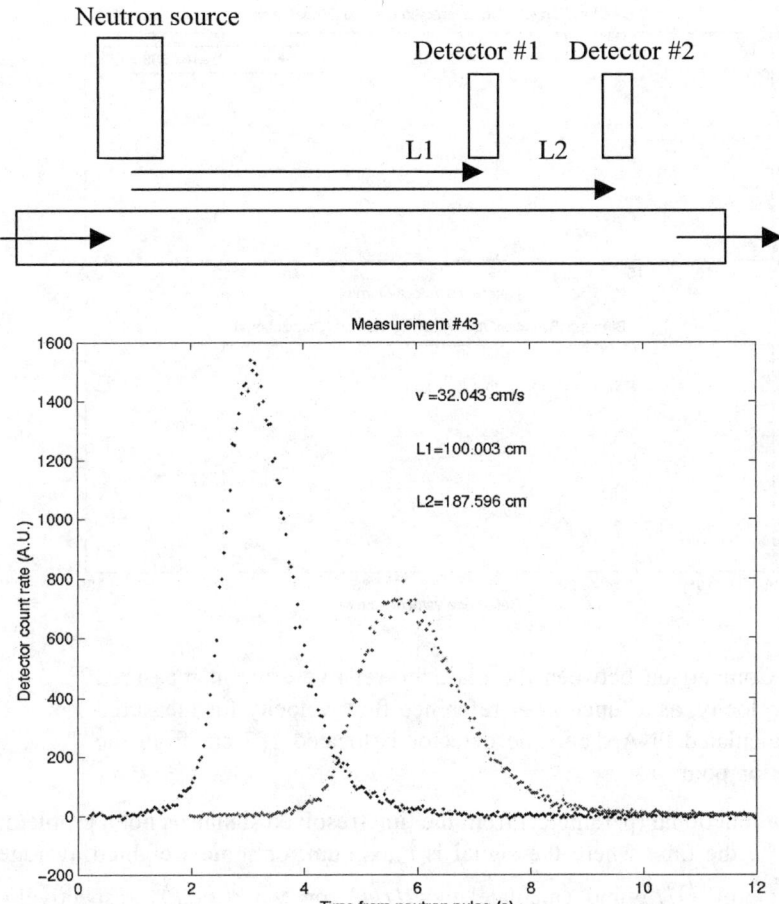

Fig. 8. The principle of the pulsed neutron activation method to measure water flow in a pipe. The water is activated by a pulsed neutron source and downstream the activation position one or more gamma detectors are located. The recorded activity time distribution from a measurement with two detectors can be seen in the lower part of the figure

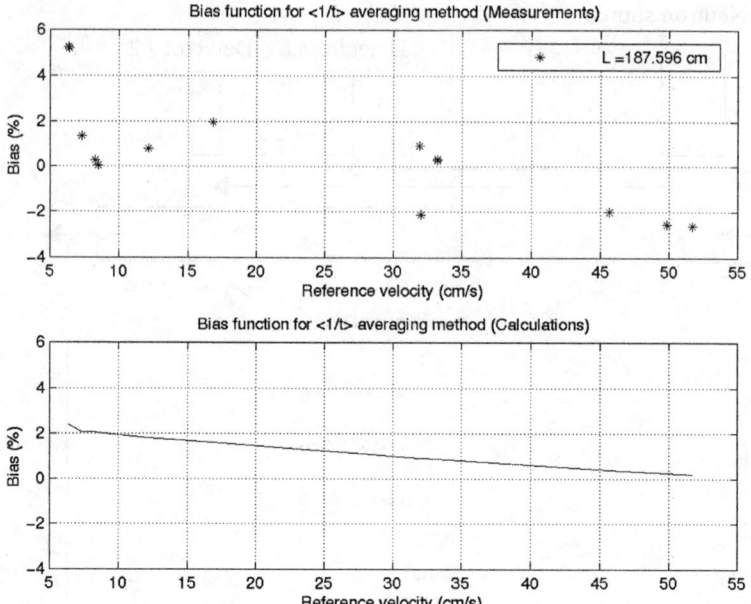

Fig. 9. Comparison between the bias, the relative error in measured flow velocity, as a function of reference flow velocity for measured and calculated PNA data. The detector is located 187 cm from the activation point

To derive a functional (parameter) from the time-resolved signals is not a problem: one can take the time where the signal is maximum, or some weighted average such as $\langle t \rangle$ or $\langle 1/t \rangle$ and calculate $v = d/\langle t \rangle$ or $v = d \langle 1/t \rangle$, respectively. Experience shows however, that the mass flow, when calculated by any of the existing time averaging methods, has a certain error in the several percent range. In addition, the error is not an absolute constant or constant fraction of the true value, rather it is a function of the flow and measurement parameters, such as flow velocity, source-detector or detector-detector distance etc. These errors, as functions of the flow or measurement parameters, were coined "bias functions" and were determined extensively recently in a project performed at Chalmers by the present authors (project FlowAct, [14,15]). An example of such measured bias functions is shown in Fig. 9. (the bias functions from calculations in the Figure will be explained later). The bias function shown corresponds to the $\langle 1/t \rangle$ type averaging method, which is generally known as the best. It is seen that the error of the measurement is still in the several percent range, much larger than the required better than 1% accuracy.

The reason for the failure of the time averaging methods is the complicated physics of diffusive transport of the activated volume of water with the existence of a velocity profile. In addition, the mass flow is defined by the area-averaged velocity profile over the pipe cross-section, whereas the PNA method determines a line-integrated average with an exponential weight function. Due to flow properties (diffusion, mixing, velocity profile etc.), the shape of the activated volume will be distorted and its axial extension widening all the time during the transport. The consequence is that the detector signals are asymmetric and become wider with increasing distance from the activation point.

The failure of the time averaging methods is just a consequence of the lack of a simple analytical (and hence invertible) relationship between the time-resolved detector signals and the mass flow. The reason is, as usual, causality; it is the flow and irradiation properties that determine the detector signals and not the other way round. This also means however, that through advanced numerical calculations, it is possible in principle to calculate the detector counts belonging to a given flow structure and irradiation geometry. The necessary tools or numerical packages are available. The neutron and γ transport can be calculated for arbitrary geometries and materials by Monte Carlo simulation codes, such as MCNP [15]. Computational fluid dynamics (CFD) methods and codes are also available, such as CFX, Fluent and Star-CD, for numerical simulation of the turbulent transport of the activated volume.

This means that we clearly have a case similar to that described in the preceding Section, determining of control rod elevation from axial flux shape through training a neural network on simulated data and applying it on measurements. The same strategy can be executed here. There is however one significant difference. In the case of rod positioning, there are high quality codes developed and tested for the calculation of the neutron flux (detector signal), i.e., exactly the same quantity that is measured. In the flow measurement method, there are no complete (and tested) codes that calculate the detector signals as measured by the PNA experiment. As mentioned above, high quality tested codes are available for the two main ingredients, i.e., neutron and γ transport and flow calculations, respectively. Their combining and benchmarking for real PNA measurements is yet to be done. This appears however fully possible and work is going on in this direction.

It appeared however worthwhile to perform a proof-of-principle type study of the applicability of the above strategy by developing a model of the PNA detector signals based on components that perform a simplified calculation of both the particle and the fluid transport part. Since a neural network trained on data from the simplified model would not be applicable to evaluate real measurement, the purpose is to study the inherent sensitivity of the method e.g., on flow conditions, external perturbations etc. The study is performed such that the trained network is tested on new data that are still produced with the same simulation model. This is again the same strategy as also was applied in a previous Section. If the conceptual study yields satisfactory results (meaning that the parameter to be determined depends sufficiently sensitively on the quantities measured, which is not granted in advance) then for the applicability of the method it only remains to develop the simulation model such that it agrees with measurements sufficiently well. Of course in order that such a study gives relevant results, it is essential that the simplified model still yields results that qualitatively reproduce all important features of the measurement.

Such a model was developed and tested by using a simple first collision and uncollided flux approximation of the neutron and γ transport, and a convective-diffusive model of turbulent flow, implemented in a finite difference scheme based on the diffusion equation for a soluble matter in a fluid. These calculations are described in more detail in [15,16], here we only give a description of the results by its use.

As said before, it was required that the model reproduces the important features of the experiments. This was verified by two separate means. A natural requirement is that by using the same flow and measurement parameters in the calculations and in a measurement, the calculated values of the detector signal time distributions must be similar to the measured ones, including the asymmetry of the curves and an increasing widening with increasing distance from the source. Such a comparison is shown in Fig. 10 below, showing a good qualitative agreement.

A second, more stringent requirement was that if the $1/t$ averaging method is applied to the simulated signals, it should show a similar trend of the bias function on flow and detector parameters as the measurement, i.e., it should produce an error similar to the measurement. If this is achieved, then one can trust that the effects of turbulent diffusion and velocity profile are faithfully reconstructed, and thus the model contains the same type of information in the signal profile as the measurement. This requirement was also fulfilled, as it is seen in Fig. 9, displaying both measured and calculated bias functions.

Thus the model was found satisfactory to conduct a feasibility study of the application of neural networks to determining mass flow from PNA measurements. This study was made along the same lines as the one reported in the previous section. The only difference compared to all previous cases is that it proved to be necessary to employ a four-layered network, with two hidden layers. The other parameters of the network were as follows. The input nodes take the detector counts in the scaler channels, thus the number of the nodes is equal to the number of channels with non-zero counts. In addition, a few extra input data were used, obtained by time averaging from the detector signals, namely $\langle t \rangle$ and $\langle 1/t \rangle$, because their use speeds up the training in certain cases. In the investigations reported here we used $N_{inp} = 300$ as the number of input nodes. The output layer consists of a single node, which is expected to give the mass flow rate. The number of hidden layers and also the number of the nodes in the hidden layers is not uniquely determined, their optimum value is usually found in an application with a trial-and-error procedure. In the calculations reported here we have selected two hidden layers with $N_{hid} = 30$ in the first hidden layer and $N_{hid} = 20$ in the second layer. A scheme of the network is shown in Fig. 11.

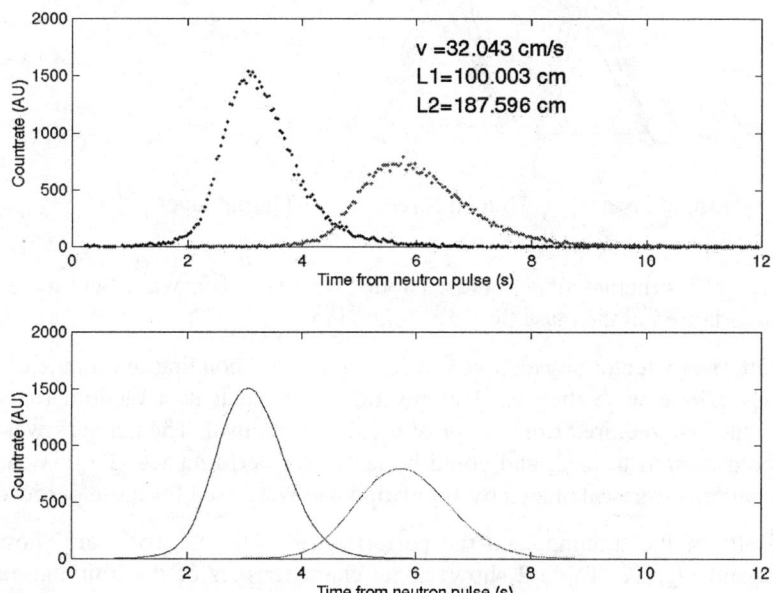

Fig. 10. Comparison between measured and calculated time-resolved detector signals. The detectors are located at 100 and 187.6 cm from the source. The reference velocity is 32.04 cm/s. The area under the calculated signal is normalised to the measured area

Although there are several parameters, most important the flow velocity and detector distance, the training was made with fixed detector distance and varying mean flow velocity. This corresponds to an experimental case where detector distance can be chosen by the experimenters, but the flow velocity is unknown and thus need to be treated as a variable. As usual, expected measurement errors can be modelled by adding suitably selected random numbers ("noise") to the "clean" simulated data. For one case, that is for a given detector distance, 44 different

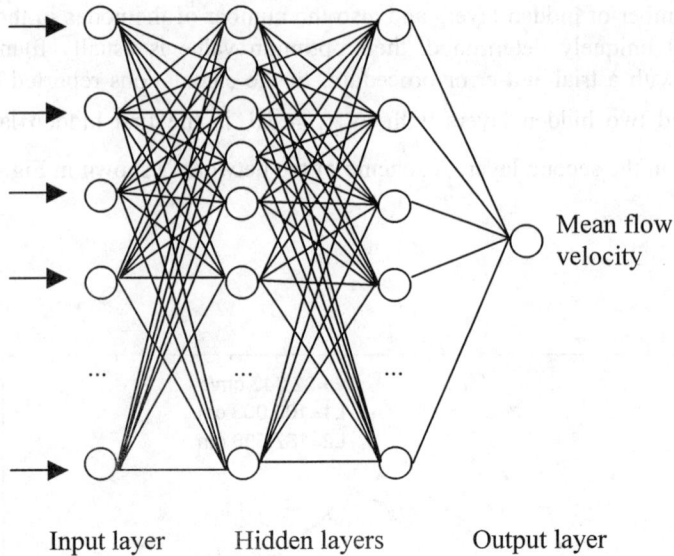

Input layer   Hidden layers   Output layer

Fig. 11. The structure of a standard four-layered feed-forward neural network used in the mass flow det·rmination

input patterns (detector signals) were calculated, corresponding to a range of flow velocities. These were then used in several cycles each in a random repetitive manner until the required r.m.s. error of 0.1% was attained. The network was then considered as well trained, and could be tested for performance. To this end, 12 further patterns were calculated by simulation and were used for these purposes.

The results of the training, and the performance of the network, are shown in Table 2 and Fig. 12. Table 2 shows some characteristics of the training, among others the training cycles necessary to achieve a r.m.s. error of 0.1%. It is seen that with noise added, the number of the training cycles increases, but the performance of the trained network deteriorates only very slightly. It is also seen that including a few extra input parameters in addition to the raw detector counts, such as $\langle t \rangle$ and/or $\langle 1/t \rangle$ in some cases speeds up the training. The table also shows the network performance in a condensed way, by displaying the standard deviation of the error taken over all test samples. It is seen that the standard deviation of the network identification is in the order of 1%, which agrees with the target value of the practical flow-meter, to be developed in the FlowAct project.

The performance of the trained network is seen in more detail in Fig. 12. As seen from the Figures, the precision of the network is about 0.5% when no noise is added, and between 0.5% and 1.5% when noise was added. There is no tendency or correlation between the magnitude of the error and the value of the flow velocity. The error is largely due to the statistical uncertainty of network performance, which is not specific to the task to be solved.

Table 2. Summary of the results from the neural network calculations

Input characteristics	Noise amplitude added (%)	# iterations	Average relative error (%)	Standard deviation of relative error (%)
Time distribution only	0	9218	0.0015	0.1402
	1	15905	-0.0634	0.3097
	3	16347	-0.0535	0.6030
	5	14963	-0.1698	0.9103
	10	9982	-0.4724	1.6173
<1/t> weighting velocity and time distribution	0	3785	-0.1874	0.1816
	1	12791	0.1145	0.2696
	3	15027	-0.3577	0.5299
	5	13473	-0.7383	0.8126
	10	11108	-1.2440	1.2550
<t> weighting velocity, <1/t> weighting velocity and time distribution	0	5487	-0.2060	0.2194
	1	10347	0.0553	0.2450
	3	15483	0.0440	0.6761
	5	12853	-0.1811	1.2515
	10	10936	-0.4476	3.0241

It was thus demonstrated that it is possible to construct and train a network such that the true flow rate can be determined with an accuracy of 0.5%. This possibility is provided mainly by the fact that the time-resolved detector signals contain information on the flow structure, and the neural network can utilise this information.

A more detailed comparison of the simulated and measured data in Fig. 10 reveals that the accuracy of the simulation is not good enough to reconstruct measurements with less than 1% error. This means that if using the network with measured data, as opposed to using it with further simulated data, the error of the network output is in the range of 1-5%. This was actually confirmed by using measured data as input to the trained network. Thus in order that the method be applicable in real measurements, the simulation method need to be improved. This is nevertheless fully possible since, as mentioned before, modules for high accuracy simulation of both particle transport as well as fluid flow are available. They need to be compiled into one dedicated package and tested for the reproduction of the PNA measurements. Work in this direction is underway.

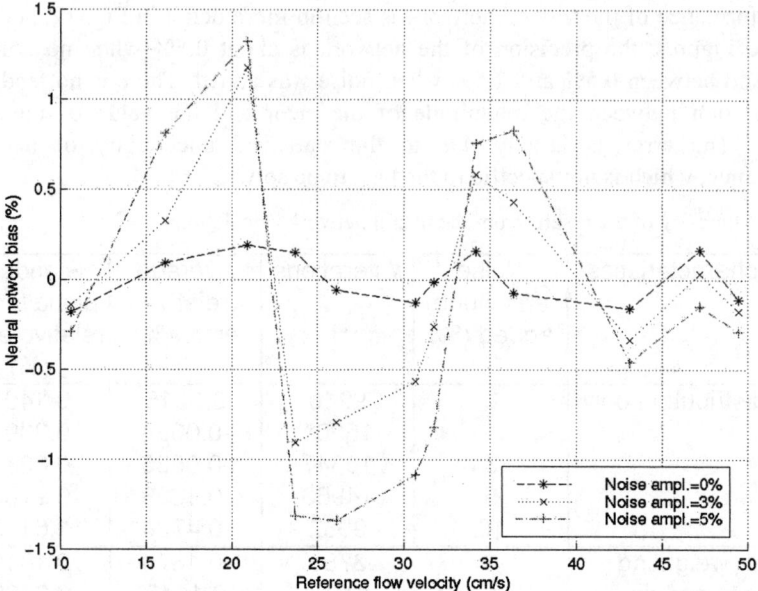

Fig. 12. Bias function for an artificial neural network trained with the time resolved detector signal only. Three cases with different noise amplitude added to the detector signal are shown. No correlation between the magnitude of the bias and the reference flow velocity can be seen

7 Conclusions

In this chapter, a few concrete applications of artificial neural network techniques in nuclear engineering were described. The problems were taken from the operational needs of pressurized and boiling water reactors. The solution in all cases relies on the fact that the measurement data, from which the neural network needs to estimate a source parameter, i.e., control rod elevation, control rod position and mass flow of water, can be simulated by high accuracy calculational codes, that take the searched parameter as input. With the combination of high accuracy numerical simulation and ANN techniques, real problems at an operating plant can be solved with minimum interference with plant operation. This strategy opens up a wide field of diagnostic and monitoring problems, which were hitherto hindered by the fact that only analytically tractable and invertible simulations could be applied. At the same time application of ANN techniques promotes the development of further high accuracy core simulations, notably the calculation of the complex dynamic transfer functions of inhomogeneous cores. Development of such codes went very slow in the past because of the difficulties of using numerical simulations in conventional diagnostic algorithms.

References

1. I. Pázsit and M. Kitamura, *The role of neural networks in reactor diagnostics and control*. Advances in Nuclear Science and Technology **24**, 95 - 130 (1996)

2. N. S. Garis, I. Pázsit, U. Sandberg and T. Andersson, *Determination of PWR control rod position by core physics and neural network methods*. Nuclear Technology **123**, pp 278 - 295 (1998)

3. I. Pázsit, Development of Core Diagnostic Methods and their Application at Swedish BWRs and PWRs. Submitted to J. Nucl. Sci. Technol.(1999)

4. M. Edenius, K. Ekberg, B. H. Forssén and D. Knott, "CASMO-4 - A Fuel Assembly Burnup Program", User's Manual. Studsvik of America, Studsvik/SOA-93/1. (Draft 2) (1993)

5. J. A. Umbarger and A. S. Digiovine, "SIMULATE-3 - Advanced Three-Dimensional Two-Group Reactor Analysis Code", User's Manual. Studsvik of America, Studsvik/SOA-92/01 (1992)

6. J. Hertz, A Krogh and R. G. Palmer, *Introduction to the theory of neural computing*, Addison-Wesley, Reading, Massachusetts (1991)

7. S. Haykin, *Neural Networks: A Comprehensive Foundation.* 2nd Edition, Prentice-Hall Inc., Upper Saddle River, New Jersey (1999.)

8. D. N. Fry, Experience in reactor malfunction diagnosis using on-line noise analysis. Nuclear Technology.**10**, p. 273, (1971)

9. I. Pázsit and O. Glöckler, On the neutron noise diagnostics of PWR control rod vibrations III. Application at power plant. Nucl. Sci. Eng. **99**, pp. 313-328 (1988)

10. J. A. Thie, *Power Reactor Noise.* ANS monograph, New York (1983)

11. I. Pázsit, N. S. Garis and O. Glöckler, *On the neutron noise diagnostics of PWR control rod vibrations IV: Application of neural networks.* Nucl. Sci. Eng. **124**, pp. 167-177 (1996)

12. J. K-H Karlsson, *A two-group dynamic transfer function for inhomogeneous reactors.* To be submitted to Ann. Nucl. Energy (1999)

13. N. S. Garis and I. Pázsit, *Control rod localisation with neural networks using a complex transfer function.* Prog. Nucl. Energy **34**, pp. 87 - 98 (1998)

14. P. Lindén, G. Grosshög and I.Pázsit, *FlowAct, Flow Rate Measurements in Pipes with the Pulsed Neutron Activation Method.* Nuclear Technology, **124**, pp. 31-51 (1998)

15. P. Lindén, *A Study of Using Pulsed Neutron Activation for Accurate Measurements Water in Pipes*, CTH-RF-136, Chalmers University of Technology, Göteborg, Sweden (1998)

16. J. Briesmeister (editor), MCNP - A general Monte Carlo N-particle transport code, Version 4A, LA-12625-M (1993)

17. P. Lindén and I. Pázsit, Study of the Possibility of Determining the Mass Flow of Water from Neutron Activation Measurements with Flow Simulations and Neural Networks. Kerntechnik **63**, pp 188-196 (1998)

13 Regularization Methods for Inferential Sensing in Nuclear Power Plants

J. Wesley Hines, Andrei V. Gribok, Ibrahim Attieh, and Robert E. Uhrig

Nuclear Engineering Department
The University of Tennessee
Knoxville, Tennessee 37996, USA
hines@utkux.utcc.utk.edu

Inferential sensing is the use of information related to a plant parameter to infer its actual value. The most common method of inferential sensing uses a mathematical model to infer a parameter value from correlated sensor values. Collinearity in the predictor variables leads to an ill-posed problem that causes inconsistent results when data based models such as linear regression and neural networks are used. This chapter presents several linear and non-linear inferential sensing methods including linear regression and neural networks. Both of these methods can be modified from their original form to solve ill-posed problems and produce more consistent results.

We will compare these techniques using data from Florida Power Corporation's Crystal River Nuclear Power Plant to predict the drift in a feedwater flow sensor. According to a report entitled "Feedwater Flow Measurement in U.S. Nuclear Power Generation Stations" that was commissioned by the Electric Power Research Institute, venturi meter fouling is "the single most frequent cause" for derating in Pressurized Water Reactors. This chapter presents several viable solutions to this problem.

1 Introduction

The safe and economical operation of Nuclear Power Plants requires knowledge of the state of the plant. This knowledge is obtained by measuring critical plant parameters with sensors and their instrument chains. The correct operation of the sensor systems must be validated to assure the safe, efficient operation of nuclear power plants.

Traditional approaches to sensor validation at nuclear power plants involve the use of redundant sensors coupled with periodic instrument calibration. Many periodic sensor calibration techniques require the process shut down, the instrument taken out of service, and the instrument loaded and calibrated. This method can lead to equipment damage, incorrect calibrations due to adjustments made under non-service conditions, increased radiation exposure to maintenance personnel, and possibly increased downtime. Since few of the sensors are actually out of calibration, the end result is that many instruments are unnecessarily maintained. While correct adjustment is vital to maintaining proper plant operation, an alternative condition based technique is desirable.

When implementing condition based calibration methods, the instruments are calibrated only when they are determined to be out of calibration. On-line, real-time sensor calibration monitoring will allow nuclear utilities to reduce the maintenance efforts necessary to assure the instruments are in calibration and increase the reliability of the components. The EPRI/Utility On-Line Monitoring Working Group estimates an industry wide cost savings of $40M to $290M over the next 20 years depending on the values applied to indirect benefits.

Inferential sensing is the prediction of a plant variable through the use of correlated plant variables. Most on-line calibration monitoring systems produce an inferred value and compare it to the sensor value to determine the sensor status. The system can be used to monitor sensors for drift or other failures making periodic instrument calibrations unnecessary. There are many methods used for inferential sensing including many types of regression, neural networks, fuzzy logic, and several statistical techniques. Several of these techniques produce believable results but can be adversely affected by collinearity of the predictor variables. This chapter investigates how collinearity adversely affects inferential sensing techniques by making the results inconsistent and unrepeatable; and presents regularization as a potential solution.

Sensor fault detection is a subset of fault detection and isolation that includes the detection of faults in other components. Several surveys on FDI technologies have been published including those by Basseville (1988), Gertler (1988), Isermann (1984), Patton (1991), and Willsky (1976). Many techniques have been studies for use in sensor fault detection in Nuclear Power Plants including expert systems [Qualls, Uhrig and Upadhyaya, 1988; Tsoukalas, 1992], model based techniques [Glockler, 1991; Grini, 1989; Hardy, 1992; Holbert, 1990; Kittamura, 1980], state estimation techniques [Black, 1998; Gross, 1997; Singer, 1997], artificial neural networks [Hines, 1997b; Kavaklioglu and Upadhyaya, 1994; Uhrig, 1998; Upadhyaya and Eryurek, 1992], fuzzy logic [Hines, 1997a; Holbert, 1995], and hybrid combinations of these techniques [Ikonomopoulos, 1992]. All of these techniques can be divided into two basic categories: physical model based techniques and data driven models. We will define physical model based techniques as those that use mathematical models developed from first principals while data driven techniques are constructed using data collected from the process.

This chapter will investigate four linear and several non-linear data driven methodologies for inferential sensing:

1. Linear Regression (LR)

2. Ridge Regression (RR)

3. Truncated Singular Value Decomposition (TSVD)

4. Partial Least Squares (PLS)

5. Neural Networks (NN)

These methodologies will be compared on the basis of prediction performance including their ability to model non-linearities, their ease of design, and their ability to handle collinear predictor variables while producing consistent results. The example used to make these comparisons will be that of inferential measurement of nuclear power plant feedwater flow. The data set is extremely ill-conditioned and will illustrate the necessity for the application of regularization techniques. Most inferential sensing problems are not this extreme and will be easier to solve, although regularization techniques should still be used.

1.1 Feedwater Flow Measurement

In the United States, a nuclear power plant's operating limit is tied to its thermal power production. The steam generators' feedwater flow rate is an input to the calculation of thermal power and therefore must be accurately known. The majority of pressurized water reactors (PWRs) utilize venturi meters to measure feedwater flow because of their ruggedness and precision. However, these meters are susceptible to fouling due corrosion products that are present in the feedwater. The fouling increases the measured pressure drop across the meters, which results in an over estimation of the flow rate. Consequently, the reactors' thermal power is also overestimated [Nuclear News, 1993]. To stay within regulatory limits, reactor operators are forced to derate their plants. According to a report entitled "Feedwater Flow Measurement in U.S. Nuclear Power Generation Stations" that was commissioned by EPRI (1992); venturi meter fouling is "the single most frequent cause" for derating in PWRs. The amount of derating, according to the report, varied from insignificant to 3% of full power. On average, the derating was between 1% and 2% of full power. It is estimated that derating an 800 MWe unit by 2% will cost the utility approximately $20,000 per day given a cost of electricity of $0.05/kWh.

Despite the susceptibility of the venturi meter to fouling, it is still the most common flow measurement instrument used in nuclear reactors. To overcome the loss of generating capacity, some utilities have developed a fouling coefficient or a correction factor to offset the degradation in the measurements' accuracy. The drawback of this method is that the flow measurement could be "corrected" when in fact there is no venturi fouling resulting in the reactor operating above its thermal power limits. The best solution to this problem would be to have an inferential sensing system that can accurately predict feedwater flow. This chapter presents several methodologies to perform that function.

1.2 Prediction Problems Caused by Collinear Data

Traditionally, the on-line prediction of instrument performance is based on the use of redundant sensors. The use of redundant sensors and other highly correlated measurements as predictor variables causes a potential problem in data based prediction models. This problem occurs due to the collinearity in the predictor variables.

Variables are collinear if the data vectors representing them lie on the same line (i.e., subspace of dimension one). More generally, k variables are collinear if the vectors that represent them lie in a subspace of dimension less than k; that is, if one of the vectors is a linear combination of the others. In practice, such "exact collinearity" rarely occurs due to process and measurement noise. A broader notion of collinearity is therefore needed to deal with the problem as it affects statistical estimation. More loosely, two variables are collinear if they lie almost on the same line; that is, if the angle between them is small. In the event that one of the variables is not constant, this is equivalent to saying that they have a high degree of correlation between them.

To show the degrading effects of collinearity on prediction, consider two redundant sensors (x_1 and x_2) that are operating in a noisy environment and being used to infer the value of a third redundant sensor using the common linear regression model of Equation 1.

$$x_3 + \varepsilon_3 = (x_1 + \varepsilon_1)w_1 + (x_2 + \varepsilon_2)w_2 \tag{1}$$

Since x_1 and x_2 are redundant sensors, their outputs are the same except for their respective noise term ε_i, and therefore, are almost perfectly collinear. Singular Value Decomposition (SVD) [Jolliffe, 1996] can be used to determine if the k=2 variables lie in a subspace of dimension less than two. This method performs a linear transformation of the data into a new coordinate system so that a maximum amount of the variation of the data is along one principal axis and the remaining variation is along the second axis. A mathematical simulation of this redundant sensor case resulted in singular values of 7.0871 and 0.0059. The square of the singular values is proportional to the amount of variance in the data. In this case over 99.9% of the variation of the original signals could be represented with a single signal. In other words, the 2 variables lie in a single dimensional subspace. This shows the high degree of collinearity in this ill-conditioned problem (the condition number is over

1,000). Solving for this prediction model using linear regression yields coefficients of: w_1 = -388.4 and w_2 = 389.4. These huge weights result in a solution that has a much larger variance than that of the predictor variables. This results in noisy and inconsistent predictions.

Using regularization, a technique presented in the next section, the following solution: w_1 = 0.47 and w_2 = 0.47 was generated. The optimal solution to the problem would be w_1 = 0.5 and w_2 = 0.5, which is an average of the two redundant predictors. Since the regularization solution is slightly biases towards a small norm resulting in weights that are slightly smaller than the optimal solution. This is the cost of regularization; the benefit of regularization is a stable inferred value with a smaller variance than that of the predictors.

2 Methodology

The development of a data based inferential sensing system consists of collecting training and testing data, preprocessing the data to remove outliers, and scaling the data to allow the use of statistical signal evaluation techniques. Once the data is collected and preprocessed, the inferential model is developed and tested. The final system is shown in Figure 1 where X is a vector of predictor signals, X_s is the preprocessed and scaled signal, and Y is the inferred signal.

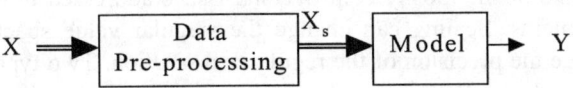

Fig. 1. Inferential Sensing System

Several inferential models can be used for inferential sensing including linear techniques such as linear regression, principal component regression, ridge regression, and partial least squares; and non-linear techniques such as non-linear regression, non-linear partial least squares, artificial neural networks, and fuzzy inference systems.

There are two primary categories of methods used to deal with the problem of collinearity. The first category transforms the predictors to a new orthogonal space thus removing the collinearity, while the second category, called direct regularization methods, deals with making the ill-conditioned problem a well-conditioned problem. The following matrix shows which techniques will be used with each inferential sensing method. We will investigate each of these methodologies in the following section.

Method	Predictor Transformations	Direct Regularization
Linear Regression	X	X
Partial Least Squares	X	
Neural Network		X

2.1 Data Preprocessing

Prior to any statistical evaluation of the data, a number of preprocessing techniques should be applied to the raw data to ensure consistent results. The most common preprocessing techniques are filtering and scaling.

2.1.1 Outliers

It is a well-known fact that least squares models are very sensitive to outliers. Just one outlier can significantly distort the estimation computed with a least squares method. In fact, a noise spike occurring at the same sample time, in two otherwise uncorrelated signals can result in a correlation coefficient of 0.95. To reduce measurement noise we suggest using a median filter because it has well known outlier rejection and fast digital implementation properties.

2.1.2 Data Scaling

Data scaling is one of the mostly controversial issues addressed in the regularization of ill-posed problems. Scaling can change the singular value spectrum of the data matrix and change the precision of the regularized solution. Two types of data scaling are commonly used: the first one is the well-known Z-score scaling which provides signals with zero mean and unit variance and the second one we term range scaling which linearly scales the data's range to between zero and one. In the following examples we will use range scaling and also mean center the data.

2.2 Regularization

The inferential measurement of feedwater flow sensor drift is based on the inference of actual feedwater flow rate. Actual flow is predicted through its relationship to other correlated plants parameters. The problem with using these parameters as predictors is that they are not only highly correlated with feedwater flow, but they are also correlated with each other. If this degree of correlation is quite high, the data matrix becomes ill-conditioned and the problem of drift detection becomes ill-posed in the sense of Hadamard (1923). Hadamard defined a well-posed problem as a problem which satisfies the three following conditions:

1. The solution for the problem exists.

2. The solution is unique.

3. The solution is stable or smooth under small perturbations of the data; i.e., small perturbations in the data should produce small perturbations in the solution.

If any of these conditions are not met, the problem is termed ill-posed and special considerations must be taken to ensure a reliable solution. To understand the essence of ill-posed problems for inferential sensing, let us consider the linear least squares model whose objective is to find a linear combination of predictor variables that accurately models the response variable.

$$\min\|Xw - y\|_2^2, \quad X \in R^{mxn}, \ m \geq n \tag{2}$$

Where X is a data matrix containing m samples of n predictor variables related to feedwater flow rate, y is a vector of measured values of feedwater flow and w is a solution of regression coefficients. A very valuable tool in the analysis of ill-posed problems is singular value decomposition (SVD) [Golub, 1996]. The SVD of data matrix X can be written as

$$X = U\Sigma V^T = \sum_{i=1}^{n} u_i \sigma_i v_i^T \tag{3}$$

where u and v are called left and right eigenvectors of X and σ_i are the singular values of the matrix X. In terms of the SVD, the solution for the least squares problem (2) can be written as:

$$w_{LS} = \sum_{i=1}^{n} \frac{u_i^T y}{\sigma_i} v_i \tag{4}$$

Here we assume that matrix X has a full rank of n. Equation (4) gives insight into the essence of ill-conditioning. The division by small singular values results in amplification of high-frequency oscillations of the right singular vectors of the data matrix X. To deal with ill-conditioned problems, several methods have been developed which essentially damp or filter out these high frequency oscillations. These methods are called regularization methods because they regularize or smooth potentially unstable least squares solutions. The simplest regularization method is the truncated SVD (TSVD) method. This method truncates the sum in Equation 2 at some value k<n, thus avoiding small singular values in the denominator of Equation 4. The two heuristics used in this method of regularization are as follows:

1. The singular values have a distinct gap in their spectrum. The location of this gap on the singular values curve can be a natural choice for the truncation parameter k.

2. The left and right singular vectors u_i and v_i tend to have more sign changes in their elements as the index i increases, i.e., as σ_i decreases [Hansen, 1997].

Heuristic 2 is only guaranteed to hold for totally positive matrices [Hansen, 1995]. A matrix is called totally positive if all its minors of any order are positive [Gantmacher, 1959]. Matrices that arise in most practical applications usually satisfy heuristic 2, but it is worth checking this property prior to the application any regularization method based on heuristic 2. This kind of regularization is especially appropriate for ill-conditioned problems having a large gap (say two orders of magnitude) between a pair of singular values σ_i and σ_{i+1}. These kinds of problems are called problems with well-determined numerical rank. Not all real-world problems have a well-determined numerical rank. If the singular value spectrum has no distinct gap, then the problem has an ill-determined numerical rank and the choice of truncation or regularization parameter is not as evident as in the former case. But as it was stressed in [Hansen, 1989], the success of TSVD depends on the satisfaction of the Discrete Picard Conditions (DPC) [Hansen, 1990] but not on the existence of a particular gap in singular value spectrum of the data matrix X. To deal with ill-conditioned problems having ill-determined numerical rank, the method of regularization due to Tikhonov (1963) can be used. In this method the minimization problem (2) is replaced by the following augmented functional:

$$
\min \left\{ \|Xw - y\|_2^2 + \lambda^2 \|Lw\|_2^2 \right\} \tag{5}
$$

where λ is a regularization parameter that controls the trade-off between smoothness of the solution and fitness to the data. L is a well-conditioned matrix; for example, a discrete approximation of the derivative operator. The main assumption behind Tikhonov regularization is that the solution should be smooth or non-oscillating. In the case of L=I, where I is identity matrix, the Tikhonov's functional (5) is said to be in standard form and is known in statistical literature as ridge regression [Hoerl, 1970]. In this case, we can write the regularized solution as:

$$
w_\lambda = \sum_{i=1}^{n} f_i \frac{u_i^T y}{\sigma_i} v_i \tag{6}
$$

where u_i and v_i are left and right singular vectors of the data matrix X, σ_i are singular values of this matrix and $f_i = \dfrac{\sigma_i^2}{\sigma_i^2 + \lambda^2}$ are filter factors. The role of filter factors is to suppress the contribution of minor components to the solution thus providing a more stable non-oscillating solution. In Tikhonov regularization the filter factors for large σ_i are close to 1 and for small σ_i they tend toward zero, thus providing necessary filtering of minor components. The heuristic 1 for Tikhonov regularization can be formulated as the singular value spectrum decays to zero without any particular gap in singular values.

Heuristic 2 is the same for Tikhonov regularization and the smoothness or stability of the regularized solution depends on satisfaction of the conditions of heuristic 2. It should be noted that in any practical situation the singular value spectrum does not decay to zero but levels off at some index "i" due to unavoidable errors in measurements or instrumentation noise. The noise level in both the right and left parts of Xw = y is a crucial factor for satisfaction of the Discrete Picard Condition and thus for the existence of a "good" regularized solution which is a reasonable approximation to a desired true solution.

With these theoretical considerations in mind we will now tackle the problem of drift detection in feed water flow instrumentation. This problem belongs to the so-called inferential or virtual measurements, which is an inverse problem, where the aim is to recover or "infer" about unknown parameters of a physical system from measurable noisy data.

2.3 Feedwater Flow Data

Twenty-four variables that were well correlated with feed water flow were selected as predictor variables for the inferential models. These variables include pump speeds, pressures, temperatures, levels, and other flow rates in the NPP secondary. A listing is included in the appendix. The inferential models are constructed using the 24 predictor variables to estimate the feed water flow rate. The "training" region for the models was chosen to be the plant start-up and the first few days of the fuel cycle. Since the feedwater flow venturi is removed, acid cleaned, and calibrated between fuel cycles, the venturi is assumed to be working correctly at the beginning of each cycle.

The training data consists of 601 samples of the 24 predictor variables recorded at 30-minute intervals. The response variable is the feed water flow rate recorded at the 601 sample times. The training data sets are median filtered and scaled. The resulting training data have linear correlation coefficients above 0.99 showing the data is very linearly correlated with each other and with feedwater flow.

A principal components analysis (PCA) [Jolliffe, 1996] shows that the first principal component contains 99.93 % of the variation of the training data. This result indicates that the data is extremely collinear because all of the predictors lie along the same axis and indicates that the prediction problem will be ill-conditioned. Therefore, this inferential prediction problem will exhibit the instability problems caused by collinearity.

3 Results

Solutions to inferential sensing problems that are highly collinear, and thus ill-conditioned, are inconsistent and unstable. The results may be dependent on the selected training data used to construct the inferential model, the architecture of the inferential model, and the initialization of the inferential model. This raises concerns about the stability of the feedwater flow drift estimation and the reliability of inferential measurements. To evaluate the consistency of the inferential drift estimation system, a bootstrap technique was used [Efron, 1982].

The bootstrap technique is a statistical method used for evaluating the stability prediction models. From a training data set of sample size n, the bootstrap technique samples values from both predictor and response variables at random with replacement, thus providing a bootstrap sample of size n but with some original values duplicated and some missing. Each bootstrap sample set is used to construct an inferential model to estimate feedwater flow. When the method is repeated a large number of times, the bootstrap procedure provides a set of solution models whose consistency can be evaluated by forming a probability density function of their drift estimates at a specific time. More specifically, 100 bootstrap samples were generated from the original data and used to develop 100 inferential models to predict the feedwater flow sensor drift value six months into the fuel cycle. This method will determine if the inferential model is influenced by small changes in instrumentation noise.

3.1 Linear Regression Models

The linear regression model's feedwater flow prediction and the probability density function for the drift estimates are shown in Figure 2. The PDF for the drift value has a mean of 39.9 klb/hour a large variance (standard deviation of 8.3 klb/hour). These results show the inconsistency of drift estimation using the ordinary least squares model. The instability of drift estimation is due to the ill-conditioned nature of the problem; and as a consequence, due to the high sensitivity of the ordinary least squares solution to small perturbations in the data. To stabilize the drift prediction, a regularization method should be used to alleviate the ill-conditioning problem.

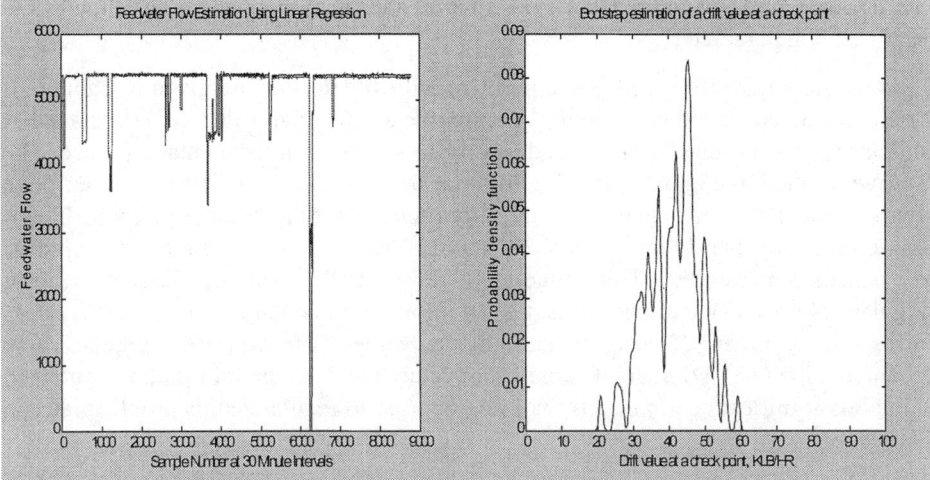

Fig. 2. PDF of feedwater drift using OLS

First, the standard form of Tikhonov regularization, ridge regression, was used to regularize the problem. Prior to applying this form of regularization, the regularization parameter λ must be chosen to resolve the subtle compromise between the smoothness of the regularized solution and solution bias. This biasing towards small regression coefficients is the "price" paid for the smoothness and stability of the regularized solution. We want to get as much smoothness as possible without significantly biasing our solution.

Several methods of such optimal "payment" have been proposed. The principle of discrepancy by Morozov (1984) requires the knowledge of the right hand side error (e) of equation Xw=y+e. When a good estimation of e is available, this method yields a good regularization solution. Two other highly regarded methods for regularization parameter selection do not assume any knowledge about the error level but are based on the extraction of information from the data. The generalized cross-validation method [Golub, 1979] is based on the assumption that if an arbitrary element y_i (of the right-hand side y) is removed, then the corresponding regularized solution should predict this observation well [Hansen, 1994]. However, the most common method of determining the regularization parameter is the L-curve method [Hansen, 1992].

The L-curve is a plot of the residual norm versus the solution norm. The residual norm is comprised of error that cannot be reduced by the model and bias due to regularization. The solution norm is a measure of the size of the regression coefficients. As the regularization parameter (λ) is increased, the regression coefficients are reduced, making the solution smoother, but bias is added to the solution. The best solution occurs when there is little bias but the solution is smooth. The optimal regularization parameter is found by locating the "corner" of the L-curve.

The L-curve method is reliable and simple and is implemented to choose the regularization parameter.

The singular values for the ill-posed problem of drift detection are given in Table 1. It should be noted that there is only one significant singular value (4.97), the other singular values are much smaller and mostly reside in the instrumentation noise. The L-curve, plotted as Figure 3, is a log-log plot because the singular values span three orders of magnitude. An analysis of the curve shows that the best λ, corresponding to the "corner" of the L-curve, is 4.37, so this value was chosen as the optimal regularization parameter. This value is slightly smaller than the least significant singular value (4.97) and tends to damp out information contained in the eigenvectors corresponding to the 23 insignificant singular values. The truncated singular value decomposition (TSVD) method, which completely removes the information contained in the less significant components, will also be used to regularize this problem later in this section.

Table 1. Singular Values

4.9737	0.0101
0.0773	0.0073
0.0461	0.0069
0.0379	0.0066
0.0307	0.0061
0.0204	0.0047
0.0194	0.0043
0.0170	0.0034
0.0164	0.0031
0.0140	0.0030
0.0121	0.0028
0.0111	0.0023

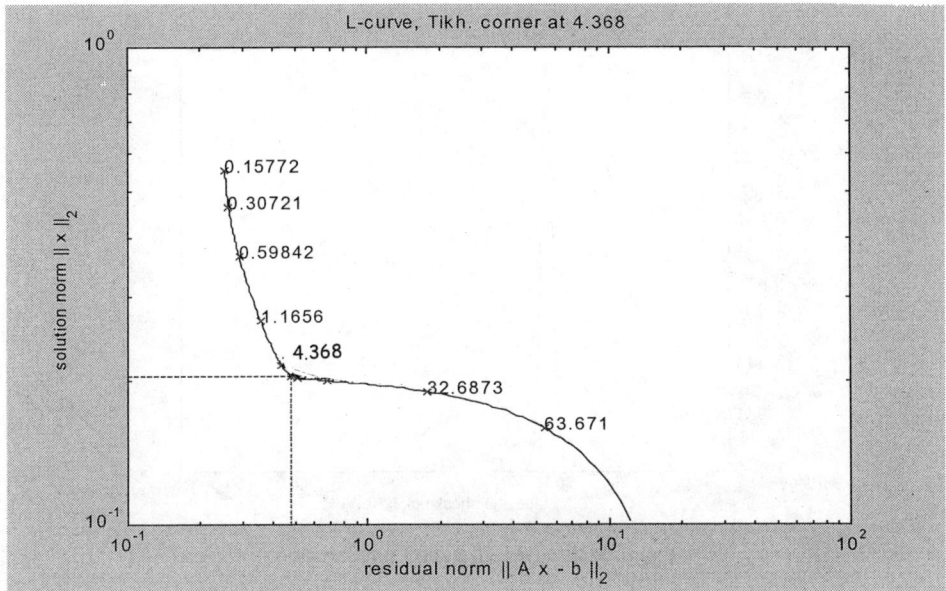

Fig. 3. L-curve using Tikhonov regularization

The optimal regularization parameter is used to regularize the solution and the bootstrap procedure is used to test the solution for stability. The results of the application of the bootstrap technique using Tickhonov regularization are shown in Figure 4. As can be seen from this figure, the variance of the drift estimation was reduced more than 50 times (standard deviation of 0.99 klb/hour). The stability of the regularized drift is clearly seen from the unimodal nature of the PDF function. The mean value of the regularized drift was found to be around 33.2 klb/hour. This value corresponds to 0.61 % drift in the six months of operation and coincides with previous feedwater flow rate drift estimation studies.

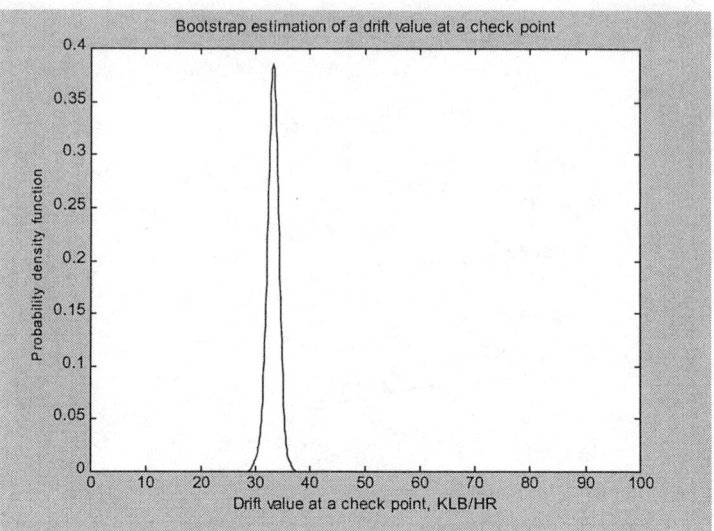

Fig. 4. PDF using Tikhonov regularization

The second method of regularization used for the linear solution is the truncated singular value decomposition (TSVD). This method linearly transforms the data into an orthogonal space using either singular value decomposition or principal components analysis. The transformed variables are called the principal components. These principal components are then evaluated to determine how many should be kept as predictor variables. Since the components are orthogonal, collinearity is no longer a problem.

In this case, only the first principal component has valuable information. It is regressed onto the response variable to calculate the single weight. The transformation matrix and weight are then used to predict feedwater flow. The results of the bootstrap procedure, shown in Figure 5, produce a mean drift or 36.1 klb/hour with a standard deviation of 0.2 klb/hour. This shows the TSVD model properly stabilizes the solution.

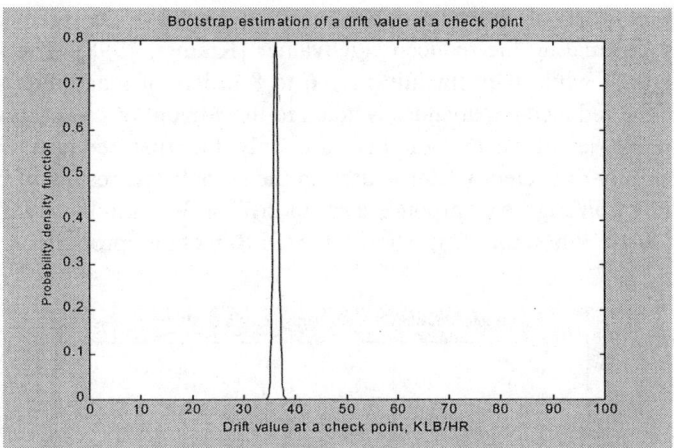

Fig. 5. PDF using Truncated Singular Value Decomposition

3.2 Partial Least Squares Model

Partial Least Squares (PLS) is a factor based technique used to perform multilinear regression [Geladi, 1986; Hoskuldsson, 1988]. Like the TSVD and principal components analysis (PCA) methods, the PLS architecture (see Figure 6) linearly transforms the input space (x) into an orthogonal space (t) thus removing the collinearity of the inputs. While PCA transforms the inputs so that the first principal component explains most of the variance of the data, PLS transforms the inputs so that the first latent vector accounts for the majority of the covariance between the inputs and the response variable (y). This method takes into consideration the response variable when performing the transformation while PCA only considers the predictor variables. The latent vectors are iteratively transformed and regressed onto the response variable through a linear weight matrix (w).

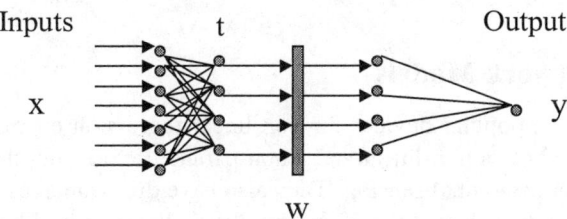

Fig. 6. PLS Inferential Model Architecture

The optimal number of latent variables used to perform the prediction can be calculated by calculating the reduced eigenvalues [Kramer, 1998]. The first reduced eigenvalue is 0.19 while the remaining are 6 to 8 orders of magnitude smaller. The magnitude of the reduced eigenvalues is equal to the amount of covariance explain by its corresponding latent vector. In this case only the first reduced eigenvalue is significant, so only one latent vector is used in the model. The results of the bootstrap procedure, shown in Figure 7, produce a mean drift or 36.2 klb/hour with a standard deviation of 0.14 klb/hour. This shows the PLS method properly stabilized the solution.

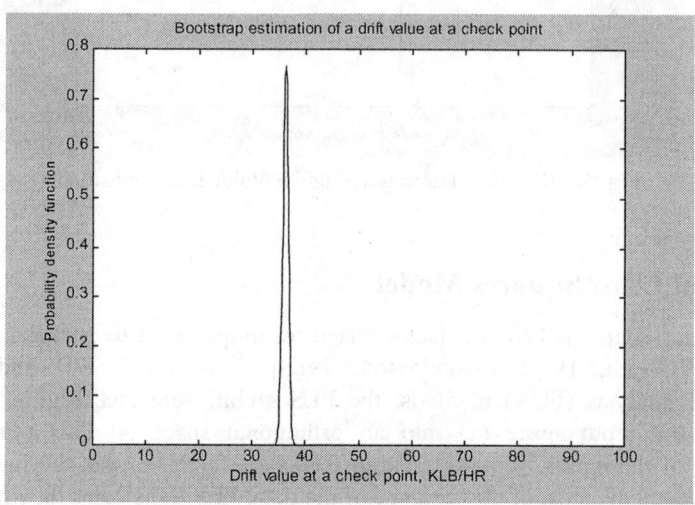

Fig. 7. PDF using Partial Least Squares

A Non-Linear Partial Least Squares (NLPLS) model [Qin and McAvoy, 1992] performs a non-linear mapping between the latent vectors and the response variable using neural networks. This technique is useful when non-linear relationships exist between the predictor variables and the response variable. Since these methods are based on the linear covariances, they do not perform well for very non-linear relationships. In this example the relationship between the predictors and response variable is highly linear and NLPLS methods are not needed.

3.3 Neural Network Models

Neural networks are popular devices for non-linear regression estimation and pattern recognition. Being both a non-linear and a non-parametric method, they are extremely flexible models for inferential sensing. They also have disadvantages that are a natural extension of their advantages. One of the mostly challenging problems in the correct application of neural networks is their regularization. Nonlinear regularization is a much more difficult problem than its linear counterpart and it has no general solution due to nonlinear error propagation and the existence of multiple local minima (multiple solutions) on the error surface. Linear regression models have only one minima. Because of multiple local minima, the training procedure may find one of several solutions that meets the error criteria. Each of these solutions will have different generalization properties that must be evaluated. Generalization is the ability of a model to respond correctly to inputs that were not used for training.

The solutions provided by neural networks depend on a number of factors including weight initialization, network architecture, stopping criteria, and the training algorithm. To insure consistent inferential measurements, the neural network's solution must be invariant under all these conditions. If we are not able to get consistency under different conditions, we at least need to estimate the reliability of our inference. A number of methods have been proposed to guarantee the stability and consistency of neural network solutions. In this chapter the mostly popular methods are evaluated: training with weight decay, the Levenberg Marquardt training algorithm, cross-validation training, small weight initialization, and Bayesian regularization. The different neural network methods are tested with the same data set to determine their dependence on initial conditions and network architecture.

The neural network used in this study was the standard multilayer perceptron (MLP) with a single nonlinear hidden layer and a linear output layer. A MLP without any regularization was first trained to a mean squared error of 10^{-3} using ordinary gradient decent and used as the reference neural network for baseline comparison. The network architecture had two hidden neurons using a hyperbolic tangent activation function. The training was performed one hundred times using different initial weights and a limited number of epochs equal to 5000. The mean value of estimated drift and its standard deviation were 25.3 klb/hour and 28.8 klb/hour correspondingly. Changes in neural network architecture and error goal did not improve the training results. The probability density function (PDF) of the drift estimation at a check point for this network is shown in Figure 8. It is obvious that this PDF is extremely broad, ranging from negative to positive values of drift estimation. Clearly, this unstable performance of the benchmark neural network is unacceptable for the task of inferential sensing. We will now investigate several neural network regularization methods.

302

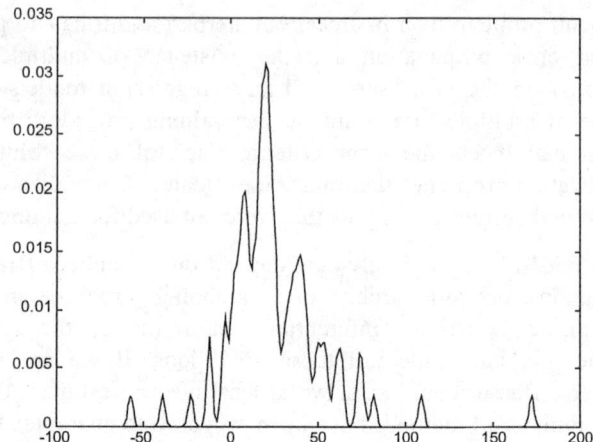

Fig. 8. Drift dependence on random start for an MLP trained by gradient decent

3.3.1 Levenberg-Marquardt Training

The Levenberg-Marquardt (LM) algorithm has inherent regularization properties as discussed in [J. Erriksson, 1996]. To check the regularization properties of the LM algorithm, we trained a number of neural networks one hundred times starting from different initial conditions for different numbers of hidden neurons and for different numbers of training points. The results of this training are presented in Tables 2-4. The mean of predicted drift denotes the mean value of predicted drift calculated for 100 runs with different initial weights for each architecture and number of training points. As can be seen from these tables, the mean value of drift estimation depends slightly on neural network architecture but largely on the number of training points. As the number of training points is increased, the network becomes more constrained and the output moves towards a more stable, correct estimate. This shows the importance of a complete training set. The large standard deviation for small training sets shows that the neural network inference is unstable under different random starts. It should be noted that the variance of drift inference was substantially reduced compared with the ordinary unregularized gradient decent solution. The LM algorithm was also able to reduce training error down to 10^{-6} which is three orders of magnitude less than that of ordinary gradient decent.

Table 2. LM Method: Number of Training Patterns = 400

Number of Hidden Neurons	Mean of Predicted Drift	Standard Deviation of Predicted Drift
2	12.0	7.62
3	11.8	5.87
4	12.6	6.67
5	13.0	6.98
10	11.6	6.98

Table 3. LM Method: Number of Training Patterns = 600

Number of Hidden Neurons	Mean of Predicted Drift	Standard Deviation of Predicted Drift
2	14.4	2.90
3	16.6	2.52
4	16.1	2.60
5	15.5	2.39

Table 4. LM Method: Number of Training Patterns = 800

Number of Hidden Neurons	Mean of Predicted Drift	Standard Deviation of Predicted Drift
2	22.8	1.70
4	21.3	1.64
10	21.3	1.72

3.3.2 Training with Weight Decay

The instability of neural network inference can be attributed to the redundant flexibility of neural networks when used as function approximators and also to the collinearity of the training data. It is known [J. Hertz, A. Krogh, R.G. Palmer, 1995] that the development of neural networks with good generalization capabilities requires some type of complexity control imposed on the neural net. Such control can be implemented by controlling the magnitude of weights and biases in the neural net. It has been shown [David J.C. MacKay, 1995] that the complexity of the function which can be implemented by a neural network depends on the magnitude of the weights: the larger the weights the more complex the function that a neural network can approximate. It has been proven that neural networks with sufficiently large numbers of hidden neurons can approximate any arbitrary complex function up to any degree of accuracy [Hornick 1989]. Without complexity controls, the neural network will approximate noise or artifacts in the training data which are not general to the entire data set. This results in a predictive model with poor generalization capabilities. By constraining the neural network's complexity a subtle compromise between fitting the data and keeping our model as simple as possible is resolved. This is in accordance with Occam's razor which states that a simple model should be preferred to a complex one provided both are consistent with the data. The easiest way to restrict the complexity of a neural net is to add a penalty term to its least square error function. This penalty term is usually the sum of squares of all network's weights and biases and is analogous to ridge regression. The penalized functional is:

$$\text{Total Performance} = \lambda E + (1-\lambda)S \qquad (7)$$

where E is the mean squares error term, S is the penalty term, and λ is the regularization parameter which controls the trade-off between E and S. One more rationale behind this type of regularization is that the mapping should be smooth or non-oscillating and the second term in formula (7) penalizes such non-smoothness. The parameter λ should be chosen prior to application of this method of regularization. This choice is a difficult problem when using linear models and even more difficult when the models are non-linear due to multiple error minima. The λ parameter is defined by the amount of noise in the data which is usually not known *a priori*. The value of this parameter can be roughly approximated through an analysis of the eigenvalues of the Hessian matrix of the neural network [Bishop, 1996].

The results of neural network training with weight decay are shown in Tables 5-7 for different numbers of hidden neurons, and different regularization parameters (λ). For each number of hidden neurons and each λ, the network was reinitialized 100 times to check the dependence on random starts. This dependence is summarized by the standard deviation of the predicted drift value. For Table 5 the regularization parameter was not appropriate, giving much more preference to penalization than to data fitting; and as a result, the network was undertrained with a large standard deviation for prediction and substantially different mean values for different architectures. Increasing the regularization parameter helps reduce the standard deviation and make the mean value of the predicted drift more consistent but still results in a less robust model than that of linear regression or the LM trained neural net.

Table 5. Training with Weight Decay Function (LM): Regularization Parameter =0.1

Number of Hidden Neurons	Mean of Predicted Drift	STD of Predicted Drift
2	75.0	71.6
4	63.1	45.8
6	55.7	41.3
10	34.9	18.2

Table 6. Training with Weight Decay Function (LM): Regularization Parameter =0.3

Number of Hidden Neurons	Mean of Predicted Drift	STD of Predicted Drift
2	47.8	46.7
4	32.5	14.9
6	30.3	9.7
10	27.5	9.3

Table 7 Training with Weight Decay Function (LM): Regularization Parameter =0.5

Number of Hidden Neurons	Mean of Predicted Drift	STD of Predicted Drift
2	24.9	7.2
3	23.3	7.3
4	25.7	7.5
5	24.9	7.6
10	21.5	6.3
20	18.9	5.1

3.3.3 Cross Validation Training

Another popular method for neural network regularization is termed early stopping or cross-validation. This regularization method is largely ad hoc and is based on dividing the training data into two sets: a training set and a validation set. The basic idea is to stop training before the neural network begins to learn the noise and spurious structures in the training data. During training, the neural network learns more and more structure from the data; however, at some point neural net begins to learn pseudo-structure or noise thus providing a more "rough" mapping. The validation set is used to provide an independent test set for verifying how well the trained network will generalize on previously unseen data. The results for this regularization method are shown in Table 8. For this test, the number of training patterns was chosen to be 600 and the number of validation patterns was 400. From this table it can be seen that the predicted drift again depends on the number of hidden neurons and also on random initialization. The standard deviation is relatively small for "large" networks (6-10 hidden neurons) but is unacceptably big for "small" networks. An obvious limitation of this type of regularization is that the final solution depends on the initial start and on the path by which system evolved to the final state. In addition, it requires division of the data into two sets thus decreasing the amount of data available for training, which in the case of scarce data, can be a serious limitation. An obvious advantage of this kind of complexity control is its simplicity.

Table 8. Training with Cross Validation (LM)

Number of Hidden Neurons	Mean of Predicted Drift	STD of Predicted Drift
2	23.3	23.6
4	18.2	5.2
6	15.8	4.1
10	16.0	4.8

3.3.4 Regularization by Small Weight Initialization

A lesser known type of regularization for neural networks is regularization by initialization. In this method the initial weights of the neural network are set to small values, thus forcing the neural network to search a smaller area of the error surface and minimizing its dependence on random starts. This kind of regularization is also ad hoc because setting initial weights to small values partly specifies the solution. The results of the application of this kind of regularization are shown in Tables 9-10. These tables show that we can reach consistent results (small standard deviation) for a specific training set but the mean values are rather different for different training sets because changing the data changes the error surface. This inconsistency with respect to the training set makes this method unreliable for inferential sensing with highly collinear data.

Table 9. Four Hidden Nonlinear Neurons, 100 Random Starts, 400 Training Patterns

Weights and Biases Initialization	Mean of Predicted Drift	Standard Deviation of Predicted Drift
0.01*randn	14.2	3.14
0.001*randn	14.9	3.54
0.00001*randn	13.5	3.60

Table 10. Four Hidden Nonlinear Neurons, 100 Random Starts, 600 Training Patterns

Weights and Biases Initialization	Mean of Predicted Drift	Standard Deviation of Predicted Drift
0.01*randn	19.2	1.64
0.001*randn	18.8	1.68
0.00001*randn	19.1	1.05

3.3.5 Bayesian Regularization

The most advanced method of neural network regularization is Bayesian regularization [David J.C. MacKay, 1992]. The Bayesian point of view on neural network training is rather different from traditional views. Traditional methods are variations of the maximum likelihood principle which states that from a variety of possible models we should pick the one that is the most probable with respect to the observed data. The maximum likelihood principle considers model parameters as unknown but fixed values and tries to estimate these parameters from the available data, providing the only set of parameters which is claimed most likely generated the observed data. Therefore, in conventional neural network training, a single set of weights is used for future inference. In contrast to the maximum likelihood principle, the Bayesian approach considers the model parameters to be random variables having an *a priori* distribution. Having this *a priori* distribution, Bayesian inference uses the application of Bayes theorem to modify the prior distribution and produce a posterior distribution which now depends on both prior information and current data.

The key to the success of Bayesian training is the correct choice of prior distributions. This is sometimes considered to be the fatal flaw of Bayesian inference because of the "subjective" nature of such a choice. However, when dealing with ill-posed problems we must use some prior information because the data underdetermines the solution. Having chosen prior distribution of weights, Bayesian training gives rise to posterior weight distributions, which in turn gives rise to a distribution of the output values during the inference with new data. The mean of this output distribution is considered to be the inferred value.

The results of application of Bayesian regularization to the problem of drift detection are shown in Table 11. The number of training points used for this method was 600. As can be seen from this table, Bayesian regularization is the only method which generates a neural network which does not depend on the random weight initialization. This is shown by the standard deviation of the drift prediction due to different random starts being zero. However, this inference method drastically depends on the initial number of hidden neurons (network architecture) and therefore provides inconsistent estimations of the drift mean value.

Table 11. Bayesian Regularization

Number of Hidden Neurons	Predicted Drift	STD
2	19.0	0
3	13.5	0
4	23.2	0
5	26.6	0
6	58.5	0

3.3.6 Neural Network Methods Summary

Neural networks are a powerful and flexible tool for non-parametric modeling and inference but their regularization and consistency with highly collinear data is challenging due to the inherent local minima. Unregularized neural networks can provide very inconsistent results, which are non-interpretable and non-repeatable.

Our results show that stability with respect to initial conditions, can be attained with Bayesian regularization. On the other hand, Bayesian training provided drastically different drift estimations for different network architectures. A very serious limitations of the Bayesian approach to neural network training is its computational burden which limits its application to small amounts of data and small networks. The use of this approach in an on-line system is probably not practical. Simpler network regularization methods such as cross-validation and weight decay provide a reasonable trade-off between stability and computational time and can be used to stabilize solutions. The Levenberg-Marquardt algorithm proved to be a rather stable technique, although not as stable as linear methods. The stability of this algorithm can be explained by its built-in regularization properties which help it dampen high frequency noise in weights in the vicinity of the solution. Regularization by initialization is a rather new technique and its validity has to be evaluated more rigorously in theoretical and practical aspects, but our results show that it can reduce dependence on initial conditions and can be effective from a computational point of view because it does not require any additional computational efforts.

4 Conclusions

Data driven inferential methods provide unstable estimation when the predictor variables are collinear. This instability is caused by the ill-conditioning of the data matrix and manifests itself as a non-smooth least squares solutions. This solution becomes sensitive to the noise in the data. Regularization methods combat this instability and provides more consistent drift predictions.

The data used in this study is extremely collinear and the relationship between the predictors and response variable (feedwater flow) is mostly linear. This data set was chosen as a worst case example and results on other more usual data sets should be significantly better. Several linear and non-linear techniques were applied to predict the feedwater flow drift six months into the fuel cycle. The results of these techniques are summarized in Table 12. This table shows that techniques that do not use some sort of regularization have unrepeatable results as evidenced by large standard deviations of the predictions. It also shows that the three linear regularization techniques give consistent results. Although the problem was linear, non-linear techniques were also used to demonstrate the potential problems of using neural networks to solve ill-posed problems.

The non-linear drift predictions were somewhat dependent on the neural network model architecture as evidenced by the different mean drifts. This is another adverse feature of ill-posed problems and can only be dealt with by knowing the solution *a priori*. The solution to the feedwater drift detection problem is known due to the availability of a redundant ultrasonic measurement device on several plants. Many inferential sensing problems are not concerned with measuring the level of drift, only upon ascertaining when a sensor has drifted so that efficient maintenance can be planned. Again, the example presented in this study is a more difficult problem than most inferential sensing problems.

Table 12. Results Summary

Method Type	Model	Mean Drift	Standard Deviation
Linear	Regression	39.9	8.3
Linear	Ridge Regression	33.2	0.99
Linear	TSVD	36.1	0.20
Linear	PLS	36.2	0.14
Non-Linear NN	Gradient Descent	25.3	28.8
Non-Linear NN	LM Training	21.3	1.64
Non-Linear NN	Weight Decay	25.7	7.5
Non-Linear NN	Cross Validation	15.8	4.1
Non-Linear NN	Small Weight Initialization	18.8	1.68
Non-Linear NN	Bayesian Regularization	26	0

The non-linear techniques did not perform as well as the linear techniques, although if there were substantial non-linearities in the mapping, the non-linear techniques would have the ability to perform the mapping while linear techniques would be limited to giving a best linear fit. When the neural network was used without any method of regularization, the results were totally unusable. The LM training algorithm performed well when large amounts of training data were used. he weight decay method was unstable. The cross validation method may have performed equal to the LM algorithm if additional training data were available since it also used the LM training algorithm. Using small weight initialization limited the solution search space and consistently found the same solution, but that solution was dependent on the training set and was not equal to the actual drift computed by the linear methods. Lastly, Bayesian Regularization was consistent but dependent on architecture. All of the non-linear techniques have disadvantages that come with their additional flexibility. The use of neural network techniques to solve ill-conditioned problems must be done with caution and the knowledge of their potential problems. The importance of thorough validation and testing techniques cannot be overstated. Many of these problems have not been properly documented in the literature giving neural network users a false sense of security.

With the proper use of regularization techniques, linear techniques can accurately and consistently predict the feedwater flow drift accurately. When using non-linear techniques, proper architecture selection, training set selection, regularization, and validation testing must be incorporated into the network design. Lastly, it should again be noted that the example presented is a worst case example of ill-conditioned prediction and most inferential problems will have several principal factors involved in constructing a reliable prediction.

References

Basseville, M. (1988). "Detecting Changes in Signals and Systems - A Survey," *Automatica*, 24, 309-326.

Bishop, Christopher M. (1996), *Neural Networks for Pattern Recognition*, Oxford University Press.

Black, C.L., R.E. Uhrig, and J.W. Hines, (1998), "System Modeling and Instrument Calibration Verification with a Non-linear State Estimation Technique," published in the proceedings of the *Maintenance and Reliability Conference (MARCON 98)*, Knoxville, TN, May 12-14.

Cherkassky, V., and F. Muller (1998), *Learning From Data*, John Wiley & Sons.

Desai, M., A. Ray (1981), "A Fault Detection and Isolation Methodology," *Proc. 20th Conf. on Decision and Control*, 1363-1369.

EPRI (1992), "Feedwater Flow Measurement in U.S. Nuclear Power Generation Stations," TR-101388, Electric Power Research Institute.

Efron, B. (1982), *The Jacknife, The Bootstrap, and Other Resampling Plans*. Philadelphia, Penn.: Society for Industrial and Applied Mathematics.

Erriksson, J., Marten Gulliksson, Per Lindsrom and Per-Ake Wedin , "Regularization Tools for Training Large-Scale Neural Networks," Technical report UMINF-96.05 Department of Computing Science, Umea University, Sweden.

Gantmacher, F.R. (1959), *Applications of the Theory of Matrices*, Interscience Publishers, New York.

Geladi, P, and B.R. Kowalski (1986), "Partial Least-Squares Regression: A Tutorial," *Analyta Chimica, Acta*, 185, pp 1-17.

Gertler, J. (1988), "Survey of Model Based Failure Detection and Isolation in Complex Plants," *IEEE Control Systems Magazine*, pp. 3-11.

Glockler, O. (1991), "Fault Detection in Nuclear Power Plants Applying the Sequential Probability Ratio Test of Mar-Based Residual Time Series," *Proceedings of the AI 91 Frontiers in Innovative Computing for the Nuclear Industry*, Jackson, Wyoming, 859-868.

Golub, G. H., M.T. Heath, and G. Wahba (1979), *Generalized cross-validation as a method for choosing a good ridge parameter*, Technometrics, 21, pp. 215-223.

Golub, G.H., and C.F. Van Loan (1996), *Matrix Computations*, Third Edition, the Johns Hopkins University Press, Baltimore, MD.

Grini, R.E., O. Naess, O. Berg (1989), "Model-Based Fault Detection and Diagnosis in Process Plant Operation," *Proceedings of the Seventh Power Plant Dynamics, Control & Testing Symposium*, Knoxville, Tennessee, 56.01-56.14.

Gross, K.C., R.M. Singer, S.W. Wegerich, J.P. Herzog, R. Van Alstine, and F. K. Bockhorst (1997), "Application of a Model-based Fault Detection System to Nuclear

Plant Signals," Proc. 9th Intl. Conf. on IntelligentSystems Applications to Power Systems, Seoul, Korea.

Gull, S.F. "Bayesian Inductive Inference and Maximum Entropy," *Maximum Entropy and Bayesian Methods in Science and Engineering*, Vol.1, Erickson and Smith, eds., pp.53-74, Kluwer, Dordrecht.

Hadamard, J. (1923), *Lectures on Cauchy's Problem in Linear Partial Differential Equations*, Yale University Press, New Haven.

Hansen, P.C. (1989), *Regularization, GSVD and Truncated GSVD*, BIT 29, pp. 491-504.

Hansen, P.C. (1990), *The Discrete Picard Condition for Discrete Ill-Posed Problems*, BIT 30, pp. 658-672.

Hansen, P.C. (1994), *Regularization Tools: A Matlab package for analysis and solution of discrete ill-posed problems*, Numer. Algorithms, vol.6, pp. 1-35.

Hansen, P.C. (1995), *Test Matrices for Regularization Methods*, SIAM J. SCI. Comput. vol. 16, No. 2, pp. 506 –512.

Hansen, P.C. (1997), *Rank –Deficient and Discrete Ill-Posed Problems. Numerical Aspects of Linear Inversion*, SIAM, Philadelphia.

Hansen, P.C. (1992), *Analysis of Discrete Ill-Posed Problems by Means of the L-curve*, SIAM Review vol. 34, No.4, pp. 561-58.

Hardy, C.R., D.W. Miller, B.K. Hajek (1992), "A Model-Based Approach to Malfunction Isolation in Interacting Systems," *Proceedings of the 8th Power Plant Control & Testing Symposium*, Knoxville, Tennessee, 37.01-37.13.

Hertz, J., A. Krogh, and R.G. Palmer (1991), *Introduction to the Theory of Neural Computation*, Addison-Wesley.

Hines, J.W. and D.J. Wrest (1997a), "Signal Validation Using an Adaptive Neural Fuzzy Inference System," *Nuclear Technology*, August, pp. 181-193.

Hines, J.W. and R.E. Uhrig (1997b), "Use of Autoassociative Neural Networks for Signal Validation," *Journal of Intelligent and Robotic Systems*, Kluwer Academic Press.

Hines, J.W., R.E., Uhrig, C. Black and X. Xu (1997c), "An Evaluation of Instrument Calibration Monitoring Using Artificial Neural Networks," published in the proceedings of the 1997 American Nuclear Society Winter Meeting, in Albuquerque, NM, November 16-20.

Hoerl, A.E., and R.W. Kennard (1970), Ridge Regression: *Biased Estimation for Nonorthogonal Problems*, Technometrics, 12, pp. 55-67.

Holbert, K.E. and B.R. Upadhyaya (1990), "An Integrated Signal Validation System For Nuclear Power Plants," Nuclear Technology, Vol. 92, pp. 411 - 427.

312

Holbert, K.E., Heger, S.A. and A.M. Ishaque, (1995) "Fuzzy Logic For Power Plant Signal Validation," *Proceedings of the 9th Power Plant Control & Testing Symposium*, Knoxville, Tennessee, May 24-26, 20.01 - 20.15.

Hornick, K., Stinchcombe, M., and H. White, 1989, Multilayer Feedforward Networks are Universal Approximators, *Neural Networks*, **2**, pp 359-366.

Hoskuldsson, A., (1988), "PLS Regression Methods," Journal of Chemometrics, Vol. 2, pp. 211-228.

Ikonomopoulos, A., R.E. Uhrig, L. H. Tsoukalas (1992), "A Methodology for Performing Virtual Measurements in a Nuclear Reactor System," *Transactions of the 1992 American Nuclear Society International Conference on Fifty Years of Controlled Nuclear Chain Reaction: Past, Present, and Future*, Chicago, Illinois, 106-107.

Isermann, R. (1984), "Process Fault Detection Based on Modeling and Estimation Methods-A Survey," *Automatica*, Vol. 20, No. 4, 381-404.

Jolliffe, I. (1986), *Principal Components Analysis*, , Springer-Verlag, New York.

Kavaklioglu, K., and Belle R. Upadhyaya (1994), "Monitoring Feedwater Flow Rate and Component Thermal Performance of Pressurized Water Reactors By Means of Artificial Neural Networks," Nuclear Technology Vol. 107.

Kittamura, M. (1980), "Detection of Sensor Failures in Nuclear Plants Using Analytical Redundancy," *Transactions of the American Nuclear Society*, 34, 581-583.

Kramer, R. (1998), *Chemometric Techniques for Quantitative Analysis*, Marcel-Dekker.

MacKay, David J.C. (1992), "Bayesian Interpolation," *Neural Computation*, Vol. 4, No. 3, pp.415-447.

MacKay, David J.C. (1992), "A Practical Bayesian Framework for Backpropagation Networks," *Neural Computation*, Vol. 4 No. 3, pp.448-472

MacKay, David J.C. (1995), "Bayesian Methods for Neural Networks:Theory and Applications," Course notes for neural network summer school. University of Cambridge programme for industry.

Morozov, V.A. (1984), *Methods for Solving Incorrectly Posed Problems*, Springer-Verlag, New York.

Nuclear News, "Flow Rate Mismeasurement Causes Unneeded Derating," *Nucl. News*, 36, 2, 39, Feb., 1993.

Patton, R.J., J. Chen (1991), "A Review of Parity Space Approaches to Fault Diagnosis," *IFAC/IMACS Safeprocess Conference*, Baden-Baden, Germany, 239-255.

Qin, S.J., and T.J. McAvoy (1992), "Nonlinear PLS Modeling Using Neural Networks," *Computers chem. Engng.*, Vol. 16, No. 4, pp.379-391.

Qualls, A.L., R.E. Uhrig, and B.R. Upadhyaya (1988), "Development of an Expert System for Signal Validation," Topical Report prepared for the U.S. Department of Energy, by The University of Tennessee, Knoxville, DOE/NE/37959-17.

Singer, R.M., K. C. Gross, J.P. Herzog, R. W. King, and S. W. Wegerich (1997), "Model-Based Nuclear Power Plant Monitoring and Fault Detection: Theoretical Foundations," Proc. 9th Intl. Conf. on Intelligent Systems Applications to Power Systems, Seoul, Korea.

Tikhonov, A.N. (1963), *Solution of incorrectly formulated problems and the regularization method*, Doklady Akad. Nauk USSR 151 (1963), pp. 501-504.

Tsoukalas, L.H. (1992), "Expert Systems for Power Plant Applications: An Overview," Proceedings of the 8th Power Plant Control & Testing Symposium, Knoxville, Tennessee, May 27-29, pp47.01-47.10.

Uhrig, R.E. (1993), "Artificial Neural Networks and Potential Applications to Nuclear Power Plants," *Proceedings of the Conference on Structural Mechanics in Reactor Technology*, Knostanz, Germany.

Uhrig, R.E., J.W. Hines, and W.R. Nelson (1998), "Integration of Artificial Intelligence Systems into a Monitoring and Diagnostic System for Nuclear Power Plants," presented at the Special Meeting on Instrumentation and Control of the Halden Research Center, Lillihammer, Norway, March 28-21.

Upadhyaya, B.R., E. Eryurek (1992), "Application of Neural Networks for Sensor Validation and Plant Monitoring," *Nuclear Technology*, Vol. 97, 170-176.

Willsky, A.S. (1976),"A Survey of Design Methods for Failure Detection in Dynamic Systems," *Automatica*, Vol. 12, 601-611.

"Flow Rate Mismeasurement Causes Unneeded Derating," *Nucl. News*, 36, 2, 39, Feb., 1993.

Appendix

Var. Num.	Description	Range	Units
1	FWP Speed	0-7500	RPM
2	'A' OTSG EFIC HIGH LEVEL	0-100	PERCENT
3	FEEDWATER PUMP A SPEED	0-7500	RPM
4	LINEAR POWER CH NI-6	0-125	PERCENT
5	HEATER 3A INLET COND TEMP	40-300	DEGF
6	HEATER 3B OUTLET COND TEMP.	40-350	DEGF
7	DEARATOR INLET COND TEMP	40-350	DEGF
8	HEATER 6A INLET FW TEMP	40-500	DEGF
9	FWP A DISCHARGE TEMP	40-500	DEGF
10	FWP A SUCTION TEMP	40-500	DEGF
11	HEATER 5B OUTLET FW TEMP	40-500	DEGF
12	STEAM GEN B INLET FW TEMP	40-600	DEGF
13	HEATER 6B OUTLET FW TEMP	40-600	DEGF
14	STEAM GEN A LEVEL (OP)	0-100	PERCENT
15	STEAM GEN A LEVEL (FULL)	40-640	INCHES
16	STEAM GEN A LEVEL (START UP)	0-250	INCHES
17	STEAM GEN B INLET FW TEMP	0-500	DEGF
18	STEAM GEN B LEVEL (START UP)	0-250	INCHES
19	STEAM GEN A INLET FW TEMP	40-600	DEGF
20	STEAM GEN B INLET FW TEMP	40-600	DEGF
21	REHEATER A COLD REHEAT PRESS.	0-200	PSIG
22	REHEATER D COLD REHEAT PRESS.	0-200	PSIG
23	REHEATER C COLD REHEAT PRESS.	0-200	PSIG
24	NO. 2A EXTR LP TURB PRESSURE	0-20	PSIA

14 Genetic Algorithms Applied to Nuclear Reactor Design Optimization

C.M.N.A.Pereira[1], R. Schirru[2], and A.S.Martinez[2]

[1]Comissão Nacional de Energia Nuclear, Instituto de Engenharia Nuclear,
P.O.Box 68550 - 21945-970, Rio de Janeiro - Brazil
[2]Universidade Federal do Rio de Janeiro, Programa de Engenharia Nuclear,
P.O.Box 68550 - 21945-970, Rio de Janeiro - Brazil
cmnap@cnen.gov.br

A genetic algorithm is a powerful search technique that simulates natural evolution in order to fit a population of computational structures to the solution of an optimization problem. This technique presents several advantages over classical ones such as linear programming based techniques, often used in nuclear engineering optimization problems. However, genetic algorithms demand some extra computational cost. Nowadays, due to the fast computers available, the use of genetic algorithms has increased and its practical application has become a reality.

In nuclear engineering there are many difficult optimization problems related to nuclear reactor design. Genetic algorithm is a suitable technique to face such kind of problems. This chapter presents applications of genetic algorithms for nuclear reactor core design optimization. A genetic algorithm has been designed to optimize the nuclear reactor cell parameters, such as array pitch, isotopic enrichment, dimensions and cells materials. Some advantages of this genetic algorithm implementation over a classical method based on linear programming are revealed through the application of both techniques to a simple optimization problem. In order to emphasize the suitability of genetic algorithms for design optimization, the technique was successfully applied to a more complex problem, where the classical method is not suitable. Results and comments about the applications are also presented.

1 Introduction

Nowadays, the most recent nuclear reactor design philosophies, such as the European Pressurized Reactor - EPR [Teichel, 1996, 1999], that embeds the safety and operation cost optimization concepts, as well as the new nuclear fuel concepts such as the Mixed Oxide - UO_2 - PuO_2 (MOX), have been motivating the nuclear reactor design techniques improvement.

The nuclear reactor core design involves a lot of constraints. During the design, a set of parameters must be adjusted in order to obtain a safe and economical reactor. Finding the best configuration in the design process includes a good representation of the phenomena related to the neutron interactions in the reactor core and an efficient optimization technique.

Due to their simplicity and low computational cost, gradient search has been used as the optimization methods. Most of them, such as the one described in Rozon (1992), use linear programming techniques. However, because of the nature of exploitation of these hill-climbing-like methods, their application to a multimodal search space can lead to a local optimum. The nuclear reactor core design optimization process includes non-linearities, discontinuities and multimodality, becoming a complex problem with large number of state parameters to be optimized subject to a large number of constraints.

In this chapter it is presented a global scope optimization approach for reactor core design based on genetic algorithms (GA) [Goldberg, 1989] in which it is considered the all-variable space and no prior knowledge to restrict the search is required. Global optimization has been explored by some researchers in the field of nuclear reactor design, such as Cacuci (1990). Genetic algorithms have been successfully applied to nuclear core reload optimization [DeChaine, 1995; Chapot, 1999], transient classification [Alvarenga, 1997; Pereira, 1998], some others specific problems in the nuclear field [Haibach, 1997; Omori, 1997], and, recently, in core designs [Pereira, 1999].

Here, results of the application of the method to a simple core configuration reveal some advantages when compared to a classical nonlinear optimization method based on linear programming. It is also shown the results of an application of the proposed method to a more complex problem to which the classical method mentioned above is not suitable.

2 The Genetic Algorithm Search Paradigm

The Holland's genetic algorithms [Holland, 1975], inspired on the species evolution theory [Darwin, 1859], manipulate a population of symbolic structures, that represent points in the search space, in order to evolve this population to its best adaptation, hence, the best solution of the problem. In GA [Goldberg, 1989; Davis, 1991; Michalewicz, 1994], the parameters to be varied in the optimization process are codified in a symbolic structure, metaphorically called genotype, that is formed by a set of genes that carry intrinsic characteristics of the symbolic individual. These characteristics dictate the adaptability of the individual in the environment, in which it may survive or die. The selection and evolution are made in such a way that stronger individuals have more chance to be selected, transferring theirs characteristics to the offsprings. This way, from generation to generation, the tendency is that strong individuals become stronger and more numerous while the weak individuals tend to be extinct.

The GA starts the adaptation process from a set of possible configurations, in other words, a random generated population of individuals. Evaluating independently each individual by an objective function, the GA assigns to each one a fitness that predicts its resistance and adaptability. Then it is simulated a natural selection process in which the selection probability of a given individual is a function of the fitness.

The crossover in a classic GA is simulated by choosing a random cross point over the binary string that represents the genotype, followed by the change of parts between the two parents, as illustrated in Figure 1.

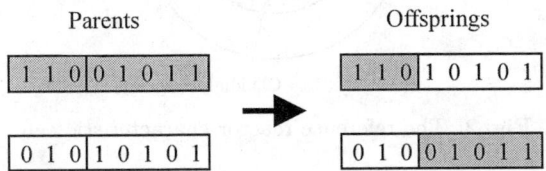

Fig. 1. An example of crossover in the classical genetic algorithm

The mutation is simulated by the inversion of one of the bits of the genotype according to the mutation probability that is a genetic parameter.

Taking into account the reproduction, crossover and mutation, it can be statistically proved, by the schemata theory, Goldberg (1989), that strong individuals may be more numerous in subsequent generations, as the population becomes more adapted, generally concentrating to near-optimum regions.

Due to the parallel search together with the genetic operators above mentioned, the GA provides a global exploitation of the search space, requiring no prior knowledge about the search space itself. Prior information about critical points, Cacuci (1990), or limits, Gray (1997), are only two examples of characteristics found in other non-linear and global optimization methods.

3 A Simple Nuclear Reactor Design Optimization Problem

In this section it is presented an optimization problem found in the scope of nuclear reactor core designs. This is a simple problem, specially proposed in order to validate the application of the application of GA for nuclear reactor optimization. Because of its simplicity it is possible to know the expected results via high computational cost methods. On the other hand, the proposed problem reveals advantages of the proposed method over a classical one. This problem aims to evaluate the robustness of the GA in finding the global optimum in the search space.

The problem is created in the context of a hypothetical reference reactor design, using a simplified model. The reactor is cylindrical, 1.63 m height and 0.86 m of radius. It has only one enrichment zone completely filled with the unit cell shown in Figure 2.

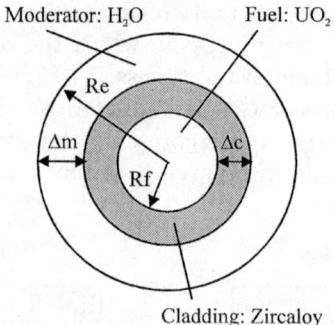

Moderator: H₂O   Fuel: UO₂

Cladding: Zircaloy

Fig. 2. The reference reactor characteristic cell

In this problem, the goal is to maximize the average flux, ϕ, in the reactor, so that the multiplication factor (k_{eff}) is 1.0 ± 0.01, varying the equivalent radius (Re) (proportional to the array pitch) and the isotopic enrichment (E), according to the ranges specified in Table 1. The cladding thickness (Δc) and fuel radius (Rf) are 0.889 cm and 0.127 cm respectively. Hence, the optimization problem can be written as:
maximize:

$$\phi(Re, E)$$

subject to:

$$\begin{aligned}
0.99 &\leq k_{eff} \leq 1.01 \\
Re_{min} &\leq Re \leq Re_{max} \\
E_{min} &\leq E \leq Re_{max}
\end{aligned} \qquad (1)$$

where k_{eff} is the neutron multiplication factor and the *min* and *max* subscript refers to the minimum and maximum values given in Table 1.

Table 1. Ranges of parameters variations for the simple problem

Parameter	Minimum	Maximum
Equivalent radius (cm)	1.041	3.048
Isotopic Enrichment (%)	3.5	5.0

4 The Genetic Algorithm Implementation

Modeling an optimization problem using genetic algorithm consists in two basic steps: choosing the appropriate genotype structure and designing a suitable objective function that must evaluate the fitness of each genotype.

4.1   The Genotype

The proposed genotype represents the set of parameters that can vary in the nuclear core design. The parameters not included in the genotype representation cannot be changed during the optimization process.

The genes in the proposed model are the binary representation of the parameters values, coded into binary fixed length strings. The length of the strings dictate the precision of the codification.

The lower limit of the gene range correspond to all bits set to zero as well as the upper limit correspond to all bits set to one. In other words, the range of each variable is discretized into 2^n values, where n is the genotype length. So a decodification of the genotype in its respective phenotype is like a digital to analog (D/A) conversion.

$$v = min + step * g \tag{2}$$

$$step = \frac{max - min}{maxint - 1} \tag{3}$$

where v is the decimal value of the parameter, min and max, the lower and upper limits, g is the gene integer value (direct conversion from binary to decimal) and $maxint$ the maximum integer represented by the number of bits used. For example consider a gene represented by 4 bits, 0101, that must be converted into a real number between 10 and 17.5. In this case, $max = 18$, $min = 10$, $g = 5$ (the binary 0101) and $maxint = 15$ (the binary 1111). So, $step$ is 0.5 and v is 12.5. The genotype is the concatenation of the genes strings.

4.2   The Fitness

To each genotype is assigned a fitness that is a function of the variables to be maximized or minimized and of the constraints of the problem, calculated by the reactor physics simulators with the parameters decoded from the genotype. The generalized fitness can be written as:

$$f = y + \sum_{i=1}^{N} F(x_i) \tag{4}$$

$$F(x_i) = \begin{cases} 0 & L_{inf} \leq x_i \leq L_{sup} \\ c_i \left| \Delta x_i \right|, & otherwise \end{cases} \tag{5}$$

where y is the objective variable that must be maximized or minimized, Δx_i is how far variable x_i is from the middle of the constraint range, L_{inf} and L_{sup} are the lower and upper limits of the constraint range, and c_i is the penalty constant.

4.3   The Interface Between the GA and the Nuclear Reactor Simulator

A schematic diagram in Figure 3 shows the link between the genetic algorithm and the nuclear reactor physics simulator.

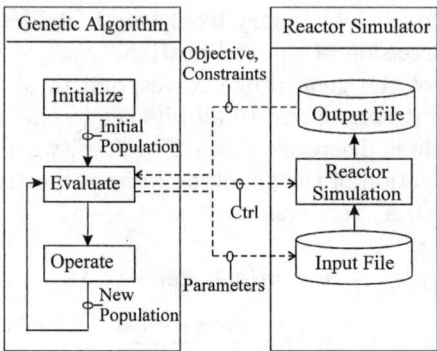

Fig. 3. Interface Between the Genetic Algorithm and the Nuclear Reactor Simulator

Based on the decodification of the genotypes the GA writes the parameters into the simulators input file and commands its execution. After running the simulator, the variable to be maximized or minimized, as well as the set of constraints variables are read from the simulator output file. When all the individuals in the population (genotypes) are evaluated, the GA applies the genetic operators, and a new generation is ready to be evaluated.

5   Application of the Genetic Algorithm to the Simple Problem

5.1   The Genetic Modeling

In the simple problem, the equivalent radius and the isotopic enrichment are encoded into binary fixed length strings representing the genes of an individual. Their concatenation forms the genotype. Figure 4 shows an example of genotype and its decodification in the respective phenotype.

The objective function is given by:

$$f = \begin{cases} \phi(Re, E), & k_{eff}(Re, E) = 1.0 \pm 0.01 \\ \phi(Re, E) + c\,|1.0 - k_{eff}(Re, E)|, & k_{eff}(Re, E) \neq 1.0 \pm 0.01 \end{cases} \quad (6)$$

where c is the penalty constant for the multiplication factor. The value of the penalty constant, c, must be adjusted so that the fitness assigned to every individual which is out of constraints is smaller than the fitness of any individual that satisfies the constraints.

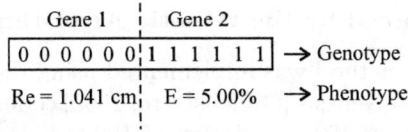

Gene 1   Gene 2

| 0 0 0 0 0 0 | 1 1 1 1 1 1 | → Genotype

Re = 1.041 cm   E = 5.00%   → Phenotype

Fig. 4. Example of genotype and the respective phenotype

5.2 Reference Results

In order to evaluate the method application to the simple problem, it was used a high computational cost method to obtain the reference optimum. It was done by discretizing the whole space using a sufficiently reduced step. The reference optimum (see Table 2) is then the best point found that satisfies the constraints of the problem. The surface obtained in the discretization is shown in Figure 5. The point where the constraints are not satisfied were set to zero only for visual effect.

Table 2. Reference optimum solution for the simple problem

Equivalent radius (cm)	Isotopic Enrichment (%)	Flux ($\times 10^{-4}$)
2.449	5.00	4.32

Note that there are two separated regions where the constraints are satisfied. Such kind of problem creates the possibility of local convergence. The difficulty imposed by this problem intend to evaluate the robustness of the genetic algorithm in finding the global optimum in a multimodal space.

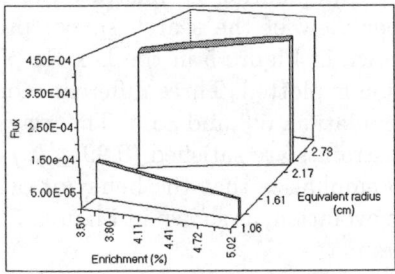

Fig. 5. The discretized surface for the simple problem, setting to zero the regions where the constraints are not satisfied

5.3 Results Obtained by the Genetic Algorithm

The above described method was implemented using the Genetic Search Implementation System - Genesis [Grefenstette, 1990] code. Using crossover rate of 60%, mutation rate of 1%, population of 100 individuals, elitism, ranking and 6 bits genes, 7 experiments were made varying the random seed and the penalty constant. The Hammer [Suich, 1967] system was used for cell and reactor core calculations. The results of the application fo the GA to the simple problem are shown in Table 3. Seven experiments varying the initial population (by using different random seeds) and the penalty coeficient were made in order to verify the GA robustness. The fitness convergence relative to the fourth experiment can be seen in Figure 6.

Table 3. Results obtained by the application of the GA to the simple problem

Experiment	c	Population Size	Random Seed	Re (cm)	E (%)	k_{eff}	ϕ ($\times 10^{-4}$)
1	1	50	123456789	2.377	4.69	0.99	4.28
2	1	50	123456789	2.380	4.74	0.99	4.26
3	1	50	987654321	2.443	4.98	0.99	4.32
4	1	50	123456789	2.410	4.83	0.99	4.30
5	10	50	123456789	2.443	4.98	0.99	4.32
6	10	50	987654321	2.443	4.98	0.99	4.32
7	10	50	123456789	2.443	4.98	0.99	4.32

Note that the best point found by a GA, sometimes is not exactly the global optimum (such as experiments 1, 2 and 4), however, it is often very close to it. That is why GA is classified as a near-optimum optimization technique.

The evolution of the population is illustrated in Figures 7 to 9. These figures show the upper view of the search space (projection of the three-dimensional graph shown in Figure 5 in the E x Re plane) where each individual of the population is plotted. Three different moments of the evolution can be seen: initial population, 5th and 35th. The gray color delimits the two regions where the constraints are satisfied ($0.99 \leq k_{eff} \leq 1.01$).

It is important to emphasize that the behavior of the genetic algorithm population during the evolution, as shown in Figures 7 to 9, is a characteristic for all the experiments made.

It can be observed in Figure 7 that the initial population is spread at the whole search space. In the 5th generation, as can be seen in Figure 8, the population is getting concentrated in the neighborhood of both regions where the constraints are satisfied. After the 5th generation the population tend to concentrate at that region which contains the global optimum, leaving the other one, but keeping points in other regions of the space, in order to provide

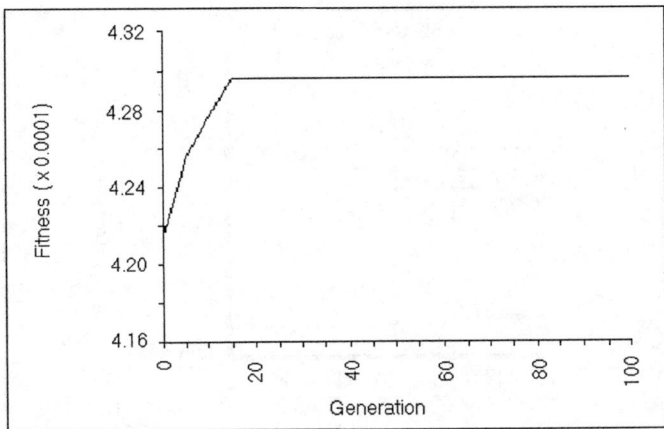

Fig. 6. Fitness convergence in the fourth experiment for the simple problem

diversity in the search. This diversity imposed by the GA is a very important mechanism in the search for the global optimum.

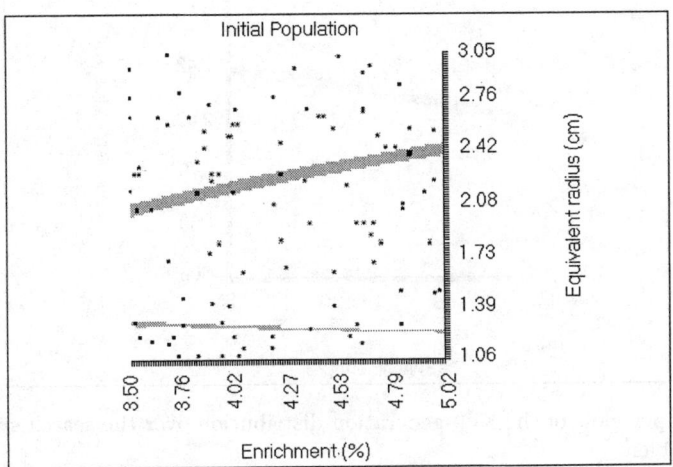

Fig. 7. Upper view of the initial population distribution over the search space of the simple problem

Finally, in the 35th generation, as can be seen in Figure 9, all the individuals of the population are concentrated at the global optimum. After this generation, no modifications in the population have occur and the GA reaches its convergence criteria.

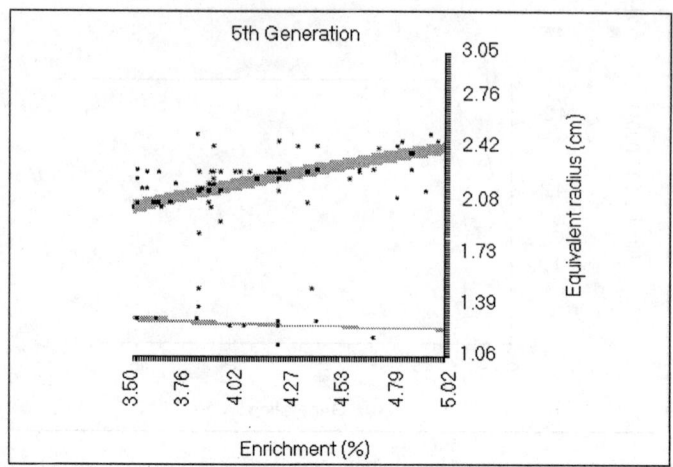

Fig. 8. Upper view of the 5th generation distribution over the search space of the simple problem

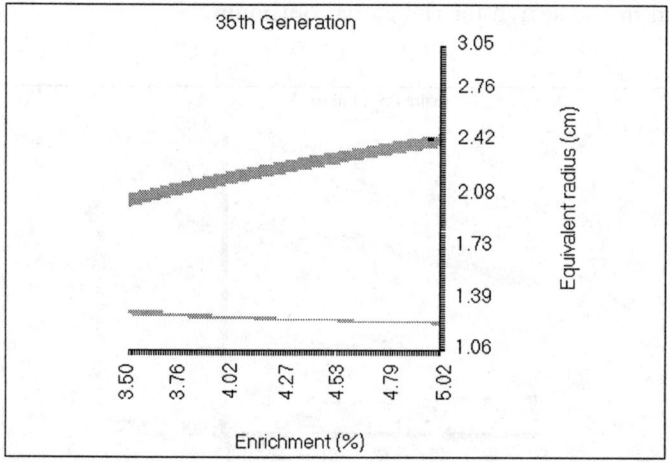

Fig. 9. Upper view of the 35th generation distribution over the search space of the simple problem

5.4 Comparisons Between the Genetic Algorithm and the Classical Method

In Section 5.3 it was mentioned that the genetic algorithm could find the global optimum. In order to emphasize the advantages of the method, an usual nonlinear optimization method, based in some concepts proposed in Rozon (1992), was used to solve the problem. This method uses linear programming (LP) techniques, specifically simplex algorithm, Gass (1975) and Ferguson (1958), to successively optimize small regions of the search space, considering them linear. In a small region around the starting point a linear problem is created in which the variations of flux per variation of the parameters are considered constant, and calculated by small perturbation of the involved parameters. The goal is to find the step length and direction that lead to maximum flux inside that region. This step leads to another point, and another iteration is ready to be made. The same procedure is repeated until the convergence criteria is reached.

The results obtained by the application of the classical method for three experiments, using different starting points, are shown in Table 4. These results reflect the difficulty of the classical method in finding the global optimum. The optimization is strongly dependent on the starting point. The flux found in the first and in the third experiment is exactly the best point of the surface that does not contain the global optimum, indicating a convergence to a local optimum.

Table 4. Results obtained by the application of the classical method to the simple problem

Experiment	E initial (%)	Re initial (cm)	E (%)	Re (cm)	k_{eff}	ϕ ($\times 10^{-4}$)
1	4.25	1.321	3.5	1.304	1.01	1.28
2	4.25	2.032	5.0	2.450	0.99	4.32
3	4.00	1.143	3.5	1.304	1.01	1.28

The poor robustness of the classical method, when applied to multimodal optimization problems, is due to its hill climbing characteristic. However, such kind of technique is certainly a good choice when the optimization problems involves an unimodal search space.

Figure 10 shows the flux convergence in the classical method for the first experiment. It can be seen in this figure that the convergence is very fast, spending few iterations. However, the overhead that is paid for this quick convergence is the susceptibility of finding local optima. The GA, on the other hand, takes more time to converge, however it could find the global (or near-global) optimum in all experiments, demonstrating robustness in finding the global optimum.

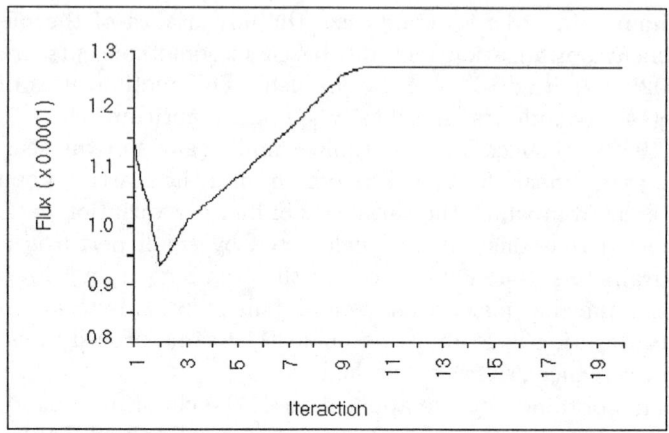

Fig. 10. Convergence of the flux in the classical method

Figure 11 shows the upper view of the search space in which the points found in each iteraction of the classical method are ploted. Once more, it can be seen that the convergence is very fast, however, when a region in which the constraints are satisfied is found, a local convergence occurs.

The overhead paid for the improvements offered by the global optimization provided by the GA is the extra computational cost. The use of the LP based method took about 30 simulation calls contrasting with the GA based method that took about 600. It must be remembered that the objective of the proposed method is to improve the results of the optimization by using a global approach.

6 The Robustness of the GA

It could be observed through the tests made that the GA was robust in finding the global optimum. One important observation is that the GA starts the search with a set of candidates to solutions (many points) contrasting with the philosophy used by the classical method that starts the search at an unique point. If this point is far from any region where the constraints are satisfied, it cannot start the search, because no points satisfying the constraints are found. Moreover, if the starting point is close to a region where constraints are satisfied, but is not connected to the region that contains the global optimum, a local convergence may occur. Hence, it is not a robust technique for this kind of problem.

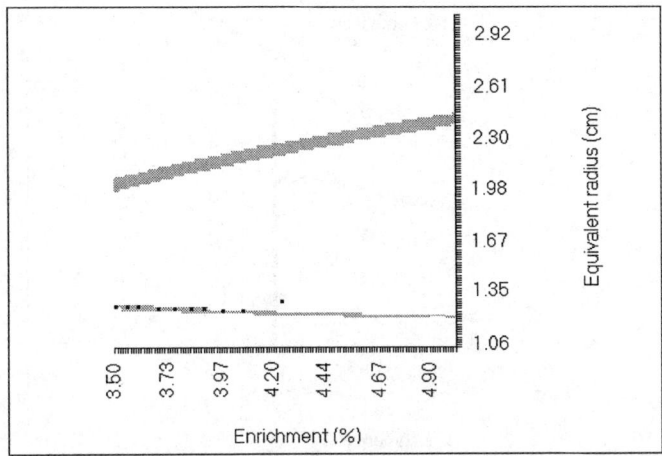

Fig. 11. Upper view of the search space in which are ploted the points found in each iteraction of the classical method

In complex optimization problems that have many parameters to be adjusted, and the computational cost involved in the simulation is high, the possibility of having a bad randomized initial populationis increased.

To verify the robustness of the GA, a bad randomized initial population was chosen. The test was made using the simple problem, with 25 individuals in population. The distribution of the population on the search space can be seen in Figure 12.

Note that the best point found in the initial population is close to the region that contains the local optima. Starting with this initial population, the GA found, in the 5th generation the local optima (see Figure 13), however, some points were getting closer to the global optimum region. Finally, in the 10th generation, the GA found the global optimum (see Figure 14).

This test validates the robustness of the GA that could find the global optimum starting from a poorly randomized population.

6.1 Application of the Genetic Algorithm to a More Complex Optimization Problem

After the tests made with sample problems in order to validate and compare the method with other technique, it is proposed a more complex optimization problem that has characteristics which make it untreatable by conventional techniques. Due to the complexity of the search space, it is practically impossible to evaluate how near are the obtained results from the global optimum. Hence, a qualitative analysis is made, observing convergence and constraint satisfactions.

328

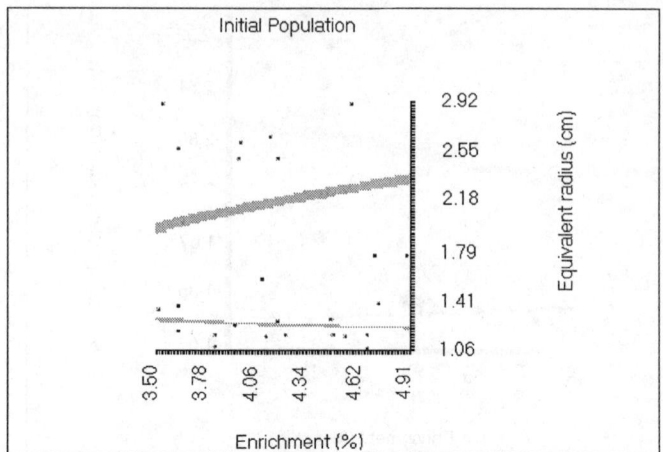

Fig. 12. Upper view of the initial population distribution over the search space of
the simple problem, for an experiment with 25 individuals in the population

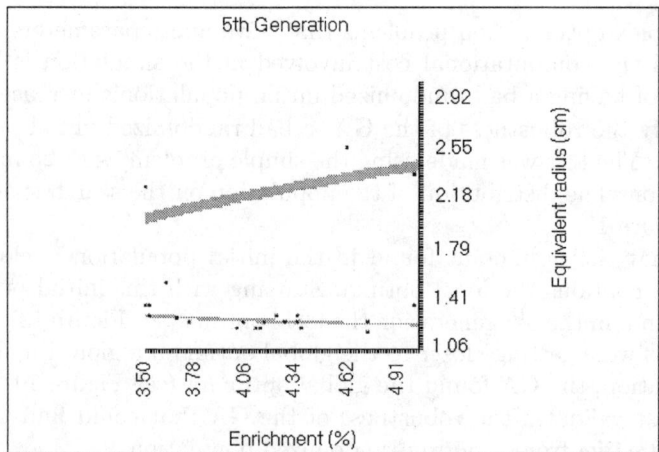

Fig. 13. Upper view of the 5th generation distribution over the search space of the
simple problem, for an experiment with 25 individuals in the population

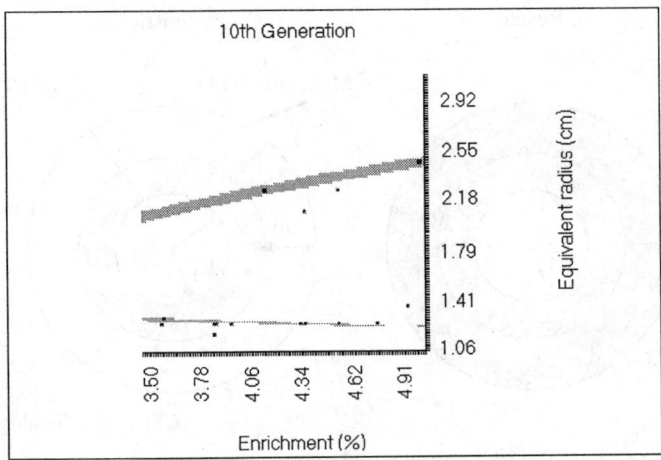

Fig. 14. Upper view of the 10th generation distribution over the search space of the simple problem, for an experiment with 25 individuals in the population

6.2 Problem Description

The objective of this problem is to minimize the average peak factor f_m for the three enrichment zone reactor shown in Figure 15, in order to obtain a given average flux $\phi = 8.0 \times 10^{-5}$ (source normalized), criticality ($k_{eff} = 1.0 \pm 0.01$) and submoderation varying the fuel radius, cladding thickness, equivalent radius, the enrichment of the three zones, cladding material and fuel material, according to the ranges specified in Table 5. The regions dimensions, R1, R2 and R3 remain constant and the values are 86 cm, 38 cm and 18 cm respectively.

Table 5. Parameter ranges for the complex problem

Parameter	Minimum	Maximum
Fuel radius (cm)	0.508	1.27
Cladding thickness (cm)	0.0254	0.254
Moderator thickness (cm)	0.0254	0.762
Isotopic enrichment, zone 1 (%)	2.0	5.0
Isotopic enrichment, zone 2 (%)	2.0	5.0
Isotopic enrichment, zone 3 (%)	2.0	5.0
Fuel material	UO_2 or U-metal	
Cladding material	Zircaloy, Al or Stainless-304	

The optimization problem can be written as:

330

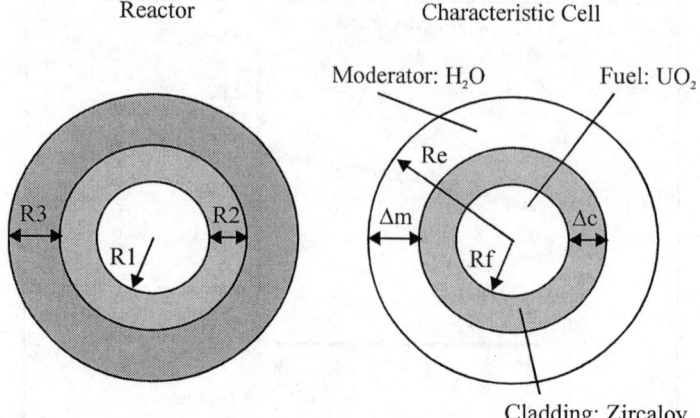

Reactor   Characteristic Cell

Fig. 15. The reactor and characteristic cell of the complex problem

minimize:
$$f_m(Rf, \Delta r, \Delta m, E1, E2, E3, Mf, Mc)$$

subject to:

$$0.99 \leq k_{eff} \leq 1.01$$
$$7.92 \times 10^{-5} \leq \phi_0 \leq 8.08 \times 10^{-5}$$
$$\frac{dk_{eff}}{dV_m} > 0$$
$$Rf_{min} \leq Rf \leq Rf_{max}$$
$$\Delta r_{min} \leq \Delta r \leq \Delta r_{max} \tag{7}$$
$$\Delta m_{min} \leq \Delta m \leq \Delta m_{max}$$
$$E1_{min} \leq E1 \leq E1_{max}$$
$$E2_{min} \leq E2 \leq E2_{max}$$
$$E3_{min} \leq E3 \leq E3_{max}$$
$$FuelMaterial = \{UO_2, U - metal\}$$
$$CladdingMaterial = \{Zircaloy, Aluminum,$$
$$Stainless - 304\}$$

where Mf is the fuel material, Mc is the cladding material V_m is the moderator volume and the *min* and *max* subscript refers to the minimum and maximum values given in Table 5.

6.3   Method Application and Obtained Results

Figure 16 presents an example of a genotype for this problem. The genes that correspond to the continuous parameter, are decoded as described in section 4.1. For the materials, enumerated decode is used, such as 0 to UO_2 and 1 to $U - metal$.

The objective function for this problem is:

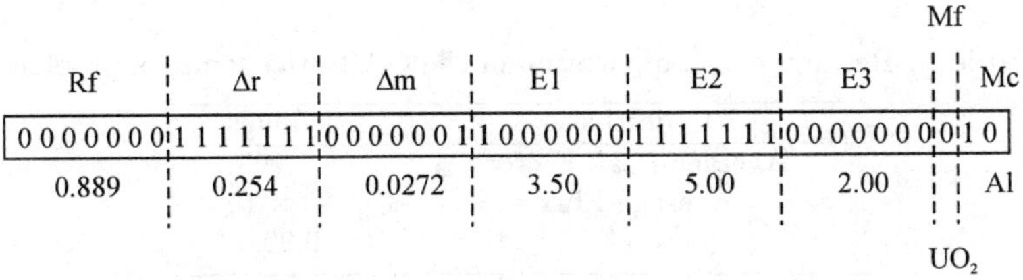

Mf

Rf	Δr	Δm	E1	E2	E3		Mc
0 0 0 0 0 0 0	1 1 1 1 1 1	0 0 0 0 0 0 1	1 0 0 0 0 0 0	1 1 1 1 1 1	0 0 0 0 0 0 0	0	1 0
0.889	0.254	0.0272	3.50	5.00	2.00		Al

UO_2

Fig. 16. Example of genotype and its phenotype representing the optimization parameters of the complex problem

$$f = \begin{cases} f_m, & all\, constraints\, satisfied \\ f_m + c_1 \Delta k_{eff}, & k_{eff}\, out\, of\, constraints \\ f_m + c_2 \Delta\phi, & \phi\, out\, of\, constraints \\ f_m + c_3 \frac{\Delta' k_{eff}}{\Delta V_m}, & \frac{\Delta' k_{eff}}{\Delta V_m}\, out\, of\, constraints \\ f_m + c_1 \Delta k_{eff} + c_2 \Delta\phi, & k_{eff},\, \phi\, out\, of\, constraints \\ f_m + c_1 \Delta k_{eff} + c_3 \frac{\Delta' k_{eff}}{\Delta V_m}, & k_{eff},\, \frac{\Delta' k_{eff}}{\Delta V_m}\, out\, of\, constraints \\ f_m + c_2 \Delta\phi + c_3 \frac{\Delta' k_{eff}}{\Delta V_m}, & \phi,\, \frac{\Delta' k_{eff}}{\Delta V_m}\, out\, of\, constraints \\ f_m + c_1 \Delta k_{eff} + c_2 \Delta\phi + c_3 \frac{\Delta' k_{eff}}{\Delta V_m}, & no\, constraint\, satisfied \end{cases}$$

(8)

where

$$\Delta k_{eff} = |1.0 - k_{eff}| \tag{9}$$

$$\Delta\phi = \frac{|\phi - \phi_0|}{\phi_0} \tag{10}$$

$$\frac{\Delta' k_{eff}}{\Delta V_m} = \frac{k_{eff} - k'_{eff}}{\Delta V_m} \tag{11}$$

c_1, c_2 and c_3 are the penalty constant for k_{eff}, ϕ and $\frac{\Delta' k_{eff}}{\Delta V_m}$ respectively. k'_{eff} is the multiplication factor obtained for ΔV_m of variation in V_m (as the fuel volume - V_f - remains constant, the rate $\frac{V_m}{V_f}$ is altered.

Results obtained by the application of the proposed method to the complex problem are shown in Table 6. Figure 17 shows the fitness convergence.

The genetic parameters used were: population size of 300 individuals, crossover and mutation rate of 60% and 1% respectively, ranked selection and elitism. Possibly because of the difficulty found in this problem, the GA took more generations to reach the stop criteria.

The convergence of the fitness has occured, and a good configuration that has generated small peak factor (with all the constraints satisfied) was found by the GA. The expert knowledge could not indicate a better configuration.

Table 6. Results of the application of the GA to the complex problem

Objestives	Values
Average Peak Factor	1.295
Average Flux	8.23×10^{-5}
k_{eff}	0.9948

Parameters	Values
Fuel radius (cm)	0.6280
Cladding thickness (cm)	0.1604
Moderator thickness (cm)	0.6808
Isotopic enrichment, zone 1 (%)	2.7087
Isotopic enrichment, zone 2 (%)	3.0394
Isotopic enrichment, zone 3 (%)	4.7638
Fuel material	U-metal
Cladding material	Stainless-304

During the genetic algorithm evolution, many others valid configurations (configurations that satisfy the constraints) could be found, however, higher peak factors were associated with them. For example, it could be found valid configurations with peak factor greater than 20. On the other hand, peak factors less than 1.5 could be found for different materials. This discontinuities and multimodalities can often lead an expert to find a local optimum.

7   Conclusions

In the simple problem, it could be clearly observed the advantage of the genetic algorithm over the classical technique. The proposed method was very precise and robust in finding the global optimum, contrasting with the other method, that has converged to local optima.

The applicability of the method to mixed continuous/discrete optimization was demonstrated by the application of the GA to the complex problem. It is important to outline that the GA contrast with others methods in the sense that it does not need any prior knowledge about the search space, fact that allow the optimization of problems without concerning about the existence of derivatives or its inverse, continuity, limits, and so on. Due to the fact that the GA does not need prior knowledge about the search space, this model can be adapted and applied to any kind of problem for which it is possible to represent the points in the search space by a binary string and evaluate these points by an objective function.

In the complex problem, a low peak factor was found. Based on expert knowledge, or during the GA convergence, it could be found several other configurations that lead to a low peak factor and also satisfy the constraints.

However, these configurations are not so good as the optimum found by the GA after convergence.

The extra computational cost imposed by the GA is accentuated by the nuclear reactor simulators that are very time consuming. However, due to the fact that each individual can be evaluated independently from the others, parallel programming [Muhlenbein, 1991] can lead to a great gain in the optimization time.

References

Cacuci, D. G. (1990): Global Optimization and Sensitivity Analisys, Nuclear Science and Engineering, **104**(1), 78-88.

Darwing, C. (1859): The Origin of Species by Means of Natural Selection, John Murray, London.

Davis, L. (1991): Handbook of Genetic Algorithms, VNR, New York.

DeChaine, M. D. and Feltus, M. A. (1995): Nuclear Fuel Management Optimization Using Genetic Algorithms, Nuclear Technology, **111**, 109-114.

Ferguson, R. O., Sargent, L. A. (1958): Linear Programming. Fundamentals and Applications, Mc Graw-Hill Book Company.

Gass, I. S. (1975): Linear Programming. Methods and Applications, Mc Graw-Hill Book Company.

Goldberg, D. E. (1989): Genetic Algorithms in Search Optimization and Machine Learning, Addison-Wesley.

Gray, P., Hart, W., Painton, L., Phillips, C., Trahan, M., Wagner, J. (1997): A Survey of Global Optimization Methods, Sandia National Laboratories, Albuquerque.

Grefenstette, J. J. (1990): A User's Guide to Genesis Version 5.0.

Holland, J. H. (1975): Adaptation in Natural and Artificial Systems, Ann Arbor, University of Michigan.

Haibach, B. V., Feltus M. A. (1997): A Study on the Optimization of Integral Fuel Burnable Absorbers Using Genetic Algorithms based on Cigaro fuel Management System, Annals of Nuclear Energy, **24**(6).

Michalewicz, Z. (1994): Genetic Algorithms + data Structures = Evolution Programs, Springer-Verlag, 2nd Extended Edition.

Muhlenbein, H., Schomisch, M. and Born, J. (1991): The Parallel Genetic Algorithm as Function Optimizer, In Proc. Of Fourth Int. Conf. On Genetic Algorithms, San Diego, CA, 271-278.

Omori, R., Sakakibara, Y. and Suzuki, A. (1997): Application of Genetic Algorithms to Optimization Problems in the Solvent Extraction Process for Spent Nuclear Fuel, Nuclear technology, **118**(1), 26 - 31.

Pereira, C. M. N. A., Schirru, R. and MArtinez, A. S. (1999): Basic Investigations Related to Genetic Algorithms in Core Designs, Annals of Nuclear Energy, **26**(3), 173-193.

Rozon D. and Beaudet, M. (1992): Canada Deuterium Uranium Reactor Design Optimization Using Three-Dimensional Generalized Perturbation Theory, Nuclear Science and Engineering, **111** (1), 1-20.

Suich, J. E. and Honec, H. C. (1967): The HAMMER System Heterogeneous Analysis by Multigroup Methods of Exponentials and Reactors, Savannah River Laboratory, Aiken South Carolina.

15 Genetic Algorithms Applied to the Nuclear Power Plant Operation

R. Schirru[1], C.M.N.A. Pereira[2], and A.S.Martinez[1]

[1]Universidade Federal do Rio de Janeiro, Programa de Engenharia Nuclear, Ilha do Fundão, Rio de Janeiro - Brazil
[2]Comissão Nacional de Energia Nuclear, Instituto de Engenharia Nuclear -Ilha do Fundão, Rio de Janeiro - Brazil
cmnap@cnen.gov.br

Nuclear power plant operation often involves very important human decisions, such as actions to be taken after a nuclear accident/transient, or finding the best core reload pattern, a complex combinatorial optimization problem which requires expert knowledge. Due to the complexity involved in the decisions to be taken, computerized systems have been intensely explored in order to aid the operator. Following hardware advances, soft computing has been improved and, nowadays, intelligent technologies, such as genetic algorithms, neural networks and fuzzy systems, are being used to support operator decisions.

In this chapter two main problems are explored: transient diagnosis and nuclear core refueling. Here, solutions to such kind of problems, based on genetic algorithms, are described. A genetic algorithm was designed to optimize the nuclear fuel reload of Angra-1 nuclear power plant. Results compared to those obtained by an expert reveal a gain in the burn-up cycle. Two other genetic algorithm approaches were used to optimize real time diagnosis systems. The first one learns partitions in the time series that represents the transients, generating a set of classification centroids. The other one involves the optimization of an adaptive vector quantization neural network. Results are shown and commented.

1 Introduction

The evolution and improvement of the intelligent soft computing (ISC) have been following the advances in the hardware technology. These techniques, many times classified as Artificial Intelligence (AI), often require extra computational cost, when compared to traditional ones, however, they present several advantages. Nowadays, because of the fast computers available, practical applications of ISC can be seen in many fields of the engineering, including nuclear engineering that requires safe and robust systems.

The ISC can be applied to the nuclear power plant operations aiding the operator in difficult tasks that requires expert knowledge or cognition, and also automating optimization processes. Two good examples are: the fuel management optimization and the transient/accident diagnosis.

2 PWR's Fuel Management Optimization Using Genetic Algorithm

2.1 The Nuclear Core Reload Problem

The nuclear fuel management optimization is a hard combinatorial optimization problem which during many years have been solved by expert knowledge. The goal is to get the best arrangement of fuel in the nuclear reactor core that lead to a maximization of the operating time. As a consequence, the operation cost will decrease and money will be saved.

In the nuclear core reload, fuel assemblies must be replaced in the core according to its isotopic enrichment, related to the cycles in the core. Besides, fresh fuel must substitute some old assemblies. In other words, fuel assemblies with different ranges of isotopic enrichment must be positioned in the reactor core so that the operating time of that load is maximized.

The simulation of the reactor must be done to evaluate each loading pattern (LP), measuring the fuel cycle length or related parameters such as the peaking factor.

The core reload problem (CRP) consists in combining fuel assemblies with different enrichment (represented by different identification - ID) in the possible positions in the nuclear core. The CRP is a combinatorial problem that can be translated to the classical Traveling Salesman Problem (TSP) which is an np-complete problem of special interest for the optimization community.

The objective in the TSP is to find the minimum distance traveled to complete a tour for a set of cities passing only once in each city and coming back to the starting one. In other words, the goal is to find the minimum hamiltonian cycle in a given graph.

In the CRP the fuel assemblies correspond to the cities in the TSP and the position in the core is related with the order of the cities to be visited. Thus, the solution of the CRP is exactly the solution of the TSP.

2.2 Genetic Algorithms and the Traveling Salesman Problem

The TSP is an np-complete combinatorial problem that has been interesting many researchers in the optimization field. Due to its difficulty, classical optimization methods have been giving place to alternative solutions, including specific heuristics, simulated annealing (SA), genetic algorithms (GA) and others evolutionary algorithms (EA). The results obtained by the use of GA in solving the TSP have encouraged researchers to the application of such technique.

The genetic modeling of a TSP has been made in such a way that the genotypes represent the tour, where the genes represent the cities. The order of the genes in the genotype is the order of the cities to be visited. Using the classical crossover, applied to such genotype representation generate invalid offspring, with some cities twice visited and others not present in the tour. For example, let A B C D E and E C A B D be the two tours (genotypes) chosen for crossover. Let also a crossover point between the second and third city. So, the offspring is A B A B D and E C C D E. Note that both tours are invalid candidates to solution. To face such problem, modifications in the crossover were proposed in order to generate valid offspring, such as Partially Mapped Crossover (PMX), Order Crossover (OX), Cycle Crossover (CX) and others described in Holland (1975). Bean (1994) proposed the Random Keys (RK) to decode the genotype instead of modify the crossover. Schirru (1997) proposed the List Model (LM) as an alternative method that does not need the modifications in crossover. The last one has been successfully applied in practical CRP [Chapot, 1999].

The decodification in the LM occurs as follow: let $G = G_1, ..., G_n$ be the genotype formed its respective genes. $G_1, ..., G_n$ can be any integer between 1 and the number of cities. Let $C = C_1, ..., C_n$ be a base city list. To decode G in a valid tour T, each G_i^{th} city element must be removed from C and pushed into T until G is empty.

For example let $G = 3, 2, 3, 2, 1$ and $C = A, B, C, D, E$. At the first step, the 3rd element will be removed from G and put into T. So, the new list C is A, B, D, E and $T = C$. In the second step, the actual 2nd element will be removed from G and put into T. Now, $C = A, D, E$ and $T = C, B$. Then, the 3rd element will be removed from G and put into T, generating $C = A, D$ and $T = C, B, E$. The same occur with the actual 2nd element, and $C = A$ and $T = C, B, E, D$. Finally, the 1st and last element of C is removed and put into T, generating the decoded tour $T = C, B, E, D, A$ that is valid.

2.3 A Practical Application of Genetic Algorithms in the Core Reload Problem

To illustrate the use of GA in the CRP, a practical application made by Chapot (1999) will be described. The reload optimization system, called AL-GER (Genetic Algorithm Applied to Reload Studies) is a GA modeled based on the Genetic Search Implementation System, Genesis [Grefesntette, 1990] and the Advanced Nodal Code, ANC [Liu, 1985], using the LM approach to the TSP.

The GA creates a random initial population representig the positioning of the fuel assemblies in the nuclear core. Each genotype decodification is a loading pattern that must be simulated by the ANC. The result of the simulation is a value corresponding to the cycle length, or peaking factor, which make part of the objective function. This value is used as fitness for each individual. Then the GA operates over the population. The process operation-evaluation is repeated until the convergence criteria is reached.

The ALGER system has been applied to optimization problems for Angra 1 nuclear power plant. It is a Brazilian 626 MW Westinghouse PWR. Angra 1 core is comprised of 121 fuel assemblies, 80 of them belonging to 8-fold symmetric type, 40 having a 4-fold symmetry and one fuel assembly located in the middle of the core.

Applying the ALGER system to the minimization of the radial peaking factor, F_{XY}, using eight-core symmetry and two dimensional geometry modeled in ANC, it was obtained, at the end of 120 generations (stop criterion) the value of 1.304 for the radial peaking factor. Figure 1 illustrate the convergence of the F_{XY}. In the third generation, the GA found $F_{XY} = 1.327$ which is much lower than the value obtained by the manual optimization $F_{XY} = 1.344$. As a result, the cycle length for this LP was 3 ppm (about one Effective Full Power Day - EFPD).

Another test with the objective of cycle length maximization was made. In this case, the results obtained by manual method, using the ANC 3D quarter-core full-cycle depletion, was Nuclear Enthalpy Power Peaking Factor $(F\Delta H) = 1.366$ and Cycle Length = 228 EFPD.

Maximizing cycle length implies maximization of end of cycle boron concentration. However, to minimize the computational costs it was used the equilibrium of Xenon C_B as the parameter to be maximized.

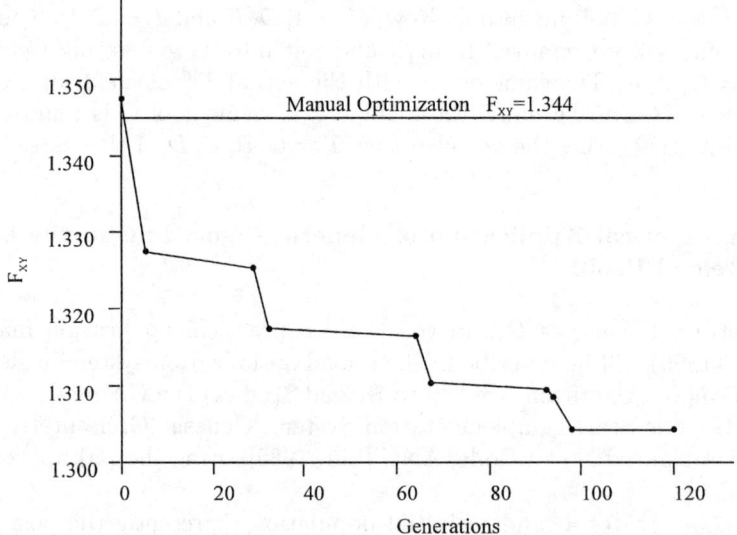

Fig. 1. F_{XY} convergence

At the end of 100 generations, the maximum C_B found was 1198 ppm, with $F_{XY} = 1365$, as shown in Figure 2.

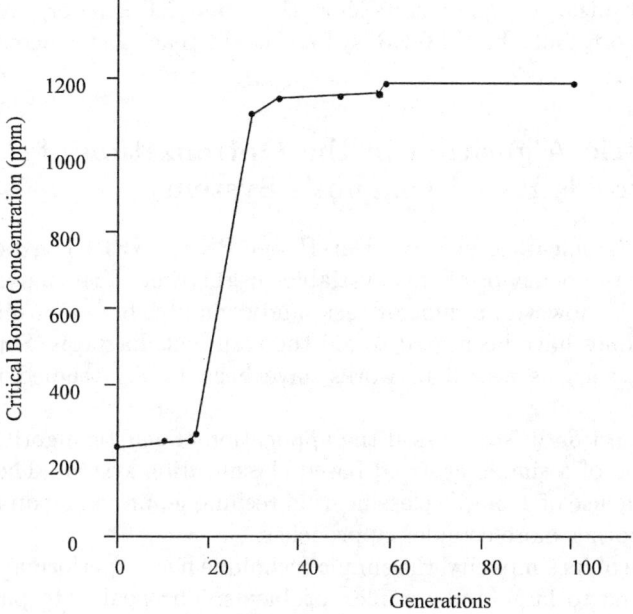

Fig. 2. C_B convergence

Figure 3 shows the LP found.

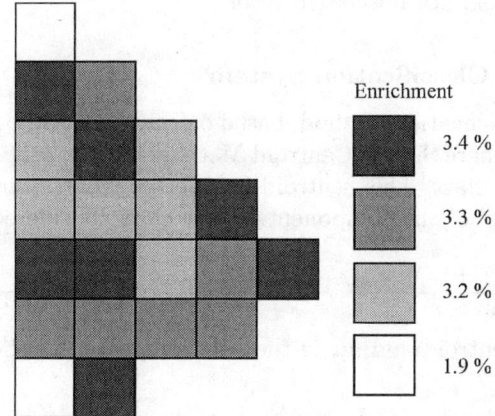

Fig. 3. Best loading pattern: eight-core symmetry

Extrapolating the critical boron concentration to 0 ppm, it was found a cycle length of 277 EFPD, 49 EFPD greater than the value obtained by manual optimization. The results show that the CRP is a very difficult task for the expert, but the ALGER system could reach better configurations automatically.

3 Genetic Algorithm in the Optimization of a Centroids Based Diagnosis System

Transient classification in a Nuclear Power Plant (NPP) can be made by monitoring the behavior of state variables in the time. The analysis of these time series is, however, a difficult task, and computer based pattern recognition techniques have been used to aid the transient diagnosis. Sophisticated techniques, such as neural networks have been hardly studied in the last years.

In this section it is discussed the application of genetic algorithms to the optimization of a simple centroid based classification system. The idea is to optimize the use of a simple classification technique, providing an alternative method to other more complex approaches.

Because of its simplicity, the simple technique has its performance reduced when applied to large and complex problems. The goal is to partition the problem domain into smaller sub-domains where the performance of a simple classification system is maximized. Besides, the minimization of the number of partitions is aimed, in order to get minimum number of classification rules (dictated by the number of centroids).

To find the best partition set implies in finding the minimum number of clusters in the training set to which the application of the simple technique can lead to minimum of misclassification.

3.1 A Simple Classification System

The simplest classification method, based on centroids will be described. This technique, called here Simple Centroid Method (SCM), proposes the calculus of a centroid per class. This centroid is a vector whose components are the arithmetic average of the components of the class considered as a cluster.

$$c = \left(\frac{a_{11} + \ldots + a_{m1}}{m}, \frac{a_{12} + \ldots + a_{m2}}{m}, \ldots, \frac{a_{1n} + \ldots + a_{mn}}{m} \right) \tag{1}$$

where c is the centroid and a_{ij} is the j-th component of the i-th point that belongs to a class.

In the classification, by the SCM, the centroid which is closer to the test point will dictate the class. The metric used by the SCM is the Euclidean distance between two points, given by:

$$d(c, p) = \left[\sum_{i=1}^{n} (c_i - p_i)_2 \right]_{0.5} \tag{2}$$

where c_i is the i-th component of the centroid c and p_i is the i-th component

of the point p.

In the example shown in Figure 4, the centroids of two clusters of points (representing two classes) can correctly classify the points p_1 and p_2.

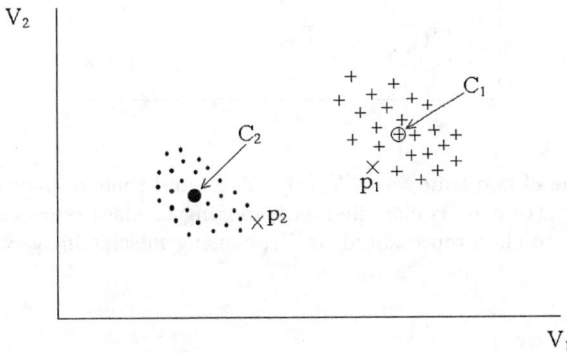

Fig. 4. Example of two clusters and their respective centroids C_1 e C_2 where the pattern p_1 is classified as belonging to class identified by C_1 and p_2 as belonging to class identified by C_2

In the time series, however, the patterns are spread along the time axis, and they can interlace themselves creating confusion areas. In this case, the application of the SCM may lead to misclassifications. An example of time series representing two classes can be shown in Figure 5.

Applying the SCM to the hypothetical example of two other time series in Figure 6, the classes centroids will be located at the middle of the time interval, that may lead to misclassification for all patterns yet to be classified.

It must be observed that the example in Figure 6 presents classes that are easy to be distinguished visually or by simple rules. However, because of its symmetry this distinction becomes a difficult task to the SCM, that will have coincident centroids for both classes.

However, if each class in Figure 6 could be partitioned into two subclasses, as shown in Figure 7, the centroids for each subclass would give the correct classification for each point. For example, if the closest centroid to a given point is the centroid of a subclass of class V_1, then the point belongs to class V_1.

Considering the patterns and centroids shown in Figure 7, the application of the simple algorithm described in Figure 8, where C(c,i) is the i-th centroid of class c and d(P,C(c,i)) is the distance between point P and the cluster C(c,i), it may correctly classify a given pattern P. This algorithm, called here

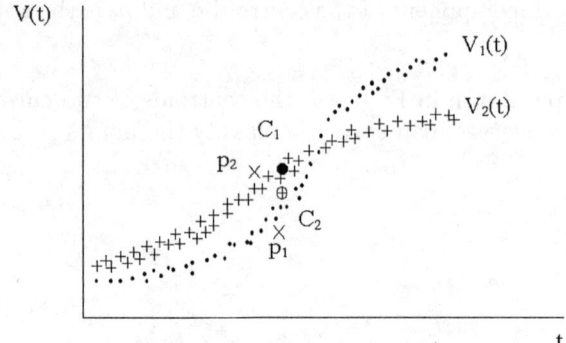

Fig. 5. Example of two time series $V_1(t)$ e $V_2(t)$ and their respective centroids C_1 and C_2 where pattern p_1 is classified as belonging to class represented by C_2 and p_2 as belonging to class represented by C_1, causing misclassifications

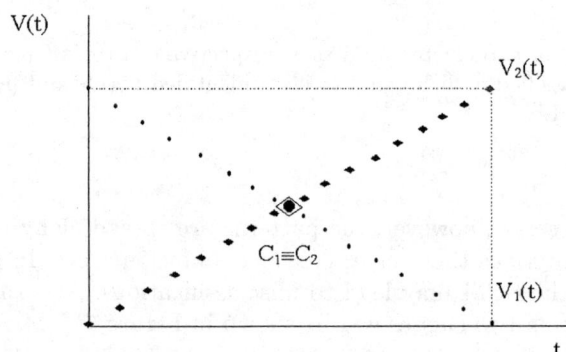

Fig. 6. Two hypothetical time series where the centroids are coincident

multiple centroids method (MCM) uses the very simple philosophy proposed in the SCM in an optimized arrangement of centroids.

In the example shown in Figure 7, it is easy to find the best centroid set that well classifies V_1 and V_2. However, if the number of variables that characterizes each pattern is high, the visualization of the best centroids may be very difficult or in some cases almost impossible. In this case, the distinction of patterns may become a very difficult task. That's why it is necessary to use a powerfull optimization technique. Here it is proposed the use of genetic algorithm [Goldberg, 1989; Davis, 1991], that is a blind and global search technique [Gray 1997; Renders 1992].

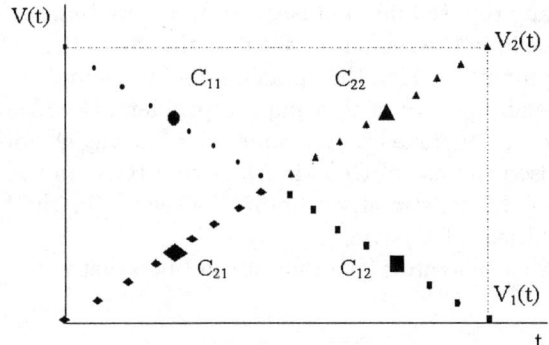

Fig. 7. Two hypothetical time series partitioned into two subclasses each one

```
MinimumDistance=d(P,C(1,1));
For c=1 to NumberOfClasses
   For i=1 to NumberOfSubclassesPerClass
      If (d(P,C(c,i)) < MinimumDistance) Then
         MinimumDistance=d(P,C(c,i));
         Class=c;
      End If;
   End For;
End For.
```

Fig. 8. The algorithm of classification on a multiple-centroid scenario

3.2 The Minimum Centroids Set Method

The goal of the proposed Minimum Centroid Set (MCS) method [Pereira, 1998] is to find the best set of centroids that better distinguishes the classes, considering one or more centroids per class. In other words, the MCS method intends to find the minimum number of time partitions (and the positions) of the classes in which the subclass centroid better classifies a set of test patterns, using the MCM of Figure 8.

It is very difficult to observe the similarity between patterns that are represented by a great number of variable time series. The class does not have any chance of being aided by visual perception. Hence, the mathematical foundations and their different kind of metrics comes to help in these cases. However, the simple methods may not be applied successfully in most real cases. But if the complex real problem is divided into simple small ones, it may be possible to apply the simple methods to each simple problem. The problem is how to find this minimum set of small problems.

The goal of the proposed method is to apply the well-known and simplest mathematical metric in the regions of the domain where they work well. But to find this subdomain in which the application of the simple method is well-succeeded may be an np-hard or np-complete problem. Because of the nature of the optimization to be done and the poor prior knowledge about the search space, it is proposed the use of Genetic Algorithm (GA) in the optimization model. The use of the genetic algorithm in the search for the best centroids set is the main subject of this work.

Figure 9 shows a schematic diagram of the MCS method.

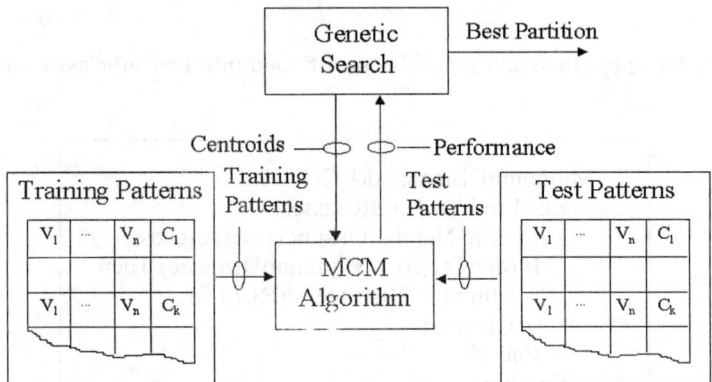

Fig. 9. Schematic Diagram of the MCS method

The genetic algorithm generates candidates for the best partitions in which centroids (of each subclass) are calculated by the SCM, that provide the centroids to the MCS algorithm. The MCS algorithm tries to classify the pattern test set, sending the number of well succeeded classifications (performance) back to the GA that, in its turn, uses it to guide the optimization search. The process is repeated until either the convergence criteria is achieved or the performance is satisfactory, when the loop may be interrupted.

3.3 The Genetic Algorithm Model

Genetic algorithms [Holland, 1975] are computational optimization techniques that simulate the species evolution theory [Darwin, 1859] using natural selection, crossover, mutation, and other genetic operators. The genetic algorithms manipulate a population of symbolic structures, that represent points in the search space, in order to evolve this population to its best adaptation, leading to the best solution for the problem. In the GA [Goldberg, 1989; Davis, 1991] a symbolic structure called genotype represents the codification of a candidate to the solution of the optimization problem. These characteristics of each candidate (individual) dictate its adaptability in the environment. The selection and evolution are made in such a way that stronger individuals have more chance to be selected, passing their characteristics to the offspring. This way, from generation to generation, the tendency is that strong individuals become stronger and more numerous while the weak individuals may be extinct.

The GA starts the adaptation process from a set of possible configurations - a random generated population of individuals. Evaluating independently each individual by an objective function, the GA assigns to each one a fitness that predicts its resistance and adaptability. Then it is simulated a natural selection process in which the selection probability of a given individual is a function of the fitness.

The crossover in a classic GA is simulated by choosing a random cross point over the binary string that represents the genotype, followed by the change of parts between the two parents. The mutation is simulated by the inversion of one of the bits of the genotype according to the mutation probability.

Taking into account the reproduction, crossover and mutation, it can be proved statistically, by the schemata theory, described in Goldberg (1989), that strong individuals may be more numerous in subsequent generations, and the population converge to the optimum adaptation.

The genotype is a binary string with length equal to the number of training patterns per accident. Each bit represents a time position. The groups of 1's (one) or 0's (zeros) together form a subclass.

Figure 10 shows an example of time partitions, and in Figure 11 it can be seen the genotype that encode the clusters of two hypothetical classes of Figure 10.

Structures like the one shown in Figure 11 are over which the GA works to guide the optimization based on the objective function or fitness.

The fitness was designed in order to reward the high percentage of correct classification and low number of centroids. Equation 3 presents the fitness of the GA for the MCS method.

$$f = K_d.P - (K_c + K_o).C \qquad (3)$$

where P is the performace (number of correct diagnosis), C is number of clusters, K_d is the weighting factor for the performance, K_c is the weighting factor for the number clusters and K_o is the offset factor.

The fitness presents three constants that must be adjusted according to the relative importance of the variables to be optimized.

In this work, it was used $K_d = 4$ and $K_c = 1$. This means that the performance is four times more important than the number of clusters. The value of K_o was 0.01 to impose a conflict resolution in cases where the variables are different and the fitness is the same (of course if Ko were zero). This value

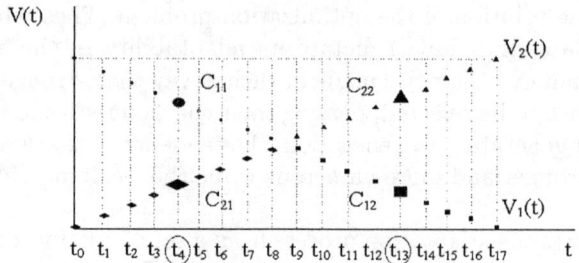

Fig. 10. Two hypothetical time series

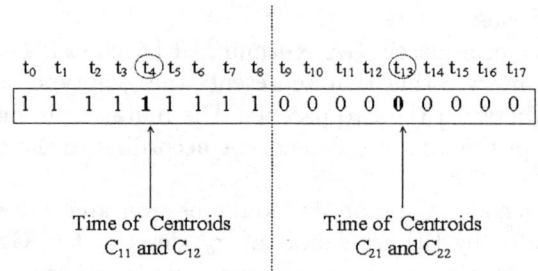

Fig. 11. Example of genotype representing the clusters

should be sufficiently small to distinguish combinations with the required precision.

For example, let $K_d = 4$, $K_c = 1$ and $K_o = 0$. If P = 98 and C = 1, then the fitness, f, is 391 (f = 4 x 98 - 1), on the other hand, if P = 99 and C = 5, the fitness should be also 391 (f = 4 x 99 - 5). To solve this conflict, it is necessary to establish who must win this competition. If K_o is set to 0.01, the fitness values would be 390.99 (f = 4 x 98 - 1.01) and 390.95 (f = 4 x 99 - 5.05) respectively, and the performance is said to be more important. Once defined the genotype shape and the fitness function, the last thing to do is to adjust the genetic parameters with the aim of optimizing the convergence process of the genetic algorithm.

3.4 Results

In order to validate the MCS method for pattern recognition two steps were proposed. At first, three simple reference cases were created, aiming to verify the precision of the proposed method, once the results (best centroids) are well-known.

The three scenarios shown in Figure 12 were used to validate the proposed method. The curves are discretized into 57 training patterns for each class with steps of 1 second.

The scenarios are very simple and the minimum centroids set can be easily identified by a simple look. However, they are good to evaluate the precision of the GA model. As the GA works only with the time axis, increasing the number of variable time series that represents each class not necessarily increases the complexity for the GA.

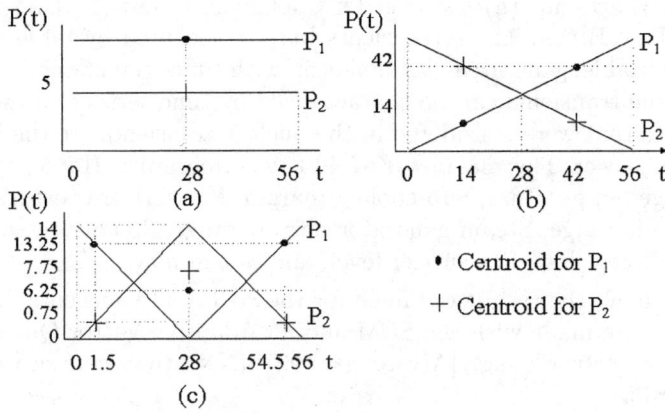

Fig. 12. The test scenarios. (a) Test-1 (b) Test-2 (c) Test-3

The system was implemented with the Genesis code [Grefenstette, 1990], and the results obtained for the centroids optimization are shown in Table 1.

Table 1. Best centroids obtained by the GA model

| | Centroid 0 | | Centroid 1 | | Centroid 2 | | |
	Teoretical	GA	Teoretical	GA	Teoretical	GA	Correct
Test-1	28.0	28.0	-	-	-	-	114/114
Test-2	14.0	14.0	42.0	42.5	-	-	112/112
Test-3	1.5	3.0	28.0	28.5	54.5	53.5	108/110

It can be observed that the minimum centroids set were found with small errors, when compared to the expected results, leading to a high performance in the classification. The small difference between the result of the genetic search and the optimum values may be due to the intrinsic characteristic of genetic algorithms methods that aim to approximate the solution with a acceptable computational cost, finding near-optimum regions. Not differently from other Artificial Intelligence methods, the Genetic Algorithm is better applied when a formal method to solve the problem is very time consuming or non-existent (np-hard and np-complete problems).

The scope of the application was extended to a real case involving three transients in a nuclear power plant (NPP). Such kind of pattern recognition have been treated using artificial neural networks, e.g., [Bartlett 1992; Bartal 1995]. It was chosen three typical transients in nuclear power plants represented by 15 variables each, and it is assumed by hypothesis that all variables are required to the transient recognition. The transients considered were (i) the Black-out, (ii) the Lost Of Coolant Accident (LOCA) and (iii) the Steam Line Break. These transients were represented by tables with 60 points per variable, generated by simulation with time-step of one second (by hypothesis the transients can be characterized by time series of 60 seconds). The variables that were considered in the nuclear transients are the Primary flux, Nuclear power, Thermal power, Cold Leg temperature, Hot leg temperature, Average temperature, Sub-cooling margin, Pressurizer pressure, Steam generator wide range, Steam generator narrow range, Steam pressure, Feed water flow, Break flow, Pressurizer level, and Steam flow.

Table 2 shows the results obtained for the NPP real transients diagnosis. Comparisons are made with the SCM and an Adaptive Vector Quantization (AVQ) Neural Network, e.g., [Alvarenga, 1997] (NN), that were submitted to the same tests.

Table 2. Results for the real nuclear transient diagnosis problem

	Correct/Trials	Correct (%)	Partitions
SMC	144/180	82	1
MCSM	171/180	95	3
AVQ-NN	155/180	86	-

The MCS method could efficiently deal with the compromise between minimization of the number of centroids (that implies in smaller rules associated) and the performance (number of correct classifications) maximization, that is the main objective this pattern recognition system.

As shown in Table 1 the GA was very precise in the optimization of the centroids for the reference scenarios, converging to a configuration that is very close to the known best solution. In Table 2 the method demonstrates the ability to optimize the relation between performance and number of centroids for the real nuclear transient identification, improving considerably the efficiency of the SCM. Besides, the results were better than the ones obtained by the AVQ NN. The fact is that the AVQ NN tries to learn the best vectors that distinguish the patterns, that is something analogue to finding simple clusters. Maybe that's why the results are not far from those obtained by the SCM.

An advantage to be considered is that the computational cost of the optimization is not necessarily increased with the number of variable, as occurs with the NN learning.

References

Alvarenga, M.A.B., Martinez, A.S. and Schirru, R. (1997): Adaptive Vector Quantization Optimized by Genetic Algorithms for Real-Time Diagnosis through Fuzzy Sets, Nuclear Technology, **120**(3), 188-197.

Bartal, Y., Lin, J. and Uhrig, R.E. (1995): Nuclear Power Plant Transient Diagnostics Using Artificial Neural Network that Allow "Don't-Know" Classification, Nuclear Technology, **110**, 436-449.

Bartlett, E.B. and Uhrig, R.E. (1992): Nuclear Power Plant Status Diagnostics Using an Artificial Neural Network, Nuclear Technology, **97**, 272-281.

Bean, J.C. (1994): Genetic Algorithms and Random Keys for Sequencing and Optimization, ORSA, **6**(2).

Chapot, J. L., Da Silva, F. C. and Schirru, R. (1999): A New Approach to the Use of Genetic Algorithms to Solve the Pressurized Water Reactor's Fuel Management Optimization Problem, Annals of Nuclear Energy, **26**(7), 641-655.

Darwing, C. (1859): The Origin of Species by Means of Natural Selection, John Murray, London.

Davis, L. (1991): Handbook of Genetic Algorithms, VNR, New York.

DeChaine, M.D. and Feltus, M. A. (1995): Nuclear Fuel Management Optimization Using Genetic Algorithms, Nuclear Technology, **111**, 109-114.

Goldberg, D.E. (1989): Genetic Algorithms in Search Optimization and Machine Learning, Addison-Wesley.

Gray, P., Hart, W., Painton, L., Phillips, C., Trahan, M., Wagner, J. (1997): A Survey of Global Optimization Methods, Sandia National Laboratories, Albuquerque.

Grefenstette, J.J. (1990): A User's Guide to Genesis Version 5.0.

Holland, J.H. (1975): Adaptation in Natural and Artificial Systems, Ann Arbor, University of Michigan.

Kim, Y.S., and Mitras, S. (1994): An Adaptive Integrated Fuzzy Clustering Model for Pattern Recognition, Fuzzy Sets and Systems, **65**, 297-310.

Kosko, B. (1992): Neural Networks and Fuzzy Systems, Prentice Hall.

Liu, Y.S. et al. (1985): ANC: A Westinghouse Advanced Nodal Computer Code. WCAP-10965, Westinghouse.

Oliver, I.M., Smith, D.J., Holland, J.R.C. (1987): A Study of Permutation Crossover Operator on th Traveling Salesman Problem. Procedings of The Second International Conference on Genetic Algorithms and Their Applicationa, 224-230.

Pereira, C.M.N.A., Schirru, R. and Martinez, A.S. (1998): Learning an Optimized Classification System From a Data Base of Time Series Patterns Using Genetic Algorithms, In: Ebecken, N.F.F. (ed), Data Mining, Computational Mechanics Publications, WIT Press, England.

Renders, J.M., Flasse, S.P., Verstraete, M.M. and Nordwik, J.P. (1992): A Comparative Study of Optimization Methods for the Retrieval of Quantitative Information from Satellite Data, EUR 14851 EN, Joint Research Centre.

Schirru, R, Pereira, C.M.N.A., Chapot, L. Carvalho (1997): F. "A Genetic Algorithm Solution For Combinatorial Problems - The Nuclear Core Reload Example," XI Encontro Nacional de Fsica de Reatores, 357-360, Brazil.

S.K. Rogers and M. Kabrisky (1991) An Introduction to Biological and Artificial Neural Networks for Pattern Recognition, Spie Optical Engineering Press.

16 Reactor Controller Design Using Genetic Algorithms with Simulated Annealing

Kadir Erkan and Erhan Bütün

Kocaeli University, Anıtpark, 41300, İzmit - Kocaeli, Turkey
{kerkan,ebutun}@kou.edu.tr

This chapter presents a digital control system for ITU TRIGA Mark-II reactor using genetic algorithms with simulated annealing. The basic principles of genetic algorithms for problem solving are inspired by the mechanism of natural selection. Natural selection is a biological process in which stronger individuals are likely be winners in a competing environment. Genetic algorithms use a direct analogy of natural evolution. Genetic algorithms are global search techniques for optimisation but they are poor at hill-climbing. Simulated annealing has the ability of probabilistic hill-climbing. Thus, the two techniques are combined here to get a fine-tuned algorithm that yields a faster convergence and a more accurate search by introducing a new mutation operator like simulated annealing or an adaptive cooling schedule. In control system design, there are currently no systematic approaches to choose the controller parameters to obtain the desired performance. The controller parameters are usually determined by test and error with simulation and experimental analysis. Genetic algorithm is used automatically and efficiently searching for a set of controller parameters for better performance.

1 Introduction

The control system is based on a second order linear model with unknown but time varying parameters. A self-tuning control algorithm with the generalised minimum variance (GMV) strategy is employed. Genetic algorithms (GAs) are applied here to the problem of estimating controller parameters. The control rod speed is selected as a control variable. The controller controls the reactor power to follow the given trajectory. Self-Tuning control is achieved by an iteration at each sample interval through the following cycle: parameter estimation, calculation of trajectory, and calculation of control signal.

The basic requirement of a research reactor is to meet power demand. This is accomplished either manually or automatically by the reactor control system.

The automatic control system of ITU TRIGA Mark-II reactor is based on the electromechanical choppers that are used to compare the deviations form setpoints such as demand power and period. When measurements deviate from the setpoints the regulating rod is moved to restore the system. It takes long time to reach the demand power and to stay constant by using current control system of the reactor [1].

In this chapter, a self-tuning digital control system is designed. The structure of the controller is determined by the GMV [2,4]. The parameters of the controller need to be optimised to meet the design objective. A genetic algorithm (GA), which is a proven search/optimisation technique, has been applied to optimal design of the controller [5,7].

2 Discrete-Time Model of the Reactor

The purpose of self-tuning controller is to control the reactor with unknown and time-varying dynamic. It is assumed that the system can be described by the discrete-time relationship [8,9]:

$$A(z^{-1})y(t) = z^{-k}B(z^{-1})u(t) + C(z^{-1})\xi(t) \tag{1}$$

where $u(t)$ is the control input (rod speed), $y(t)$ is the measured output (power), $\xi(t)$ is a noise, z^{-1} is a backward shift operator, t is the sampling instant, k is the system delay time. A, B and C are polynomials in the backward shift operator defined as follows:

$$A(z^{-1}) = 1 + a_1 z^{-1} + \ldots + a_{na} z^{-na}$$

$$B(z^{-1}) = b_0 + b_1 z^{-1} + \ldots + b_{nb} z^{-nb} \tag{2}$$

$$C(z^{-1}) = c_0 + c_1 z^{-1} + \ldots + c_{nc} z^{-nc}$$

Although the order of the polynomials is determined by that of the model which describes the reactor, an acceptable control can be achieved with self-tuning controllers of lower order than suggested by the theoretical design. Therefore, the following values were chosen:

$$na = 2, nb = 1, nc = 1, k = 1 \tag{3}$$

3 Controller Design

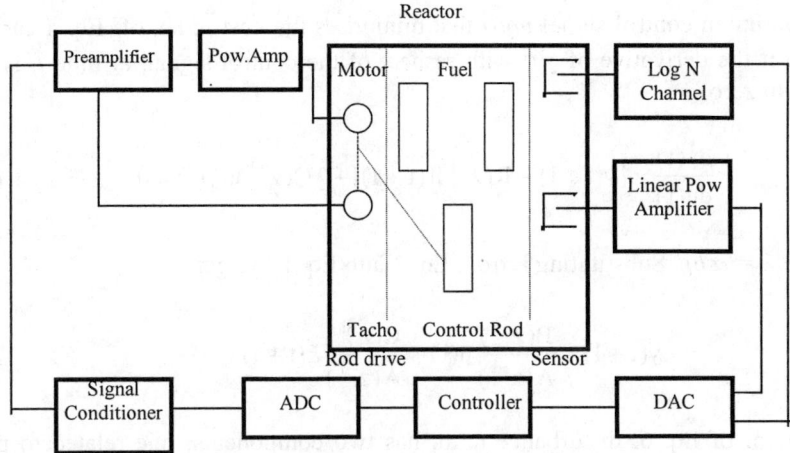

Fig. 1. Self-tuning control system

The proposed self-tuning control system is shown in Figure 1. The signal conditioner, the Analog/Digital converter (ADC), Digital/Analog Converter (DAC), and personal computer are used in control system in addition to the present reactor hardware. ADC converts analog signals coming from Log N channel and Linear Power Amplifier to digital form. Self-tuning controller software generates a control signal that will be converted to analog form using DAC depending on these signals and set points. The signal conditioner circuits are used to apply this control signal to the reactor. Signal conditioner output drives a preamplifier to move the regulating rod. The amplitude of this signal changes the rod speed while its phases determine the moving direction. The purpose of self-tuning controller is to control systems with unknown parameters. Self-tuning controller can be designed by starting with known system and a design method. The control algorithm is obtained by introducing a recursive parameter estimator. The parameters are replaced with their estimated values. Self-tuning control algorithm uses generalised minimum variance strategy that evolves from minimisation of a single-stage cost function that includes control costing. In order to calculate the optimal control signal the following cost function was chosen:

$$J = E\{\gamma(t)\} = E\left\{[y(t+1) - R(z^{-1})f(t+1)]^2 + \alpha[Q(z^{-1})u(t)]^2 \big| t\right\} \qquad (4)$$

where $E\{.\}$ is the expectation at time $(t+k)$, which is conditioned on data acquired up to time t, $f(t)$ is the specified trajectory, α is weight coefficient to adjust the trade off between the tracking accuracy and the magnitude of the control signal.

$R(z^{-1})$ and $Q(z^{-1})$ are the polynomials to eliminate the steady-state error and to limit abrupt changes in the control signal $u(t)$, respectively.

The optimum control signal $u_0(t)$ that minimises the cost index of Eq. 4 can be found if the derivative of $\gamma(t)$ with respect of the control signal, at time t, is set equal to zero.

$$\frac{d\gamma(t)}{du(t)} = [y(t+1) - R(z^{-1})f(t+1)] + \lambda Q(z^{-1})u_0(t) = 0 \tag{5}$$

where $\lambda = \alpha b_0$. Substituting k from Eq. 3 into Eq. 1, we get

$$y(t+1) = \frac{B(z^{-1})}{A(z^{-1})}u(t) + \frac{C(z^{-1})}{A(z^{-1})}\xi(t+1) \tag{6}$$

Last term of Eq. 6, disturbance term, has two components: one related to past disturbance ($\xi(t)$, $\xi(t-1)$, ...); one related to future disturbances ($\xi(t+1)$, $\xi(t+2)$, ..). Using the following Diphontine equation can separate these components:

$$\frac{C(z^{-1})}{A(z^{-1})} = E(z^{-1}) + z^{-1}\frac{G(z^{-1})}{A(z^{-1})} \tag{7}$$

Ignoring the future disturbance term in the minimisation of Eq. 4, we get

$$y(t+1|t) = \frac{G(z^{-1})}{C(z^{-1})}y(t) + \frac{B(z^{-1})}{C(z^{-1})}u(t) \tag{8}$$

Substituting Eq. 8 into Eq. 5, the optimum control signal is written as

$$u_0(t) = \frac{C(z^{-1})R(z^{-1})f(t+1) - G(z^{-1})y(t)}{B(z^{-1}) + \lambda Q(z^{-1})C(z^{-1})} \tag{9}$$

In order to limit sudden changes in the control signal $u(t)$, Q can be selected as follows

$$Q(z^{-1}) = 1 - z^{-1} \tag{10}$$

From Eq. 1 and Eq. 9, it can be shown that the closed-loop equation is

$$y(t) = z^{-1}\frac{B(z^{-1})}{B(z^{-1}) + \lambda Q(z^{-1})A(z^{-1})}R(z^{-1})f(t) + \frac{B(z^{-1}) + \lambda Q(z^{-1})C(z^{-1})}{B(z^{-1}) + \lambda Q(z^{-1})A(z^{-1})}\xi(t) \tag{11}$$

For zero steady-state offset between $y(t)$ and $f(t)$, we can set [3]

$$R(z^{-1}) = \left. \frac{B(z^{-1}) + \lambda Q(z^{-1})A(z^{-1})}{B(z^{-1})} \right|_{z=1} \tag{12}$$

The coefficients of the G, B and C polynomials of Eq. 9 need to be estimated in order to calculate optimal control signal. Therefore using Eq. 1, Eq.11 is written as

$$y(t+1|t) = g_0 y(t) + g_1 y(t-1) + b_0 u(t) + b_1 u(t-1) - c_1 y(t|t-1) \tag{13}$$

Introducing the following measurement vector $\phi(t)$ and parameter vector $\theta(t)$

$$\theta(t) = [g_0 \; g_1 \; b_0 \; b_1 \; c_1]^T$$

$$\phi(t) = [y(t) \; y(t-1) \; u(t) \; u(t-1) - y(t|t-1)] \tag{14}$$

then, Eq. 13 can be rewritten as

$$y(t|t+1) = \phi(t)\theta(t) \tag{15}$$

GASS is applied here to the problem of estimating unknown parameters in $\theta(t)$.

4 Trajectory

Operating procedures often dictate that the power be raised on a constant period [10,12]. In this chapter, power changes are made on constant period until power level reaches 90% of the demand power. Then, period becomes longer and approaches infinity as the power level reaches the demand level. Equations to calculate the trajectory are,

$$f(t+1) = \left(\frac{T+\tau}{\tau} \right) y(t) \;\; , y(t) < 0.9 y_s \tag{16}$$

$$f(t+1) = (1-\beta)\, y_s + \beta y(t), \;\; y(t) \geq 0.9 y_s \tag{17}$$

where T is sampling interval, τ is the desired period, y_s is demand power, and β can be calculated by setting Eq. 16 equal to Eq. 17 for $y(t)=0.9y_s$.

5 Genetic Algorithms for Parameter Estimation

Calculus-based search problems such as point to point method used in many optimisation methods from a singular point to the next in decision space, using some transition rule to find next point, are risky because it is a perfect prescription for locating false peaks in many peaked search spaces. Some problems may be expressed by a parameters set. These parameters are regarded as chromosome genes and can be structured with a string of values in binary form. GAs work from a highly valued database of points simultaneously climbing many peaks in parallel; so, the probability of finding a false peak is reduced over methods that move point to point. Many search techniques need some auxiliary information to work acceptably. For example, gradient techniques need derivatives calculated analytically or numerically to be able to climb the peak. In contrast, GAs do not need this kind of secondary information, they only need objective function values. This characteristic makes GAs more canonical. In fact GAs are not considered as a mathematically guided algorithm. GAs are stochastic and a non-linear processes and a final product containing the best or strongest elements of the previous generations tends to be carried forward into the following generation. In other words, the rule is survival of the fittest. GAs use random choice as a mechanism to guide a search toward regions of the search spaces and probabilistic transition rules to get a trajectory for search.

GAs have some different aspects from the other optimisation methods:

1. GAs work with a coding of the parameter set, not the parameters themselves.

2. GAs search from a population of points, not a single point.

3. GAs use objective function information, not derivatives or other auxiliary knowledge.

4. GAs use probabilistic transition rules, not deterministic rules [6].

The initial population of size N is generated randomly. The user can define several strings if the user guesses the controller parameters for the different operating region of the system. User defined strings may be kept constant between generations so that actual parameters neighbourhood of these is found faster by GAs. In this chapter no user-defined string is used. GAs work between sampling instants. Several generations can be made for each sample. The proposed algorithm can be summarised as follows.

1. Determine the parents according to fitness value.

2. Reproduce the children from parents.

3. Apply simulated annealing to new population.

4. Calculate fitness value of each string,

5. Select fittest string as a controller parameters.

The mechanics of genetic algorithms consist of copying and swapping partial strings. Simple GAs are composed of three operators: *reproduction, cross-over,* and *mutation.*

5.1 Reproduction

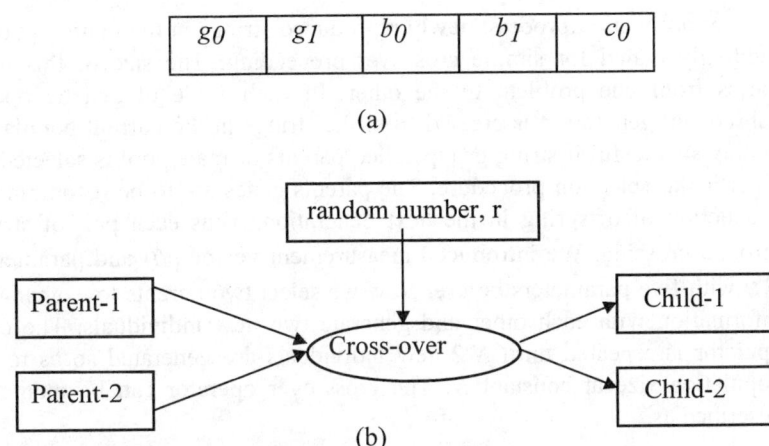

g_0	g_1	b_0	b_1	c_0

(a)

(b)

Fig. 2. a. Chromosome-structure; b. Cross-over operator

In reproduction process, individual strings are copied according to their objective function values. This function is called as fitness function. It means that fitness function is an artificial adaptation of natural selection, a measure of profit, utility or goodness that we want to maximise. This relative copying process means that higher valued strings have a higher probability of contributing one or more offsprings in the next generation.

The fitness function to be maximised is chosen as

$$g(t) = \left[\sum_{i=0}^{w} \gamma^{t-i} [y(t-i) - \overline{y}(t-i)]^2 \right]^{-1} \tag{18}$$

where $\overline{y}$ is the estimated output, γ is the forgetting factor to give more weight to recent data, w is the window size.

The reproduction operator may be implemented in algorithmic form in a number of ways. One of them is to give a bias number for each current string in population according to their proportion to their total fitness. A roulette wheel can be designed where each current string in the population has a roulette wheel slot sized in proportion to its fitness. Each time we need another offspring, a simple spin of the weighted roulette wheel yields the reproduction candidate. More highly fit strings have a higher number of offsprings in the subsequent generation in this method. An exact copy of string is made, after a string has been selected for reproduction. This copied string is then taken into a mating pool for following genetic operator actions.

5.2 Crossover

After reproduction process, newly reproduced strings in the mating pool are first randomly mated for simple crossover proceeding. The size of this population varies from one problem to the other. In each cycle of genetic operation, a subsequent generation is created from the strings in the current population. This is only successful if string group called parents or mate pool is selected by using a particular selection procedure. The parents genes are to be recombined for the production of offspring in the next generation. Thus each pair of strings goes through crossing. We introduced measurement vector $\phi(t)$ and parameter vector $\theta(t)$ with five parameters before. Now we select two parents to exchange genetic information with each other and generate two new individuals. The cross-over operator is repeated until $N/2$ new individuals are generated so as to keep the population size at constant N. The cross-over operator can be mathematically described as,

$$Child1 \quad : \quad \begin{cases} g_0 = r \cdot g_{01} + (1-r) \cdot g_{02} \\ g_1 = r \cdot g_{11} + (1-r) \cdot g_{12} \\ b_0 = r \cdot b_{01} + (1-r) \cdot b_{02} \\ b_1 = r \cdot b_{11} + (1-r) \cdot b_{12} \\ c_1 = r \cdot c_{01} + (1-r) \cdot c_{02} \end{cases}$$

$$Child2 \quad : \quad \begin{cases} g_0 = (1-r) \cdot g_{01} + r \cdot g_{02} \\ g_1 = (1-r) \cdot g_{11} + r \cdot g_{12} \\ b_0 = (1-r) \cdot b_{01} + r \cdot b_{02} \\ b_1 = (1-r) \cdot b_{11} + r \cdot b_{12} \\ c_1 = (1-r) \cdot c_{01} + r \cdot c_{02} \end{cases}$$

where $r \in (0,1)$ is a random number. Such a cross-over operator for real numbers is called weighted average operator.

5.3 Mutation

In artificial genetic systems mutation operator can effectively protect against an irrecoverable loss like even reproduction and crossover, search and recombine existing concepts. Sometimes they lose potentially useful genetic material. Mutation is seldom random alteration of the value of string position for the simple GAs and a random walk through the string space. Mutation takes place with a certain probability. Thus the genetic information of the selected individual changes and genetic information introduced. In our simulations, mutation rate is set to be 0.01 and the $(g_0, g_1, b_0, b_1, c_0)$ values of the selected individual encounter the following changes:

$$g_0 = g_0 + (r_1 - 0.5) \cdot 2 \cdot g_{0\_max}$$
$$g_1 = g_1 + (r_2 - 0.5) \cdot 2 \cdot g_{1\_max}$$
$$b_0 = b_0 + (r_3 - 0.5) \cdot 2 \cdot b_{0\_max} \qquad (19)$$
$$b_1 = b_1 + (r_4 - 0.5) \cdot 2 \cdot b_{1\_max}$$
$$c_1 = c_1 + (r_5 - 0.5) \cdot 2 \cdot c_{1\_max}$$

where $r_1, r_2, r_3, r_4, r_5 \in (0,1)$ are five random numbers and $g_{0\_max}, g_{1\_max}, b_{0\_max}, b_{1\_max}, c_{0\_max}$ are maximum changes of g_0, g_1, b_0, b_1 and c_0 under mutation. If the mutation rate is too low to move the search space to other domains, it will yield the result that the population becomes more homogeneous.

There are adjacent chromosomes with higher fitness values at a search point in every generation. Consequently, all new chromosomes are replaced by their adjacent chromosomes to get faster convergence with finer tuning before next generation is evolved. This may be accomplished by mutation operator in a standard operator in standard GAs with SA.

As we stated before GAs are stochastic optimisation algorithms based on natural selection and genetics. The data processed by the GAs is set (population) of binary coded strings (chromosomes). Each chromosome has its own fitness value determined by an objective function. Standard GAs are weak in fine-tuning the parameters to arrive at exact optima. Therefore, in this chapter, the simulated annealing genetic algorithms (SAGA) was used to achieve more accurate search and a faster convergence. SAGA has been discussed elsewhere [15].

Real coding [16] is used in the chapter to reduce the chromosome length and to avoid problems that are often encountered in binary coded strings.

6 Simulated Annealing

Simulated Annealing (SA) is a stochastic computational technique derived from statistical mechanics to find near global minimum cost solutions to large optimisation problems. It is very important and powerful algorithm in optimisation and high order problems although it is very slow. Behaviour of very large systems of interacting systems in a thermal equilibrium at a finite temperature is studied in statistical mechanics. SA searches minimal energy states by using random processes. To improve simulated annealing, GAs can merge with SA. This improved algorithm uses simulated annealing crossover and simulated annealing mutation operators instead of standard ones.

The set of system components spatial positions represents system configuration. If a system is in thermal equilibrium at temperature T, then the probability $\pi_T(s)$ that the system is in a given configuration s depends on the energy $E(s)$, and follows the Boltzman distribution

$$\pi_T(s) = \frac{\exp\left[\dfrac{-E(s)}{kT}\right]}{\displaystyle\sum_{w \in S} \exp\left[\dfrac{-E(w)}{kT}\right]} \tag{20}$$

where k is Boltzman's constant and S is the set of all possible configurations.

Let at time i the system be in configuration q for the simulation of the behaviour of a system in thermal equilibrium. A candidate r for the configuration at time $i+1$ generated randomly and accepted according to:

$$p = \frac{\pi_T(r)}{\pi_T(q)} = \exp\left[\frac{-(E(r) - E(q))}{kT}\right] \tag{21}$$

If p is greater than 1, that means energy of r is less than energy of q, then configuration r is accepted as the new configuration for time $i+1$. If p is less than or equal to 1, then configuration r is accepted as the new configuration, with the probability p. Thus configurations of the higher energy states may be attained [15].

7 Simulation Results

The proposed self-tuning controller is first simulated and its properties are determined. To maintain stability during the initial stages, the controller must either possess a reasonable estimate of the reactor or another controller must control the reactor. In this chapter, $u(t) = f(t) - y(t)$ is used for the first w iterations.

The research reactor was simulated using YAVCAN code [17]. The self-tuning controller and genetic algorithms with simulated annealing are programmed in Pascal. Simulation was done for a population size of 100 and four generations per sample. Simulation values are: $y_0 = 5$ kW, $y_s = 250$ kW, $\tau = 10$ sec., $T = 0.2$ sec., $N = 100$, $w = 10$, $\gamma = 0.95$.

Figures 3-6 show the results obtained using simulator. As shown in Figure 3 controller is able to bring the reactor power to the desired level without overshoot by following the given trajectory. There is no steady state error. So, simulation results demonstrate the effectiveness of the proposed genetic algorithms with simulated annealing approach.

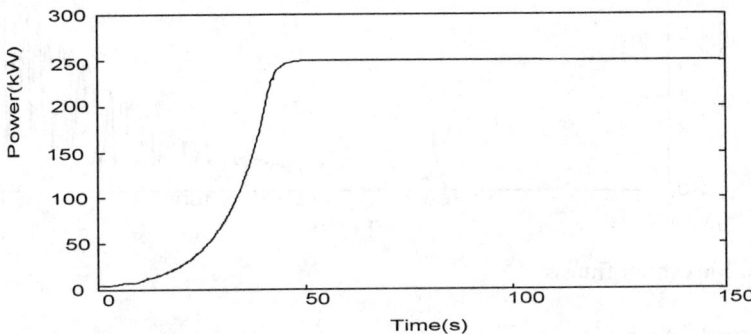

Fig. 3. Reactor Power

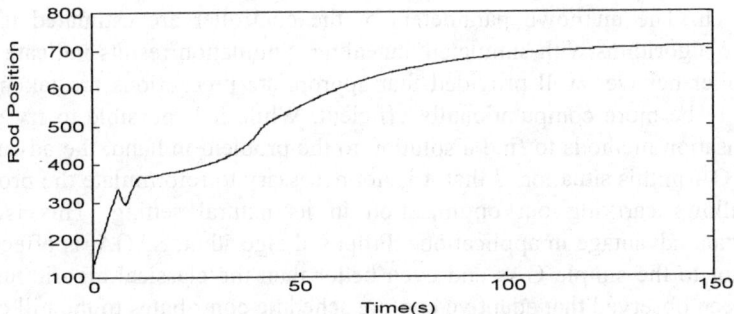

Fig. 4. Rod position

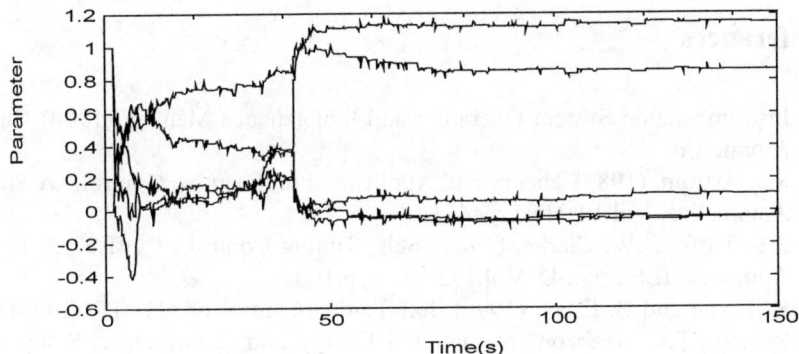

Fig. 5. Estimated parameters

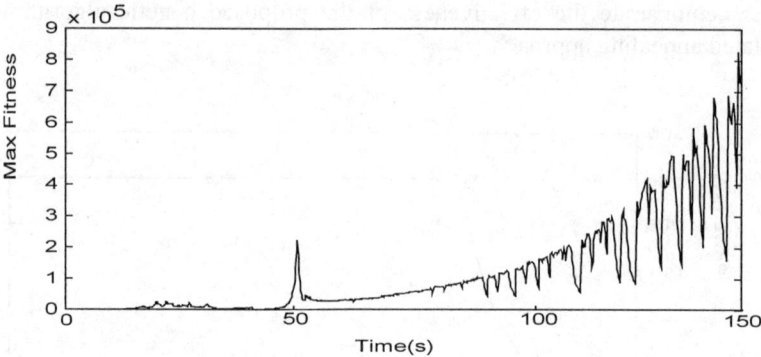

Fig 6. Maximum fitness

8 Conclusion

A self tuning controller for ITU TRIGA Mark-II Reactor has been derived in this chapter. The self-tuning controller is based on generalised minimum variance approach. The unknown parameters of the controller are estimated using the genetic algorithms with simulated annealing. Simulation results indicate that the controller behaves well provided that appropriate precautions are taken. GA is likely to be more computationally efficient. While it is possible to try classical optimisation methods to find a solution to the problem in hand, the advantage of using GA in this situation is that it is not necessary to reformulate the problem as GA allows carrying out optimisation in its natural setting. This is a very important advantage in applications. Proposed algorithm, SAGA, is effective and superior to the simple GAs and even better than the classical algorithms. It has also been observed that adaptive cooling schedule contributes to the hill climbing and faster convergence.

References

1. Instrumentation System Operation and Maintenance Manuel, (1976) General Atomic Co.
2. K.J. Astrom, (1983) Theory and Application of Adaptive Control - A Survey, Automatica, Vol.19, No.5, p. 471.
3. P.S. Tuffs, D.W. Clarke, (1985) Self -Tuning Control of Offset: A Unified Approach, IEE Proc.-D, Vol.132, No:3, p.100.
4. K. Erkan and B. CAN, (1996) Self Tuning Control of ITU TRIGA Mark-II Reactor, First Trabzon International Energy and Environment Symposium, Karadeniz Technical University, Turkey, p 285.
5. D.E. Goldberg, (1989) Genetic Algorithms in Search, Optimisation and Machine Learning, Addision Wesley.

6. J.H. Holland, (1992) Adaptation in Natural and Artificial Systems: An Introductory Analysis with Applications to Biology, control and Artificial Intelligence, Cambridge: The MIT Press.

7. K.F. Man, K.S. Tang, S. Kwong, (1996) Genetic Algorithms: Concepts and Applications, IEEE Trans. On Industrial Electronics, Vol. 43, No.5, p.519.

8. M.A. El-Sharkawi, S. Weerasooriya, (1990) Development and Implementation of Self-Tuning Tracking Controller for DC Motors, IEEE Trans. on Energy Conversation, Vol.5, No.1, p.122.

9. L. Ljung, (1987) System Identification - Theory for the User, Prentice-Hall, Englewood Cliffs, N.J.

10. J.A. Bernard, D.D. Lanning, A. Ray, (1985) Use of Reactivity Constraints for the Automatic Control of Reactor Power, IEEE Trans. On Nuclear Science, Vol.32, No.1, p.1036.

11. J.A. Bernard, D.D. Lanning, (1985) Issues in the Closed-Loop Digital control of Reactor Power, IEEE Trans. On Nuclear Science, Vol.33, No.1, p.992.

12. K. Erkan and M. Inal, (1996) Neural-Net Controller for ITU TRIGA Mark-II Reactor, First Trabzon International Energy and Environment Symposium, Karadeniz Technical University, Turkey, p 279.

13. B. Can, (1992) Optimal Control of ITU TRIGA Mark-II Reactor, 12th European TRIGA Users Conference, NRI Bucaresti-Pitesti, Romania.

14. K. Krinstinsson and G.A. Dumont, (1992), System Identification and Control using Genetic Algorithms, IEEE Trans. Syst. Man. And Cyber., p.1033.

15. I.K. Jeong and J.J. Lee, (1996) Adaptive Simulated Annealing Genetic Algorithm for System Identification, Eng.Appl.Artif.Intell., Vol.9, No.5, p.523.

16. S.S. Ge, T.H. Lee, G. Zhu, (1996) Genetic Algorithm Tuning of Lypunov-Based Controllers: An Application to a single-Link Flexible Robot System, IEEE Trans. On Industrial Electronics, Vol.43, No.5, p.567.

17. H. Yavuz, B. Can, E. Akbay, (1990) The Investigation of Nonlinear Dynamics Behaviour of ITU TRIGA Mark-II Reactor, 11th European Triga Users Conf., Hiedelberg, Germany, pp 2.39-2.53.

17 Logic-Based Hierarchies for Modeling Behavior of Complex Dynamic Systems with Applications

Y.-S. Hu and M. Modarres

Center for Technology Risk Studies
University of Maryland, College Park, MD 20742, USA
modarres@eng.umd.edu

Most complex systems are best represented in the form of a hierarchy. The Goal Tree Success Tree and Master Logic Diagram (GTST-MLD) are proven powerful hierarchic methods to represent complex snap-shot of plant knowledge. To represent dynamic behaviors of complex systems, fuzzy logic is applied to replace binary logic to extend the power of GTST-MLD. Such a fuzzy-logic-based hierarchy is called Dynamic Master Logic Diagram (DMLD). This chapter discusses comparison of the use of GTST-DMLD when applied as a modeling tool for systems whose relationships are modeled by either physical, binary logical or fuzzy logical relationships. This is shown by applying GTST-DMLD to the Direct Containment Heating (DCH) phenomenon at pressurized water reactors which is an important safety issue being addressed by the nuclear industry.

1 Hierarchy – the Nature of Complexity

Most complex systems are in the form of a hierarchy. As suggested by Courtois (1985), building upon the pioneering work of Simon (1982),

> Frequently, complexity takes the form of a hierarchy, whereby a complex system is composed of interrelated subsystems that have in turn their own subsystems, and so on, until some lowest level of elementary components is reached.

Complex systems in a hierarchic model may be represented in form of hierarchy of "objects" and "relationships" between these objects along with attributes of objects. The objects may represent functions, behaviors, properties, goals, states or events relevant to the system. The relationships show the nature and degree to which objects influence each other.

The hierarchic decomposition usually proceeds to a point where system functions/sub-functions have been sufficiently described such that the purpose of each element of the system (i.e., equipment, human, software) can be explicitly and unambiguously described. For each element, pertinent properties such as functions, behavior, events, parameters may also be described. Similarly, relationships between various objects describing systems elements and their properties should be described. It is important to recognize that, ideally, if one precisely knows all basic elements of a hierarchy, their properties and their causal relationships, then no uncertainty would have remained in the analysis of the system. Unfortunately, behaviors of complex systems carry a considerable amount uncertainty. Therefore, a hierarchic model reflects the state of our understanding about the system. This is perfectly acceptable, so long as the state of knowledge about the system is completely and accurately represented.

Hierarchic models may center on one or more properties of the system objects. For example they may be functional, structural, behavioral, goal or event driven hierarchic model. Depending on the nature of the uses of these models it may be more beneficial that a hybrid combination of them be used. One such modeling paradigm discussed in this chapter is the Goal Tree Success Tree (GTST) framework. This framework is used to model system objects and their relationships primarily based on their functions. However, related properties of states, goals, events, etc. may also be incorporated. Master Logic Diagram (MLD) is also used in connection with GTST (called GTST-MLD) to provide a better way of modeling complex relationships.

To represent the dynamic behavior of complex systems, as shown in Fig. 1, Modarres (1992) classifies the hierarchic connectivity relationship into three types: binary, physical (temporal) and fuzzy. Although it may be argued that theoretically binary logic is a subset of fuzzy logic, for simplicity and for all practical purposes the division provided above is a practical and reasonable categorization. Each object in a binary logic relationship may be in one of two possible states or may take one of two possible values. The relationship between two objects (e.g. parent – child relationships) describes that when the child object is in one of its two possible states what would be the state of the parent object. Fuzzy logic relationships on the other hand describe relationships between two objects when each object can be or take values from a grade of different states or values. Physical relation is referred to object relations that are described by some mathematical expression such that each object may be or take values from a continuous state or continuum of variable. When the relationships are not fully known (i.e., uncertain) or are linguistically imprecise, a fuzzy relation may be most appropriate. For different relations, separate representations are necessary to best describe the system behavior.

It has been shown by Hu and Modarres (1994, 1996, and 1999) that both logical and physical relationships can be adequately represented by fuzzy relations; thus, to extend the ability of a GTST-MLD to represent system dynamics, time-dependent fuzzy logic is applied to full-scale GTST-MLD modeling. Such a fuzzy-logic-based approach is called Dynamic Master Logic Diagram (DMLD).

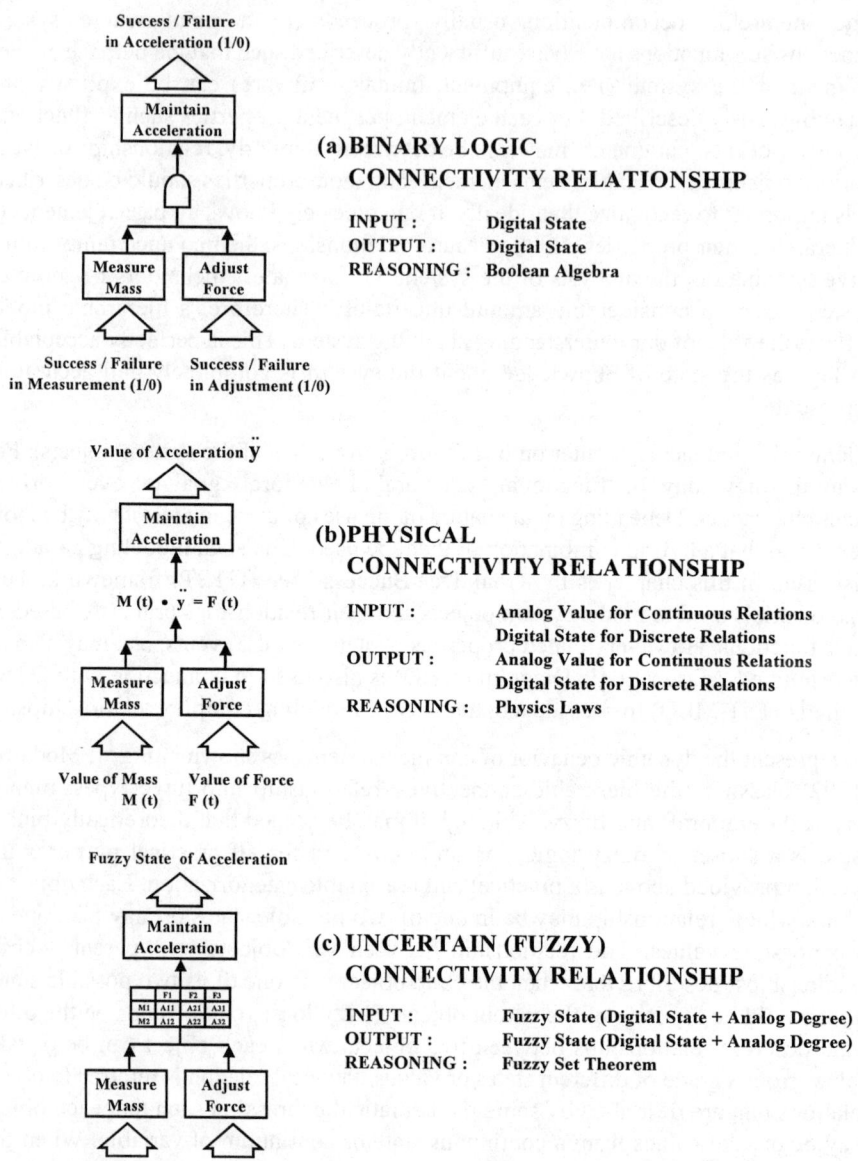

(a) BINARY LOGIC CONNECTIVITY RELATIONSHIP

INPUT :   Digital State
OUTPUT :   Digital State
REASONING :   Boolean Algebra

(b) PHYSICAL CONNECTIVITY RELATIONSHIP

INPUT :   Analog Value for Continuous Relations
               Digital State for Discrete Relations
OUTPUT :   Analog Value for Continuous Relations
               Digital State for Discrete Relations
REASONING :   Physics Laws

(c) UNCERTAIN (FUZZY) CONNECTIVITY RELATIONSHIP

INPUT :   Fuzzy State (Digital State + Analog Degree)
OUTPUT :   Fuzzy State (Digital State + Analog Degree)
REASONING :   Fuzzy Set Theorem

Fig. 1. A Classification of Relationship for Connecting Hierarchies

2 The GTST-DMLD Hierarchy Modeling

2.1 The Goal Tree Success Tree (GTST)

The GTST structure was initially developed as part of a DOE sponsored research (Hunt and Modarres 1984) to assess and improve information systems in a nuclear plant. The Goal Tree Success Tree (GTST) is a functional decomposition framework for modeling complex physical systems. The GTST framework is conceptually illustrated in Fig. 2. A GTST is a functional hierarchy of a system organized in levels starting with a "Functional Objective" at the top. The functional objective describes, in an unambiguous term, the principal purpose of the system. For example, in the case of a whole nuclear power plant the objective can be "Economical, Safe and Reliable Generation of Electric Power." While the objective should be carefully selected to limit the scope of the decomposition to the areas of interests, the modeler has total freedom in defining the scope of the decomposition.

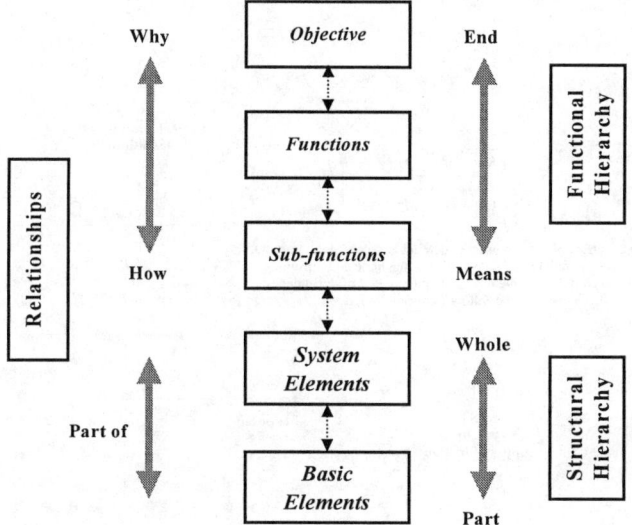

Fig. 2. A Conceptual Goal Tree Success Tree (GTST) Framework

2.2 The Master Logic Diagram (MLD)

The MLD clearly shows the interrelationships among the independent parts of the systems, including all of the support items. For example consider a conceptual MLD shown in Fig. 3. Similar to the GTST framework, the MLD is developed and displayed hierarchically. In essence, the dependency matrix in form of a lattice displays the hierarchy of the MLD. For each object (subsystem, hardware, human, software) shown on the top of the MLD (i.e., the main elements of the complex system that realizes the lowest level physical functions in the GTST model) are shown. On the left side of the matrix, the set of supporting elements needed (to start, control, cool, power, and monitor) the main elements have been displayed. Since among these supporting elements there may be some relationships, another hierarchy may depict and display this complexity. For example, in Fig. 3, element T needs X, and element B needs T. These are shown by character •.

Because of its hierarchy, the MLD of a complex system provides an excellent model to describe causal effects of a failure or disturbance in a complex system. There are two important causal relations that can be extracted from a MLD. The first is to know the ultimate effect of a failure, and the second is to determine the ways that a function can be achieved or a system (subsystem) would successfully work. For example, if one desires to use it inductively (that is to answer "what happens if..." type questions), he/she should follow the causal path A shown in Fig. 3. In order to describe the effect of losing a supporting item, the loss should be followed from left to right and from bottom to top.

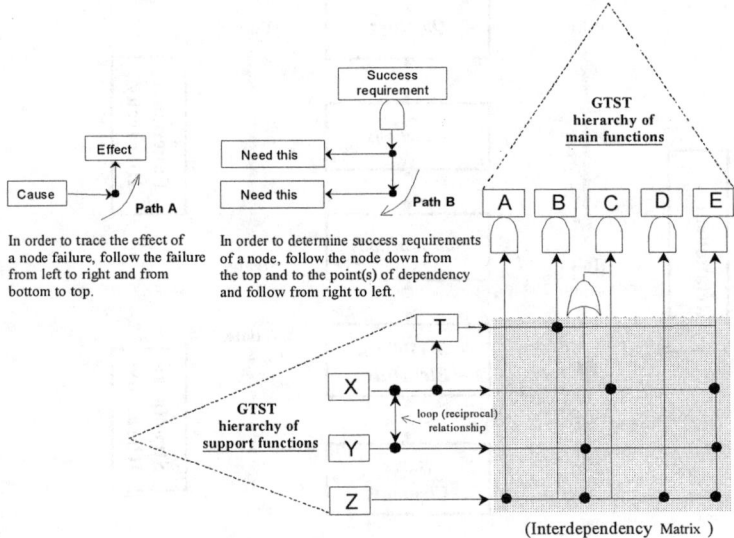

Fig. 3. A Conceptual Master Logic Diagram (MLD)

2.3 The Combined GTST-MLD Hierarchy

The combined GTST-MLD provides a powerful functional/structural description method. The combined framework is conceptually depicted in Fig. 4. Applications of the GTST-MLD framework have been widespread. They can be classified into two general categories of static and dynamic applications. In the static applications, normally the logic of a GTST-MLD representation is used as a logical representation of a complex system to perform the analyses. No time variable can be entered into the calculations. The GTST may be applied to show not only "how" the system works, but also "how well" it works. The performance of the system may be classified into capability (i.e., ability to realize the main functions), availability, reliability, and efficiency. These analyses include applications in reliability engineering [Modarres et al. 1985], maintenance [Modarres et al. 1992; Roush et al. 1985; Hunt and Modarres 1985; Hadavi et al. 1996], evaluating operating events [Kim et al. 1990; O'Brien et al. 1997; Modarres and Cadman 1986], and risk assessment and management [Zamanali et al. 1991; Modarres 1992; Ni and Modarres 1996; Dezfuli et al. 1994].

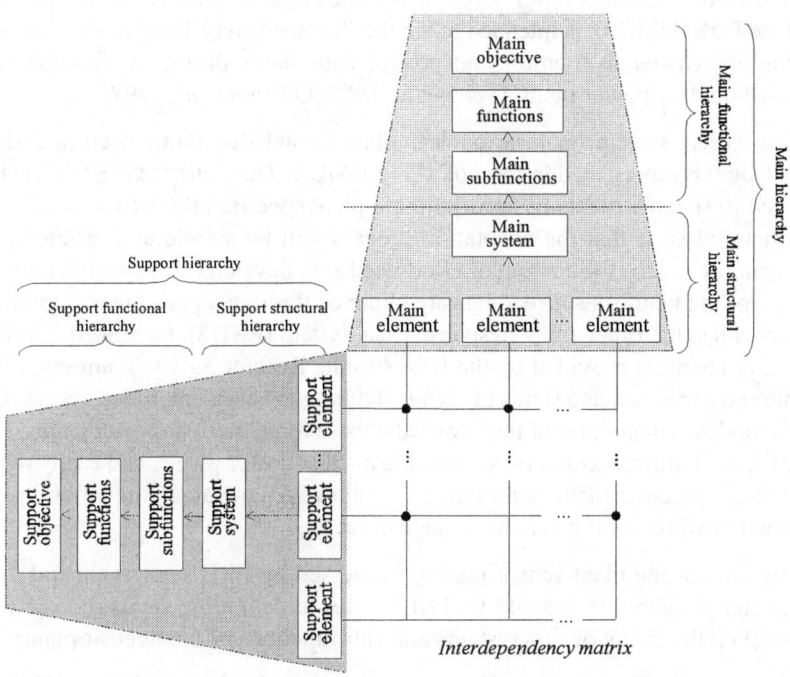

Fig. 4. A Combined GTST-MLD Framework

In the dynamic applications, the time-dependent changes should be considered. Most of the dynamic applications of the GTST-MLD method have been in the area of knowledge-based systems. These can be further classified into two subcategories: "snapshot" type (i.e., representing the knowledge base of a complex interacting physical system) and "temporal behavior" type (i.e., modeling physical behaviors of complex systems) knowledge-based systems.

The applications to the snapshot-type systems include Root Cause Failure Workstation [Modarres et al. 1989], GOTRES [Chung et al. 1989], MOAS-II [Kim et al. 1990], FAX [Chen and Modarres 1992], RSAS [Sebo et al. 1989], FORMENTOR [Nordvik et al. 1993], and STARS [Nordvik et al. 1996]. For the temporal behavior type application, Hu and Modarres (1994, 1996, and 1999) proposed the use of fuzzy logic method to represent dynamic behavior in a GTST-MLD model and the behavior of their parents. More details are discussed in Section 2.4 of this chapter.

Example: Reactor Safety Assessment System (RSAS)

The Reactor Safety Assessment System (RSAS) software is one example of the GTST-MLD snapshot-type applications. Started in 1987, the RSAS software developed for the United States Nuclear Regulatory Commission (U.S.NRC), by the Idaho National Engineering Laboratory (INEL) and the University of Maryland at College Park (UMCP), is intended to aid the Reactor Safety Team at the U.S. NCR's Operations Center to monitor and project core status during an emergency at a licensed nuclear power plant [Sebo et al. 1989; O'Brien et al. 1997].

In this expert system, generic nuclear plant knowledge (both deep and shallow knowledge) is represented in form of a GTST-MLD. This generic GTST-MLD is then adapted to specific plants by incorporating plant specific information. GTST-MLD is structured such that the adaptation process will be simple and efficient. Thus, substantially reduce the amount of effort needed to develop plant-specific knowledge base. On-line plant sensor readings are obtained through an on-line communication system called the Emergency Response Data System (ERDS). Integrated information and advisement is provided by the RSAS to the Reactor Safety Team regarding to monitored parameter and status of the plant during the accident. Since generic GTST-MLD models a major part of the knowledge-base for all nuclear power plants, experts may focus on differences between specific nuclear power plants, and easily reuse the knowledge on other plants with limited changes. In general, as shown in Fig. 5, the on-line RSAS reasoning is a three-stage process.

Firstly the on-line plant sensor readings accessed by ERDS are evaluated through parameter threshold tables and IF-THEN rules to determine status of GTST-MLD nodes (i.e., the status of sub-systems and sub-functions of the nuclear plant).

Secondly the status of the nuclear plant is determined through the GTST-MLD hierarchies that represent knowledge of the plant from both functional and structural point of views; thus, the safety of the nuclear plant is assessed through the GTST-MLD hierarchies dynamically. In the RSAS, the GTST-MLD-hierarchy displays are also used as a tool to show the propagation of failures.

Finally trend plots, plant diagrams, summary tables and other graphic or linguistic displays are used to provide safety assessment of the nuclear plant.

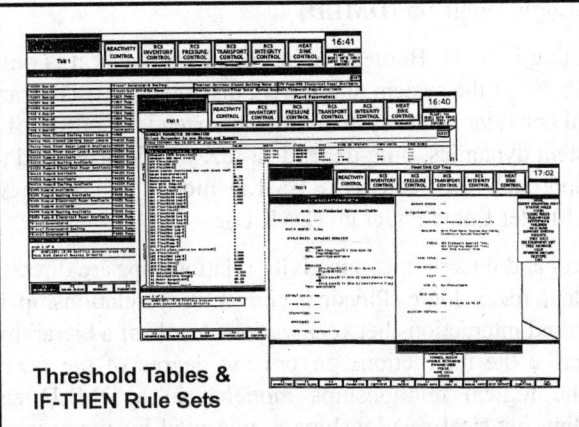

Stage One:

Process ERDS assessed on-line sensor readings and operator's setting through RSAS threshold tables and IF-THEN rule sets to determine current status of sub-systems and sub-functions.

Threshold Tables & IF-THEN Rule Sets

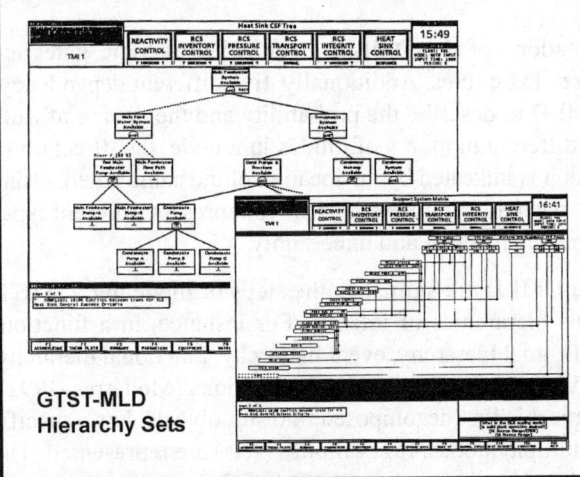

Stage Two:

Propagate integrated sub-systems and sub-functions information through GTST-MLD hierarchies to infer current status of the nuclear plant.

GTST-MLD Hierarchy Sets

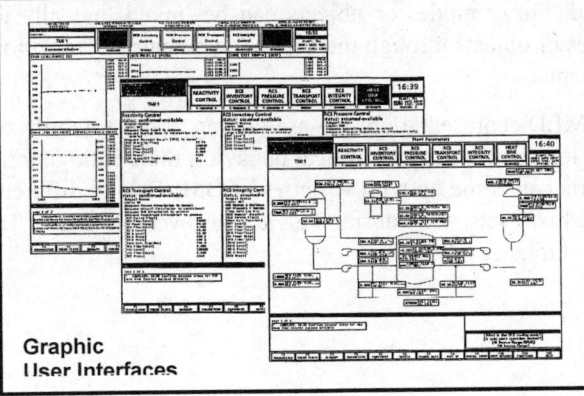

Stage Three:

Design graphic user interfaces to display and to advise root causes, potential failures, critical subsystems and other desired information of the nuclear plant.

Graphic User Interfaces

Fig. 5. Examples of RSAS Data Processes and Displays

2.4   The Dynamic Master Logic Diagram (DMLD)

Since the GTST-MLD modeling is static, Boolean-logic-based approach, it is only effective for showing "snapshots" of the system at different discrete times. They are limited to model the temporal behavior of the system; thus, to extend the power of a GTST-MLD to represent system dynamics, time-dependent fuzzy logic is applied to replace Boolean logic to approach full-scale GTST-MLD modeling. This new approach is called Dynamic Master Logic Diagram (DMLD).

In this DMLD concept, logical and uncertain connectivity relationships are directly represented by time-dependent fuzzy logic. Physical connectivity relationship is represented by fuzzy-logic-based interactions between various levels of a hierarchy. The physical laws that describe the interactions govern the degree of the fuzzy integration. Accordingly, the logical relationships modeled in a DMLD are accompanied by corresponding physical relationships represented by fuzzy logic relationships.

Table 1 summarizes the notations of the DMLD. Four types of logic gates are designed to represent the fuzzy logic rules. Additionally, five different dependency-matrix nodes are used in DMLD to describe the probability and the degree of truth in relationships. To convert different numbers of modes in a node, the direction of fuzzification and defuzzification is indicated by the location of the name noted. Using these notations and symbols, we may organize a DMLD to represent different types of connectivity relations, time dependency and uncertainty.

The steps of implementing a DMLD are similar to the steps of modeling a GTST-MLD. We first determine the hierarchies of interest. For instance, in a function-centered-model-based DMLD, goal hierarchy, event hierarchy, functional hierarchy, structure hierarchy and behavioral hierarchy are classical options [Modarres, 1992]. For each hierarchy, the top object is first decomposed into sub-objects. For a specific sub-object, which requires multiple modes, fuzzy modes (sets) are represented. The decomposition process is repeated until some lowest level of elementary components or fuzzy modes are reached. Fuzzy modes or objects can be linked logically to support specific fuzzy modes or objects through the dependency matrix. Transition gates represent time dependency.

When we talk about the DMLD representation, we are talking about a family of models (diagrams). That is, for representing a physical behavior, there is no unique DMLD model; rather, experts can come up with varieties of DMLDs and different numbers of nodes and layers, fuzzy sets, and transition logic. However, these DMLDs should yield approximately similar results.

Table 1. Notations of DMLD Based on Time-dependent Fuzzy Logic

Notation	Description
HIGH MID LOW / State / State / HIGH MID LOW	**Fuzzy Sets** of a node: the location, name, staple address, direction of fuzzification and defuzzification is shown Takagi and Sugeno's weighted average fuzzy/defuzzy system (Takagi and Sugeno) may be applied.
	AND gate: The minimum value of inputs will be the output value.
	OR gate: The maximum value of inputs will be the output value.
dt	**AND Transition** gate: The minimum value of inputs will be the output value with a delay **dt**.
dt	**OR Transition** gate: The maximum value of inputs will be the output value with a delay **dt**.
0.9	**Uncertain** node: The number inside the node represents the uncertainty of relationship. The minimum between the input degree of membership function (**dmf**) and the uncertainty (i.e., the degree of true relationship) is selected as the output.
	Certain node: represents a certain relationship that directly propagates the input **dmf** to the output **dmf**.
	NOT node: means certain negation in which the output dmf is (1 − input **dmf**).
0.9	**Uncertain Negation** node: a hollow node with a bar above the degree of certainty inside. The output dmf is the minimum between the uncertainty (i.e., the degree of true relationship) and (1 − input **dmf**).
	Independent node: means no relationship.
D	**Dependent** node: represents that more detailed DMLD hierarchy may be used to model the system behavior in details.

Table 2. Basic Structures in DMLD

Name	Example Hierarchy	Description
Static		A static DMLD represents a linear or nonlinear physical relation between X and Y which is certain and not varied with time.
Uncertain (Weighted) Input		Sometimes, we are uncertain about the effect of specific input, or the input has only partial effects. Uncertain nodes can be applied on the left lattice of a DMLD to represent an uncertain or weighted input. For example, in a light rain, experience of a driver would not affect safety. For heavier rain, degree of safety depends on the driver's experience. Assuming that a threshold exists at almost 0.7 (degree of experience).
Uncertain (Priority) Output		When the output is not unique for given inputs, we take advantage of uncertain nodes in the lattice of the DMLD. For example one may turn on a fan, or open a window while feeling hot. And, of course, for different people, the comfort range is may vary, so the fan speed is also uncertain.
Scheduled		Some outputs happen after a required time has elapsed. As such, transition logic gates can be applied to represent such effect. For example, an automatic dishwasher takes 15 minutes to wash, 15 minutes to rinse, and 30 minutes to dry dishes. (One can also use transparency of rinsing water and humidity of drying air to control the timer. Such a DMLD is static.). At t=0, the washer is turned on to either start a wash, a rinse or dry.

Table 2. Basic Structures in DMLD (con't)

Name	Example Hierarchy	Description
Time-lagged	$a(t) = \dfrac{dV(t)}{dt} = \dfrac{V(t) - V(t-1)}{1}$ Acceleration a(t) — NH NL ZE LW HI ZE LW HI V(t-1) ZE : Zero NH : Negative High LW : Low NL : Negative Low HI : High ZE LW HI V(t) Velocity	Sometimes, the output depends not only on the current inputs but also on the past inputs. For example, the rate of acceleration can be calculated from the current and the past velocity. Differ to a scheduled DMLD, the inputs of a time-lagged DMLD are not at the same time "snap shot ".
Auto-correlated	O(t) O(t -2ΔT) — D ΔT O(t - ΔT) — D — D ΔT	Autocorrelation addresses an output which depends on its own values in past. Sometimes, while we have difficulty to identify the inputs describing a specific behavior (e.g., stocked), we may assume the hidden input-output relation may be represented via an autocorrelation. For example, this system's output depends on its previous two outputs. Here, the AND transition gate performs a continuous full-scale state change.
Feedback	O(t) I(t) — D O(t) — D	When the output is affected by the inputs and in turn, the inputs are affected by the output, a feedback relation exists. The structure of a DMLD with feedback is similar to an autocorrelated DMLD but the output can return to itself (as autocorrelated) or other variables.
Comparison	Degree of B > A Degree of A > B -1 0 +1 -1 0 +1 A B	This is a DMLD comparison gate that shows the degree of difference between two nodes. A comparison gate is useful when possibility of an event occurrence is based on a required parameter value and an available parameter value. It can tell how close these two values are, and determine the potential for the occurrence of the event.

The basic structures in DMLD can be grouped into eight basic classes (templates): static, uncertain (or priority) output, uncertain (or weighted) input, scheduled, time-lagged, autocorrelated, feedback, and comparison. Table 2 summarizes these different structures with descriptions and examples. A DMLD model (diagram) is assembled from these basic structures.

Generally, a DMLD approach can improve the GTST-MLD in various ways. From state determining point of view, since DMLD applies fuzzy sets that overlap and full-scale membership functions, it allows floating threshold with fault tolerance and preventive warning. For connectivity relationship representation, DMLD can model not only full-scale physical and logical connectivity but also probabilistic, linguistic and resolution uncertainty. Transition effects of a system (e.g., partial success/failure, auto-correlation, feedback, schedule and time-lagged dynamics) can also be well represented in a DMLD hierarchy.

On one hand, since the DMLD estimation is based on logic, the speed of the estimation is much faster than a numerical simulator. On the other hand, DMLD provides a full-scale logical reasoning information that cannot be concluded in the classical logic-based systems. As such, a DMLD-based expert system, which has capability of full-scale logical reasoning and rapid simulation, can be implemented efficiently and economically.

DMLD for Representing Physical Models

One of the major advantages of the DMLD modeling is physical connectivity representation. Physics plays a fundamental role in most important fields of science and engineering. Since most of the known physical models can be represented by mathematical relations. A DMLD must be able to describe mathematical relations. Two major types of physical models are discussed in this section: solved model and raw model.

- Solved models are represented by functions in which the solution of the target variable can be computed in a straightforward manner, for example, using polynomials. As shown in Fig. 6(a), for a known solution of solved model, critical points of the relation should be chosen as fuzzy modes. The states between modes will be approached by membership functions automatically and linearly. To adapt the minor curve shape difference, membership functions can be relocated. Optimization algorithm, such as Least Squares Method, can be applied to identify the best fuzzy-set family. Theoretically, by adjusting the number of fuzzy sets and the shape of membership function, one might approach a curve to any acceptable accurancy. However, because of the uncertain nature of fuzzy logic modeling, not all objects requires such an accurancy.

- Raw models are represented by unsolved equations in which the target variable (i.e., the supported DMLD node) cannot be computed a priori. For example, a connectivity includes linear/nonlinear equations, integration, differentiation, or differential equation. The DMLD and its estimations of a **simple harmonic motion** example are shown in Fig. 6(b). The known equation of raw model is represented in a DMLD. Initial values are given to trigger the temporal behavior of the model. For instance, when $c<0$, the amplitude of the motion is increasing because there is a pulling force equal to c times the velocity in the motion direction. When $c=0$, it is a **simple harmonic motion**. The amplitude of the motion is decreasing because of the damping force when $c>0$.

For more discussions about specific applications of the DMLD, see papers of Hu and Modarres (1994, 1996, and 1998).

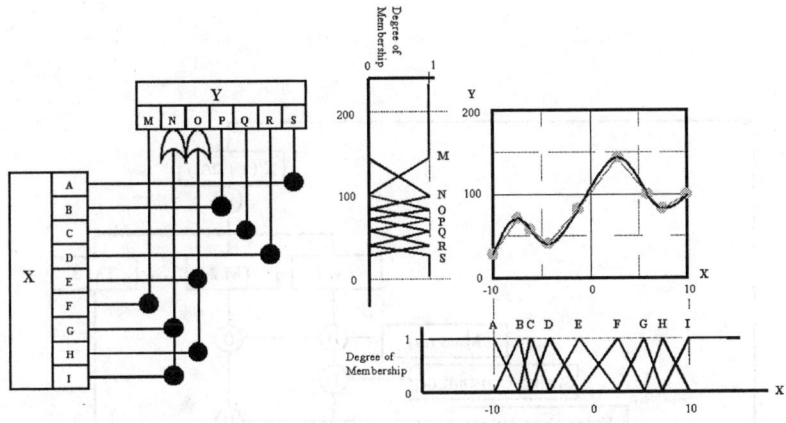

Fig. 6(a). DMLD as a Representation of Physical Solved Models

3 Application of DMLD to Nuclear Plant Direct Containment Heating Phenomenon

Direct Containment Heating (DCH) phenomenon in a Pressurized Water Reactor (PWR) nuclear power plant refers to the process whereby, under certain accident scenarios, molten core debris is ejected under high pressure from the reactor vessel into the containment atmosphere. The subsequent rapid heating of the containment atmosphere, in conjunction with possible hydrogen combustion, can lead to early containment failure. The DCH phenomenon was identified as one of the important contributors to early containment failure for Parse in NUREG-1150 (1985) and has also been identified as one of the contributors to early containment failure for PWRs in subsequent studies such as the nuclear industry's individual plant examination (IPE) studies. The results of previous research into the characteristics of debris dispersal and resultant containment loadings has led to closure of the DCH issue for all Westinghouse plants with large dry or sub-atmospheric containment, excluding ice condenser plants.

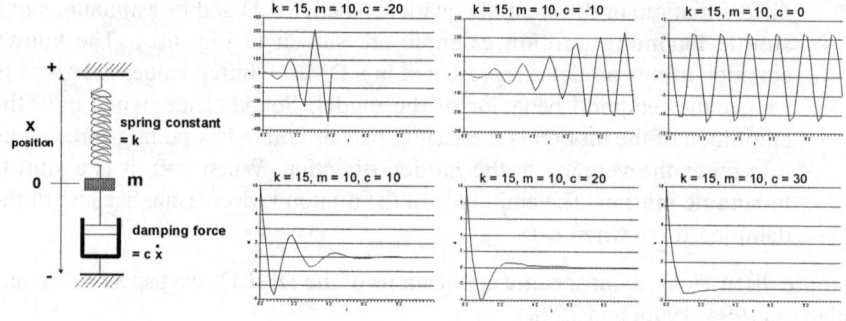

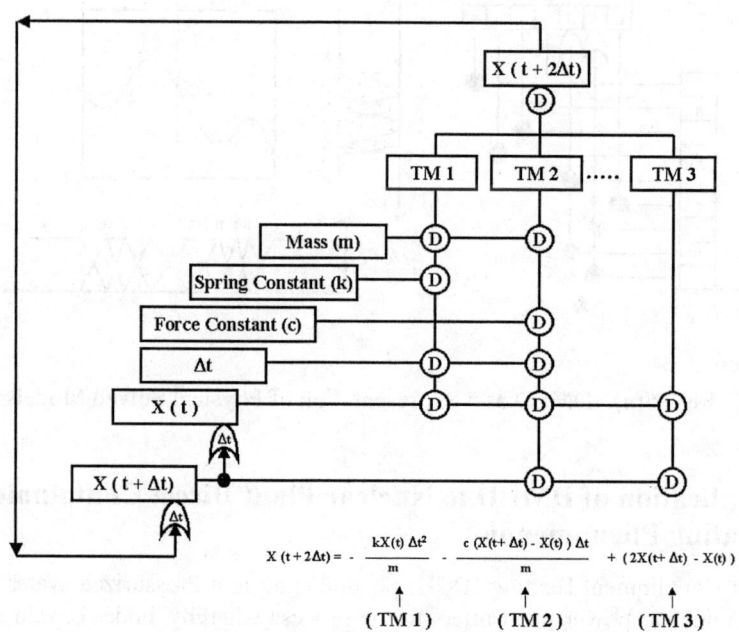

$$X(t+2\Delta t) = - \frac{kX(t)\Delta t^2}{m} - \frac{c(X(t+\Delta t) - X(t))\Delta t}{m} + (2X(t+\Delta t) - X(t))$$

(TM 1)   (TM 2)   (TM 3)

Fig. 6(b). DMLD as a Representation of Physical Row Models

Fig. 7 shows the basic energy and mass balance process applied to DCH. This figure shows that as a result of DCH the mass inventory in the vessel may be transferred to the containment. Similarly the energy content of the mass transferred (as a result of a blowdown) as well as the other source of energy (mainly, hydrogen combustion, debris thermal energy and oxidation energies) may also be added to the containment resulting in a pressure spike. If the pressure spike is high enough that leads to a load which exceeds the strength of the containment, then the containment fails.

Using the principles described in Fig. 7 a hierarchic model of the dynamic of the DCH is shown in Fig. 8. In this hierarchy each layer specifies a class of information that is either calculated or obtained. These calculations are necessary so that the magnitude of energies added to the containment may be computed.

3.1 Discussion of the Underlying Physics

Nuclear Plant containment failure occurs when applied stress (load) is greater than containment strength. Containment strength may be represented by a single value or, since it is uncertain as to whether it fails at a given stress level, by a probability distribution function. Applied stress is due to the pressure inside the containment that translates to a load applied to the containment structure. Therefore, one needs to measure the pressure inside the containment after the DCH occurs. This pressure has two components: the original pressure prior to DCH (p^0) and the pressure rise due to DCH (i.e., Δp). Therefore, the final pressure (load), p, inside the containment is obtained from,

$$p = p^0 + \Delta p \tag{1}$$

Pressure rise is proportional to the added energy into the containment atmosphere. As such

$$\Delta p \propto \Delta U / U^0$$

Therefore,

$$\Delta p = p^0 (\Delta U / U^0) \tag{2}$$

where

Δp = Pressure rise inside the containment
ΔU = Change in the internal energy (added) to the containment
U^0 = Internal energy of the containment prior to DCH
p^0 = Containment pressure prior to DCH

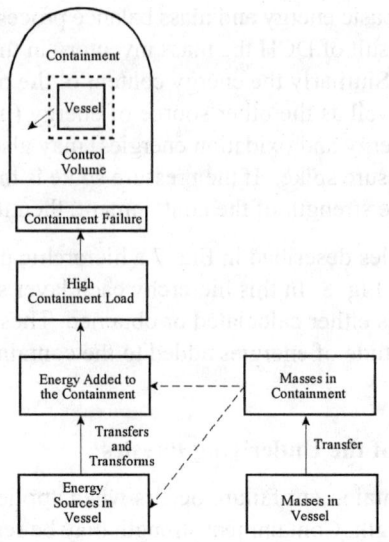

Fig. 7. Basic Energy and Mass Balance Applied to DCH

The internal energy added to the containment may be calculated assuming that the whole containment atmosphere constitutes a single cell. This may be a major, but not unrealistic approximation. Therefore,

$$U^0 = N_{cont}*C_{cont}*T_{cont} \qquad (3)$$

$$N_{cont} = P_{cont}*V_{cont} / (R*T_{cont})$$

Where,

N_{cont} = moles of air in containment
C_{cont} = containment molar heat capacity
P_{cont} = containment original pressure
V_{cont} = containment volume
T_{cont} = containment atmospheric pressure
R = Universal gas constant (8.31 Joules/^{0}K-mole)

Since all the terms in this equation 2 are known except ΔU, this value should be calculated from the energies added to the containment.

The change (added) internal energy is proportional to the sum of four major sources of energy that may be added to the containment following a DCH event. Therefore,

$$\Delta U \propto \Sigma \Delta E_i / (1+ \Psi) \quad (I = 1 \text{ to } 4) \qquad (4)$$

The four most significant sources of energy are:

ΔE_1 = Energy added due to the blowdown of the reactor coolant system
ΔE_2 = Energy added due to the thermal content of the debris dispersed
ΔE_3 = Energy added due to the oxidation energy of the debris
ΔE_4 = Energy added due to the combustion of H_2 in the debris generated as a result of Zr oxidation prior to reactor vessel breach
Ψ = Heat capacitance ratio

One may ignore Ψ (capacitance ratio- that is the ratio of the energy due to added debris to the total internal energy of the containment). Note that ΔU is inversely proportional to $(1+ \Psi)$ and since Ψ is usually much smaller then 1 (because the energy added is small in comparison to the internal energy of the containment) then,

$$\Delta U \approx \Sigma \Delta E_i \qquad (i = 1 \text{ to } 4). \qquad (5)$$

The one-cell model provides the simplest approximation for estimating the various sources energy. The models for calculating each source of the energy added are simplified versions of the form described by (Piltch et al, 1994, 1995)

Blowdown Energy

This energy is calculated from the energy content of coolant in the vessel and may be obtained from :

$$\Delta E_b = \frac{V_{rcs} \, P_{RCS}^o}{\gamma - 1} \left(1 - \frac{P^o}{P_{RCS}^o} \right) \qquad (6)$$

where

V_{rcs} = RCS Volume
P_{RCS} = RCS Pressure
γ = Isentropic exponent of blowdown gas ($\gamma \approx 1.33$)
P^o = Initial containment pressure
P_{RCS}^o = Initial RCS pressure

Since RCS Pressure is much greater than containment pressure then

$$\frac{P^o}{P_{RCS}^o} \ll 1.$$

Therefore

$$\Delta E_b \approx \frac{V_{rcs} \, P_{RCS}^o}{\gamma - 1} \qquad (7)$$

Debris Thermal Energy

The debris thermal energy depends on the amount of various species in the debris, the specific heat capacity of the debris, and the temperature of the debris,

$$\Delta E_d = N_d \, C_d \, (\, T^0_d - T_r \,) \qquad (8)$$

where

N_d = The number of moles of debris in the blowdown
C_d = The specific molar heat capacity of debris (which is a function of the mixture of chemical species in the debris)
T^0_d = Debris temperature
T_r = Reference temperature = 298 °K   (constant)

The value of N_d can be calculated by adding the number of moles of all the species in the debris. That is,

$$N_d = N_{Uo2} + N_{Zro2} + N_{Zrm} + N_{Fe} + N_{Cr} + N_{Ni} \qquad (9)$$

where

N_{Uo2} = $Mass_{UO2}/MW_{Uo2}$ (where MW_{Uo2} is the molecular weight of Uo2,
$Mass_{Uo2}$ = Uo2 mass
N_{Zro2} = Moles of Zr oxidized, where
N_{Zro2} = $f_{zro2} * N_{zr}$
f_{Zrox} = Fraction of Zr oxidized
N_{Zrm} = Moles of Zr not oxidized $(1 - f_{Zrox}) \, N_{Zr}$
N_{Zr} = Total Moles of Zr (Mass Zr / Molecular weight of Zr)
$N_{Fe} = N_{Cr}$ = N_{Ni} (molecular weight of other species ~ 0)

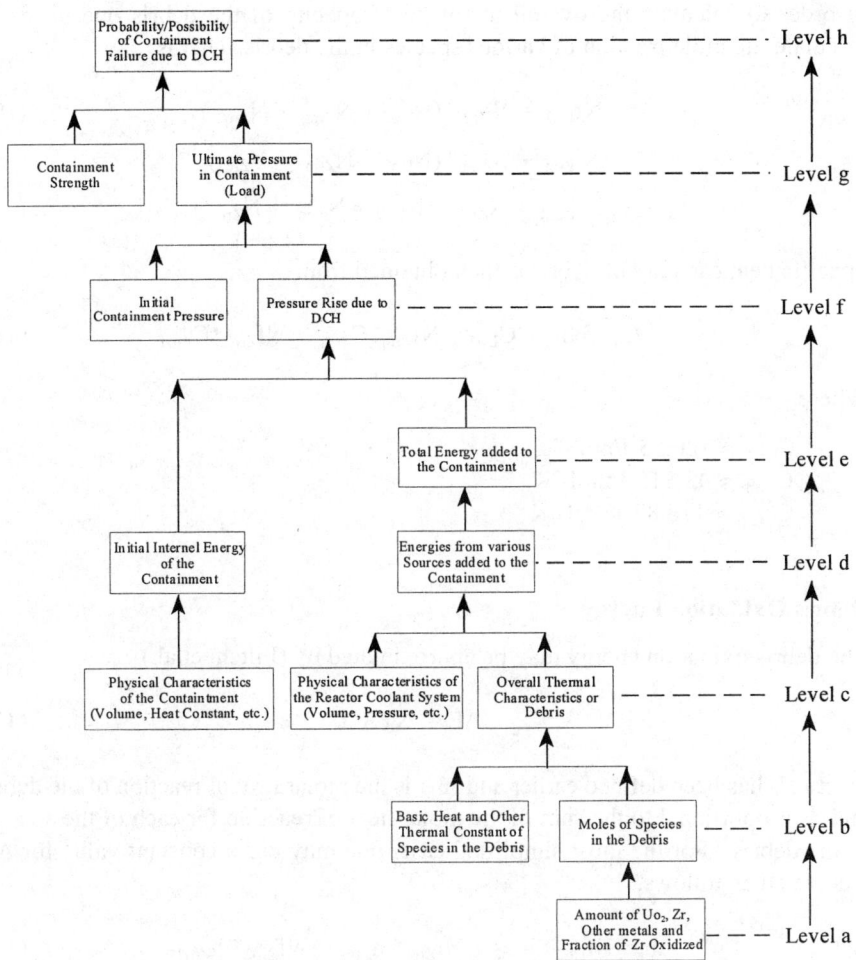

Fig. 8. Hierarchical Levels in the DCH Model

In order to calculate the overall molar heat capacity of the debris one needs to calculate the mole fraction of various species in the debris. That is

$$Nf_{Uo2} = N_{Uo2} / (N_{Uo2} + N_{Zrm} + N_{Zro2})   \qquad (10)$$

$$Nf_{Zrm} = N_{Zrm} / (N_{Uo2} + N_{Zrm} + N_{Zro2})$$

$$Nf_{Zro2} = N_{Zro2} / (N_{Uo2} + N_{Zrm} + N_{Zro2})$$

Specific heat capacity of debris is then obtained from

$$C_d = Nf_{Uo2}*C_{Uo2} + Nf_{Zrm}*C_{Zrm} + Nf_{Zro2}*C_{Zro2}   \qquad (11)$$

Where

$$C_{Uo2} \approx 126.55 \text{ J/mol-}^\circ K$$
$$C_{Zrm} \approx 42.315 \text{ J/mol-}^\circ K$$
$$C_{Zro2} \approx 116.87 \text{ J/mol-}^\circ K$$

Debris Oxidation Energy

The debris oxidation energy may be approximated by (Piltch, et al.)

$$\Delta E_r = N_d \Delta h_r   \qquad (12)$$

where N_d has been defined earlier and Δh_r is the molar heat of reaction of the debris and is proportional to the sum of the molar heat of reaction for each of the species in the debris. For the most simplified case, one may use a constant value for Δh_r calculated as follows.

$$\Delta h_r = Nf_{Uo2}* h_{Uo2} + Nf_{Zrm}* h_{Zrm} + Nf_{Zro2}* h_{Zro2}   \qquad (13)$$

Typically, $h_{Uo2} = 0$, $h_{Zrm} = 5.98 \times 10^5$ J/mol, $h_{Zro2} = 0$.

Hydrogen Combustion Energy

For simplicity consider the case where all of the hydrogen in RCS is transported into the containment. Then,

$$\Delta E_{H2} = (N_{Zroz}) \Delta h_{H2}   \qquad (14)$$

where

N_{Zroz} = number of moles of zirconium oxidized
Δh_{H2} = Heat of combustion of H_2

3.2 DMLD for DCH Modeling

Binary Logic Model

Fig. 9 shows the DMLD as a binary logic model for the DCH phenomenon. This model is a special case of the fuzzy-logic-based DMLD when there are only two states possible and the two states are mutually exclusive. Different from the physical model, a binary logic model can only estimate the interaction of the systems, subsystems and components involved and show full success or failure at various levels.

Temporal / Physical Model

To describe the physical relations of the DCH phenomenon, mathematical equations can be integrated into the hierarchy. As shown in Fig. 9, in a physical model for the DCH phenomenon, the DMLD is the expansion of the concept described in the binary logic model into specific nodes (parameters) and relationships that describe the physical processes involved in DCH process as discussed in section 3.1. Starting from the top down, one can see how each node in the model is calculated or obtained using the lower levels of the hierarchy. For example the ultimate containment pressure at *level g* of the hierarchy is obtained by summing the initial containment pressure and the pressure rise resulted from DCH (i.e., $g_2 = f_1 + f_2$) as described by equation 1 in Section 3.1.

In a physical hierarchy, all major parameters and relations involved in the model must be included to represent dynamic behavior of the system. The benefit is that transition behavior of the system caused by specific parameters can be identified easily, numerically and accurately. However, large scales of hierarchies with non-standardized mathematical connectivity relationships are required.

Fuzzy Logic Model

Fig. 10 shows the DMLD as a fuzzy logic model for the DCH phenomenon. This model only describes the nodes (parameters) that are uncertain and/or the nodes that highly influence the DCH load. The relationships are representative of fuzzy rules that experts may use to describe effect of one or more parameters (shown in each node of the diagram) being in a given state (e.g., high, low) on other state(s) (e.g., causing another state to be low, medium, etc.). Fig. 11 shows the relevant fuzzy sets for the DMLD structure shown in Fig. 10.

The nature of a fuzzy logic hierarchy is a linguistic computation that combines both logical state reasoning and numerical degree-of-membership estimation. The interaction status of the system can be represented by multiple fuzzy states. In advanced, numerical results can be approached by estimating the degree of memberships between fuzzy states linearly. The design of a fuzzy-logic hierarchy is to have a draft-rapid simulation to the detail level that engineers need to perform real-time diagnosis or judgement during design or operation.

386

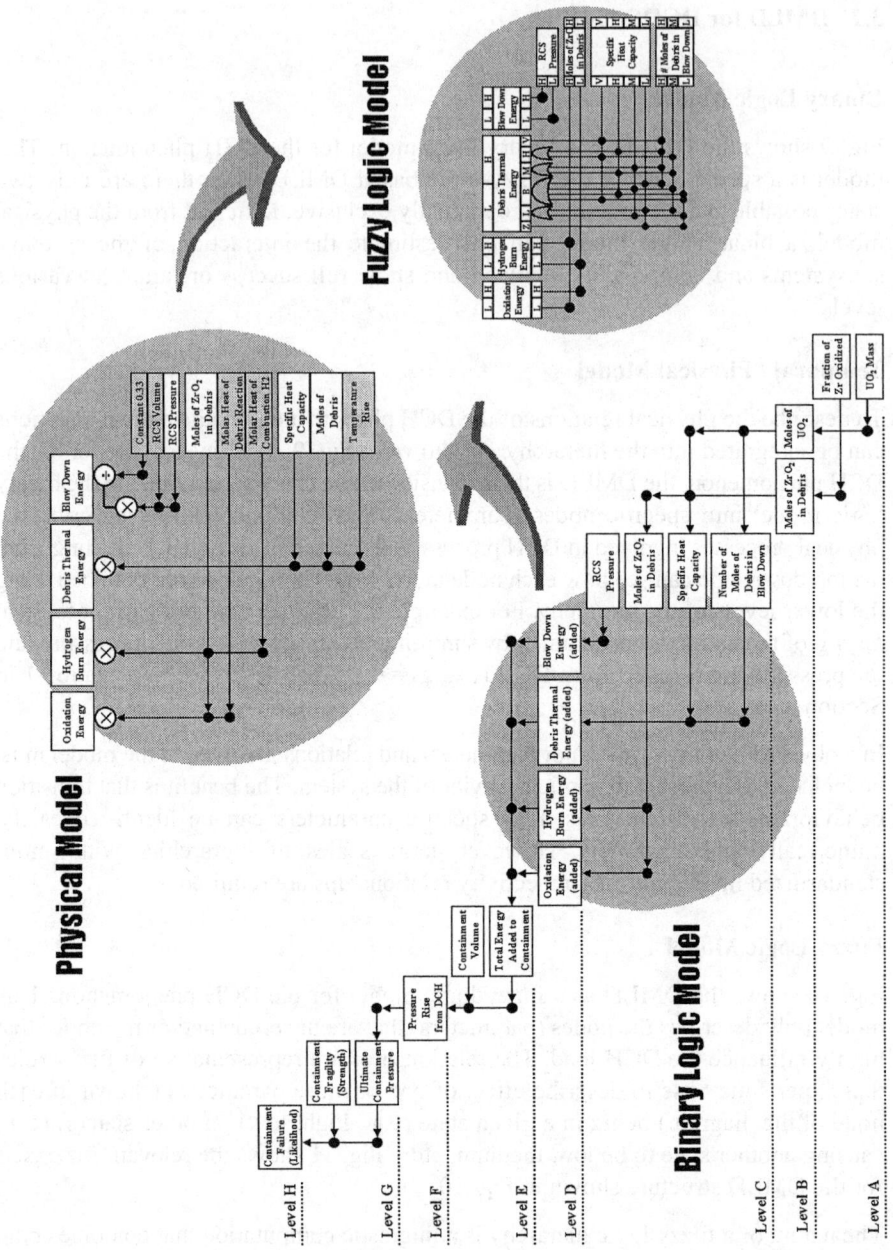

Fig. 9. DMLD for DCH as Binary Logic, Physical, and Fuzzy Logic Models

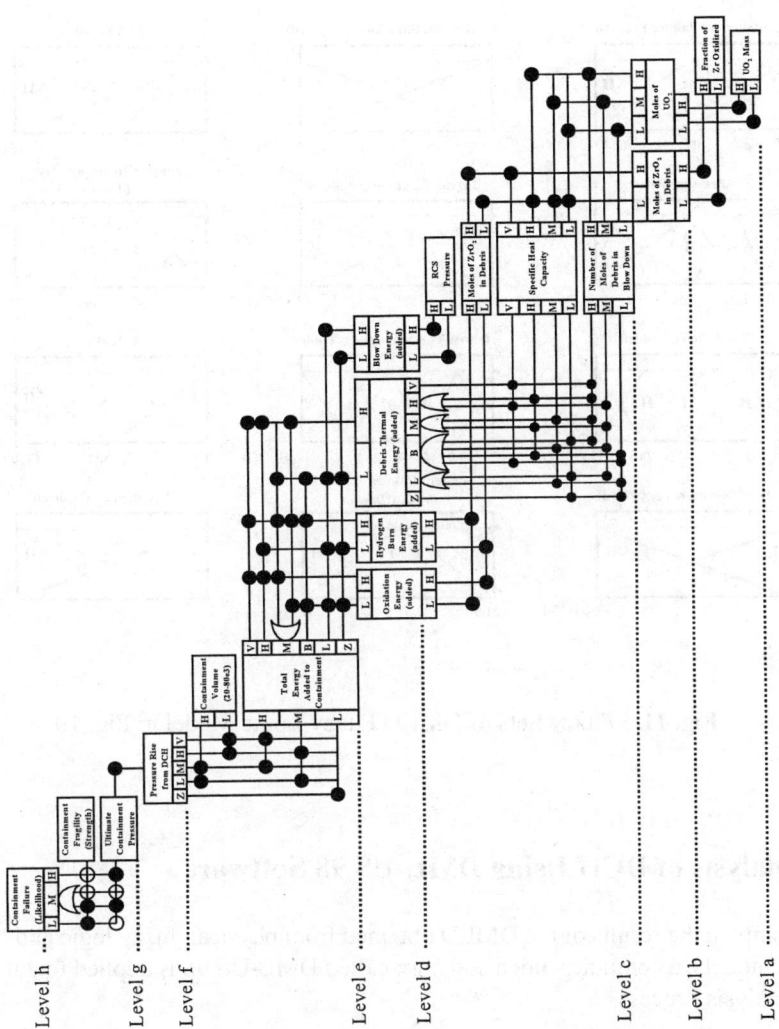

Fig. 10. DMLD for DCH as a Fuzzy Logic Model

388

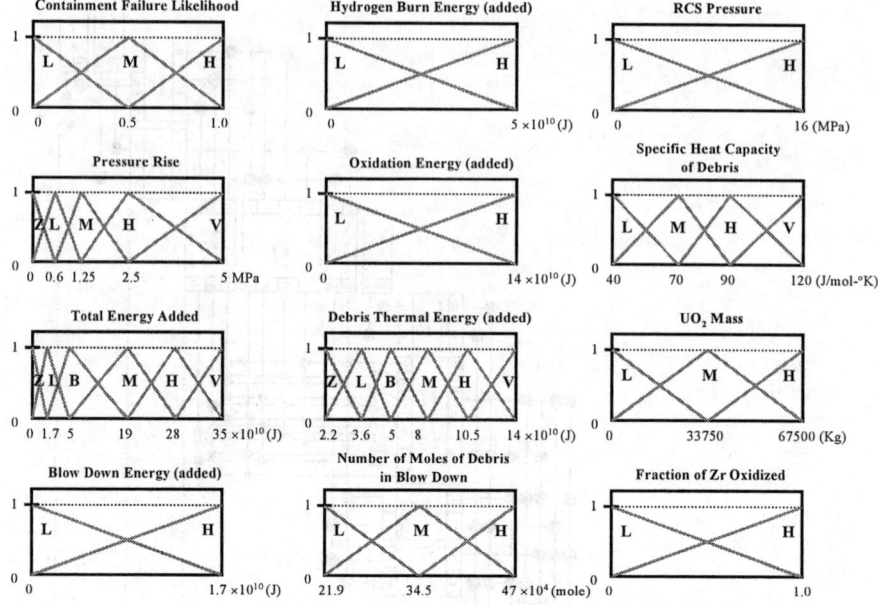

Fig. 11. Fuzzy Sets of DMLD Fuzzy Logic Model in Fig. 10

4 Analysis of DCH Using DML-US 98 Software

To compare the results of the DMLD obtained from physical, fuzzy logic and binary logic models, a computer-aided software called DML-US 98 is applied to automate the analysis process.

DML-Unified Software (DML-US) 98

As a generic hierarchy-based system building tool, the design concept of DML-US 98 is to allows the users to customize and configure critical elements of hierarchies (i.e., objects, relationships, attribute of objects and relationships) to meet requirements of different methodologies. As shown in Fig. 12, DML-US 98 provides user-friendly tools for

- implementing hierarchies,
- customizing binary, fuzzy and physical connectivity relationships of hierarchies,
- customizing fuzzy sets, IF…THEN rules and physical attributes for objects of hierarchies,

- advising simulation, inferrence and diagnosis results in user-defined graphic and user interfaces, and
- accessing data through ODBC and WWW interfaces.

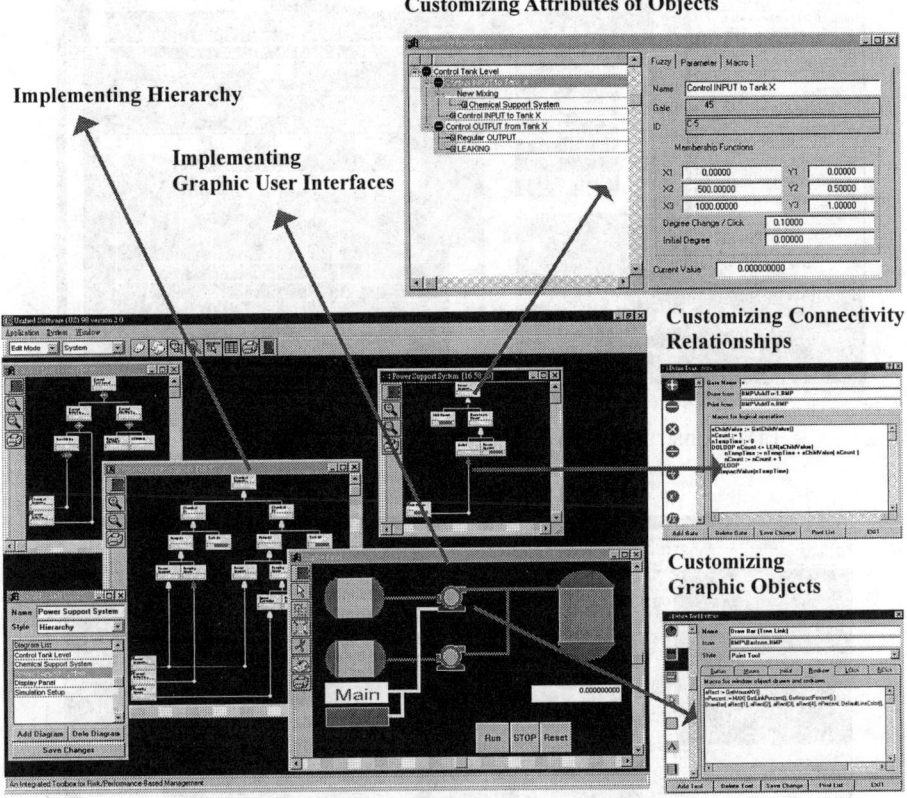

Fig. 12. Major Features in the DML-US 98 Software

Implementing DMLD for DCH on DML-US 98

As shown in Fig. 13(a), to implement DMLD for DCH on DML-US 98, different connectivity relationships can be defined to implement hierarchies. Physical connectivity relationships are defined to represent the underlying physics to connect different layers of hierarchies. Binary Boolean logic hierarchy is decomposed by pure AND/OR logic gates. Fuzzy sets are defined to represent fuzzy-logic-based hierarchies. The examples of graphic user interfaces are designed as Fig. 13(b)

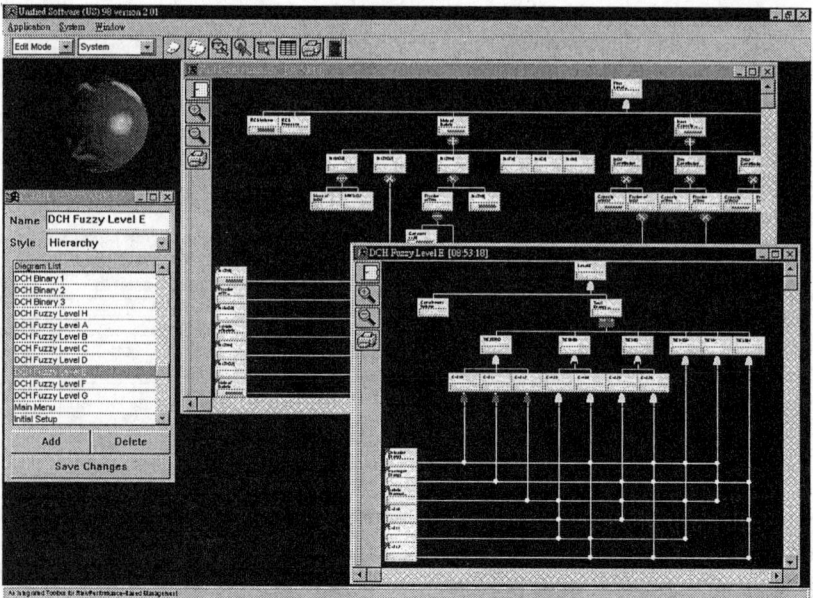

(a) Constructing DMLD Hierarchies on DML-US 98

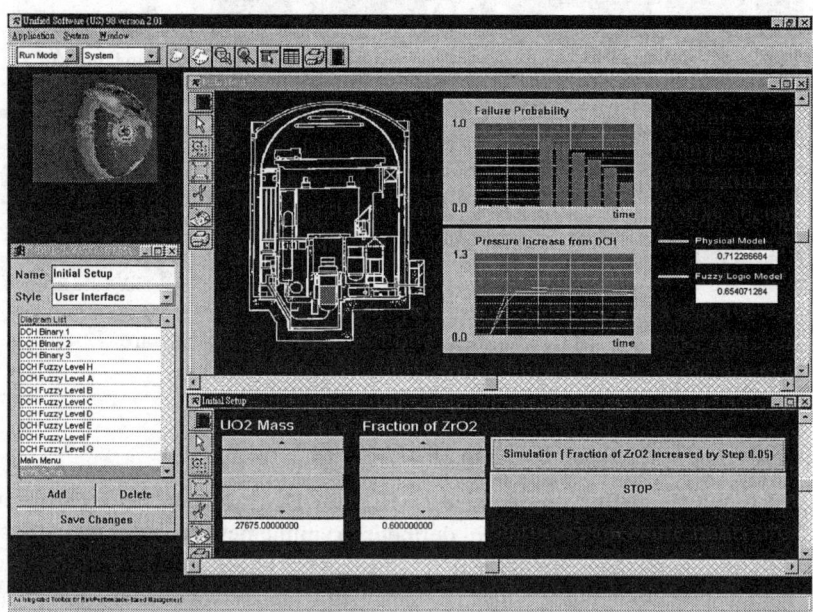

(b) Graphic User Interface for the DCH Example

Fig. 13. Implementing DMLD for DCH on DML-US 98

A comparison of the DMLD results is shown in Fig. 14. Clearly the results of fuzzy-logic-based hierarchy and physical hierarchy matched closely. Which confirms that the fuzzy logic model accurately depicts and models the underlying physics represented by the physical model.

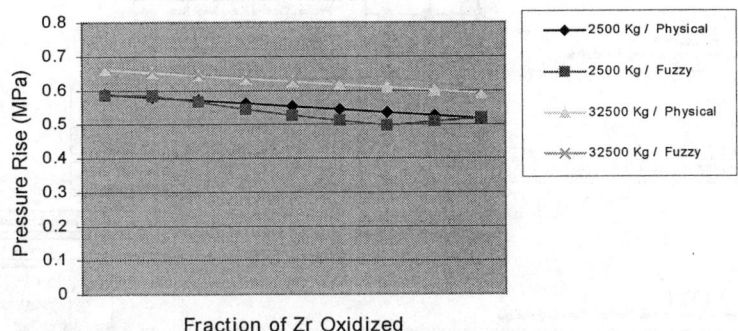

Fig. 14. A Comparison of the DMLD Results

5 Other Areas of GTST-DMLD Applications

The GTST-DMLD has also been applied to many other research areas, such as semiconductor fabrication process diagnosis and control [Chang, Y.-J. et al. 1999], spacecraft mission operation supervising, software development configuration management, aircraft maintenance management [Hu and Modarres, 1999]. Fig. 15 shows a GTST-DMLD example for semiconductor fabrication process supervising. A real-time run-to-run process control is decomposed into seven major hierarchic levels. For each hierarchy level, sub-hierarchy objects can be determined. To represent the detail behaviors of hierarchies, as shown in Fig. 16, the top object is first decomposed into sub-objects; then, for a specific sub-object, which requires multiple modes, fuzzy modes (sets) are represented. The decomposition process is repeated until some lowest level of elementary components or fuzzy modes are reached. Fuzzy modes or objects can be linked logically to support specific fuzzy modes or objects through the dependency matrix. In this example, the relationship between [Process Anomaly] and [Raw Database] has been decomposed to show details. As shown in Fig. 16, different process anomalies are classified on the top of the hierarchy. Major process measurements are represented on the left side of the hierarchy. The fuzzy rules are represented in the center of the hierarchy in the form of a matrix. For instance, a fuzzy rule "IF process measurements [RSP+ is HIGH], [RCN+ is NORMAL], [RCP+ is HIGH], [RCP1 is NORMAL] THEN [P+ Imp Anomaly]" is represented.

392

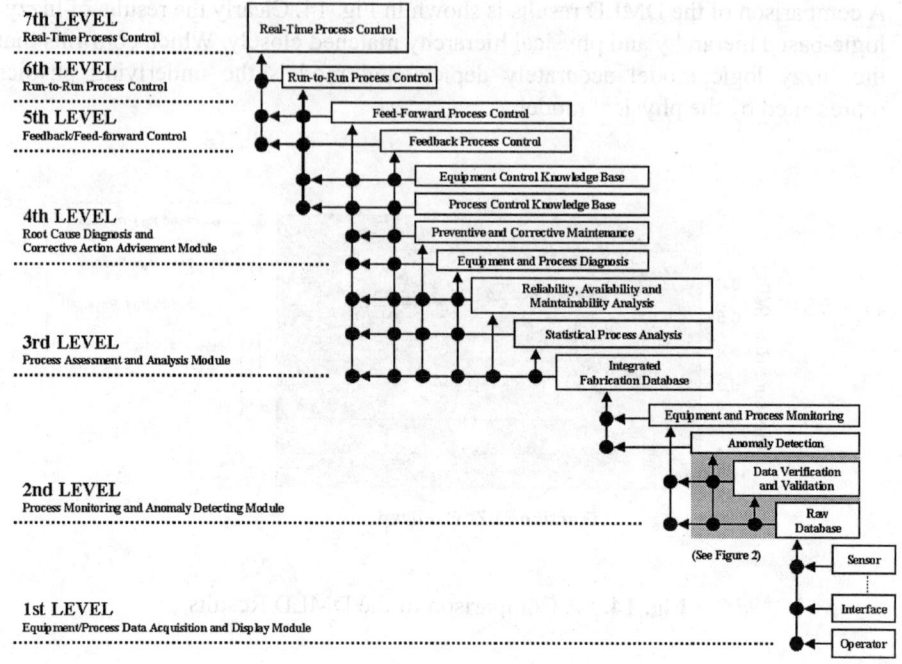

7th LEVEL
Real-Time Process Control

Real-Time Process Control

6th LEVEL
Run-to-Run Process Control

Run-to-Run Process Control

5th LEVEL
Feedback/Feed-forward Control

Feed-Forward Process Control

Feedback Process Control

4th LEVEL
Root Cause Diagnosis and
Corrective Action Advisement Module

Equipment Control Knowledge Base

Process Control Knowledge Base

Preventive and Corrective Maintenance

Equipment and Process Diagnosis

3rd LEVEL
Process Assessment and Analysis Module

Reliability, Availability and
Maintainability Analysis

Statistical Process Analysis

Integrated
Fabrication Database

2nd LEVEL
Process Monitoring and Anomaly Detecting Module

Equipment and Process Monitoring

Anomaly Detection

Data Verification
and Validation

Raw
Database

(See Figure 2)

Sensor

1st LEVEL
Equipment/Process Data Acquisition and Display Module

Interface

Operator

Fig. 15. A DMLD Hierarchy for Semiconductor Fabrication Process Control

6 Conclusions

The GTST-DMLD is a powerful technique for modeling physical processes associated with complex dynamic systems. This chapter describes the nature of these models and applies them to an important topic of interest. (i.e., the Direct Containment Heating in PWR nuclear power plants.) The use of DMLD approach in modeling complex systems is a practical reality. The DCH example has been demonstrated this claim by developing physical, fuzzy, and binary (Boolean) logic. These models provide similar results but offer different emphasis when used.

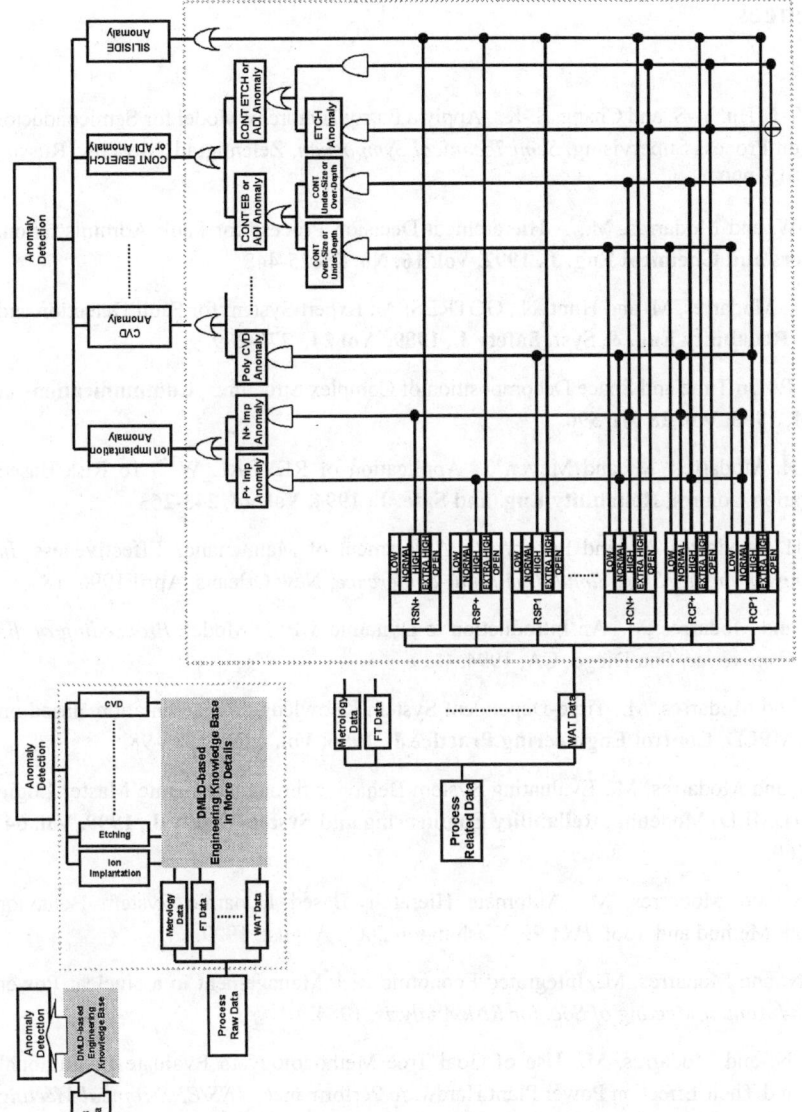

Fig. 16. A DMLD Hierarchy for Semiconductor Fabrication Process Anomaly Detection

References

Chang, Y.-J., Hu, Y.-S. and Chang, S.-K. , Apply a Fuzzy Hierarchy Model for Semiconductor Fabrication Process Supervising, *Semi Technical Symposium*, Zelenograd, Moscow, Russia, May 17-19, 1999.

Chen, L-W. and Modarres, M., A Hierarchical Decision Process for Fault Administration, **Computers and Chemical Eng. J., 1992, Vol. 16, No. 5**, 425-448.

Chung, D., Modarres, M. and Hunt, N., GOTRES: An Expert System for Fault Detection and Analysis. **Reliability Eng. & Syst. Safety J., 1989, Vol.24**, 276-289.

Courtois, P., On Time and Space Decomposition of Complex Structures, **Communications of the ACM, 1985, Vol. 28 (6)**, 596.

Dezfuli, H. Modarres, M. and Meyer, J., Application of REVEAL_W^{TM} To Risk-Based Configuration Control, **Reliability Eng. and Safe. J., 1994, Vol. 44**, 243-263.

Hadavi, H., Modarres, M. and Fakory, R., Assessment of Maintenance Effectiveness. *In Proceedings of the 6th SMC Simulation Multi-conference*, New Orleans, April 1996. 18.

Hu, Y-S. and Modarres, M., An Introduction to Dynamic MPLD Model, *Proceeding of the PSAM-II Conference,* San Diego, CA, 1994.

Hu, Y-S. and Modarres, M., Time-Dependent System Knowledge Representation Based on Dynamic MPLD, **Control Engineering Practice J., 1996, Vol. 4, No 1**, 89-98.

Hu, Y.-S. and Modarres, M., Evaluating System Behavior through Dynamic Master Logic Diagram (DMLD) Modeling, **Reliability Engineering and System Safety J., 1999, Vol. 64**, pp. 241-269.

Hu, Y.-S. and Modarres, M., Automate Hierarchy-Based Dynamic System Behavior Estimation: Method and Tool, *PSA'99*, Washington D.C., August, 1999.

Hunt, R.N. and Modarres, M., Integrated Economic Risk Management in a Nuclear Power Plant, *1984 Annual Meeting of Soc. for Risk Analysis*, 1984.

Hunt, R. N. and Modarres, M., Use of Goal Tree Methodology to Evaluate Institutional Practices and Their Effect on Power Plant Hardware Performance. *ANS/ENS Topical Meeting on Probabilistic Safety Methods and Applications*, San Francisco, Feb. 1985. 17.

Kim, I., Modarres, M. and Hunt, N., A Model-based Approach to On-line Process Disturbance Management: the Applications. **Reliability Eng. & Syst. Safety J., 1990, 28**, 185-239. 19.

Modarres, M., Application of the Master Plant Logic Diagram in PSAs During Design, *ANS Topical Meeting on Risk Management-Expanding Horizons*, Boston, 1992.

Modarres, M. and Cadman, T., A Method of Alarm System Analysis for Process Plants. **Computers and Chemical Eng. J., 1986, 10**, 557-565. 21.

Modarres, M., Chen, L. and Danner, M., A Knowledge-based Approach to Root-cause Failure Analysis. *Proc. of the Expert Systems Applications for the Electric Power Research Industry Conf.*, Orlando, FL, June 1989.

Modarres, M., Roush, M. L. and Hunt, R. N., Application of Goal Trees in Reliability Allocations for Systems and Components of Nuclear Power Plants. *12th INTER-RAM Conf.*, Baltimore, 1985.

Ni, T. and Modarres, M., Using Dynamic Master Logic Diagram for Component Partial Failure Analysis. *Proc. of the ASME Pressure Vessels & Piping & ICPVT-8*, Montreal, Canada, 1996.

Nordvik, J-P., Mitchison, N. and Wilikens, M., The Role of the Goal Tree-Success Tree Model in the Real-time Supervision of Hazardous Plant. *Proc. of the First International Workshop on Functional Modeling of Complex Technical Systems*, Ispra, Italy, May 12-14, 1993, 127-141. 31.

Nordvik, J-P., Atkinson, M. and Carpignano, A., Advances with STARS: Applications to Safety Problems. *Proc. of the Fourth International Workshop on Functional Modeling of Complex Technical Systems*, Athens, Greece, May 19-20, 1996, 91-100. 32.

NUREG-1150 (1990) Severe Accident Risks: An Assessment for Five U.S. Nuclear Power Plants, U.S. Nuclear Regulatory Commission.

O'Brien, J., Marksberry, D. and Modarres, M., Reactor Safety Assessment System (RSAS) Development and Lessons Learned. *Second OECD Specialist Meeting on Operator Aids for Severe Accident Management*, Lyon, France, Sept. 1997. 20.

Piltch, M., Yan, H and Theofanous, T. (1994), "The Probability of Containment Failure by Direct Containment Heating in Zion", NUREG/CR-6075, U.S. Nuclear Regulatory Commission.

Piltch. M. (1995), The Probability of Containment Failure by Direct Containment Heating in Surry, NUREG/CR-6109, U.S. Regulatory Commission.

Roush, M. L., Modarres, M. and Hunt, R. N., Application of Goal Trees to Evaluation of the Impact of Information upon Plant Availability. *ANS/ENS Topical Meeting on Probabilistic Safety Methods and Applications*, San Francisco, CA, Feb. 1985. 16.

Sebo, D. Marksberry, D. And Modarres, M., RSAS: A Reactor Safety Assessment System, *Proc. Of the 7th Power Plant Dynamics, Control and Testing Symposium*, Knoxville, 1989.

Simon, H., *The sciences of the artificial*, **The MIT Press,** 217-221, Cambridge, MA, 1982.

Takagi, T. And Sugeno, M., Fuzzy Identification of Systems and Its Application to Modeling and Control, **IEEE Trans. On Systems, Mans, and Cybern., Vol. SMC-15(1)**, 116-132.

Zamanali, J. H., Modarres, M. and Wang, J., Application of Master Plant Logic Diagram (MPLD) PC-based Program in Probabilistic Risk Assessment. *Proc. of the Int. Conf. on Probabilistic Safety Assessment and Management (PSAM)*, Beverly Hills, 1991.

18 Continued Fractions in Time Series Forecsting

Andrew Zardecki

Los Alamos National Laboratory, MS F645, Los Alamos, NM 87545, USA
azz@lanl.gov

Through their ergodic properties, continued fractions provide a fascinating example of a dynamical system whose properties we attempt to encode in a library of rules. For an irrational number, randomly selected from a unit interval, the probability distribution of partial quotients can be derived from the ergodic theorem, allowing one to distinguish a random time series from a time series whose elements are drawn from the probability distribution resulting from the ergodic hypothesis. Applications of ergodicity to modular transformations and chaotic cosmology are sketched. In addition to being an object of study, continued fractions are used as a tool to overcome the curse of dimensionality in rule-based forecasting. To this end, we encode the successive (possibly rescaled) values of a time series, as the partial quotients of a continued fraction, resulting in a number from the unit interval. The accuracy of a ruled-based system utilizing this coding is investigated to some extent. Qualitative criteria for the applicability of the algorithm are formulated.

1 Introduction

Fuzzy logic control is an effective approach to utilizing linguistic rules, whereas neural control is suited for using numerical data pairs. Fuzzy basis functions, which are algebraic superpositions of fuzzy membership functions, can combine both numerical data and linguistic information. In parallel with neural net numerical techniques, an increasing effort has been devoted to rule-based forecasting by employing fuzzy logic controllers. Wang and Mendel [13] developed a general method to generate fuzzy rules from numerical data and used their method for time series prediction. Subsequently, Mendel and coworkers also represented fuzzy systems as series expansions of fuzzy basis functions (FBF), obtained as algebraic superpositions of fuzzy membership functions [14]. The FBF method avoids the combinatorial explosion problem associated with fuzzy logic systems having a large number of antecedents in the rule base [15]. The price one needs to pay, though, is long running times that are needed for the algorithm to converge. An alternative fuzzy rule configuration that avoids the combinatorial explosion has recently been advanced by Combs and Andrews [3].

The goal of this chapter is twofold. First, we apply the predictive ability of a fuzzy controller to distinguish chaotic and noisy behavior in a one-dimensional time series. We achieve this by using well-known examples from the number theory pertaining to the continued fractions. Any sequence of natural numbers drawn from the probability distribution of the quotients of the continued fraction corresponding to an irrational number represents a typical sequence, in the sense that almost all sequences of quotients have this distribution. On the other hand, a sequence of numbers drawn from a uniform probability distribution does not have this property. The implications of this distinction will be illustrated through applications of the continued fractions to transformations of the modular group and to chaotic relativity.

The second objective of this chapter is to provide a rapid prediction method, in which a larger number of antecedents are accounted for than are currently considered. To this end, we encode the successive (possibly rescaled) values of a time series as the partial quotients of a continued fraction. (We recall that a set of natural numbers, called partial quotients, determines a simple continued fraction whose value belongs to the unit interval; conversely, given the value of a continued fraction, its partial quotients are readily recovered.) Within the 64-bit representation of the double precision numbers, we can thus encode up to 40 antecedents in the continued fraction form. When this representation is processed by a standard fuzzy logic controller, one obtains the rule base including the historical data for each term of the time series. One can speak about the dressed rules, in which not only the values, but also their history, are encapsulated. We study to some extent the accuracy of the continued fraction representation. We also investigate the different decoding schemes that lead to different forecasting accuracy. In the simplest case, when the continued fraction is expressed in terms of its first partial quotient, a considerable increase in the forecasting power, as compared to the standard fuzzy controller-based technique, is achieved. This method can be viewed as a data-compression technique for the time series. In the context of nuclear safeguards, we consider it a refinement of the anomaly detection algorithm proposed earlier [16].

2 Continued Fractions

Let x be a real number from the interval $(0, 1)$. An expression of the form

$$x = \cfrac{1}{q_1 + \cfrac{1}{q_2 + \cfrac{1}{q_3 + \ldots}}} \tag{1}$$

is called a (simple) continued fraction [4] representation of x; for reasons of technical convenience, Eq. (1) is often written as

$$x = [q_1, q_2, q_3, ...] \tag{2}$$

The integers $q_1, q_2, q_3, ...$ are called the partial quotients, whereas the successive approximations of x, obtained by retaining an increasing number of partial quotients, are referred to as complete quotients or convergents.

For our purpose, the most important results of the theory of continued fractions can be expressed as two theorems [8].

Theorem 1. To every $x \in [0, 1]$, there corresponds a unique continued fraction with value equal to x. This fraction is finite if x is rational and infinite if x is irrational.

Theorem 2. Let us agree to call a rational fraction a/b ($b > 0$) a best approximation of a real number x if every other rational fraction with the same or smaller denominator differs from x by a greater amount. Then every best approximation of x is a convergent or an intermediate fraction of the continued fraction representing that number.

The one-to-one correspondence between x and the partial quotients of the continued fraction representing x is readily obtained. In fact, given a number x in the unit interval, we can write $x = 1/(q_1 + x')$ so that q_1 is the integer part $[1/x]$ of $1/x$, and x' is its fractional part $\{1/x\}$. If we define the functions

$$T(x) = \{1/x\} \tag{3}$$

and

$$q(x) = [1/x] \tag{4}$$

then $q(x) = q(T^{n-1}(x))$, $n = 1, 2, ...$, represent the partial quotients of the continued fraction expansion of x. Conversely, given the partial quotients, $q_1, q_2, q_3, \ldots$, the number x represented by them can be recovered through the successive approximations $x_n = A_n/B_n$, where, for $n \geq 1$, the A_n and B_n are given recursively as

$$\begin{aligned}
A_n &= q_n A_{n-1} + A_{n-2}, \\
B_n &= q_n B_{n-1} + B_{n-2}.
\end{aligned} \tag{5}$$

The initial values, for $n = -1$, are $A_{-1} = 1$, $B_{-1} = 0$; for $n = 0$, we have $A_0 = 0$, $B_0 = 1$.

The numbers $B_1, B_2, B_3, \ldots$ are strictly positive increasing natural numbers; hence B_n increases indefinitely with n. It readily follows [4] from Eq. (5) that

$$\left| x - \frac{A_n}{B_n} \right| < \frac{1}{B_n B_{n+1}}, \tag{6}$$

which proves that A_n/B_n has the limit x as n increases indefinitely. This is the property which makes the word "convergent" appropriate; A_n/B_n converges to the value of the original number x as n increases indefinitely.

As an example illustrating our considerations, the rational number $x = 11/31$ has the following finite expansion in continued fractions

$$\frac{11}{31} = \cfrac{1}{2 + \cfrac{1}{1 + \cfrac{1}{4 + \cfrac{1}{2}}}} \; . \tag{7}$$

In this case, the recurrence relation given by Eq. (5) yields successively

$$A_1 = 2 \times 0 + 1 \,,$$
$$B_1 = 2 \times 1 + 0 \,,$$

resulting in $x_1 = 1/2$. For $n = 2$, we get

$$A_2 = 1 \times 1 + 0 \,,$$
$$B_2 = 1 \times 2 + 1 \,,$$

Similarly, for $n = 3$ we have

$$A_3 = 4 \times 1 + 1 \,,$$
$$B_3 = 4 \times 3 + 2 \,,$$

giving $x_3 = 5/14$. Finally, for $n = 4$ the equations

$$A_4 = 2 \times 5 + 1 \,,$$
$$B_4 = 2 \times 14 + 2 \,,$$

reproduce the number $x = A_4/B_4$.

The irrational number $\sqrt{3} - 1$ has an infinite periodic expansion

$$\sqrt{3} - 1 = \cfrac{1}{1 + \cfrac{1}{2 + \cfrac{1}{1 + \cfrac{1}{2 + \cdots}}}} \; . \tag{8}$$

The recurrence relation produces the best successive approximants to $\sqrt{3} - 1$, as described in Theorem 2.

3 Ergodic Properties of Continued Fractions

If m is a measure on space E, consider transformations T of E into itself. T is said to be measure preserving, if $m(T^{-1}A) = m(A)$. A is invariant under T in case $T^{-1}A = A$. T is ergodic if it is measure preserving and if each invariant set is trivial in the sense of having a measure of either 0 or 1.

As a simple example, consider the five-point space $E = (a, b, c, d, e)$ on which the transformation T acts as a permutation $(a, b, c)(d, e)$, a product of two cycles. If T preserves measure m, then m must assign equal masses to a, b, and c, and equal masses to d and e. In this case, T is not ergodic for the set $A = \{a, b, c\}$ is sent into itself by T and so is the complimentary set $B = \{d, e\}$; that is, $T^{-1}A = A$, and $T^{-1}B = B$, with nontrivial A and B. On the other hand, taking the same space E with equal masses of $1/5$ an the cyclic permutation $T = (a, b, c, d, e)$, we see that, for any $x \in E$, every orbit $\{x, Tx, T^2x, \ldots\}$ resembles E in the primitive sense of being equal to E as a set.

For ergodic transformations, the sequence of iterates $T^i(p)$ of a point p is uniformly dense in space. This means that, starting with a point p at time 0, one asks for the frequency with which the iterates of p fall into A. In the limit of infinitely many iterates, for almost every point p this frequency is equal to the relative measure of the region. To state the ergodic theorem in mathematical terms, let us first call a measurable function f invariant if $f(Tp) = f(p)$. Suppose that T is a measure-preserving transformation and that f is measurable and integrable. Then there exists an integrable, invariant function $\widehat{f}$ such that $\int \widehat{f}\, dm = \int f\, dm$ and

$$\lim_{N \to \infty} \frac{1}{N} \sum_{n=1}^{N} f(T^i(p)) = \widehat{f}(p) \tag{9}$$

with probability 1. If T is ergodic, then $\widehat{f} = \int f dm$ with probability 1. In other words, if T is ergodic, then each invariant function is almost everywhere constant.

As a special case of the ergodic theorem, we obtain

$$\lim_{N \to \infty} \frac{1}{N} \sum_{n=1}^{N} I_A(T^i(p)) = \frac{m(A)}{m(E)} , \tag{10}$$

where I_A is the indicator function of A. This equation details the meaning of the statement about the frequency with which the iterates of p fall into A: Every orbit $\{x, Tx, T^2x, \ldots\}$ of x is a sort of replica of E itself. In physics, one often says that each ensemble average is equal to the corresponding time average involving a typical member of the ensemble.

The transformation $T(x)$, introduced in Eq. (3), can be rewritten as

$$T(x) = \frac{1}{x} - \left[\frac{1}{x}\right] . \tag{11}$$

It maps the unit interval $[0, 1]$ into $[0, 1]$, and possesses an infinite number of discontinuities at the points of the form of $1/k$, where k is a natural number larger than 1. T preserves the Gauss measure

$$m(A) = \frac{1}{\log 2} \int_A \frac{dx}{1 + x} , \tag{12}$$

which means that

$$m(T^{-1}A) = \int_A m(x)dx = m(A) . \tag{13}$$

To prove that T preserves m, it is enough to prove that it preserves measures of intervals $[a, b] \in [0, 1]$. Since the inverse image of $[a, b]$ is a disjoint union of intervals of the form

$$T^{-1}[a, b] = \bigcup_{k=1}^{\infty} \left[\frac{1}{k + b}, \frac{1}{k + a} \right] , \tag{14}$$

we need only verify the equation

$$\int_a^b \frac{dx}{1 + x} = \sum_{k=1}^{\infty} \int_{1/(k+b)}^{1/(k+a)} \frac{dx}{1 + x} . \tag{15}$$

The kth term on the right-hand side of Eq. (15), however, is

$$\log \frac{(k + a + 1)/(k + a)}{(k + b + 1)/(k + b)} , \tag{16}$$

which, on summation, yields the right-hand side.

Furthermore, T is measure-preserving and ergodic with respect to the Gauss measure. The probability of finding an integer k in the sequence of partial quotients $q_1, q_2, q_3, \ldots$ corresponding to x is given as

$$p(k) = \frac{1}{\log 2} \log \left[\frac{(k + 1)^2}{k(k + 2)} \right] . \tag{17}$$

Equation (17) is obtained from the ergodic theorem, stated in Eq. (10), by taking E to be interval $(0, 1)$; A to be the interval $(1/(k + 1), 1/k)$, that is the set on which $q_1(x) = k$, and the measure to be Gauss measure, given by Eq. (12). The first five cases of Eq. (17) are: $p(1) = 0.4150$, $p(2) = 0.1690$, $p(3) = 0.0931$, $p(4) = 0.0589$, and $p(5) = 0.0406$.

Any sequence of natural numbers drawn from the probability distribution of the quotients of the continued fraction corresponding to an irrational number represents a typical sequence, in the sense that almost all sequences of quotients have this distribution. On the other hand, some numbers lead to sequences of quotients that do not have this property. For example, the quotients corresponding to $(1 + \sqrt{5})/2 - 1$ are all equal to 1. In the same vein, a sequence of uniformly distributed integers will have more large numbers than allowed by the probability distribution of quotients. We note in passing that periodic sequences of quotients correspond to quadratic numbers; that is, numbers which are solutions to quadratic equations with integer coefficients.

These comments lead to a method allowing one to distinguish the iterates of an ergodic transformation from a random sequence. After constructing the rules based on the ergodic sequence, we register the forecast error for ergodic and random sequences, thus obtaining a classification tool.

In forecasting error using fuzzy rule-based system (FRBS) with lag vector of length 6 [15], rules are obtained for ergodic and random sequences, drawn from a set of random numbers with the probability distribution given by Eq. (17) and from a uniformly distributed set of numbers, respectively.

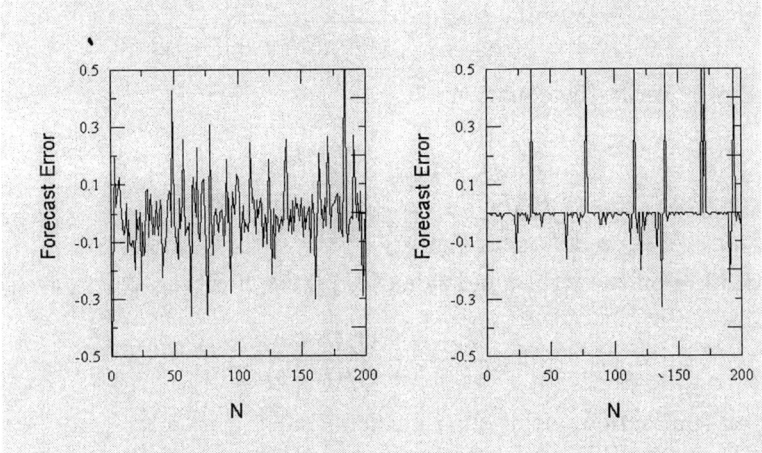

Fig. 1. Ergodic sequence of numbers with probability function given by Eq. 10 (left) and chaotic time series drawn from a random number distribution (right)

In Fig. 1, the ergodic sequence (left), provides the time series that is too irregular for FRBS to be captured in rules. The random sequence (right) shows anomalies where large integers occur.

3.1 Modular Group

A link between continued fractions and ergodic theory has been first established by Artin [1] in 1924. A modern, and more accessible, exposition of Artin's ideas is contained in the article written by Series [11]. In his study, Artin was concerned with ergodic properties of geodesic curves on a Riemann surface; the proof of ergodicity was obtained by the use of elementary results from the theory of continued fractions. In this section, we discuss certain properties of modular transformations, which play a role in Artin's result.

Consider two complex points w and w in the upper half-plane, related by the equation

$$w = \frac{az + b}{cz + d} \, , \tag{18}$$

where a, b, c, d are integer numbers satisfying $ad - bc = 1$. The transformations defined by these conditions are called modular transformations, often denoted by $PSL(2, Z)$ when the positive value for determinant is chosen.

A similarity between the recursive relation for continued fractions, Eq. (5), and modular transformations now emerges. In fact, let a real number x be expressed in the form

$$x = [q_0, q_1, \dots, q_n, x'] \, , \tag{19}$$

where x' is the complete quotient corresponding to q_n; that is, $q_n + 1/x'$ is last term in the continued fraction representation. The numbers x and x' are related by

$$x = \frac{A_{n-1} x' + A_{n-2}}{B_{n-1} x' + B_{n-2}} \, , \tag{20}$$

where A_{n-1} and B_{n-1} define the approximants to x, introduced by Eq. (5).

In terms of the generators $T(z) = -1/z$ and $S(z) = z + b$, with an integer b, any modular transformation $M(z)$, given by Eq. (18), can be written as

$$M(z) = S^{q_1} T S^{q_1} S^{q_2} T S^{q_2} T \cdots T S^{q_n} \, . \tag{21}$$

Equation (21) implies that there is an isomorphism between modular transformations and finite continued fractions $[q_1, q_2, \dots q_n]$.

We start with a number z_0 in the upper half-plane and form its orbit with the aid of modular transformations; that is, the set M_{z_0}, where M are the modular transformations. By virtue of Eqs. (20) and (21), we see that the orbit set is equivalent to an infinite continued fraction. Since the limit point of every orbit of the modular group is a real number (possibly infinite), in the limit $n \longrightarrow \infty$, we obtain

$$x = \lim_{n \longrightarrow \infty} S^{q_1} T S^{q_1} S^{q_2} T S^{q_2} T \cdots T S^{q_n} z_0 \, , \tag{22}$$

where x is a real number independent of the choice of z_0.

The considerations following Eq. (11) tell us that a sequence of modular transformations should be characterized by the numbers whose probability distribution is given by Eq. (17). In principle, then, we should be able to distinguish a typical subset of real numbers belonging to the limit set of the modular group from other subsets. The example summarized in Fig. 1

illustrates this distinction by displaying large forecast errors corresponding to the events that do not satisfy the probability distribution given by Eq. (17).

3.2 Chaotic Relativity

An example of chaotic dynamics in Einstein's equations of general relativity is provided by the Bianchi type-IX or mixmaster universe [2]. The evolution of the three orthogonal expansion factors is characterized by the expansion rates p_1, p_2, and p_3, which can be parametrized as $p_i(u)$, $i = 1, 2, 3$, for $u \in (1, \infty)$. The process of evolution of the model toward the singular point is made up of a successive series of oscillations during which the distances along two of the space axes oscillate, while they fall off monotonically along the third. In going from one series to the next, the direction along which there is a monotonic drop-off of distances shifts from one direction to another. Asymptotically, the order of these shifts takes on the character of a random process. The order of succession of length of successive series of oscillations also takes on the stochastic character [10]. In Fig. 2 we sketch the qualitative form of the scale factors.

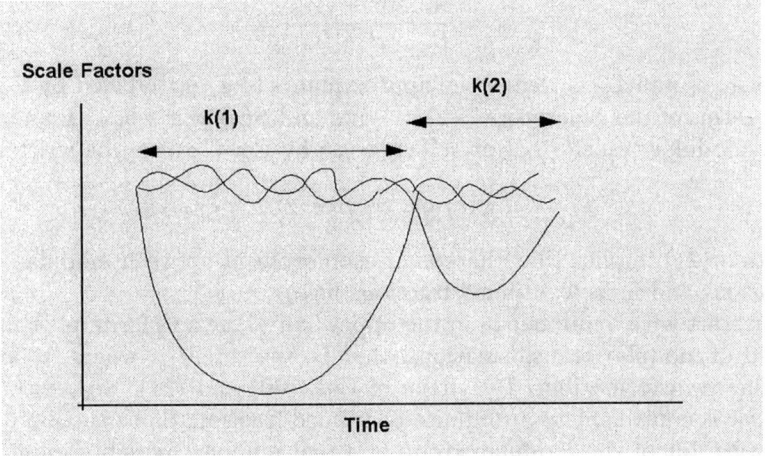

Fig. 2. Qualitative behavior of the three scale factors in chaotic relativity

If the initial state is given by an irrational number u_0, then the evolution proceeds via successive small oscillations coded by $u_0 - 1$, $u_0 - 2$, ..., etc., until the integer part of u_0 is exhausted. A new cycle of small oscillations then commences in which the initial state is coded by $u_1 = 1/(u_0 - [u_0])$. The number of small oscillations k_r occurring within the rth cycle is given by the rth partial quotient in the infinite continued-fraction expansion of

u_0. In other words, any initial (irrational) u_0 value will have a unique continued-fraction expansion and encode an infinite sequence of mixmaster cycles; similarly, any infinite set of cycle lengths determine a unique initial value of u_0.

The successive series of oscillations crowd together as we approach the cosmological singularity. An infinite number of oscillations are contained between any finite world time t and the moment $t = 0$ [10]. Although the mixmaster universe exhibits stochastic behavior, its underlying dynamics is ergodic for it is described by the now well-known transformation defined by Eq. (11). In particular, the probability distribution of the mixmaster cycle length is given by Eq. (17). This distribution implies that the mixmaster cycles are most likely to be short; over 41% of them will involve a single oscillation. Finally, if we choose in the ergodic theorem, Eq. (9), the function

$$f(x) = \log(q_1(x)) = \log[x^{-1}], \tag{23}$$

in combination with Eq. (10), we obtain

$$\lim_{N \longrightarrow \infty} \sum_{n=1}^{N} \log(q_n(x)) = \frac{1}{\log 2} \int_0^1 \log\left[\frac{1}{x}\right] \frac{1}{1+x} = \kappa. \tag{24}$$

In Eq. (24), is the universal Khintchine constant whose representation in terms of an infinite product is

$$\kappa = \prod_{n=1}^{\infty} \left(1 + \frac{1}{n^2 + 2n}\right)^{\log n / \log 2}. \tag{25}$$

The approximate value of κ is 2.68542. Results of numerical simulation related to chaotic relativity are contained in Ref. [17].

4   Time Series Coding

We encode a mapping $X : Z \longrightarrow \Re$ from integers to real numbers representing a time series by viewing each time series element X_i as a partial fraction in the continued fraction expansion. In the following, to emphasize the difference between the elements X_i of a time series and their continued fractions transforms, we use the lower case letters to denote the continued fractions, and the upper case letters to denote the elements of a time series. For reasons of accuracy, we employ a moving window covering about 40 elements of the time series. In the double precision arithmetics, taking a larger number elements does not increase the accuracy of a real number representation given by Eq. (1). If X_i are restricted to interval $(0, 1)$, we scale them by a factor s ranging from 50 to 1000. Thus for $i = k_1, \ldots, k_{k_{\max}}$, we make the assignments $q_i = sX_i$, leading to the continued fraction expansion given by Eq. (1) through the correspondence

$$[X_{k_1}, X_{k_2}, \ldots X_{k_{\max}}] \longrightarrow x \tag{26}$$

between an aggregate of the time series elements and its continued fraction representation. The aggregation of the time series elements can be viewed as a mechanism of data compression. An interesting application of continued fractions to cryptography is described by Jan and Kowng [6].

The accuracy of this assignment depends on the relative values of the partial quotients. For example, the fractional part of π has the following continued fraction expansion

$$\pi = [7, 15, 1, 292, 1, 1, 1, 2, 2, 3, \ldots] . \tag{27}$$

When $\{\pi\}$ is replaced by its decimal approximation 0.141592654, only the first four partial quotients are correctly recovered when the numerical algorithm of Sec. 2 is used. Because of the large value of q_4, the first three partial quotients lead already to a good approximation of $\{\pi\}$. On the other hand, the partial quotients of the number $(1 + \sqrt{5})/2 - 1$ are all equal to 1; in the double precision arithmetic, their numeric assignment is correct up to q_{37}.

In traditional rule-based systems, a library of rules with n antecedents $X_1, \ldots, X_n$ and output Y is constructed from input-output data pairs of the form

$$(X_1^{(m)}, X_2^{(m)}, \ldots, X_n^{(m)}; Y^{(m)}) , \tag{28}$$

where the index m labels the rules. Under the continued fraction encoding scheme, the rules can, similarly, have more than one antecedent. For example, with $k_{\max} = 5$, we transform the first 5 elements of X_i, $i = 1, \ldots 5$, into x_1; the elements X_i, $i = 2, \ldots 6$, are transformed into x_2; the elements X_i, $i = 3, \ldots 7$, are transformed into x_3, etc., as shown schematically in Fig. 3.

Once the time series has been encoded, fuzzy rules are generated from examples according the scheme of Wang and Mendel [13]. The five steps of their algorithm are well known and will not be reproduced here. In the last step, which determines a mapping from the combined fuzzy rule base, the output is generated by adopting a centroid or center of gravity defuzzification scheme. The defuzzified numeric output for the continued fraction transform allows one to decode the actual, nontransformed value. The two possible encoding-decoding schemes correspond to the order in which the elements X_i are aggregated into a continued fraction. For the natural order, as exemplified by Eq. (28), the most ancient element is the dominant partial fraction q_1; when the reverse order is used, the most recent element $X_{k_{\max}}$ is dominant.

5 Numerical Results

The Lorentz model [9] provides a well-known example of the chaotic motion; the solution to the system of differential equations

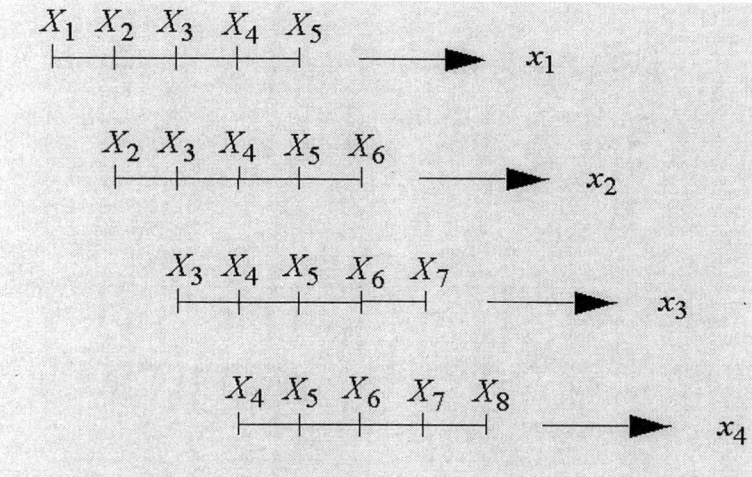

Fig. 3. Aggregating the elements of a time series into continued fractions

$$\dot{X} = -a(X - Y) \,,$$
$$\dot{Y} = -XZ + bX + Y \,,$$
$$\dot{Z} = XZ - cZ \,, \tag{29}$$

in which a, b, c are parameters, exhibits a strange attractor.

Setting $a = 10$, $b = 28$, $c = 8/3$, the computed values of the x-component of the Lorenz system is displayed in the upper part of Fig. 4, whereas the forecast error is shown in the lower part of the figure. We used the continued fraction encoding with three antecedents.

To optimize the algorithm of Sec. 4, we investigated the two-dimensional parameter space, spanned by the scale and stretch factors. Here the scale factor s refers to a multiplicative factor in the assignment $q_i = sX_i$; the stretch factor t maps the continued fraction representation x to the interval $(1 - t, t)$. As a convenient measure of the forecast accuracy we adopt the root mean square forecast error per time step, defined as

$$e = \sqrt{\frac{\sum_{n=1}^{N}(x_n - \widehat{x}_n)^2}{N}} \,, \tag{30}$$

where x_n and $\widehat{x}_n$ refer to the observed and forecasted elements of the time series, respectively. Two types of fuzzy controller were utilized: a simple fuzzy controller with center of gravity defuzzification, and an adaptive neuro-fuzzy interference system (ANFIS) of Jang implemented through a MATLAB toolbox [5]. Fundamentally, ANFIS is a neural network representation of Sugeno-type fuzzy systems, endowed with neural learning capabilities [12].

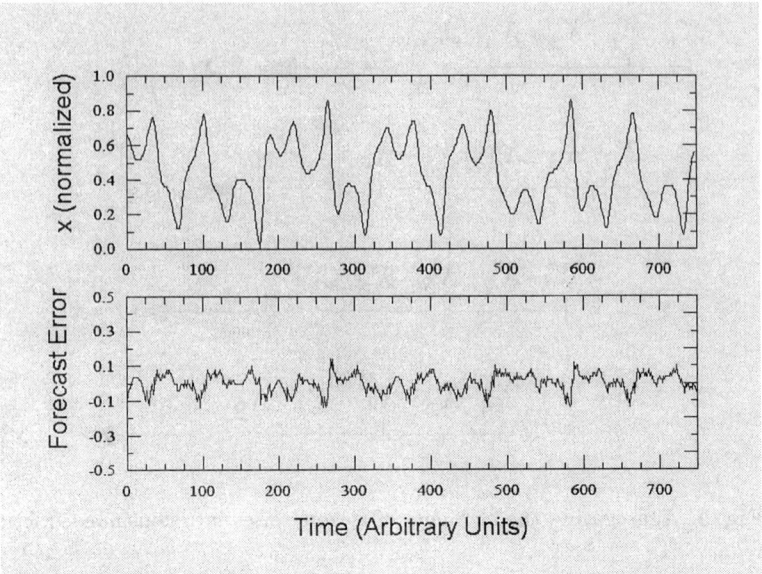

Fig. 4. The normalized x component of the Lorenz attractor, together with the forecast error. Simple fuzzy controller, $s = 50$, $t = 0.95$

When the transformed data are used, the forecast error becomes much smaller. In Fig. 5 we show the situation after the times series is written in terms of the continuous fractions, using the algorithm of the preceding section.

As a quantitative measure of the forecasting efficiency, we use the square root deviation per time step, Eq. (30); with the continued fraction encoding, we gain about 10% in the efficiency as compared to the standard fuzzy controller.

Somehow surprisingly, the ANFIS model attains the optimum conditions for $s = 10$, $t = 0.99$. Because the scale factor is five times smaller than in the case of the simple fuzzy controller, the resolution of the data is poorer. The coarse-grained rendition of the forecasted time series and the forecast error is shown in Fig. 6.

For highly chaotic time series, exhibiting large oscillations between neighboring values, the rule system fails to capture the time evolution to a satisfactory degree of accuracy. For example, using the water flow data of Kasabov [7], we observe in Fig. 7 sizable forecast errors when the water flow rate changes abruptly. In the continued fractions interpretation, this is due to differences in the dominant partial fraction of the time series encoding.

The last example shows that satisfactory prediction for the transformed time series does not necessarily imply successful reconstruction of the original

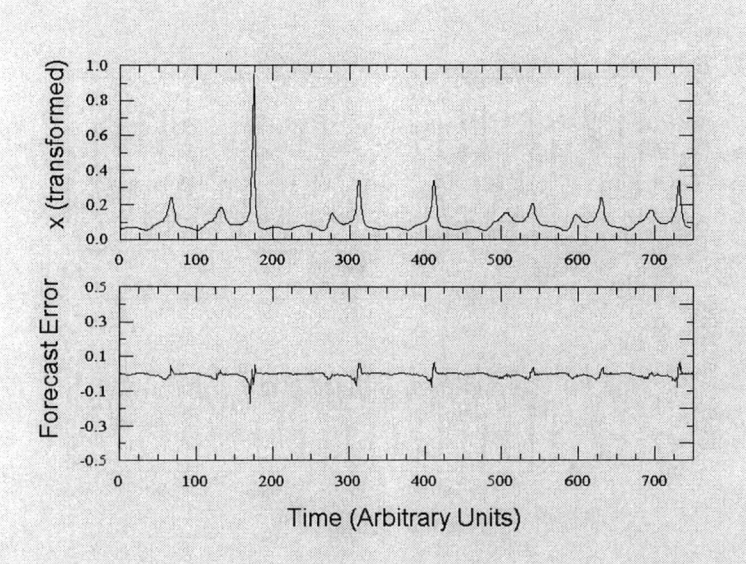

Fig. 5. The continued fraction transformation of the x component of the Lorenz system, displayed in Fig. 5.1, and its forecast error

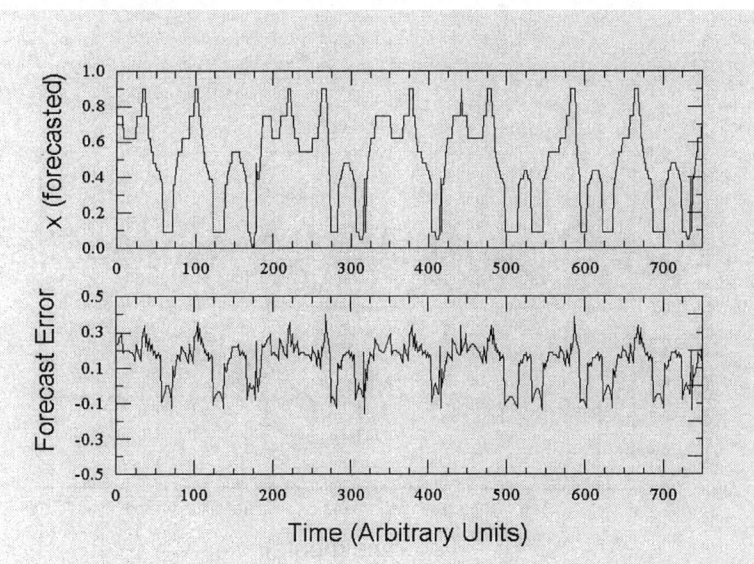

Fig. 6. The normalized x component of the Lorenz attractor, together with the forecast error. ANFIS controller with $s = 10$, $s = 0.99$

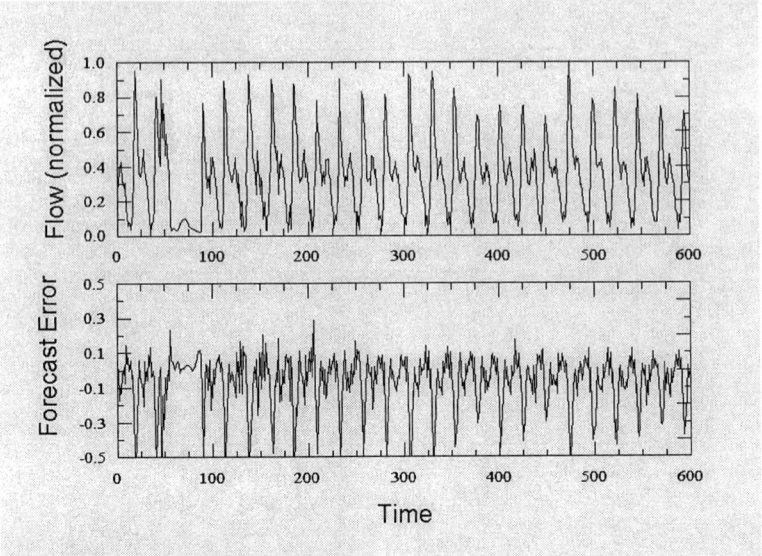

Fig. 7. Water flow data and the forecast error obtained by using the simple fuzzy controller, $s = 100$, $t = 0.9$

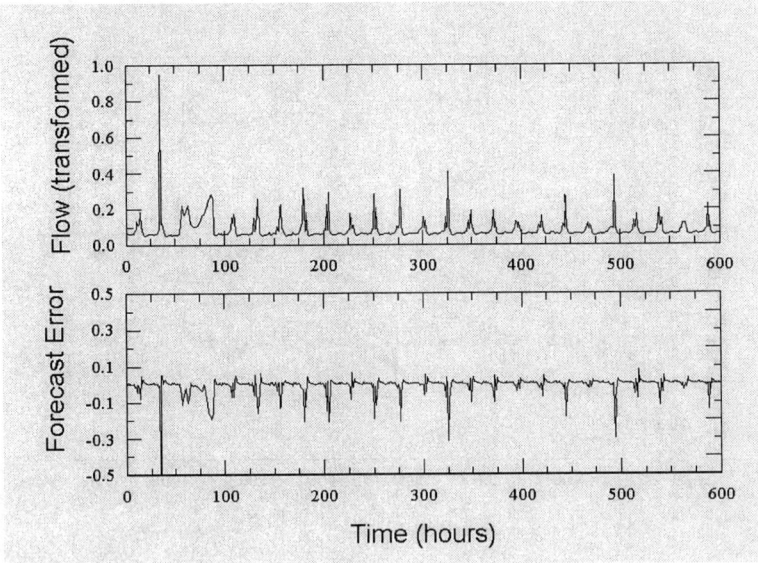

Fig. 8. The transformed time series and its forecast error for the water flow data of Ref. [7]

time series. It is only through optimization of the algorithm parameters that the original time series can be properly reconstructed.

6 Conclusions

To account for the past information contained in a time series and, simultaneously, to avoid the combinatorial explosion in the rule library, we have encoded the overlapping measurement windows into continued fractions through a one-to-one transformation. The algorithm potentially captures the history of a large number of past events, leading to quantitatively better prediction results than the simple fuzzy controller. When the time series exhibits rapid oscillations, the algorithm is less successful. To some extent, we have studied the impact of two parameters on the predictive power of the algorithm. They were the scaling factor s, which maps the original time series from the range $(0, 1)$ to the range $(0, s)$, and the stretch factor mapping the resulting continued fraction from the interval $(0, 1)$ to $(1 - t, t)$. The other parameters of interest are the number k_{max} of the time series elements included into the time series history, and the number of continued fractions that are used to build the library of rules. Future research will explore the applicability of this encoding scheme to radial basis functions approach [14]; we will also study the optimization of parameters through genetic algorithms.

References

1. Artin, E. (1965): Ein mechanisches System mit quasiergodischen Bahnen, in *The Collected Papers of Emil Artin*, edited by S. Lang and J. T. Tate (Addison-Wesley, Reading), pp. 499–504
2. Barrow, J.D. (1981): Chaos in the Einstein equations, Phys. Rev. Lett. **46**, 963–966
3. Combs, W.E. and Andrews, J.E. (1998): Combinatorial rule explosion eliminated by a fuzzy rule configuration, IEEE Trans. Fuzzy Systems **6**, 111
4. Davenport, H. (1983): *The Higher Arithmetic: An Introduction to the Theory of Numbers* (Dover, New York), Ch. IV
5. Jang, J.-S.R. (1993): ANFIS: Adaptive-Network based Fuzzy interference System, IEEE Trasactions on Systems, Man, and Cybernetics, **25** 665–685
6. Jan J.K. and Kowng, H.C. (1993): A cryptographic system based upon the continued fractions, Proc. 1993 International Conference on Security Technology (IEEE, Piscataway, NJ), pp. 219–223.
7. Kasabov, N.K. (1996): *Foundations of Neural Networks, Fuzzy systems, and Knowledge Engineering* (The MIT Press, Cambridge, MA), Ch. 7
8. Khinchin, A.Ya. (1997): *Continued Fractions* (Dover, New York)
9. Lorenz, E.N. (1963): Deterministic nonperiodic flow, J. Atmos. Sc. **20**, 130–141
10. Landau, L.D. and Lifshitz, E.M. (1975) *The Classical Theory of Fields* (Pergamon Press, New York), Sec. 118
11. Series, C (1981): Non-euclidean geometry, continued fractions, and ergodic theory, Math. Intelligencer **2**, 24–31

412

12. Tsoukalas, L.H. and Uhrig, R.E. (1997): *Fuzzy and Neural Approaches in Engineering* (Wiley, New York), Sec. 13.6
13. Wang, L.X. and Mendel, J.M. (1992): Generating fuzzy numbers by learning from examples. IEEE Trans. Systems, Man and Cybernetics **22**, 1414–1427
14. Wang, L.X. and Mendel, J.M. (1992): Fuzzy basis functions, universal approximation, and orthogonal least-squares learning, IEEE Trans. Neural Networks, **3**, 807–813
15. Zardecki, A. (1996): *Rule-based forecasting, in Fuzzy Modelling: Paradigms and Practice*, edited by W. Pedrycz (Kluver, Boston), pp. 375–391
16. Zardecki, A. (1995): Fuzzy controllers in nuclear material accounting, Fuzzy Sets and Systems **74**, 73–79
17. Zardecki, A. (1983): Modeling in chaotic relativity, Phys. Rev. **D28**, 1235–1242

19 A Possibilistic Approach to Target Classification

Albert G. Huizing[*] and Frans C.A. Groen[+]

[*]TNO Physics and Electronics Laboratory
Oude Waalsdorperweg 63, 2597 AK The Hague, The Netherlands
[+] Department of Computer Science, University of Amsterdam
Kruislaan 403, 1098 SJ Amsterdam, The Netherlands
groen@wins.uva.nl

This chapter describes an alternative to the Bayesian approach to target classification that is based on possibility theory. A possibilistic classifier minimizes the maximum cost of the classification decision taking into account the a posteriori possibilities of the target classes given the measured target attributes. The advantage of a possibilistic classifier when compared with a Bayesian classifier is that it requires only an ordinal ranking of the costs associated with the classification decisions and the uncertainty about the target class. Owing to its qualitative character, a possibilistic classifier is less sensitive to inaccuracies in a priori knowledge than a Bayesian classifier at the expense of a degraded performance in situations where accurate a priori knowledge is available. This robustness of the possibilistic classifier to inaccuracies in a priori knowledge is demonstrated in a case study where an average cost criterion is used to compare the performance of a possibilistic and a Bayesian classifier. It is shown that when the characteristics of the measured target attributes deviate strongly from the expected characteristics, the possibilistic classifier provides a lower average cost than a Bayesian classifier.

1 Introduction

Target classification is an important function in civilian and military observation systems to support the situation assessment process. A well-known approach to the problem of target classification is a Bayesian classifier [1]. A Bayesian classifier estimates the target class from the measured target attributes and the available *a priori* knowledge. This *a priori* knowledge comprises estimates of the *a priori* probabilities of the target classes, estimates of the likelihood of measured attribute values and estimates of the costs of the decisions taken by the classifier. Accurate probability and cost estimates are sometimes difficult to obtain because some target classes and attribute values may never have been observed. This situation particularly occurs in military scenarios and competitive games where we deal with intelligent adversaries that try to influence the decision of the classifier in such a way that they can complete their own mission successfully. Given these difficulties, the Bayesian classifier may not always be the best solution to the target classification problem.

The classification of targets can be regarded as a decision-theoretic problem in which decisions have to made in situations where the state of the world is uncertain. Recently, Dubois and Prade [3,4,5] developed an approach based on possibility theory which addresses some of the problems associated with other approaches to decision making under uncertainty. A possibilistic decision-theoretic approach requires an ordinal ranking of the world states with respect to their uncertainty and an ordinal ranking of the cost of decisions with respect to their consequences given a certain world state. This qualitative nature of the possibilistic approach is in contrast with conventional approaches to decision making such as a Bayesian classifier that require the uncertainties in the possible world states and the cost of decisions to be quantified. The purpose of this chapter is to compare a possibilistic and Bayesian approach to target classification and to demonstrate that the possibilistic approach can provide a better performance in situations where targets have to be classified that try to avert a correct classification.

The organization of this chapter is as follows. First, the problem of target classification is formulated and a description of a Bayesian and a possibilistic classifier is given. We then define a performance criterion so that the performance of the two classifiers can be compared quantitatively. The difference in performance of the two classifiers in an environment with an intelligent adversary is illustrated by a case study and finally some conclusions are drawn.

2   Problem Formulation

In this chapter we consider the following problem. A target classification system measures with its sensor a set of attributes of a single target in its environment and subsequently selects the best hypothesis about the target class given this set of attribute measurements and the available *a priori* knowledge about the sensor, the target and the environment (see figure 1). The best hypothesis as viewed from the perspective of the classifier depends on the class of the target which can only indirectly be observed through a measurement of the target attributes. When targets belonging to different classes can have the same attribute values there will be uncertainty about the target class. Owing to imperfections of the sensor (e.g., due to noise, clutter and jamming) the measured attribute values are only an approximation of the true attribute values and this adds to the uncertainty about the target class.

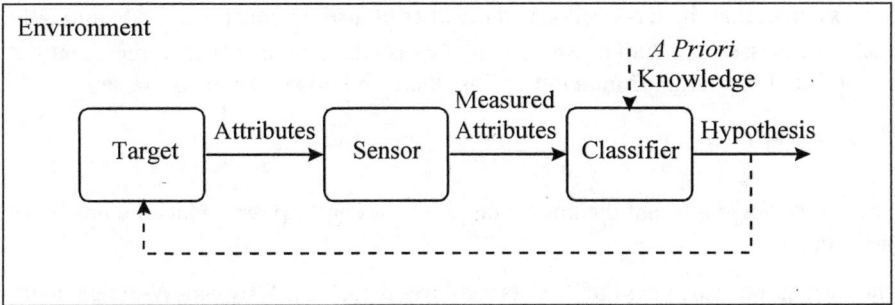

Fig. 1. Target Classification

To optimize the performance of the classifier, accurate information is required about the occurrence of target classes and the associated attribute values and the costs of the various hypotheses given a target class. Because some target classes and attribute values may never have been observed before, it is not always possible to obtain the required information and educated guesses must be made, particularly in military scenarios. Another problem is that targets with hostile intentions may try to avert a correct classification. This can for example be achieved by avoiding detection by the sensor through application of camouflage or stealth techniques, jamming of the sensor, or deception of the classifier by imitating the characteristics or behaviour of other targets. This type of behaviour is here referred to as non-cooperative. Cooperative targets on the other hand do not try to avert a correct classification and behave in the same way (in a statistical sense) as they used to behave in the past. The classifier does not know in advance if a target will behave cooperatively or non-cooperatively.

In summary, it can be concluded that the classifier has to deal with the following uncertainties:

1. uncertainty about the target class given the true attribute values,

2. uncertainty about the true attribute values given the measured attribute values,

3. uncertainty about the costs of hypotheses,

4. uncertainty about the strategy of the target (cooperative or non-cooperative).

A Bayesian (or probabilistic) and a possibilistic approach to this problem are discussed in the next section.

3 Target Classification

It is assumed that the target class is a member of a set Ω consisting of M mutually exclusive classes. An "unknown" target class is added to the set Ω to represent the possibility that a target from another class than the known classes is present:

$$\Omega = \left\{ \omega_0, \omega_1, \cdots, \omega_M \right\}$$

where $\omega_1 \ldots \omega_M$ represent the known target classes and ω_0 represents the unknown target class.

The measured attributes of the target are represented by an N-dimensional attribute vector $\hat{\mathbf{x}}$ which is an estimate of the true attribute vector $\mathbf{x}$. From the measured attribute vector and the *a priori* knowledge, the classifier estimates the target class. Owing to the uncertainty in the target class given the measured attribute vector, the classifier cannot always estimate the target class unequivocally. As a consequence, the set of hypotheses that is available to the classifier for the classification process not only contains unambiguous hypotheses that associate the origin of the measured attribute vector with a single target class, i.e., $h_i = \{\omega_i\}$, $i = 0, \ldots, M$, but also ambiguous hypotheses that associate the measured attribute vector with a subset of the entire set of classes which contains more than one target class, i.e., $h_i \subset \Omega$, $|h_i| > 1$, $i > M$. The set of hypotheses, H, used by the classifier may therefore include up to $2^{M+1} - 1$ hypotheses [1]:

$$H = \left\{ h_0, h_1, \cdots, h_L \right\} \qquad M \leq L \leq 2^{M+1} - 1$$

We will now successively discuss how a Bayesian classifier and a possibilistic classifier select the best hypothesis h from the set H given the measured attribute vector $\hat{\mathbf{x}}$.

3.1 Bayesian Classifier

A well-known approach to the problem of target classification is provided by Bayesian decision theory. A Bayesian classifier, $f_B(\hat{\mathbf{x}})$, selects the hypothesis $h \in H$ that minimizes the average cost of the decision given the measured attribute vector $\hat{\mathbf{x}}$ [5]:

---

[1] The empty set is excluded because the hypothesis that the measured attribute vector does not originate from a known class is represented by the unknown class hypothesis h_0.

$$h = f_B(\hat{\mathbf{x}}) = \arg\min_{i=0}^{L} \sum_{j=0}^{M} C(h_i, \omega_j) \cdot \Pr(\omega_j | \hat{\mathbf{x}})$$ (1)

where $C(h_i, \omega_j)$ represents the cost associated with a hypothesis h_i given the target class ω_j and $\Pr(\omega_j | \hat{\mathbf{x}})$ represents the *a posteriori* probability of the target class ω_j given the measured attribute vector $\hat{\mathbf{x}}$.

The cost function $C(h_i, \omega_j)$ maps each combination of a hypothesis h_i and a target class ω_j to a real value C_{ij} that is determined by the preferences of the Bayesian classifier. If the Bayesian classifier prefers hypothesis h_m to h_n given a target class ω_j, then the cost C_{mj} is smaller than the cost C_{nj}. The cost C_{nj} is equal to the cost C_{mj} if the Bayesian classifier considers the hypotheses h_n and h_m equivalent given a target class ω_j.

A special case is the situation where only unambiguous hypotheses, $h_i = \{\omega_i\}$, $i = 0, \ldots, M$ are considered by the Bayesian classifier and the cost function C is a zero-unit function which assigns a zero cost to the correct hypothesis and a unit cost to the wrong hypotheses:

$$C_{ij} = 1 - \delta_{ij} \quad i, j = 0, \cdots, M$$ (2)

where δ_{ij} is the Kronecker delta function. The Bayesian classifier chooses in this case the hypothesis with the maximum *a posteriori* probability:

$$h = \arg\min_{i=0}^{M} \sum_{j=0}^{M} (1 - \delta_{ij}) \cdot \Pr(\omega_j | \hat{\mathbf{x}}) = \arg\max_{j=0}^{M} \Pr(\omega_j | \hat{\mathbf{x}})$$ (3)

The *a posteriori* probability of a target class given the measured attribute vector $\hat{\mathbf{x}}$ can be calculated with Bayes' rule from the available probabilistic *a priori* knowledge about the target and the sensor according to:

$$\Pr(\omega_j | \hat{\mathbf{x}}) = \frac{p(\hat{\mathbf{x}} | \omega_j) \cdot \Pr(\omega_j)}{p(\hat{\mathbf{x}})}$$ (4)

where $\Pr(\omega_j)$ is the *a priori* probability of the target class, $p(\hat{\mathbf{x}} | \omega_j)$ is the likelihood function of the measured attribute vector $\hat{\mathbf{x}}$ given the target class ω_j and $p(\hat{\mathbf{x}})$ is the marginal probability density function.

Using the total probability theorem, the likelihood function of the measured attribute vector can be written as the convolution of the probabilistic sensor transfer function $p(\hat{\mathbf{x}} | \mathbf{x})$ and the likelihood function of the true attribute vector $p(\mathbf{x} | \omega_j)$:

$$p(\hat{\mathbf{x}}|\omega_j) = \int p(\hat{\mathbf{x}}, \mathbf{x}|\omega_j) \cdot d\mathbf{x} = \int p(\hat{\mathbf{x}}|\mathbf{x}) \cdot p(\mathbf{x}|\omega_j) \cdot d\mathbf{x} \qquad (5)$$

where we have assumed that the measurement of the attribute vector does not depend on the target class, i.e., $p(\hat{\mathbf{x}}|\mathbf{x},\omega_j) = p(\hat{\mathbf{x}}|\mathbf{x})$.

A problem for the Bayesian classifier is that *a priori* probabilities of the target classes (joint) likelihood functions of the measured attributes must be available. Unfortunately, *a priori* probabilities of target classes and likelihood functions of the measured target attributes are difficult to obtain for all target classes and scenarios that are of interest because some target classes and attribute values may never be observed. This lack of information means that educated guesses must be made about the *a priori* probabilities and likelihood functions which may prove to be inaccurate, particularly in military scenarios or competitive games. Given these difficulties, the Bayesian classifier may not always be the best solution to the target classification problem.

3.2 Possibilistic Classifier

An alternative to the Bayesian classifier is a possibilistic classifier, $f_\pi(\hat{\mathbf{x}})$, which selects the hypothesis h that minimizes the maximum possible cost of the decision given the measured attribute vector $\hat{\mathbf{x}}$ [6]:

$$h = f_\pi(\hat{\mathbf{x}}) = \arg \min_{\substack{i=0 \\ }}^{L} \max_{\substack{j=0}}^{M} \min\left(C(h_i,\omega_j), \Pi(\omega_j|\hat{\mathbf{x}})\right) \qquad (6)$$

where $\Pi(\omega_j|\hat{\mathbf{x}})$ is the *a posteriori* possibility of the target class ω_j given the measured attribute vector $\hat{\mathbf{x}}$. The cost function $C(h_i,\omega_j)$ represents the preferences of the possibilistic classifier for the hypotheses given a target class. This is analogous to the role of the cost function for the Bayesian classifier. The difference is, however, that the domain of the cost function $C(h_i,\omega_j)$ must be the same as the interval of the *a posteriori* possibilities of the target classes, i.e., $C(h_i,\omega_j) \in [0,1]$. The requirement of commensurate domains for the cost function and the *a posteriori* possibility of a target class seems to introduce a limitation when we want to compare the performance of a Bayesian and a possibilistic classifier with the same cost function. However, because the decisions taken by the Bayesian classifier are scale and translation invariant [5], we can choose the cost function of the Bayesian classifier in such a way that it also fits in the interval $[0,1]$.

The cost minimized by the possibilistic classifier is a weighted maximum which is the possibilistic counterpart of the probabilistically weighted mean cost that is minimized by the Bayesian classifier [7]. An important property is that the possibilistic classifier uses the minimum and maximum operators while the

Bayesian classifier uses the product and sum operators. Consider for example a situation where the *a posteriori* possibilities of the target classes are either zero or one. In this case the maximization is limited to the classes with an *a posteriori* possibility equal to one and the possibilistic classifier selects the hypothesis with the minimal maximum cost. This minimax decision rule is well-known from game theory where we deal with an intelligent opponent, see for example Luce and Raiffa [8].

We consider again a special case where the classifier can only select a hypothesis from the set of unambiguous hypotheses, $h_i = \{\omega_i\}$, $i = 0,...,M$ and a zero-unit cost function C. In this case, the possibilistic classifier chooses the hypothesis with the maximum *a posteriori* necessity, $N(\omega_j|\hat{\mathbf{x}})$, of a target class ω_j given the measured attribute vector $\hat{\mathbf{x}}$ [2]:

$$h = \arg\min_{i=0}^{M} \max_{j=0, j\neq i}^{M} \Pi(\omega_j|\hat{\mathbf{x}}) = \arg\max_{i=0}^{M} N(\omega_i|\hat{\mathbf{x}}) \qquad (7)$$

The *a posteriori* possibility of the target class given the measured attribute vector $\hat{\mathbf{x}}$ is calculated according to (for a derivation, see the Appendix):

$$\Pi(\omega_j|\hat{\mathbf{x}}) = \min\left(\pi(\hat{\mathbf{x}}|\omega_j), \Pi(\omega_j)\right) \qquad \max_{j=0}^{M} \Pi(\omega_j|\hat{\mathbf{x}}) = 1 \qquad (8)$$

where $\Pi(\omega_j)$ is the *a priori* possibility of the target class, $\pi(\hat{\mathbf{x}}|\omega_j)$ is the possibility distribution of the measured attribute vector $\hat{\mathbf{x}}$ given the target class ω_j and $\pi(\hat{\mathbf{x}})$ is the marginal possibility distribution. This update rule is analogous to Bayes' rule that is used to calculate the *a posteriori* probability of a target class except that the product operator is replaced by a minimum operator and the normalization is carried out in such a way that there is always a target class with an *a posteriori* possibility equal to one.

The possibility distribution of the measured attribute vector is calculated from the possibilistic sensor transfer function $\pi(\hat{\mathbf{x}}|\mathbf{x})$ which represents the possibility that an attribute vector $\hat{\mathbf{x}}$ is measured given a true attribute vector $\mathbf{x}$ and the possibility distribution of the true attribute vector $\pi(\mathbf{x}|\omega_j)$:

$$\pi(\hat{\mathbf{x}}|\omega_j) = \sup_{\mathbf{x}} \pi(\hat{\mathbf{x}}, \mathbf{x}|\omega_j) = \sup_{\mathbf{x}} \min\left(\pi(\hat{\mathbf{x}}|\mathbf{x}), \pi(\mathbf{x}|\omega_j)\right) \qquad (9)$$

---

[2] The *a posteriori* necessity of a target class is defined as the impossibility that the measured attribute vector originates from another target class:

$$N(\omega_i|\hat{\mathbf{x}}) = 1 - \Pi(\overline{\omega}_i|\hat{\mathbf{x}}) = 1 - \max_{j=0, j\neq i}^{M} \Pi(\omega_j|\hat{\mathbf{x}})$$

where we have assumed that the measurement of the attribute vector does not depend on the target class, i.e., $\pi(\hat{\mathbf{x}}|\mathbf{x},\omega_j) = \pi(\hat{\mathbf{x}}|\mathbf{x})$.

4   Performance Criterion

In this section we define an average cost criterion to compute the sensitivity of a Bayesian and possibilistic classifier to inaccuracies in *a priori* knowledge. This criterion averages the costs of the selected hypotheses over all attribute vectors that can be measured by the sensor. The true *a priori* probabilities of the target classes and the likelihood functions of the measured attributes are generally different from the expected *a priori* probabilities and likelihood functions. If the true *a priori* probabilities and likelihood functions are represented by Pr*(ω_j) and p*($\hat{\mathbf{x}}|\omega_j$), respectively, the average cost of a classifier f($\hat{\mathbf{x}}$) is given by:

$$E\left\{C\left(f(\hat{\mathbf{x}})\right)\right\} = \int C\left(f(\hat{\mathbf{x}})\right)\cdot p^*(\hat{\mathbf{x}})\cdot d\hat{\mathbf{x}} = \sum_{j=0}^{M} \int C\left(f(\hat{\mathbf{x}}),\omega_j\right)\cdot Pr^*\left(\hat{\mathbf{x}},\omega_j\right)\cdot d\hat{\mathbf{x}} =$$
$$= \sum_{j=0}^{M} \int C\left(f(\hat{\mathbf{x}}),\omega_j\right)\cdot p^*\left(\hat{\mathbf{x}}|\omega_j\right)\cdot Pr^*\left(\omega_j\right)\cdot d\hat{\mathbf{x}}$$

(10)

It will be clear that the Bayesian classifier provides the minimum average cost in situations where the true *a priori* probabilities and likelihood functions are equal or very close to the expected *a priori* probabilities and likelihood functions. However, if the true *a priori* probabilities and likelihoods differ significantly from the expected *a priori* probabilities and likelihoods, the average cost of the Bayesian classifier may be larger than the average cost of another classifier such as the possibilistic classifier. These conditions occur in a game-theoretic situation where an intelligent opponent knows the preferences and the strategy of the classifier and is able to manipulate the *a priori* probabilities of the target classes and likelihoods of the attributes in such a way that the cost of the decision taken by the classifier is maximized. In this case, the possibilistic classifier will provide a lower cost than the Bayes classifier because it minimizes the maximum possible cost.

In practical scenarios, neither of these extreme situations will occur but an intermediate case in which friendly and neutral (i.e., cooperative) targets behave more or less like expected while hostile (i.e., non-cooperative) targets try to maximize the average cost of the decisions made by the classifier. Consider for example a crisis scenario in which a jet fighter with hostile intentions is flying along a commercial air way at the same velocity and altitude as a typical airliner and the fighter replies to interrogations with the codes that correspond with a scheduled commercial flight. In this case, the classifier may wrongly conclude that the target is an airliner.

5   A Case Study

5.1   Perfect Sensor

The difference between the Bayesian and possibilistic classifier will be illustrated with a case study in which a sensor measures a single attribute x of a target. We will first assume that the sensor is perfect, i.e., $p(\hat{x}|x) = \delta(\hat{x} - x)$ where δ represents the Dirac function. The set Ω of target classes consists of two known classes, ω_1 and ω_2, and an unknown class, ω_0. The set H of hypotheses consists of the three unambiguous hypotheses $h_0 = \{\omega_0\}$, $h_1 = \{\omega_1\}$, $h_2 = \{\omega_2\}$, and two ambiguous hypotheses $h_3 = \{\omega_1,\omega_2\}$ and $h_4 = \Omega = \{\omega_0,\omega_1,\omega_2\}$. The hypothesis h_4 is here referred to as the ignorant hypothesis because it indicates that the measured attribute $\hat{x}$ provides insufficient information to preclude any target class. Table 1 shows the cost matrix that contains the costs associated with the various hypotheses given the target classes. This cost matrix is valid for both the Bayesian and the possibilistic classifier. C_a represent the cost associated with hypothesis h_3 when the true target class is either ω_1 or ω_2 while C_i denotes the cost when the ignorant hypothesis h_4 is declared in the presence of ω_0, ω_1 or ω_2.

Table 1. Cost Matrix

Hypothesis	Subset	Target Class		
		ω_0	ω_1	ω_2
h_0	$\{\omega_0\}$	0	1	1
h_1	$\{\omega_1\}$	1	0	1
h_2	$\{\omega_2\}$	1	1	0
h_3	$\{\omega_1,\omega_2\}$	1	C_a	C_a
h_4	$\{\omega_0,\omega_1,\omega_2\}$	C_i	C_i	C_i

In addition to the cost matrix, we need to define the *a priori* probabilities and *a priori* possibilities of the target classes and the likelihood functions and possibility distributions of the measured attribute. In this case study, the Bayesian classifier assumes an equal *a priori* probability of the two known classes which is higher than the *a priori* probability of the unknown class. The possibilistic classifier assigns the *a priori* possibilities of the two known classes a value equal to one because they have been observed before. The *a priori* possibility of the unknown class is assigned a value between zero and one because this event is considered to be not impossible but less possible than the occurrence of a known target class.

To allow a fair comparison between the two classifiers, the Bayesian and possibilistic classifier both use the same information about the occurrence of attribute values. This information indicates that the attribute for class ω_i can attain values in an interval with a minimum value $\mu_i - 0.5 \cdot \eta_i$ and a maximum value $\mu_i + 0.5 \cdot \eta_i$. From this information and the normalization constraints[3], we can derive the (uniform) likelihood functions and possibility distributions of the measured attribute value. Figure 2 shows the likelihood functions for the three target classes as a function of the measured attribute value when we assume that the mean, μ_i, of the attribute values for the three classes is equal to 0.5 and the width, η_i, of the likelihood functions is equal to 1, 0.4 and 0.1 for ω_0, ω_1, and ω_2, respectively.

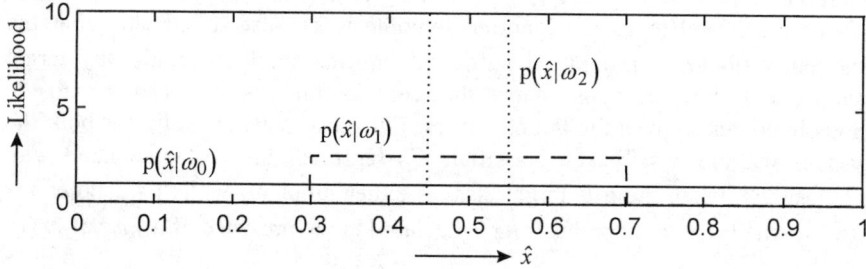

Fig. 2. Likelihood functions for the three target classes as a function of the measured attribute when the sensor is assumed to be perfect

Figure 3 shows the decisions taken by the Bayesian (top diagram) and possibilistic classifier (bottom diagram) as a function of the measured attribute value. The cost C_a of the ambiguous hypothesis h_3 is 0.3 and the cost C_i of the ignorant hypothesis h_4 is 0.8. The Bayesian classifier assumes that the *a priori* probability of the unknown class is 0.1 (implying an *a priori* probability of 0.45 for the two known classes). The possibilistic classifier assumes that the *a priori* possibility of the unknown class is equal to 0.2.

It can be observed in Figure 3 that only for attribute values in the overlap region (from 0.45 to 0.55) of the likelihood functions (and possibility distributions) of the two known classes a different decision is taken by the two classifiers: the

---

[3] The normalization constraint for a likelihood function implies $\int p(\hat{x}|\omega) \cdot d\hat{x} = 1$ while the normalization constraint for the possibility distribution implies $\max_{\hat{x}} \pi(\hat{x}|\omega) = 1$

Bayesian classifier selects the unambiguous hypothesis h_2 that is associated with the second target class, ω_2, while the possibilistic classifier chooses the ambiguous hypothesis h_3 that is associated with the two known target classes, ω_1 and ω_2.

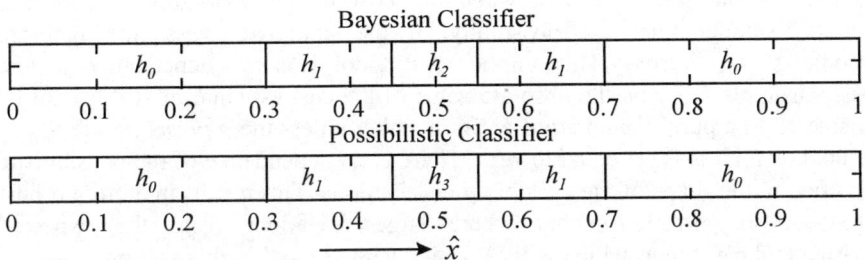

Fig. 3. Decisions taken by the Bayesian and possibilistic classifier as a function of the measured attribute value assuming a perfect sensor

Decisions in the Overlap Region

Figure 4 illustrates the decisions taken by the two classifiers in the overlap region of the likelihood functions of the two known classes as a function of the costs of the ambiguous and ignorant hypotheses, C_a and C_i, respectively. Three decision areas can be discerned for the Bayesian classifier with boundaries that are determined by the variables α, β and γ.

$$\alpha = \frac{1}{1 + \Pr(\omega_2|\hat{x})/\Pr(\omega_1|\hat{x})} = \frac{1}{1 + \eta_1/\eta_2} = 0.2 \tag{11}$$

$$\beta = 1 - \Pr(\omega_2|\hat{x}) = 0.214 \tag{12}$$

$$\gamma = \Pr(\omega_0|\hat{x}) = 0.0175 \tag{13}$$

The shapes of the decision areas illustrate the quantitative nature of the Bayesian classifier because its decisions depend not only the order of the costs of the ambiguous and ignorant hypotheses but also on their absolute values. In order to obtain the desired (i.e., robust) behaviour from the Bayesian classifier, a sensitivity analysis must be performed of the effects of (small) changes in the cost estimates and the estimates of the *a priori* probabilities and likelihood functions. This sensitivity analysis is relatively easy to perform when the number of target classes is small but it may become intractable when the number of targets classes becomes large.

424

For the possibilistic classifier there are only two decision regions. If the cost of the ignorant hypothesis is larger than the cost of the ambiguous hypothesis ánd the *a priori* possibility of the unknown class, then the possibilistic classifier decides for the ambiguous hypothesis, otherwise the ignorant hypothesis is preferred. The nature of the possibilistic classifier is mainly qualitative because the decision depends for a large part only on the order of the costs of the ambiguous and ignorant hypotheses. However, when the cost of the ambiguous hypothesis becomes smaller than the possibility of the unknown class, the ignorant hypothesis is preferred. This implies that there is some dependence of the possibilistic classifier on the absolute value of the cost estimates and it cannot be considered as a purely qualitative decision maker unless the *a priori* possibility of the unknown class is set equal to zero. There is no dependence of the possibilistic classifier on the shape of the possibility distributions. This results in a more robust behaviour in an environment where large deviations from the expected environment can occur, which will be shown in the sequel of this section.

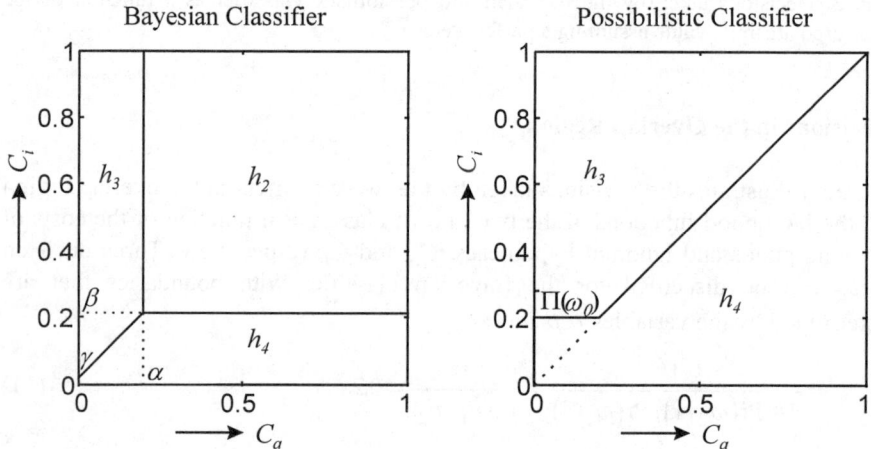

Fig. 4. Decisions taken by the Bayesian and possibilistic classifier in the overlap region of the likelihood functions of the two known classes as a function of the costs of the ambiguous and ignorant hypotheses

Sensitivity to Inaccuracies in *A Priori* Knowledge

The sensitivity of the Bayesian and possibilistic classifier to inaccuracies in the *a priori* knowledge is analyzed by calculating the average cost of the decisions taken by the classifier in an environment where we deal with an intelligent adversary, i.e., a non-cooperative target. We assume that a non-cooperative target belonging to the first known target class, ω_1, is capable of adapting its behaviour in such a way that the measured attribute value $\hat{x}$ will most likely fall in the overlap region of the likelihood functions for the two known target classes. The true likelihood functions of the unknown class and the second known class as well as the true *a priori* probabilities of the classes are all equal to the expected values.

Figure 5 shows the average cost of the decisions taken by the Bayesian and possibilistic classifier as a function of the width of the true (uniform) likelihood function of the first target class, η_1^*. As a reference, the average cost of the optimum (Bayesian) classifier is shown which knows in advance the true width of the likelihood function for the non-cooperative target. When the true width is equal to the expected width of 0.4, the average cost of the Bayesian classifier is equal to the average cost of the optimum classifier which is lower than the average cost of the possibilistic classifier. As the true width of the likelihood function of the non-cooperative target decreases, the average cost of the Bayesian classifier increases more rapidly than the average cost of the possibilistic classifier. The cross-over point occurs at the width where the optimum classifier changes the decision in the overlap region from the unambiguous hypothesis, h_2, to the ambiguous hypothesis, h_3. This cross-over point is determined by the cost of the ambiguous hypothesis, C_a, and the width of the likelihood function of the second known class, η_2:

$$\eta_1^* = \frac{(1 - C_a)}{C_a} \cdot \eta_2 = 0.23$$

As the true width decreases further, the average costs of the Bayesian and possibilistic classifier become constant when η_1^* becomes smaller than η_2 (0.1).

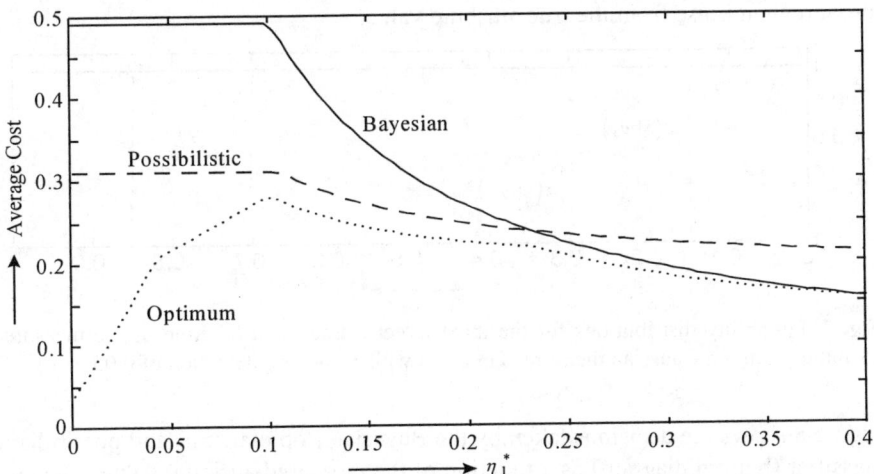

Fig. 5. Average cost of a Bayesian, possibilistic and optimum classifier as a function of the true width of the likelihood function of the non-cooperative target. The cost C_a of the ambiguous hypothesis (h_3) is 0.3, the cost C_i of the ignorant hypothesis (h_4) is 0.8 and the sensor is assumed to be perfect

426

5.2 Measurement Noise

We now examine a case where the sensor adds Gaussian noise with a standard deviation of 0.05 to the true attribute value. The characteristics of the measurement noise are known by the Bayesian and possibilistic classifier. Figure 6 shows that the likelihood functions of the measured attribute value for the two known classes become smoothed at the edges while the likelihood function of the unknown class stays uniform because we assume that the measured attribute can only attain values in the domain [0,1].

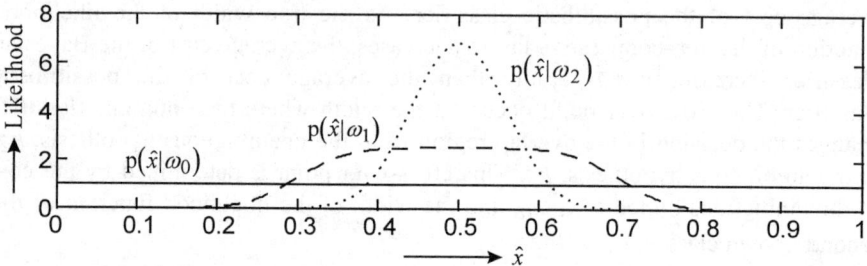

Fig. 6. Likelihood functions for the three target classes as a function of the measured attribute assuming Gaussian measurement noise with a standard deviation of 0.05

The (uniform) possibility distributions of the measured attribute value for the three classes are shown in Figure 7. We have assumed that the measured attribute value can never attain a value more than three times the standard deviation of the measurement noise from the true attribute value.

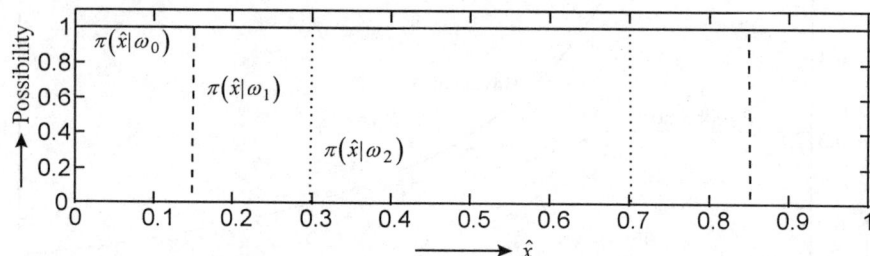

Fig. 7. Possibility distributions for the three target classes as a function of the measured attribute assuming Gaussian measurement noise with a standard deviation of 0.05

Figure 8 shows the decisions taken by the Bayesian (top diagram) and possibilistic classifier (bottom diagram) as a function of the measured attribute value. The cost C_a of the ambiguous hypothesis h_3 is 0.3. and the cost C_i of the ignorant hypothesis h_4 is 0.8. The a priori probabilities and possibilities are the same as in the case of a perfect sensor. In the centre of the overlap region of the likelihood functions, the Bayesian classifier decides for the unambiguous hypothesis h_2. At the edges of the overlap region the Bayesian classifier chooses the ambiguous hypothesis h_3 when the ratio of the *a posteriori* probabilities of the two known classes becomes smaller than a value given by $(1-C_a)/C_a = 7/3$. The possibilistic classifier decides for the ambiguous hypothesis h_3 in the overlap region of the possibility distributions for the two known classes. When compared with the perfect sensor case, this region is extended on both sides with three times the standard deviation of the measurement noise.

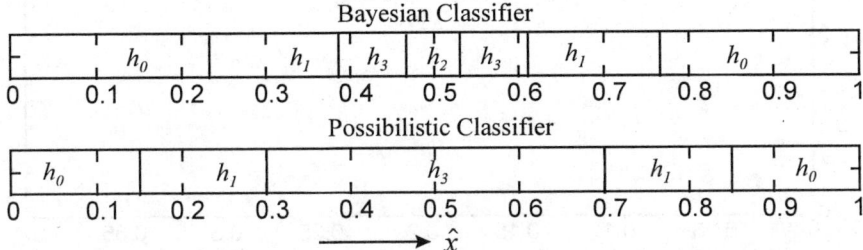

Fig. 8. Decisions taken by the Bayesian and possibilistic classifier as a function of the measured attribute value assuming Gaussian measurement noise with a standard deviation of 0.05

Sensitivity to Inaccuracies in *A Priori* Knowledge

Figure 9 illustrates the average cost of the Bayesian, possibilistic and optimum classifier as a function of the width of the true likelihood function of the non-cooperative target, η_1^*. The cost C_a of the ambiguous hypothesis h_3 is 0.3 and the cost C_i of the ignorant hypothesis h_4 is 0.8, and we assume that the characteristics of the sensor measurement noise are known. It can be observed in figure 9 that the average costs of the possibilistic and optimum classifier are almost independent of η_1^* while the average cost of the Bayesian classifier increases significantly as η_1^* decreases. The cross-over point of the average cost for the Bayesian and possibilistic classifier is close to the cross-over point for the perfect sensor case.

Figure 10 shows the average cost of the classifiers when we assume that the cost of the ambiguous hypothesis C_a is 0.2 instead of 0.3. In this case, the Bayesian classifier always decides for the ambiguous hypothesis h_3 in the overlap region because the ratio of the *a posteriori* probabilities of the two known classes never exceeds the value $(1-C_a)/C_a = 4$. The consequence is that the average cost of the Bayesian classifier is now always less than the average cost of the possibilistic classifier. This shows that when the cost of the ambiguous hypothesis is low enough and the characteristics of the measurement noise are known, the performance of the Bayesian classifier is always better than the performance of the possibilistic classifier, even in a scenario with a non-cooperative target.

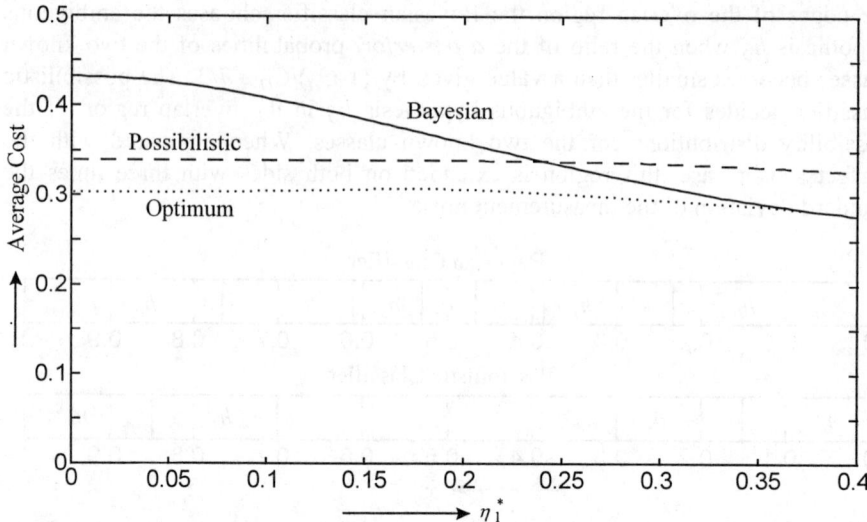

Fig. 9. Average cost of a Bayesian, possibilistic and optimum classifier as a function of the true width of the likelihood function of a non-cooperative target. The cost C_a of the ambiguous hypothesis (h_3) is 0.3, the cost C_i of the ignorant hypothesis (h_4) is 0.8 and the measurement noise is assumed to be Gaussian with a standard deviation of 0.05

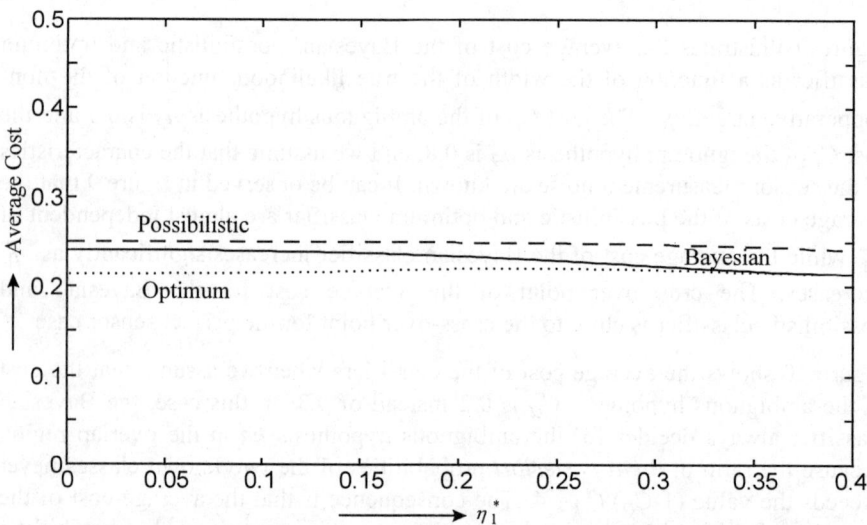

Fig. 10. Average cost of a Bayesian, possibilistic and optimum classifier as a function of the true width of the likelihood function of a non-cooperative target. The cost C_a of the ambiguous hypothesis (h_3) is 0.2, the cost C_i of the ignorant hypothesis (h_4) is 0.8 and the measurement noise is assumed to be Gaussian with a standard deviation of 0.05

6 Conclusions

In this chapter, a possibilistic alternative to the Bayesian approach to target classification has been presented. Both approaches to target classification have been expressed in a decision-theoretic framework in which a decision about the target class is taken after a measurement of a set of target attributes. Owing to the uncertainty in the target class given the measured attributes, the classifier cannot always estimate the target class unequivocally. As a consequence, the set of hypotheses that is available to the classifier for the classification process not only contains unambiguous hypotheses that associate the origin of the measured attributes with a single target class, but also ambiguous hypotheses that associate the measured attributes with multiple target classes.

The Bayesian classifier represents the uncertainties about the target class and the associated attributes with probabilities and selects the hypothesis which minimizes the average cost. The possibilistic classifier represents the uncertainties with possibilities and selects the hypothesis which minimizes the maximum possible cost. This minimax decision rule is well-known from game theory where we deal with an intelligent opponent. The use of the minimum and maximum operators instead of the product and sum operators used by the Bayesian classifier implies that the possibilistic classifier is only sensitive to the order of the uncertainties about the target class and the costs of hypotheses.

The sensitivity of the Bayesian and possibilistic classifier to inaccuracies in *a priori* knowledge has been analyzed by calculating the average cost of the classification decisions in an environment where we deal with an intelligent opponent. When the true likelihood functions of the measured attribute are close to the expected likelihood functions, the average cost of the possibilistic classifier is higher than the average cost of the Bayesian classifier. However, when the intelligent opponent is trying to maximize the average cost of the classifier by changing its attribute likelihood function, the average cost of the possibilistic classifier is smaller than the average cost of the Bayesian classifier under the condition that the cost of an ambiguous hypothesis is not small. In general, it can be concluded that the possibilistic classifier provides a more robust performance than the Bayesian classifier at the expense of a higher average cost when the true likelihood functions are close to the expected likelihood functions.

References

1   J.T. Tou and R.C. Gonzalez, Pattern Recognition Principles, Addison-Wesley, Reading 1974

2   D. Dubois, H. Prade, *Possibility Theory as a Basis for Qualitative Decision Theory*, Proc. IJCAI'95, Montreal, Aug. 2-25 1995, pp. 1925-1930.

3   D. Dubois, H. Prade, and R. Sabbadin, *A Possibilistic Logic Machinery for Qualitative Decision*, Proc. AAAI'97 Workshop on Qualitative Preferences in Deliberation and Practical Reasoning.

4   D. Dubois, H. Prade and R. Sabbadin, *Qualitative Decision Theory with Sugeno Integrals*, Proc. 14th Conference on Uncertainty in Artificial Intelligence, 1998, pp. 121-128.

5   J.O. Berger, *Statistical Decision Theory and Bayesian Analysis*, Springer-Verlag, New York, 1980.

6   T. Whalen, *Decision Making under Uncertainty with Various Assumptions about Available Information*, IEEE Trans. on Systems, Man and Cybernetics, Vol. SMC-14, No. 6, November/December 1984, pp. 888-900.

7   D. Dubois and H. Prade, *Weighted minimum and maximum operations in fuzzy set theory*, Information Sciences, 39, pp. 205-210.

8   R.D. Luce and H. Raiffa, *Games and Decisions*, Wiley, New York, 1957.

9   E. Hisdal, *Conditional Possibilities, Independence and Noninteraction*, Fuzzy Sets and Systems 1, 1978, pp. 283-297.

Appendix   Possibilistic Update Rule

The *a posteriori* possibility of a target class ω given a measured attribute vector $\hat{\mathbf{x}}$ is derived from the joint possibility of the target class and the measured attribute vector and the marginal possibility distribution of the measured attribute vector, see Hisdal [9]:

$$\min\left(\Pi(\omega|\hat{\mathbf{x}}), \pi(\hat{\mathbf{x}})\right) = \Pi(\omega \wedge \hat{\mathbf{x}}) \tag{A.1}$$

Since $\hat{\mathbf{x}}$ has been measured it must be concluded that $\hat{\mathbf{x}}$ is possible and as a consequence $\pi(\hat{\mathbf{x}}) = 1$. This implies that the *a posteriori* possibility is equal to the joint possibility of the target class and the measured attribute vector:

$$\Pi(\omega|\hat{\mathbf{x}}) = \Pi(\omega \wedge \hat{\mathbf{x}}) \tag{A.2}$$

On the other hand, the joint possibility of the target class and the measured attribute vector is also equal to the minimum of the possibility distribution of the measured attribute vector given the target class and the *a priori* possibility of the target class:

$$\Pi(\omega, \hat{\mathbf{x}}) = \min(\pi(\hat{\mathbf{x}}|\omega), \Pi(\omega)) \tag{A.3}$$

The possibilistic update rule follows from the two expressions for the joint possibility:

$$\Pi(\omega|\hat{\mathbf{x}}) = \min(\pi(\hat{\mathbf{x}}|\omega), \Pi(\omega)) \qquad \max_{\omega} \Pi(\omega|\hat{\mathbf{x}}) = 1 \tag{A.4}$$

The normalization is carried out to ensure that there is at least one target class for which the *a posteriori* possibility is equal to one.

20 From FUELCON to FUELGEN: Tools for Fuel Reload Pattern Design

E. Nissan,* A. Galperin,** J. Zhao,*
B. Knight,* and A. Soper*

(*) School of Computing and Information Technology,
The University of Greenwich, Wellington Street,
Woolwich, London SE18 6PF, England, U.K.
E-mail: E.Nissan@greenwich.ac.uk

(**) Department of Nuclear Engineering,
Ben-Gurion University, Beer-Sheva, Israel.

FUELGEN is an effective tool for refuelling design, i.e., for solving the in-core fuel management problem at nuclear power plants. Devising good fuel-allocations for reloading the core of a given nuclear reactor, for a given operation cycle, is crucial for keeping down operation costs at plants. Fuel comes in different types, and is positioned in a grid representing the core of a reactor. The starting point was Galperin and Nissan's prototype which eventually led to FUELCON, a rule-based expert system with the same task. FUELGEN, instead, is based on a genetic algorithm for optimization, and is at the current forefront of research in refuelling design, where genetic techniques are now getting increasing recognition. The end result of over a decade of research within this sequence of projects yielded a set of alternative, partly overlapping architectures. Nodal algorithms to carry out parameter prediction by simulation, heuristic rules in FUELCON's ruleset and meta-level refinement ergonomic techniques by which the ruleset can be refined during a session with FUELCON, attempts with neural computation on top of the latter, and then, replacing the ruleset altogether by resorting to genetic algorithms, are the sequence of techniques that were in turn applied, in the development of FUELCON and then FUELGEN. This actually reflects the sequence of emergence of expert systems and then neural computation methods, then genetic and hybrid methods, in knowledge engineering in general and in its application to nuclear engineering in particular.

1 A Description of the Domain and of the Task

Inside the vaguely water-boiler-like outer vessel of a nuclear reactor, the reactor core has a grid-like, symmetric planar section (see Figure 1) where fuel is inserted, in the form of assemblies of 200 to 250 parallel 350 cm rods. In the core geometry, important regions include the border of the core, the main

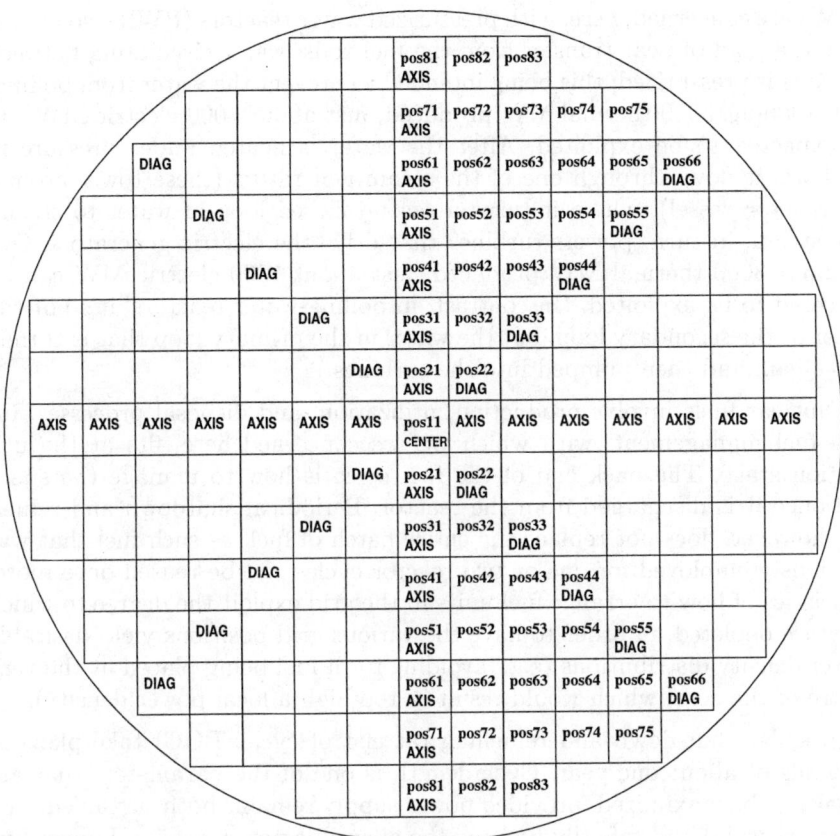

Fig. 1. A planar section of a reactor core

axis, and the diagonal. It is usual to reason on a slice of the grid in one-eighth symmetry, when devising how to rearrange and partly replace the depleted fuel at shut-down; a reactor is indeed to be stopped either for its regular shut-down period (as intended for refuelling and maintenance), or because of an emergency. Stopping is done, in both cases, by bringing down into the fuel assemblies among the fuel rods the control rods that are positioned above the assemblies in some grid positions; once the control rods are lowered, they stop the fission.

We are concerned, here, with pressurized water reactors (PWR), so named after the kind of heat transfer from the fuel rods; water circulating between the rods is pressurized, this being intended to prevent the water from boiling. Out of about 3000 thermal MW produced, just about 1000 electric MW can be expected to be exploited. After the water is heated under pressure in the core, it flows through one of the steam generators (these tower around the reactor vessel), where it turns a second closed loop of water to steam. The steam, in turn, powers turbines that drive the electric generators. Out of about 3000 thermal MW produced, just about 1000 electric MW can be expected to be exploited. (In contrast, in boiling-water reactors, it's not the water in the secondary loop, but the water in the primary loop that is turned into steam and then pumped into the turbines.)

Nuclear fuels involve production, utilization, and disposal processes. In-core fuel management, with which we are concerned here, fits at the utilization stage. The back end of the fuel cycle is how to manage the spent fuel once it is discharged from the reactor. Periodical shutdown and refueling, however, does not replace the entire batch of fuel, as such fuel that was previously employed for one or two reactor cycles can be reused once more. Heuristics of how to arrange fuel units in the grid exploit the degree to which they are depleted, in order to have the various grid positions yield desirable power density distributions (e.g., avoiding fresh fuel being placed at the very centre of the core, which would result in too high a local power density).

Regular shut-down and refuelling (at end of cycle: EOC) take place at intervals of about one year. Cycle length is one of the parameters that are usually to be maximized, provided power supply remains both sustained (i.e., "kept critical") and safe (by limiting the allowable power peaks). Longer fuel cycles and shorter refueling shutdown periods reduce the need, for a company, to supply replacement power by purchasing it from competitors (in a power market economy such as in the United States)—which the company would be required to do if, notwithstanding its own utilities' reserve capacity, it finds itself unable to provide adequate supply when demand is high. Utilities may have reserve capacity, but if this is not adequate, the company would have to purchase the power.

Cycle length is less important in monopolistic power market systems (e.g., in France), so shorter cycles are more feasible; yet, longer cycles result in less frequent licensing submittals, and having shut-down periods more spaced between reduces the load on organization [1]: downtime periods at plants typically take a few weeks per years, and are costly, in terms of energy not supplied (and the need, for a company, to provide an alternative supply). How feasible it is to invest in optimizing the yield of an operation cycle at a given plant, depends on the degree of autonomy of the management of individual plants, and on whether there is such extensive integration within a company that as the latter also produces fuel, it is less motivated to utilize it sparingly.

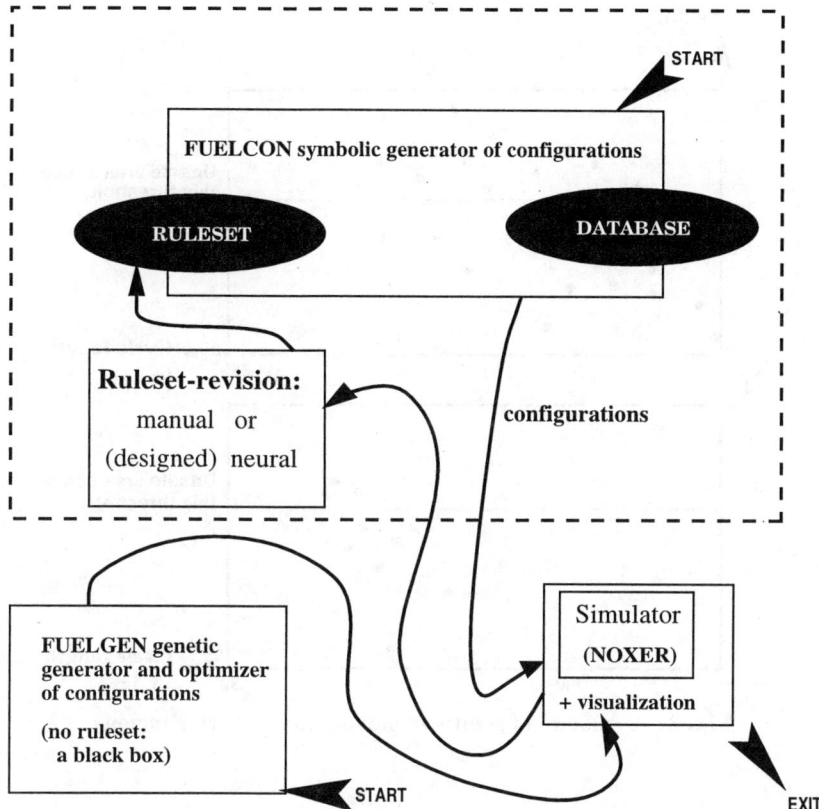

Fig. 2. Using FUELCON versus using FUELGEN

Most grid positions in a reactor core have no control rods, so the tube inside assemblies that is primarily designed for a control rod to be inserted, remains available for inserting there another kind of rods. It is indeed usual for the rate of burnup (and thus for the profile of power density distributions) during a reactor cycle to be manipulated, by inserting so-called burnable poison rods inside such assemblies that are positioned in squares of the grid with no overhanging control rods. Pattern reload design techniques which take burnable poisons into account are handled by the tools we have developed, namely, the FUELGEN genetic-algorithm optimization tool, and the earlier, rule-based FUELCON expert system. FUELCON is a tool where optimization criteria are transparent and can be tuned by the user, thus being particularly suited for a very skillful fuel engineer; FUELGEN, instead, works like a "black box" and is suitable for the average practitioner. (See Figure 2.)

436

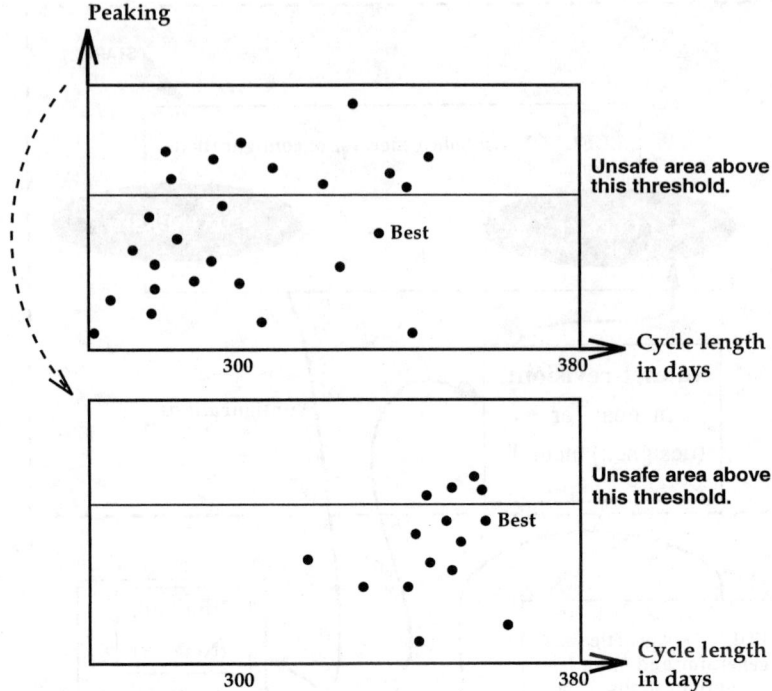

Fig. 3. A "cloud" of solutions moving into a better region

With both tools—FUELCON, but also FUELGEN—sessions gradually generate populations of solutions that outperform a previous population, which obtained at the preceding iteration; see Figure 3.

By providing such effective tools that moreover run relatively quickly, shut-down periods can be shortened. This is especially important as fuel reload pattern design, by its very nature, cannot be mainly done before EOC, as because of the very nature of the problem, such design is not robust in terms of post-optimality analysis.

That is to say, unforeseen variations (colder weather and thus higher demand, or, then, a damaged fuel unit) are prone to quite significantly affect the advantage margin of a given solution over other solutions, or even whether the given solution remains admissible. The latter factor, in particular, is why a solution that was adopted at a previous cycle at the given reactor cannot be reapplied "as is" at some subsequent EOC.

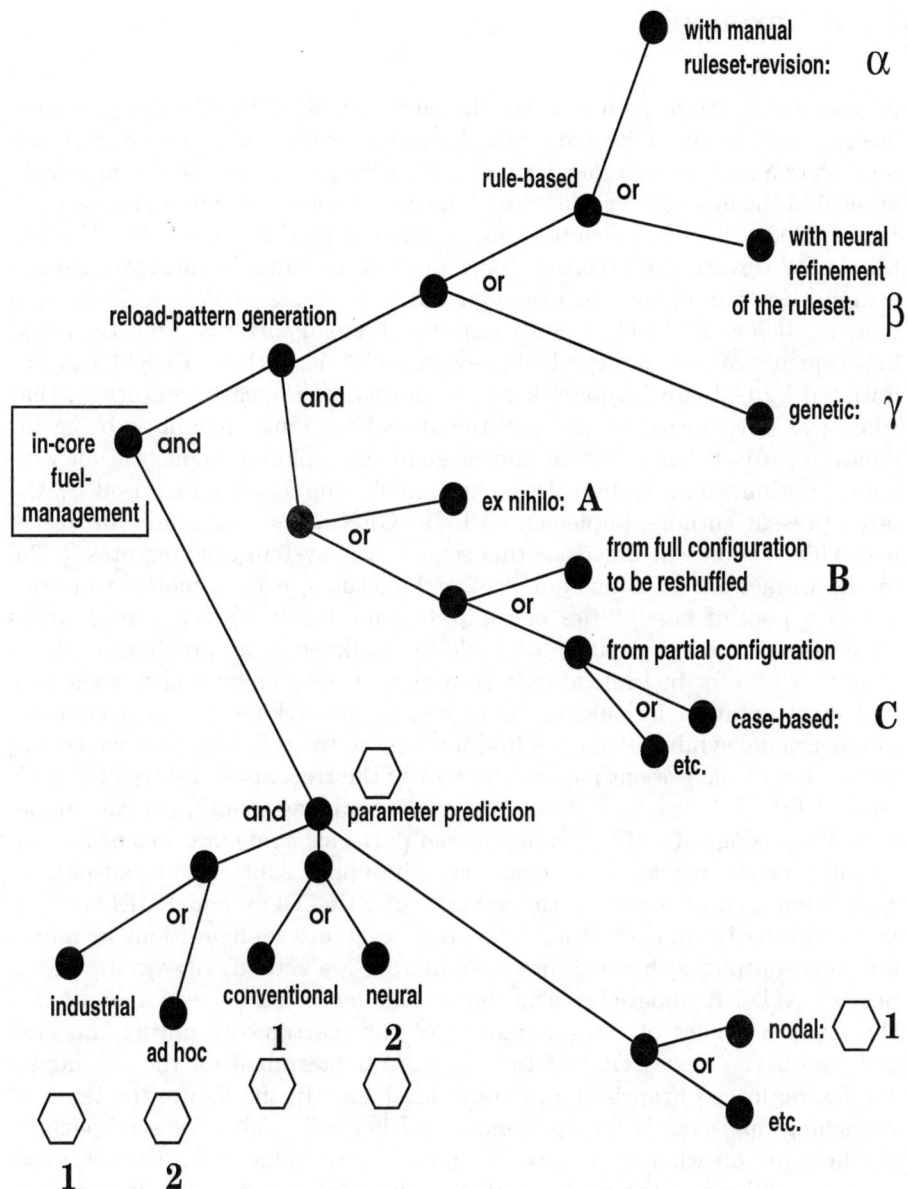

Fig. 4. An AND/OR tree of features of refuelling design tools

438

2   A Taxonomy

As mentioned, from the mid-1980s the development of FUELCON as a rule-based expert system reflected a broader picture where rule-based expert systems were a mature technology that in those very years was becoming widely applied in the most diverse domains. The neural wave was soon to follow [2,3]. In the 1990s, there has also been increasing recognition for hybrids, the emphasis for researchers involved being on how to suitably integrate several computation paradigms when carrying out a given task in a given application domain [4]. The FUELGEN generator of fuel reload patterns is but the latest in a sequence of tools designed and developed by the authors. FUELCON was initiated by Galperin, who with Nissan developed the early prototype, that when was developed into an operational tool by Shuky Kimhi as being his doctoral project. Then Nissan and Siegelmann explored augmentation with neural optimisation of the ruleset, and finally Jun Zhao, supervised by the other present authors, implemented FUELGEN into a tool ready for licensing. This section will illustrate this sequence of system architectures [5–25] in the framework of a taxonomy of architectures, each combining options out of a pool of possibilities of computational treatments. Figure 4 shows an AND/OR tree of features available for malking up an architecture for a computer tool for fuel reload pattern design. The terminal nodes in the tree introduce notational elements for capturing an architecture in a compact bidimensional symbol. This was first introduced by us in Ref. [26], which was at the core of the present paper. The root of the tree, at the centre of the left edge of Fig. 4, is the task itself: "in-core fuel management", i.e., the global task of devising safe, effective fuel reload patterns: configurations of fuel assemblies in the reactor core geometry. The upper subtree, "reload-pattern generation", corresponds to the subtask of FUELCON and FUELGEN as being alternative to each other. They both generate configurations *ex nihilo,* and their output is then fed into a simulator—we actually always used Meir Segev's NOXER nodal-algorithm based simulation package—for predicting the numeric values of the parameters of the reactor core during the next cycle until the next EOC. In the tree, the nonterminal on the second (in the figure: lower) branch at the upper level (i.e., in the figure: the leftmost branching out) stands for "parameter prediction", which we symbolize by the hexagon on whose perimeter we position the values (i.e., the terminals on the periphery of the subtree) of three variables represented by the three nonterminals of the subtree considered. In this compact notation, NOXER's features are captured by:

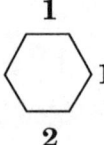

Using NOXER (which was developed by Segev *ad hoc* but can be replaced with some industrial simulator when either FUELCON or FUELGEN is adopted at a given nuclear power plant) for simulation, FUELCON was developed first. The characteristics of NOXER are notated above the hexagon (1 as it's a conventional rather than a neural simulator, the latter being a fast kind that is now gaining popularity [2] for in-core fuel management); below the hexagon (2 as it's an *ad hoc* tool developed in academia rather than an industrial tool); and on the right of the hexagon (1 because the technique embodied is nodal algorithms, which are popular for this kind of simulations from reactor physics [1]).

Actually, FUELCON arguably was the archetype of *rule-based* generation of fuel-configurations, since its earliest publication [6], which reported about the prototype that had been developed in 1986–8. It was subsequently further developed [7,9,11,5,8] into a full-fledged operational prototype, tested for effectiveness and accuracy on several case studies from the published literature of the discipline. This constituted indeed the subject of Kimhi's doctoral dissertation [10].

The ergonomics of operating FUELCON effectively [11], can be conceived of in terms of a control loop, combining automated and manual stages: there is a pre-stored ruleset in the expert system, with mandatory rules (safety constraints), and optional rules (embodying a search strategy); once the generator of configurations yields as output a batch of (typically) hundreds reload patterns, these are fed into NOXER, which then visualises them as dots in the plane of power-peaking and cycle-length. The "best" solution (peaking high but under a threshold, and a long cycle) can be then picked, if good enough and falling under the peaking safety threshold. However, the expert user may well decide to revise the ruleset, either to explore new strategies, or to improve on the features of the few very best solutions (or of the family of solutions: by zooming on a region, or moving the "cloud" of dots in the plane). This revision stage is performed manually. This proved to be, at the same time, the strength and the limit of FUELCON: like a racing car, being able to play with the transparently represented ruleset is elatingly empowering for a user who is very knowledgeable of in-core fuel management (starting with Galperin, who was more than just the domain expert for the expert system), but the average practitioner (to be expected at an average plant) cannot be expected to possess that level of competence that would allow the fuel engineer to get the very best out of the tool.

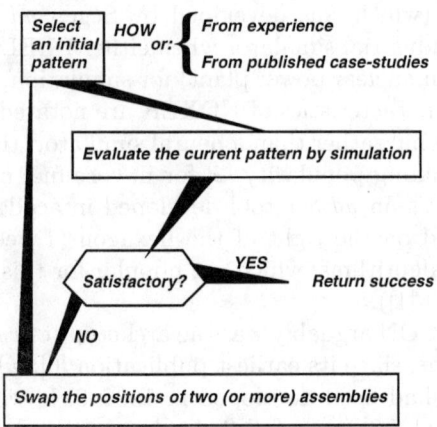

Fig. 5. Traditional reload pattern design by shuffling

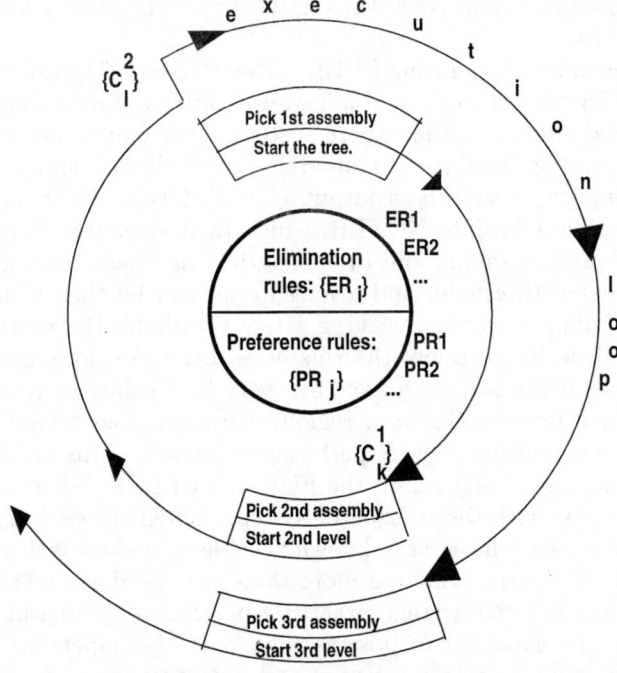

Fig. 6. FUELCON's generating *ex nihilo,* level by level, a tree of alternative reload patterns, these being the tree's terminals. FUELCON's ruleset of mandatory and optional rules is executed for each level in turn, resulting in the allocation of one fuel assembly per tree level

FUELCON's characteristics are:

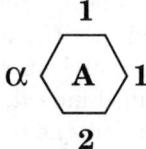

which incorporates NOXER's hexagon (because we have been using NOXER indeed, for all of FUELCON's portability to another simulator), and where in addition the α on the left of the hexagon symbolizes the fact that FUELCON has a ruleset which the user, if wishing for example to explore new heuristics, can revise manually—which resulted in a specifically tailored ergonomics of how to run sessions with FUELCON.

Instead the **A** inside FUELCON's hexagon represents the fact that this expert system generates configurations *ex nihilo,* instead of shuffling core geometry grid positions in a given reload pattern until a more satisfactory pattern obtains (which indeed is the traditional practice of fuel managers when designing a new fuel allocation either by hand, or with the help of a computer tool which sticks to the schema of the manual procedure: see Figure 5).

Generating reload patterns *ex nihilo* starts from an empty core, filling it gradually; each complete pattern is a terminal in a tree, which in turn is a family of alternative patterns. Figure 6 shows how, given a loading sequence for the available pool of fuel assemblies, FUELCON constructs a family of configurations by constructing, level by level (where one level corresponds to one fuel assembly), a tree of alternative reload patterns. (The figure shows the generation of the first three levels in the tree. The first run of the ruleset yields $\{C_k^1\}$, the set of the generated partial core-reload configurations with just one position filled in the one-eighth symmetry slice of the reactor-core geometry. The second run of the ruleset yields $\{C_l^2\}$, the set of partial configurations with two positions filled in symmetric slice of the grid; and so on).

Another possibility, adopted by others, is to look for only one reload pattern at a time, either by shuffling or possibly *ex nihilo,* the difference with respect to FUELCON being that the latter generates a family of alternatives instead of just one reload pattern. Actually option **B** in the AND/OR tree of Fig. 4 stands for the customary fuel manager's manual practice, of selecting— out of experience, the plant's record, or the literature—such an earlier reload pattern that provided a suitable solution for a problem that looks similar to the one at hand, and then to make this initial solution admissible and hopefully convenient, by shuffling positions in the grid (i.e., performing binary or more complex exchanges). There are tools which assist such a manual practice, and there exist also such tools which perform such a procedure automatically, by emulating shuffling as performed by humans [30,31].

However, FUELCON's feature of generating a "cloud" of reload pattenrs that then NOXER visualizes as dots scattered in the plane of the power peak and cycle length, challenges the traditional conception significantly. FUELCON's ruleset includes mandatory rules that are there for safety reasons, and these are not modified; it's the rest of the ruleset that shapes a search strategy by directing the search in regions of the search space, and these optional rules can be modified indeed, not necessarily in order to optimize, but even in order to just explore the behaviour of heuristics by zooming on particular regions put in evidence by the previous iteration of the tool.

The size itself of the output set of configurations per session, is not affected by the ruleset revision, as there is an arbitrarily threshold on the number of solutions (thus, dots) generated. There is no requirement for ruleset revision, as performed by the expert, to involve substantial change in the size of the set of configurations. If the desired effect of the revision is to have the generator *move* the "cloud" of configurations (in the plane where they are visually simulated as dots) towards a better region, then the size of the set of configurations can be expected to remain about the same.

If, instead, the improved version of the ruleset is meant to achieve *zooming,* then the family of generated configurations becomes smaller: the generated set of configurations converges on a smaller region of the search space, i.e. the generator looks for solutions that are less apart from each other. No identical configurations are generated inside the tree, because this is explicitly prevented by a suitable constraint in the tree-generation algorithm.

Yet, when the ruleset is revised for zooming, the configurations in the output set will tend to be more similar, both in respect of the way fuel-assemblies are positioned inside the core-geometry, and of their location in the space of simulation parameters. Visual simulation is in the plane of just two out of those parameters, and the *spread* of the configurations dotting that plane, i.e. how much they are apart, is reduced by zooming; moreover, two or more distinct configurations will be more likely to be visualized as one dot, or as a heap of overlapping dots. Far from being inconvenient, this just reflects success at achieving zooming as a goal of ruleset revision.

We turn now to our attempt to hybridize FUELCON with neural computation. The early 1990s saw an attempt on our part to integrate a *neural* component in the FUELGEN loop of operation (i.e., running the ruleset, getting a family of configurations, feeding it into the simulator, then possibly revising the ruleset by hand and running it all over again), for the purposes of revising the ruleset automatically [12–16].

This is a topic of considerable interest in principle, but whereas the design phase was carried out, we eventually chose to invest instead in the development of the genetic tool, FUELGEN, as we perceived it to be a practically more promising option. The attempts by Nissan, Siegelmann and Galperin to devise an automated ruleset-revision stage within sessions with FUELCON— such an option, NeuroFUELCON, was designed, by representing the rules from the ruleset in a special programming language devised by Siegelmann for transforming a program coded in symbolic computation into a neural internal representation—can be represented as follows, where β replaces α on the left side of the FUELCON hexagon:

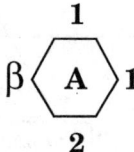

Next, the genetic-algorithm option, γ, resulted in FUELGEN. *Genetic* computation for the purposes of generating reload patterns is, indeed, an emerging paradigm, with several teams reporting on ongoing projects [27–29], or known to be experimenting with it. FUELGEN, a robust tool [17–20], is now ready for its industrial adoption phase, and negotiations are ongoing.

FUELGEN's yields results that are better than those published for comparable cases as far as they are known to us. For example, when applied to a well-researched real case [18], FUELGEN yielded a cycle length of 378.8 days and a power peak of 1.30, which is significantly better than the best result previously found by other systems, i.e., cycle length of 360 with power peak of 1.28.

The simulator used with the FUELGEN prototype, NOXER, is to be replaced with an industrial simulator upon industrial adoption of FUELGEN. FUELGEN, just like FUELCON, also effectively handles cases where burnable poison rods must be accounted for.

In the notation we are using in this paper, we represent FUELGEN as follows:

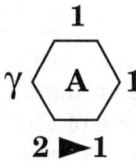

As to the structure of FUELGEN in terms of its control flow, Fig. 7 below shows the inner workings of FUELGEN within its architectural schema.

444

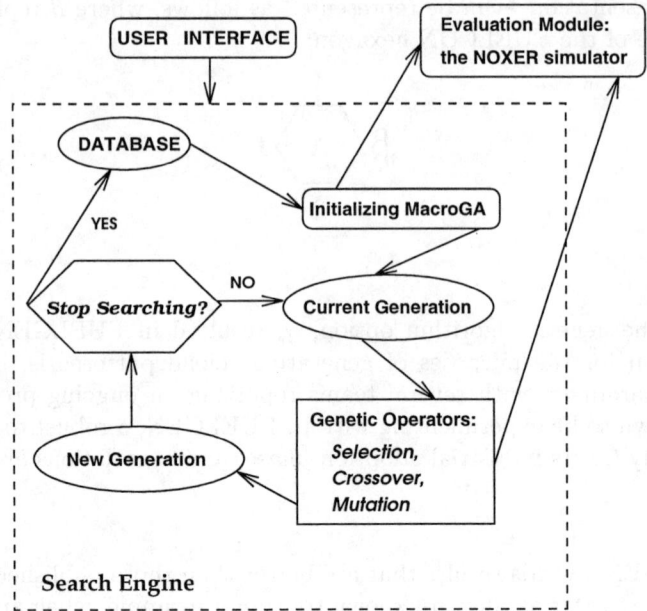

Fig. 7. The architecture of FUELGEN

In FUELGEN (as well as in other reported projects in the field), a loading pattern is treated as a chromosome, with fitness based on design criteria such as cycle length and power peak. At the initial stage, a population of valid loading patterns is generated. They are each assigned a fitness. The evolutionary operators —selection, crossover and mutation— are then applied to the population to produce a new generation of loading patterns. This process is repeated until a required loading pattern is found.

Note that in FUELGEN, reload pattern generation is *ex nihilo,* for all of a genetic algorithm obviously requiring a population to start with. (By the way, option **C** in the AND/OR tree of Fig. 4 stands for an alternative way of starting, other than by shuffling, yet from a partial configuration instead of *ex nihilo;* we are envisaging to explore a variant of FUELGEN which would start from a database of precedents, and select one by means of *case-based reasoning:* it's interesting in principle, and is a renewed attempt at coping with the expert's heuristics directly—something avoided in FUELGEN, which runs as a black box from the user's viewpoint.)

3   Further Considerations on FUELGEN

Each candidate reload pattern is represented as a set of genes, in FUELGEN.[1]
These correspond to the cells in the reactor core. The particular values (alleles) that each gene can take are the fuel assemblies. The choice of such
a minimal representation is deliberate, even though it restricts the possible
crossover and mutation operators that can be constructed, because of the
limits on information available to them. Nuclear physics dictates that our
genes, which represent cells, will interact strongly, so that fitness variance
of gene sets will be large (the power peak, being a local constraint makes it
easy to produce configurations with low fitness by giving the remainder of
the genes appropriate alleles). The fact that adjacent assemblies interact most
strongly, suggests a natural linkage for the core reload problem: Parks [27]
used a bidimensional chromosome, with structure identical to the sector of
the core represented, so that adjacent cells correspond to adjacent genes.
However, there are no groups of linked genes, that are unlinked or weakly
linked to others. The exchange of a linked set of genes cannot proceed without disruption to others. The local nature of the power peak measure implies
that disruption will normally produce undesirable loading patterns with a
high power peak and hence poor fitness.

The exchange of a linked set of genes will in most cases produce offspring
containing more than one copy of an assembly. Adjustments need to be made
to the offspring to restore the assembly set. This is further evidence against
the existence of independent groups of linked genes suitable for recombination, since restoration will disrupt linked gene combinations.

Because of these considerations, in FUELGEN it was considered prefer-
able not use any additional structure defining linkage for recombination. Instead the minimal representation was used and appropriate mutation strategies defined. We use an assorting crossover. This kind of crossover is highly
explorative, as it takes those genes that are common to both parents into the
offspring and fills in the remaining genes randomly using all available allele
values, rather than only those residing in the parents. In FUELGEN, it consists of these steps: given two selected parent loading maps, create two empty,
child loading maps with no assemblies assigned to cells; copy the contents of
all cells for which the parents have identical assemblies to the offspring; fill
the rest of the cells in the children randomly from the available assemblies,
subject to the symmetry restrictions. Since our search strategy is based on
mutation, we use higher mutation rates, than are normally used with a recombining crossover (where the mutation operator traditionally plays a secondary
role); this way, search is given a strong emphasis on hill climbing.

Population structure in FUELGEN (we adopt the so-called Island Model),
consists of many small subpopulations, each evolving under the action of a

---

[1] This section is slightly extended from an earlier version presented at
FLINS98 [26].

separate GA, and exchanging individuals infrequently with other subpopulations. As to the selection operator, we use ranking selection. Our approach to the fitness function was to use a multiobjective strategy [27]. Here population members are placed in ranked groups according to whether they are dominated by other individuals. Ours is a strategy of gradually increasing the penalisation of the fitness of solutions for violating the constraint as the search progresses. The initial stages of the search will be directed towards increasing cycle length while less hampered by the presence of local minima arising from the constraint on the power peak. As evolution progresses and the average fitness of the population increases, the search will be more directed to reducing the power peak. In the fitness function

$$f = C/P$$

C is the cycle length and P the power peak. This can be seen by calculating the ratio of the partial derivatives of f with respect to C and and P, as in the following formula:

$$\frac{\frac{\partial f}{\partial P}}{\frac{\partial f}{\partial C}} = -\frac{C}{P}$$

The latter equation is independent of the fitness scaling chosen, in our case ranking, since it holds for any differentiable function $F(C/P)$. Owing to how the genetic algorithm matches the physics of the problem, this fitness function worked well, and not for instance C^2/P which would have directed the population less strongly to areas of lower power peak.

Acknowledgments

We are grateful to Shuky Kimhi, Meir Segev, and Hava Siegelmann for their role at various stages of this long project.

References

1. R.G. Cochran and N. Tsoulfanidis. *The Nuclear Fuel Cycle: Analysis and Management.* American Nuclear Society, La Grange Park, Illinois, 1990.
2. E. Nissan, Intelligent Technologies for Nuclear Power Systems: Heuristic and Neural Tools. *Expert Systems with Applications,* **14**(4), pp. 443–460 (1998).
3. E. Nissan (ed.), *Intelligent Technologies for Electric and Nuclear Power Systems,* special issue, *Computers and Artificial Intelligence,* **17**(2/3), 1998.
4. E. Nissan, Hybrid Techniques. *Computers and Artificial Intelligence,* **17**(2/3), pp. 251–256 (1998). [Five books reviewed.]

5. A. Galperin, Exploration of the Search Space of the In-Core Fuel Management Problem by Knowledge-Based Techniques. *Nuclear Science and Engineering,* **119**(2), p. 144 ff (1995).

6. A. Galperin and E. Nissan, Application of a Heuristic Search Method for Generation of Fuel Reload Configurations. *Nuclear Science and Engineering,* **99**(4), pp. 343–352 (1988).

7. A. Galperin, S. Kimhi and M. Segev, A Knowledge-Based System for Optimization of Fuel Reload Configurations. *Nuclear Science and Engineering,* **102**, p. 43 ff. (1989).

8. E. Nissan and A. Galperin, Refueling in Nuclear Engineering: The FUELCON Project. *Computers in Industry,* **37**(1), pp. 43–54 (1988).

9. A. Galperin and S. Kimhi, Application of Knowledge-Based Methods to In-Core Fuel Management. *Nuclear Science and Engineering,* **109,** pp. 103–110 (1991).

10. Y. [=S.] Kimhi, A Non-Algorithmic Approach to the In-Core Fuel Management Problem of a PWR Core. Ph.D. Dissertation, Nuclear Engineering, Ben-Gurion Univ. of the Negev, Beer-Sheva (1992). In Hebrew.

11. A. Galperin, S. Kimhi and E. Nissan, FUELCON: An Expert System for Assisting the Practice and Research of In-Core Fuel Management and Optimal Design in Nuclear Engineering. *Computers and Artificial Intelligence,* **12**(4), pp. 369–415 (1993).

12. E. Nissan, H. Siegelmann, A. Galperin and S. Kimhi, Towards Full Automation of the Discovery of Heuristics in a Nuclear Engineering Project, by Combining Symbolic and Subsymbolic Computation. *Proc. 8th Int. Symp. on Methodologies for Intelligent Systems: ISMIS'94,* (LNAI 869, Springer Verlag, Heidelberg, pp. 427–436, 1994).

13. E. Nissan, H. Siegelmann and A. Galperin, An Integrated Symbolic and Neural Network Architecture for Machine Learning in the Domain of Nuclear Engineering. *Proc. of the 12th IAPR Int. Conf. on Pattern Recognition,* Vol. 2, pp. 494–496 (1994).

14. A. Galperin, S. Kimhi, E. Nissan, H. Siegelmann and J. Zhao, Symbolic and Subsymbolic Integration in Prediction and Rule-Revision Tasks for Fuel Allocation in Nuclear Reactors. *Proc. 3rd European Congress on Intelligent Techniques and Soft Computing (EUFIT'95),* Aachen, Vol. 3, pp. 1546–1550 (1995).

15. E. Nissan, H. Siegelmann, A. Galperin and S. Kimhi, Upgrading Automation for Nuclear Fuel In-Core Management: From the Symbolic Generation of Configurations, to the Neural Adaptation of Heuristics. *Engineering with Computers,* **13**(1), pp. 1–19 (1997).

16. H. Siegelmann, E. Nissan and A. Galperin, A Novel Neural/Symbolic Hybrid Approach to Heuristically Optimized Fuel-Allocation and Automated Fuel-Allocation in Nuclear Engineering. *Advances in Engineering Software,* **28**(9), pp. 581–592 (1997).

17. J. Zhao, An Examination of the Macro Genetic Algorithm and its Application to Loading Pattern Design in Nuclear Fuel Management. Ph.D. Dissertation, Computer Science, The University of Greenwich, London (discussed Nov. 1996).

18. J. Zhao, B. Knight, E.Nissan and A. Soper, FUELGEN: A Genetic-Algorithm Based System for Fuel Loading Pattern Design in Nuclear Power Reactors. *Expert Systems with Applications,* **14**(4), pp. 461–470 (1997).

19. J. Zhao, B. Knight, E. Nissan, A. Soper, FUELGEN: Effective Evolutionary Design of Refuellings for Pressurized Water Reactors. In Ref. 3, *Computers and Artificial Intelligence,* **17**(2/3), pp. 105–125 (1998).

20. E. Nissan (ed.), *Forum on Refuelling Techniques for Nuclear Power Plants: One Decade with FUELCON.* Thematic section (5 papers), *New Review of Applied Expert Systems,* **4,** pp. 139–194 (1998).

21. E. Nissan, The FUELCON Meta-Architecture, in the Landscape of Intelligent Technologies for Refuelling. In Ref. 20.

22. A. Galperin and E. Nissan, The FUELCON Meta-Architecture, II: Alternatives for Parameter Prediction. In Ref. 20.

23. A. Galperin S. Kimhi, E. Nissan and H. Siegelmann, FUELCON's Heuristics, Their Rationale, and Their Representations. In Ref. 20.

24. J. Zhao, B. Knight, E. Nissan, M. Petridis and A. Soper, The FUELGEN Alternative: An Evolutionary Approach. The Architecture. In Ref. 20.

25. A. Soper, Exploring Genetic Alternative Concepts for FUELGEN. In Ref. 20.

26. E. Nissan, A. Soper, J. Zhao, B. Knight, M. Petridis, Fuel Reload Pattern Design Within a Family of Hybrid Architectures. *Proc. FLINS'98: Third International FLINS Workshop on Fuzzy Logic and Intelligent Technologies for Nuclear Science and Industry (FLINS'98),* Antwerp, Belgium, Sept. 1998, pp. 408-415.

27. G.T. Parks, Multiobjective Pressurized Water Reactor Reload Core Design by Nondominated Genetic Algorithm Search. *Nuclear Science and Engineering,* **124**(1), 1996.

28. E. Tanker and A.Z. Tanker, Application of a Genetic Algorithm to Core Reload Pattern Optimization. *Proc. Int. Conf. on Mathematics and Computations, Reactor Physics, and Environmental Analysis,* Vol. 1, Portland, Oregon, 1995.

29. M.D. DeChaine and M.A. Feltus, Fuel Management Optimization Using Genetic Algorithms and Expert Knowledge. *Nuclear Science and Engineering,* **124**(1), 1996.

30. B.M. Rothleder, G.R. Poetschhat, W.S. Faught and W.J. Eich, The Potential for Expert System Support in Solving the Pressurized Water Reactor Fuel Shuffling Problem. *Nuclear Science and Engineering,* **100:** p. 440 ff (1988).

31. Y. Tahara, K. Hamamoto and M. Takase, Computer Aided System for Generating Fuel Shuffling Configuration Based on Knowledge Engineering. *Journal of Nuclear Science and Technology,* **28**(5): pp. 399–408 (1991).

21 Diagnosis of Unanticipated Plant Component Faults in a Portable Expert System

Jaques Reifman and Thomas Y. C. Wei

Argonne National Laboratory
9700 S. Cass Ave., Argonne, Illinois 60439, USA
{jreifman, tycwei}@anl.gov

We describe the first-principles-based PRODIAG expert system for on-line plant-level diagnosis of component faults in thermal-hydraulic processes. This diagnostic system combines the concepts of fundamental physical principles and function-oriented diagnosis in a qualitative reasoning framework and structures these concepts into three independent knowledge bases. PRODIAG has the unique ability to diagnose unanticipated (unforeseen) component faults and can be ported across different processes/plants through modifications of only input data files containing the appropriate process layout information. Simulation tests for two plant systems with transient data generated with the Braidwood Nuclear Power Plant full-scope training simulator confirm the unique capabilities of PRODIAG.

1 Introduction

Expert systems first emerged in the late 60's and early 70's as the result of unsuccessful attempts to create a truly intelligent machine capable of solving generic problems. For the diagnosis of component faults in nuclear power plants, one of the first expert systems was REACTOR [1]. This diagnostic expert system like the majority of the vast body of systems that followed [2], including the ones applying artificial neural networks and other soft computing tools, has significant limitations and inflexible design characteristics. These systems are plant- and process-dependent and can diagnose only previously defined fault scenarios. With these design characteristics, a diagnostic system designed for one particular thermal-hydraulic (T-H) process cannot be reused for identifying faults in other processes/plants and an incorrect diagnosis or no diagnosis is made when the system is presented with an unanticipated (unforeseen) fault.

Here, we describe a first-principles-based expert system for on-line plant-level diagnosis of component faults in T-H processes which does not have these limitations. This system, PRODIAG, is generically applicable and event- and T-H process-independent. PRODIAG advances the state-of-the-art of soft computing tools for process diagnostics because:

1. It is fully *portable* and only requires modification of the input data files containing the appropriate process layout information to be able to diagnose component faults in different processes/plants, and

2. It diagnoses *unanticipated* component faults.

These two unique attributes are achieved by combining the concepts of fundamental *physical principles* and *function-oriented* diagnosis in a *qualitative reasoning* framework. The fundamental *physical principles* of conservation of mass, momentum, and energy that govern the behavior of the process are, of course, T-H process-independent. When represented in a qualitative reasoning framework the conservation equations can be applied to different processes to detect behavioral changes, i.e., mass, momentum, and energy imbalances, due to component faults. Hence, the "diagnostic rules" derived from the conservation equations are generic and portable from T-H process to T-H process without requiring any code-related modifications. Modifications are only required for the information relating to the type and connectivity of the process components and these are provided entirely through input. This feature provides the additional advantage that verification and validation of the diagnostic rules are T-H process independent and need to be performed only "once." With each different application, verification would only be performed for the input data files.

In *function-oriented* diagnosis, the relationship between the behavior and the design T-H function objective of each component of the process is used to identify component faults. In this approach, faults are identified through the detection of an imbalance in one of the three process functions (mass, momentum, or energy transfer) using the diagnostic rules for the conservation equations. The relatively small number of generic types of components (e.g., pump, valve, heat exchanger) included in T-H processes and the fact that each component is designed to perform essentially only one key function guarantees functional completeness of the knowledge base. There is no need to explicitly enumerate scenarios or anticipate specific failures that may occur in the process. As long as these functions are properly represented, any component fault, even an unanticipated one, can be diagnosed.

In *qualitative reasoning*, a small number of qualitative values, such as increasing, decreasing, and unchanging trends, are used to represent the values of continuous real-valued variables [3]. By taking a qualitative reasoning approach diagnostics can be performed with less information. For example, there is no need to know the specific quantitative characteristics of a pump to functionally classify it as a source of momentum and associate a detected momentum decrease with a pump fault. Reasoning with reduced information may, however, lead to ambiguous inference which can only be resolved by providing more specific, quantitative information.

PRODIAG is designed for early diagnosis of transient events prior to actuation of automatic control systems or operator response. Diagnosis is based exclusively on the T-H instrumentation signals of the process which are assumed to have been validated. Also, in its current implementation, the range of applicability of PRODIAG is limited to: (1) single-component faults, (2) single-phase liquid plus noncondensable gas T-H processes, (3) coolant with bulk moduli and thermal expansion coefficients similar to those of water, and (4) use of instrumentation signal data which have been filtered for noise. These are not inherent limitations of the approach, but rather implementation simplifications.

2 Diagnostic Methodology

2.1 Overview

When components of a T-H process malfunction, they cease, either partially or completely, to perform their design functions. This change in the component behavior causes the process T-H signals, e.g., pressure, flow, temperature, and level, to vary or trend from their expected values, producing a set of symptoms. Diagnosis is then performed by reversing this causality and mapping the observed symptoms into a list of possible component faults.

This mapping can take different forms. The traditional form employed by most expert systems and neural networks is to map the relationships between process symptoms and component faults directly, with no dependence on a functional understanding of the underlying principles of the process. A good example of direct mapping is the use of "if (symptoms) then (fault)" production rules prevalent in event-oriented expert systems.

In PRODIAG, process symptoms are mapped into component faults in a three-step procedure through the use of three distinct but interacting knowledge bases: (1) physical rule database (PRD), (2) component classification dictionary (CCD), and (3) piping and instrumentation database (P&IDB). Figure 1 illustrates the three-step procedure where at each step the corresponding knowledge base performs the following function:

1. In step 1, the PRD maps the qualitative trends in the T-H signals (e.g., increasing pressure and decreasing flow) into function imbalance trends in the three conservation types of mass, momentum, and energy. For example, it detects a decreasing mass inventory or an increasing energy inventory.

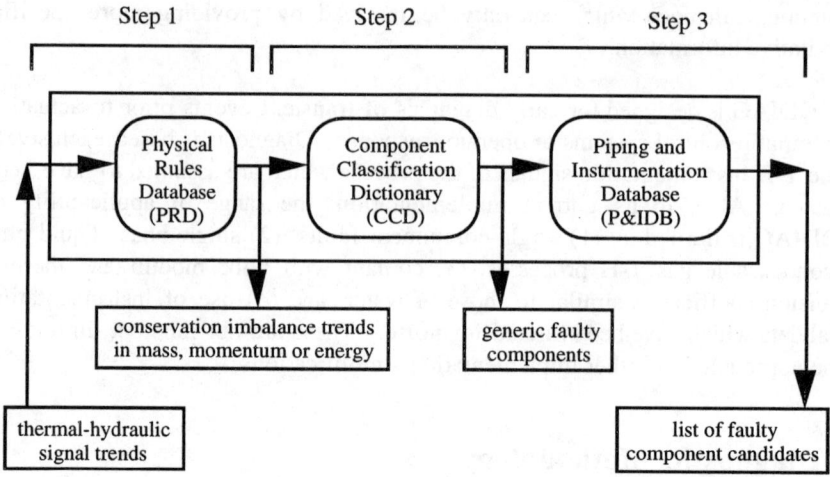

Figure 1: Three-step mapping of process symptoms into component faults.

2. In step 2, the CCD maps the identified function imbalance type and trend into generic faulty component types (e.g., closed valve, pump, and electric heater) whose failure could have been responsible for the identified imbalance.

3. In step 3, the P&IDB containing the process layout information, i.e., the piping and instrumentation diagram (P&ID) information, is applied to identify specific component candidates (e.g., closed valve CV-121, pump A, and electric heater B) from the generic faulty component types.

In this structuring of the knowledge base under a qualitative reasoning framework, we confine the process-dependent information solely to the P&IDB. The PRD (representing the physical principles of conservation) and the CCD (expressing the concept of function-oriented diagnostics through the classification of generic components) are event- and T-H process-independent. Hence, there is no need to prespecify the possible fault scenarios and the diagnostic system can be ported across different processes/plants by incorporating the appropriate P&IDB through modifications only of input data files.

2.2 Simple Example

The diagnosis of a tube break in the letdown heat exchanger shown in the simplified P&ID in Fig. 2 is used for illustration of the methodology. The tube break causes a prompt decrease in the letdown line pressure measured by PT-131, a decrease in mass flow rate from the break location up to the volume control tank (VCT) measured by FT-132, and an increase in mass flow rate from the reactor coolant system (RCS) up to the break location measured by FV-XYZ. This increase in the regenerative heat exchanger hot-side flow causes an increase in its

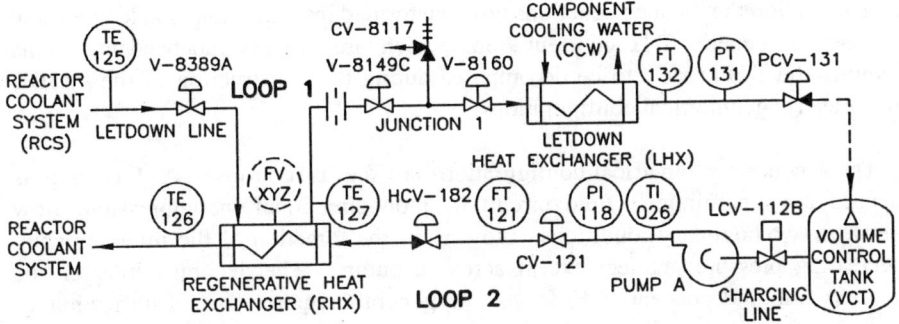

Figure 2: Simplified letdown line and charging line of a chemical and volume control system for a pressurized water reactor.

outlet temperature measured by TE-127. However, due to the heat exchanger thermal inertia, there is a time delay before the heat exchanger outlet temperature increases to reflect the tubing break. FV-XYZ is a "virtual" flow transmitter discussed in Sect. 3.1.2.

Based on the three-step mapping, the letdown heat exchanger tube break is diagnosed by PRODIAG through the following procedure (a more in-depth diagnosis in discussed in Sect. 4.2). First, the PRD is used to correlate the trends in the flow meters, FT-132 and FV-XYZ, to detect a decreasing mass imbalance in the control volume defined by the flow meters, i.e., a "mass imbalanced" control volume is identified between the inlet of the regenerative heat exchanger and the outlet of the letdown heat exchanger. Then, the CCD is searched to infer generic component types, such as closed valve, junction, and break, whose failure could have caused the detected decrease in mass inventory. The abstraction used to classify a junction and a break as a "component" responsible for mass imbalances is discussed in Sect. 3.2. With the generic component types identified as closed valve, junction, or break, the components within the imbalanced control volume are filtered through the P&IDB to narrow down and identify the possible specific components. Because there are no closed valves in the imbalanced control volume and the junction is not "malfunctioning" (see Sect. 4.2), a line break encompassing the two heat exchangers is inferred as the possible fault.

Hence, through the proposed three-step mapping used in PRODIAG, there is no need to prespecify the possible fault scenarios and unanticipated faults can be diagnosed. Furthermore, other breaks in this or other processes/plants would be diagnosed through the same procedure without the need to develop additional rules.

2.3 Spatial Loop Decomposition

The PRD consists of generic qualitative physics rules that can be applied to different T-H processes without any need for rule customization. To allow for this generalization the balance equations are transformed into rules applicable to a finite number of generic T-H configurations which are process independent. This requires the T-H process to be decomposed during the construction of the P&IDB into generic geometrical configurations.

These generic geometrical configurations are the T-H loops. A T-H loop is defined as a continuous fluid circuit of monodirectional incompressible flow between two boundary conditions. Flow is in the direction of the monotonically decreasing pressure gradient except across a pump. The decomposition of the process into independent T-H loops, in general, requires the identification of boundary conditions with flows and pressures which are essentially constant during a transient, and allows a loop-by-loop decoupling such that the rules of the PRD are applied only for one loop at a time.

There are two types of T-H loops: open and closed. An open loop starts and ends at two different boundary locations while a closed loop starts and ends at the same location. For example, the P&ID in Fig. 2 would be decomposed into two open loops. Loop 1, representing the letdown line, would start at the RCS and would end at the VCT and loop 2, representing the charging line, would start at the VCT and end at the RCS.

There can be hydraulic and thermal connections between loops. Loops can be hydraulically connected to other loops through valves that are closed during normal operations, piping with zero interconnecting flow during steady state, and junctions. Hydraulic connections permit the diagnostic search to move from loop-to-loop (through the hydraulic connection) in search of the faulty component. Loops can be thermally connected to other loops through heat exchangers where there is a thermal but not a hydraulic interface during normal operations. Thermal connections permit the exchange of information between loops (across heat exchangers) during diagnostics. For example, information about the state of the regenerative heat exchanger in Fig. 2 during the diagnosis of loop 1 can be obtained through the use of the PRD rules in loop 2.

Open and closed loops are further decomposed into control volumes of two types: separated volume and nonseparated volume. A separated volume consists of only one component, such as the VCT in Fig. 2, where the fluid state is a non-condensable gas space over a liquid volume. A nonsepararated volume consists of a specific component or group of connected components where the fluid state is entirely subcooled liquid.

Unlike T-H loops which are predefined during the decomposition of the process in the construction of the P&IDB, control volumes are defined as the diagnostics in a loop proceeds. Control volumes are defined by the location of the T-H signal instrumentation in the loop during the application of the PRD rules in that loop. Control volumes identify the location of the imbalance and hence the location of the component fault in the T-H loop. For instance, in the case of the tube break example discussed in Sect. 2.2, the location of the two flow meters, FT-132 and FV-XYZ, and the specific PRD rule utilized define the nonseparated "imbalanced" control volume to be the set of connected components starting at the regenerative heat exchanger and ending at the letdown heat exchanger.

2.4 Treatment of Dynamic Effects

The qualitative reasoning framework requires the establishment of the T-H signal trends, such as increasing pressure and decreasing flow, at each sampling time before they are utilized by the quasi-static version of the balance-equation rules in the PRD. In our current implementation, transient induced signal trends are established based on (1) signal deviations from their expected values and (2) consideration of the process dynamics.

Based on our assumption that the signal data have already been filtered by a low frequency bandpass device to filter out noise in the data, the changes in the T-H signals from their expected values are currently detected through simple threshold logic. If the value of the signal is within $x_0 \pm \varepsilon$, where x_0 is the expected value of T-H signal x and ε is the threshold, the signal is said to have an unchanging trend (-), otherwise the signal is said to have an increasing ($\uparrow$) trend or a decreasing ($\downarrow$) trend. In detail, our approach employs two thresholds, a primary threshold ε_p and a secondary threshold ε_s, with $\varepsilon_p > \varepsilon_s$. The primary threshold ε_p is used to establish the onset of a transient and the smaller ε_s is used to determine the signal trends once the occurrence of a transient has been established. This two-tier strategy provides an increased probability in correctly detecting the occurrence of a transient while allowing a large number of signals to be considered in the diagnostics.

In the tube break example discussed earlier, due to the thermal inertia of the regenerative heat exchanger, there is a time delay before the heat exchanger outlet temperature TE-127 responds to the letdown flow change. The use of this temperature information before its trend is fully developed could lead to incorrect diagnosis. Hence, consideration of the time constants associated with thermal inertia and other natural feedback phenomena, such as mass inertia and transport delay, as well as other dynamic effects, such as instrumentation response and control system action, need to be properly accounted for in establishing the T-H signal trends before they are used in the PRD.

We account for the dynamic effects of natural feedback phenomena by grouping thermal-hydraulically coupled signals and associating with each group a time window [4]. A time window is an abstraction that specifies a time interval Δt in which changes in the signal trends associated with natural feedback and instrumentation response are accepted as diagnosis input. When any one T-H signal from a prespecified group of signals reaches its primary threshold value

$x_0\pm\varepsilon_p$, the time window for those signals opens. Once the time window for a group of signals is open, changes in signal trends can be accepted for any signal of the group if the signal reaches its secondary threshold ε_s. After a time interval Δt has passed, the time window for that group of signals is closed. The closing of the time window prevents the effects of control system action and other undesirable effects with long time delays to be considered in the diagnosis. Automatic control actions can mask signals or drive signals from their expected values which, in turn, could lead to incorrect diagnoses. To avoid using these undesirable effects which have long time delays, once a time window is closed, additional trend changes in any one signal of the group are not accepted and cause the diagnosis to be halted.

For natural feedback phenomena, analysis of the three balance equations of mass, momentum, and energy identifies the coupling of the T-H signals of flow (W), pressure (P), temperature (T), and level (L) and the division of time couplings into fast and slow classes. Instrumentation time response is treated as additive delays incorporated into the time lags. Table 1 shows the classification and the numerical values of thresholds and time intervals Δt for the signal couplings of nonseparated control volumes. Integrated or cumulative couplings, such as the coupling of temperature and flow {T,W}, are slow while instantaneous couplings, such as of pressure and flow {P,W}, are fast. Each T-H signal coupling defines a time window and each time window actuates independently of the others. Mass and momentum dynamic effects are decoupled from energy effects because the small coefficient of thermal expansion for the coolant (water) and the selection of the specific threshold for temperature deviations prevent temperature driven expansion of the coolant to change the coolant pressures significantly.

Table 1: Dynamic coupling between thermal-hydraulic signals based on natural feedback response

Coupled Variables	{P,W}	{T,W}
Correlating Equation	mass and momentum balance	energy balance
Time Coupling	fast	slow
Primary Threshold ε_p (%)	{5.0,5.0}	{2.5,5.0}
Secondary Threshold ε_s (%)	{0.4,0.4}	{0.3,0.4}
Time interval Δt (s)	2.0	∞

The thresholds ε_p and ε_s and the time intervals Δt presented in Table 1 are based on analysis of simulator data for the Chemical and Volume Control System (CVCS) of a pressurized water reactor (PWR). For example, with the fast {P,W} time coupling, ε_p and ε_s for either variable is 5.0% and 0.4%, respectively, and Δt is 2.0 s. As soon as any P or W signal change in the entire T-H process reaches 5.0%, the time window for the {P,W} group opens allowing trend changes larger than 0.4% in any P or W signals to be considered. Two seconds later the time window is closed and any trend change in this group of signals after the time window has closed causes a halt in the diagnostics. Slow time couplings are treated differently. For example, in the slow {T,W} time coupling, temperature T is decoupled from flow W. Temperature T is treated separately and the dynamic behavior of the flow W is considered only in the fast {P,W} coupling. This approximation is reasonable because it does not cause the inadvertent application of the rules in the PRD and allows for process-independent treatment of these signals as it eliminates the need for selecting process-dependent time delays between T and W signals. Therefore, we use an infinite (∞) time interval for temperature signals. Once the time window opens at the 2.5% threshold level it never closes. Trend changes in temperature signals larger than 0.3% are always considered after the time window has opened except if oscillatory trends are detected, in which case, the diagnosis is halted. A similar approach is used for pairing the T-H signals in separated control volumes. The interested reader should refer to Reifman and Wei [4] for an in-depth treatment of dynamic effects.

3 Knowledge Bases

In this section we describe the three distinct but interacting knowledge bases of PRODIAG used to map process symptoms into component faults.

3.1 Physical Rules Database

The qualitative quasi-static balance equation rules that we derive in this section are applicable to generic process-independent configurations, i.e., the T-H loops discussed in Sect. 2.3 into which T-H processes are decomposed, without any need for rule customization. This approach differs from the qualitative physics reasoning methods based on De Kleer and Brown's work [3], where a set of balance equations or confluences that model the process are customized for each device (process) based on the device topology.

The qualitative physics rules of the PRD are of two classes: Q rules and CV rules. A Q rule indicates the type and trend of the imbalance in a control volume inferred from the trends in the T-H signals. Corresponding to the three balance equations of mass, momentum, and energy, we have three types of Q rules, Q_{mass}, Q_{mom}, and Q_{eng}, respectively, to infer the Q status (or imbalance indicator) of a control volume. The Q status can have one of three trends or qualitative values, increasing($\uparrow$), decreasing ($\downarrow$), and unchanging (-). Therefore, if a control volume is experiencing a loss of mass, a Q rule identifying such imbalance would characterize the Q status of the control volume as $Q_{mass}^{\downarrow}$. A CV rule infers the trend status of nonmeasured or "virtual" T-H signals, such as pressure P, flow W, temperature T, and level L, in a process component from the Q status and other T-H signals of the component. For example, the qualitative trend of the flow W through a heat exchanger can be inferred from the heat exchanger inlet and outlet temperature trends and its Q_{eng} status.

Different Q rules with different signal requirements apply equally well to open and closed loops, however, each Q rule is applied to only one loop at a time. Different CV rules apply to different individual generic components in separated and nonseparated control volumes. Implicit in the derivation of many of these two classes of rules is the assumption of single-component faults in T-H processes consisting of single-phase coolant with bulk moduli and thermal expansion coefficient similar to those of water and separated volumes containing noncondensable gas over single-phase liquid.

3.1.1 Derivation of Q Rules

Distinct classes of Q rules with varying degree of diagnostic precision can be derived as a function of the type, trend, and number of T-H signals. Specific groups of three-signal variables with specific trends are required to form the minimum set for *unique* identification of an increasing or decreasing Q status in a control volume. For instance, rules that employ the three-signal variables [$P^{\downarrow}$ $W_{in}^{\uparrow}$ $W_{out}^{\downarrow}$], can uniquely identify a decrease in the mass inventory ($Q_{mass}^{\downarrow}$) in the control volume defined by the two flow measurements in both open and closed T-H loop configurations. In the above notation, $P^{\downarrow}$ indicates a pressure decrease measurement anywhere in the T-H loop, $W_{in}^{\uparrow}$ represents an increase in the control volume inlet flow and $W_{out}^{\downarrow}$ represents a decrease in the control volume outlet flow. The control volume is defined during the diagnostics by matching the T-H signals of the rule with the location of these signals in the process.

Unique Q status identification can also be obtained for Q_{mom} and Q_{eng} if the specific variable trends are available for the sets [W P_{in} P_{out}] and [W T_{in} T_{out}], respectively. However, in many practical situations the available instrumentation set of the T-H loop is insufficient to provide this minimum set. There are cases where only two- or one-signal variables are available in a loop. In such cases, Q rules can also be constructed to provide some malfunction Q diagnostics. But as can be expected, the precision of the diagnostics decreases with a larger number of possible Q malfunctions being inferred. For instance, if only the two-signal variable set [P W] is available in the loop, then a Q rule would indicate both Q_{mass} and Q_{mom} problems.

3.1.1.1   Two-Variable Rules Using Flow Signals

Let us first derive two-variable rules using flow W signals that infer imbalances in the conservation of mass inventory. For the control volume shown in Fig. 3, associated with one or a group of connected components, the macroscopic quasi-static mass conservation equation is given by

$$Q_{mass} = W_{out} - W_{in} \, , \tag{1}$$

where W_{in} and W_{out} are the control volume inlet and outlet mass flow rates, respectively, and Q_{mass} is the mass source/sink term in the mass balance. Transforming Eq. (1) into qualitative differential expressions using De Kleer and Brown's [3] methodology and notation, gives the following confluence:

$$[dW_{in}] - [dW_{out}] = - [dQ_{mass}], \tag{2}$$

where the square brackets [·] represents the qualitative value or trend ($\uparrow$, $\downarrow$, -), of the argument, i.e., W_{in}, W_{out}, and Q_{mass}. Equation (2) represents the general confluence from which Q rules characterizing imbalances in Q_{mass} can be derived by applying the different trend combinations of W_{in} and W_{out} and using the operations of qualitative algebra.

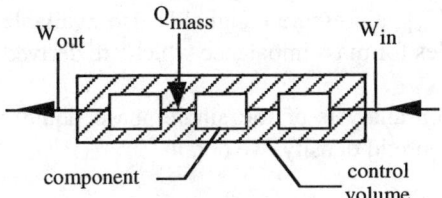

Figure 3: Control volume defined by flow signals W_{in} and W_{out}.

For the case where inlet flow into the control volume is increasing and outlet flow is decreasing, the confluence in Eq. (2) infers that Q_{mass} is decreasing ($Q_{mass}^{\downarrow}$), or, equivalently, that the control volume is loosing mass, represented through the rule:

$$\text{rule (A)} \qquad \text{If} \quad W_{in}^{\uparrow} \text{ and } W_{out}^{\downarrow}, \quad \text{Then } Q_{mass}^{\downarrow}. \tag{3}$$

This rule is applied to a T-H loop by defining a control volume where $W_{in}^{\uparrow}$ matches an upstream flow signal in the loop which has an increasing trend and $W_{out}^{\downarrow}$ matches a downstream flow signal which has a decreasing trend. While Q_{mass} in the actual balance equation is the source/sink term, in the qualitative analysis rule in Eq. (3) it should be thought of as a conservation imbalance indicator, viz. a malfunction status indicator characterizing the fact that one of the components within the control volume is malfunctioning and causing the mass inventory to decrease. Trend combinations that cause ambiguous inference, e.g., both W_{in} and W_{out} increasing, are not represented in the PRD because they would require the assumption of all three trend combinations which would cause an exponential increase in the diagnostic search procedure.

Similarly, we can also derive a rule from the general confluence in Eq. (2) corresponding to rule (A), for the case where Q_{mass} is increasing,

$$\text{rule (B)} \qquad \text{If} \quad W_{in}^{\downarrow} \text{ and } W_{out}^{\uparrow}, \quad \text{Then } Q_{mass}^{\uparrow}. \tag{4}$$

These two rules, (A) and (B), formed with two variables of the same type, i.e., W, in the condition part of the rule, uniquely identify Q_{mass} imbalances in open loops.

3.1.1.2 Three-Variable Rules

If the control volume of Fig. 3 is in a closed loop, ambiguities arise. In a closed loop, the definition of "in" (upstream) and "out" (downstream) has two possible combinations. If a fault causing a mass imbalance occurred, both rules (A) and (B) would be simultaneously activated regardless of the fault location and type ($\uparrow$ or $\downarrow$) of mass problem. This undesirable situation can be eliminated if in addition to the two flows, W_{in} and W_{out}, a pressure P signal is also available in the loop, giving rise to three-signal rules for mass imbalance which are derived as follows.

Through perturbation analysis of the single-phase liquid equation of state $P= P(\rho,T)$, where ρ is the liquid density, we obtain

$$dP = \frac{\partial P}{\partial \rho} d\rho + \frac{\partial P}{\partial T} dT. \tag{5}$$

By initiating the fault diagnosis when dT is small ($dT \ ^{\circ}F < 10^3 \ d\rho/\rho$) and using the fact that the bulk modulus ($\rho \partial P/\partial \rho$) for liquid water is positive, the qualitative differential equation for Eq. (5) becomes

$$[dP] = [dM], \tag{6}$$

where M is the liquid water mass inventory in a control volume V with density ρ. Instantiating the confluence in Eq. (6) with a decreasing pressure trend, translates into the rule

$$\text{rule (C)} \qquad \text{If } P^{\downarrow}, \quad \text{Then } M^{\downarrow}. \tag{7}$$

If a P transducer is available in a closed (or open) loop, an indication of $P^\downarrow$ anywhere in the loop, and hence $M^\downarrow$, would contradict an inference made by rule (B). Thus, the logical combination of rule (C) with rules (A) and (B) eliminates the possibility of applying rule (B) and uniquely identifies the $W^\uparrow$ instrument as being upstream of the malfunction and the $W^\downarrow$ instrument as being downstream of the malfunction. The logical intersection of rules (A) and (C) is therefore

$$\text{rule (D)} \qquad \text{If} \quad P^\downarrow \text{ and } W_{in}^\uparrow \text{ and } W_{out}^\downarrow, \quad \text{Then } Q_{mass}^\downarrow. \qquad (8)$$

Analogously, there is a corresponding rule for Q_{mom} when three-signal variables, $[W\ P_{in}\ P_{out}]$, are available,

$$\text{rule (E)} \qquad \text{If} \quad W^\uparrow \text{ and } P_{in}^\downarrow \text{ and } P_{out}^\uparrow, \quad \text{Then } Q_{mom}^\uparrow, \qquad (9)$$

where $Q_{mom}^\uparrow$ is downstream of P_{in} and upstream of P_{out}, and W is measured anywhere in the loop. Rules (D) and (E) are examples which illustrate that only three-signal variables, $[P\ W_{in}\ W_{out}]$ or $[W\ P_{in}\ P_{out}]$, are required to form the minimum set for *unique* $Q_{mass}^{\uparrow\downarrow}$ or $Q_{mom}^{\uparrow\downarrow}$ identification, respectively, for both types of loops.

3.1.1.3 Two-Variable Rules Using Flow and Pressure Signals

When only two signals of different types are available in a loop, rules can also be constructed to provide some malfunction Q diagnostics. But as can be expected, the precision of the diagnostics decreases with a larger number of possible Q malfunctions being inferred. We provide an illustration of one such rule for the case where only two-signal variables [P W] are available and show how the rule is constructed.

For the case where $[P^\downarrow\ W^\uparrow]$, rule (D) could be activated if another flow meter downstream of W was present with a decreasing trend, or rule (E) could be activated if another pressure meter downstream of P was present with an increasing trend. Since either rule could be activated in this $[P^\downarrow\ W^\uparrow]$ combination, then the *logical union* of rules (D) and (E), could be applied to obtain

$$\text{rule (F)} \qquad \text{If} \quad P^\downarrow \text{ and } W^\uparrow, \quad \text{Then } Q_{mass}^\downarrow \text{ or } Q_{mom}^\uparrow, \qquad (10)$$

where $Q_{mass}^\downarrow$ is located downstream of the W instrument and $Q_{mom}^\uparrow$ is located downstream of the P instrument. Thus, when two-variable rules employing different T-H signal types are activated, the location of one of the signals (W for Q_{mass} and P for Q_{mom}) is used to define one boundary of the imbalanced control volume with the other boundary set at the beginning of the loop if the imbalance is upstream of the signal, or set at the end of the loop if the imbalance is downstream of the signal.

The construction of rule (F) shows that there is a systematic procedure using logical union to derive Q rules with two-signal variables from the set of rules which uses the minimum three-variable sets [P W_{in} W_{out}] and [W P_{in} P_{out}]. Moreover, the two-variable rules can also be used to reconstruct the three-variable rules if the signal variables can be grouped in blocks of two. For instance, if we consider a two-signal rule analogous to rule (F), i.e.,

$$\text{rule (G)} \qquad \text{If } P^{\downarrow} \text{ and } W^{\downarrow}, \quad \text{Then } Q_{mass}^{\downarrow} \text{ or } Q_{mom}^{\downarrow}, \tag{11}$$

where $Q_{mass}^{\downarrow}$ and $Q_{mom}^{\downarrow}$ are located upstream of the W and P signal locations, respectively, and a signal set [$P^{\downarrow}$ $W_{in}^{\uparrow}$ $W_{out}^{\downarrow}$] is available, the signal set can be grouped as two two-variable sets [$P^{\downarrow}$ $W^{\uparrow}$] and [$P^{\downarrow}$ $W^{\downarrow}$]. This would mean the activation of both rules (F) and (G), where the *logical intersection* of these rules is $Q_{mass}^{\downarrow}$, which is the identical conclusion obtained with rule (D). This shows the logical consistency between the derivation of the sets of the different-variable-number rules. We apply logical union when we construct two-variable rules from two three-variable rules and logical intersection when we construct three-variable rules from two two-variable rules.

3.1.2   Derivation of CV Rules

CV rules infer the trend status of nonmeasured or virtual signals of a process component based on the Q status and trends of other T-H signals of the component. We illustrate the derivation of a CV rule that infers the trend of the flow, W_a, through the hot side of a counter current heat exchanger. For the control volume shown in Fig. 4, the quasi-static energy conservation equation is given by

$$Q_{eng} = W_a (h_{out} - h_{in}) = W_a c_p (T_{out} - T_{in}), \tag{12}$$

where h_{in} (T_{in}) and h_{out} (T_{out}) are the control volume inlet and outlet enthalpy (temperature), respectively, c_p is the specific heat and Q_{eng} is the energy source/sink term in the energy balance. Transforming Eq. (12) into a qualitative differential expression valid for $Q_{eng} < 0$ and $T_{out} < T_{in}$ and solving for W_a yields the confluence

$$- [dQ_{eng}] - [dT_{in}] + [dT_{out}] = [dW_a]. \tag{13}$$

For the case where the energy source/sink term into the control volume is not increasing, the inlet temperature is not increasing, and the outlet temperature is increasing, the confluence infers that the flow rate through the hot side of the heat exchanger is increasing, represented through the rule

rule (H)   If $\ Q_{eng}^{/\uparrow}$ and $T_{in}^{/\uparrow}$ and $T_{out}^{\uparrow}$,   Then $W_a^{\uparrow}$,   (14)

where the symbol "/" indicates negation. The trend of Q_{eng} can be obtained by applying Q_{eng} rules to the cold side of the heat exchanger [5] and the trends of T_{in} and T_{out} are obtained from the instrumentation signals. Other CV rules for inference of W_a can be obtained by instantiating the quantities in the left hand side of the confluence in Eq. (13) with different trend combinations. A similar procedure is used to derive CV rules for other types of T-H signals and components.

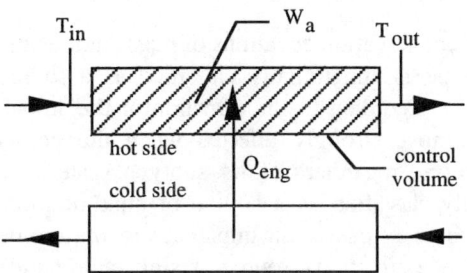

Figure 4: Control volume representation of the hot side of a counter current heat exchanger where the energy balance equation allows for the inference of the trend of the unmeasured heat exchanger flow W_a.

A special type of CV rule that we make extensive use of is the one based on enthalpy (or temperature) transport. This type of CV rule is used to infer temperature trends throughout a T-H loop (instead of just one component) based on the trend status of an upstream thermocouple signal. The trend status of the upstream temperature signal is transported to the downstream components and used as the temperature trend at the inlet and outlet of the downstream components. Temperature, however, is not transported through energy source/sink components, such as electric heaters and heat exchangers, or through junctions which result in fluid mixing downstream of the junction in the T-H loop where temperature is being transported. The underlying assumption is that only energy source/sink components and junctions with incoming flow can affect the temperature trend of components in T-H loops.

The virtual signal trends obtained through the CV rules extract the most information possible from the process and increase the total number of instruments (virtual and real) available for diagnostics. Once inferred, virtual signals can then be combined with actual signals and used to determine the Q status of components through the Q rules, which otherwise would not be determined. The additional signal information provided by the CV rules tends to increase the diagnostic precision obtained with PRODIAG because the number of faulty component candidates tends to decrease with increasing signal information.

3.2 Component Classification Dictionary

The CCD is a living dictionary where the small number of generic component types (e.g., closed valve, open valve, and electric heater) included in T-H processes are classified into predefined classes according to their design function. Through the proper functional classification of generic components we can then link the Q status (or imbalance indicator) of control volumes detected by the rules of the PRD with components whose failure could be responsible for the detected imbalance. In contrast to other approaches [6,7], the classification of generic component in the CCD is event- and process-independent.

Based on the three conservation equations of mass, momentum, and energy, each generic component type is functionally classified as a source or sink of mass, momentum, or energy according to the direction of the imbalance and to which balance equation is most strongly affected when the component fails. For example, an open valve, regardless of its subtype (gate valve or globe valve), should be functionally classified as a source or sink of momentum because an open valve failure primarily causes an imbalance in the momentum conservation equation. The imbalance directions, source or sink, automatically account for the possible failure modes of the component fault. For instance, valve blockages or unexpected closures are covered by the classification of the open valve as a sink of momentum because any of these failure modes would cause a negative imbalance in the momentum conservation equation. An unexpected increase in valve opening is covered by the classification of the valve as a source of momentum for analogous reasons.

The small number of generic component types classified in Table 2 are currently represented in the CCD knowledge base. This small set includes the most often found types of components which are used repeatedly in nuclear power plant systems. As other types of components not included here are encountered, they can be classified and assigned a location in the CCD.

Under the Q_{mass} components for nonseparated volume, we have closed valve, junction, and break. A closed valve is naturally a mass source or sink as its T-H operating function is to either inject inventory into or drain inventory from a process, respectively. The classification of junction and break invoke a certain level of abstraction. A junction is classified as a source or sink of Q_{mass} because for a control volume containing a junction, a pressure disturbance initiator in the loop across the junction, outside the control volume, would lead to the same spatial flow distribution in the control volume as a mass problem within the control volume. For example, a disturbance initiator, such as a component fault that causes a mass or a momentum imbalance, downstream of junction 1 in the loop across loop 1 in Fig. 2, (i.e., in the loop that includes valve CV-8117), would be reflected as a mass imbalance in a control volume including the junction

Table 2: Functional classification of generic components as sinks $(Q^{\downarrow})$ or sources $(Q^{\uparrow})$ or mass, momentum, and energy

Control Volume	Imbalance Type	Trend Type	Generic Component
Nonseparated	Q_{mass}	Sink $(Q_{mass}^{\downarrow})$ or Source $(Q_{mass}^{\uparrow})$	closed valve junction break
	Q_{mom}	Sink $(Q_{mom}^{\downarrow})$	pump filter pipe demineralizer open valve
		Source $(Q_{mom}^{\uparrow})$	pump open valve
	Q_{eng}	Sink $(Q_{eng}^{\downarrow})$ or Source $(Q_{eng}^{\uparrow})$	heater heat exchanger
Separated	Q_{mass}	Sink $(Q_{mass}^{\downarrow})$ or Source $(Q_{mass}^{\uparrow})$	tank pressurizer
	Q_{eng}	Sink $(Q_{eng}^{\downarrow})$ or Source $(Q_{eng}^{\uparrow})$	pressurizer

in loop 1. The classification of junctions as sources or sinks of mass allows T-H loops to be considered separately and independently for the application of the Q_{mass} PRD rules even though the loops do hydraulically intersect at the junction and are, therefore, hydraulically coupled. A "malfunctioning" junction then points the diagnosis in the direction of the intersecting loops. Physically speaking, a break is not a process component. However, defining a break as a component eliminates the need to classify every generic component type as a source or sink of mass since any component break would cause a mass imbalance. Note that in our nomenclature the term break is used to represent a specific component malfunction, namely, a leak, and not a generic component fault.

Under the Q_{mom} components for nonseparated volume, we classify a pump and an open valve as a source of momentum and a pump, filter, pipe, demineralizer, and open valve as a sink of momentum. Pumps and open valves are sources or sinks of momentum depending on the mode of the failure. Filters, pipes, and demineralizers are classified as sinks of momentum because these components are resistances in the momentum equation and an obstruction to flow in a filter, pipe, or demineralizer would cause a decrease in momentum. Only two types of Q_{eng} components have been identified so far. These are heaters, where the energy input is through external sources, such as electric heater, and heat exchangers where the energy input is through heat exchange between two T-H loops.

Separated volume components have been restricted to volumes where internal flow fields are not a major factor in the component response. For those components where the internal flow fields are a major factor, additional rules will have to be derived. They are, therefore, essentially end condition volumes for loops. Examples are tanks and pressurizers. As such, there are, thus, no Q_{mom} components, but just Q_{mass} and Q_{eng} components for the class of separated volume components.

3.3 Piping and Instrumentation Database

The P&IDB contains the information about the type and structural arrangement or connectivity of the process components expressed in the process P&ID. By representing the process P&ID information in a separate knowledge base which is independent of the diagnosis methodology, PRODIAG can be readily ported across different processes/plants through changes only in the P&IDB to accommodate customization of the program for particular diagnostic applications.

The structural domain knowledge of a process P&ID is represented in the P&IDB through directed graph structures [8] consistent with the T-H process decomposition described in Sect. 2.3. The T-H process is represented by a collection of T-H loops where each loop is characterized by a single directed graph. The mapping of a T-H loop into a graph structure is achieved by taking each physical component or component part in the loop as a node of the graph, while each connection between two components, i.e., the piping, is taken as an arc. When the arcs are directed, i.e., represented by ordered pairs of components defining the direction of fluid flow, the graph is a directed graph. Through this representation, search procedures can be readily applied to retrieve intra- and inter-loop information, such as the flow paths between any two components in the process.

In addition, in the P&IDB, each T-H loop and each component in a loop is represented as an object where each object is characterized by attributes. Loop attributes include loop name, type, and imbalance status indicator, while component attributes include component name, type, fluid phase, imbalance status indicator, and signal instrumentation trend.

The P&IDB together with the PRD and the CCD form the three knowledge bases of PRODIAG. All three knowledge bases are independent of each other, but unlike the other two, the P&IDB is process-dependent and decoupled from the assumptions of the driving reasoning algorithm.

4 Simulation Tests

To provide a realistic environment for demonstrating the unique capabilities of PRODIAG to identify unanticipated component faults and to be ported across different processes/plants through simple modifications of input data files, we test PRODIAG for two plant systems of a PWR with transient data generated using the full-scope operator training Braidwood Nuclear Power Plant (BNPP) simulator. In each test, the sole information provided to PRODIAG, through input only, consists of the simulated transient T-H instrumentation signal data and the process P&ID. The first validation test, consists of a blind test using the Braidwood Chemical and Volume Control System (CVCS) where the identities of the simulated component faults were not disclosed to us until after completion of the test. However, in this test we had prior knowledge of which component faults of the CVCS could be modeled with the BNPP simulator. The second validation test, consists of a double-blind test with simulated component faults in the Braidwood Component Cooling Water (CCW) system. In the double-blind test, in addition to having no prior knowledge of the identities of the simulated faults, we also had no prior information regarding the type of CCW component faults which could be modeled with the BNPP simulator.

4.1 The Chemical and Volume Control System

The simplified P&ID for the CVCS of the BNPP simulator used in the blind test is shown in Fig. 5. The CVCS is a water system whose primary purposes are to regulate reactor coolant chemistry for reactivity and corrosion control and to maintain the water level in the pressurizer [9]. During the normal charging/letdown mode of operation, letdown water leaves the RCS, shown in the upper-left part of Fig. 5, and flows through the shell side of the regenerative heat exchanger where it gives up its heat to makeup water being returned to the RCS. From there, letdown water proceeds through a series of valves and the letdown heat exchanger in order to reduce system pressure and temperature, until it reaches the VCT. Then, a charging pump (pump A in the lower-right part of the figure) takes the coolant from the VCT to a junction (junc_7 in the figure), where the streams divide. Charging water flows back to the RCS through the tube side of the regenerative heat exchanger and the remaining water flows to the seals of the reactor coolant pumps (RCP) where one portion returns to the RCS and the other recirculates to the suction side of pump A.

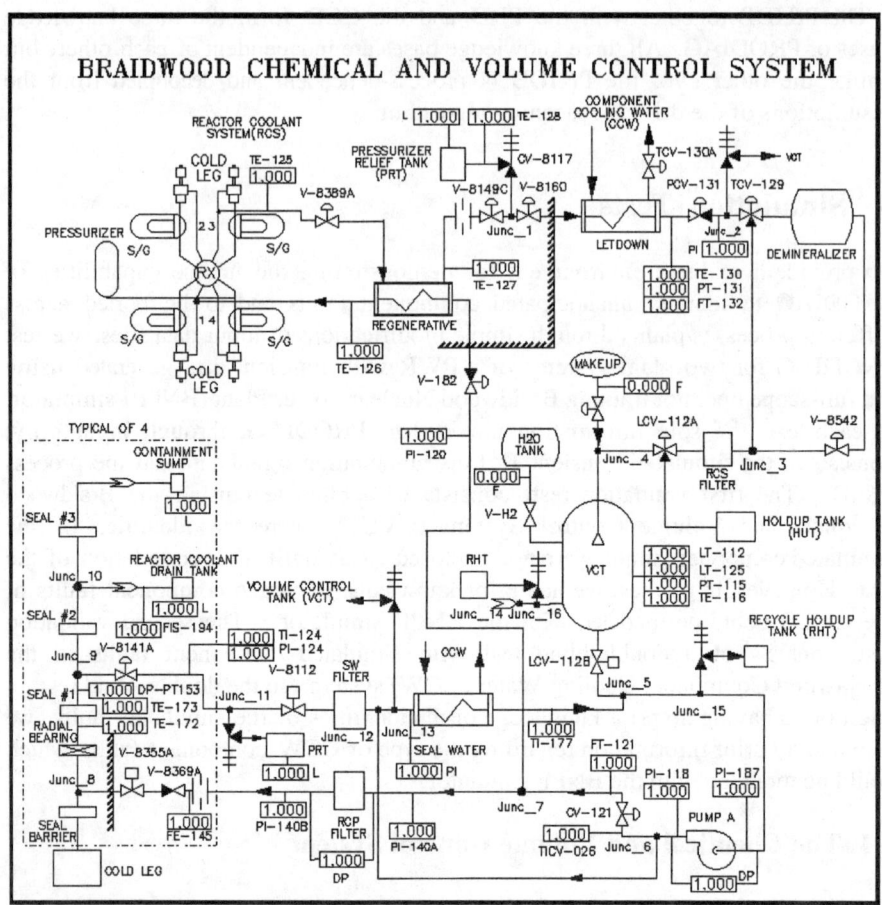

Figure 5: Simplified piping and instrumentation diagram for the chemical and volume control system of the Braidwood simulator.

A total of 39 CVCS transient events separately simulated by the BNPP simulator were used in the blind test. Data for these transients were generated through repeated simulations of 20 distinct component faults, including pump trips, valve failures, tube breaks, and clogged filters, where for each simulation different severity levels and failure modes were selected. In this blind test, the identities, severity levels, and failure modes of the simulated events were only disclosed to us after the completion of the test. However, we did know the types of the 20 transients that composed the transient test matrix.

Each single-fault transient event was simulated for 180 s, including at least 3 s of null transient, starting from the steady state normal charging/letdown mode of the CVCS operation with the plant at 100% of nominal power. Data from each simulated transient were stored in a separate file consisting of recorded values of the T-H signal measurements (flows, pressure, temperature, and level) the locations of which are illustrated in Fig. 5. The signals were recorded at 1 s sampling intervals.

In the nomenclature of Fig. 5, flow sensor indicators start with the letter F, pressure indicators with the letter P, temperature indicators with the letter T, and level indicators with the letter L. The symbol DP is used to represent pressure difference across pump A and across the RCP filter. The signal values have been normalized to 1.0 with respect to their expected steady state values. Hence, together with considerations of the dynamic transient effects, signal indications > $1.0+\varepsilon$ characterize increasing ($\uparrow$) signal trends, where ε is the signal threshold, while signal indications < $1.0-\varepsilon$ characterize decreasing ($\downarrow$) signal trends, and signal indications between $1.0\pm\varepsilon$ characterize unchanging (-) signal trends. Table 1 illustrates the numerical values used for the primary threshold ε_p, secondary threshold ε_s, and time interval Δt, for signals associated with components in nonseparated control volumes.

Of the 39 simulated transients, PRODIAG correctly diagnosed 95% (37/39) of the transients with varying precision within the first 40 s into the transient, and did not diagnose 5% (2/39). No transients were misclassified. The second row of Table 3 summarizes the diagnostic results and indicates the distribution of the diagnostic precision obtained at 40 s into the transient. Nineteen transients, corresponding to 49% of the simulated transients, were uniquely identified, 3 transients or 8% were identified as one out of two possibilities, 8 transients or 20% were identified as one out of three possibilities, 7 transients or 18% were identified as one out of more than three possibilities, and 2 transients or 5% were not identified. The two unidentified transients correspond to instrumentation malfunctions (as opposed to a component malfunction) which are beyond the scope of PRODIAG.

Table 3: Summary of the two test results

System	Uniquely Identified	Identified as One of Two Possibilities	Identified as One of Three Possibilities	Identified as One of More than Three Possibilities	No Diagnosis	Incorrect Diagnosis
CVCS	19 (49.0%)	3 (8.0%)	8 (20.0%)	7 (18.0%)	2 (5.0%)	-
CCW	3 (21.4%)	3 (21.4%)	2 (14.4%)	6 (42.8%)	-	-

4.2 Letdown Heat Exchanger Tube Break Diagnosis

Let us reconsider the diagnosis of a tube break in the letdown heat exchanger discussed in Sect. 2.2 where loop 1 and loop 2 in Fig. 2 correspond to the letdown line and a portion of the charging line, receptively, of the CVCS P&ID in Fig. 5. The dynamic response of the CVCS T-H signals affected by the heat exchanger tube break scenario is illustrated in Table 4. At 5 s into the transient, PT-131 (P) and FT-132 (W) in loop 1 have decreased beyond their primary threshold ε_p of 5% (see Table 1), causing the {P,W} time window to open and PRODIAG's fault identification algorithms to be activated. The Q rules of the PRD are searched to identify a first-principles rule which correlates the decreasing ($\downarrow$) trends in P and W with imbalance indicators. With only the two-signal variable set [P W] available for diagnostics in the loop, rule (G) in Eq. (11) is activated to infer that there is either a $Q_{mass}^{\downarrow}$ or a $Q_{mom}^{\downarrow}$ problem in loop 1 and that the imbalanced control volume is located upstream of the signals, from the beginning of the letdown line up to the outlet of the letdown heat exchanger. Then, the CCD is searched to infer that closed valve, junction, or break problems could have been responsible for the $Q_{mass}^{\downarrow}$ inference, and that pump, filter, pipe, demineralizer, or open valve problems could have been responsible for the $Q_{mom}^{\downarrow}$ inference (see Table 2).

Table 4: Dynamic response and PRODIAG treatment of the thermal-hydraulic signals for the letdown heat exchanger tube break

Transient Time (s)	Signal Change (%) and Inferred Trend				Time Window Status for Nonseparated Volumes	
	PT-131	FT-132	TE-127	FV-XYZ	{P,W}	{T,W}
5	-7.0 ($\downarrow$)	-9.0 ($\downarrow$)	0.0 (-)	-	open	closed
6	-32.0 ($\downarrow$)	-27.0 ($\downarrow$)	0.0 (-)	-	open	closed
7	-38.0 ($\downarrow$)	-33.0 ($\downarrow$)	0.0 (-)	-	open	closed
8	-38.0 ($\downarrow$)	-35.0 ($\downarrow$)	0.2 (-)	-	closed	closed
9	-37.0 ($\downarrow$)	-36.0 ($\downarrow$)	0.3 ($\uparrow$)	($\uparrow$)	closed[a]	open[a]

[a]Status will not change until the end of the diagnostic session.

Next, the P&IDB is searched to determine the existence and identity of these generic components in the imbalanced control volume. Since there are no closed valves, pumps, filters, and demineralizers in the control volume they are ruled out. There is one junction (junction 1) between valves V-8149C and V-8160 which hydraulically connects loop 1 with an open loop that starts at the junction and continues through relief valve CV-8117 and the subsequent downstream components. Because a disturbance initiator in this open loop would be reflected as a mass imbalance in loop 1, the rules of the PRD and additional logical statements are applied using T-H signals not shown in the figure to rule out any possible component fault in this loop. PRODIAG then hypothesizes that the detected imbalances are due to a break or a pipe blockage anywhere in the imbalanced control volume or that one of the three open valves in the control volume, V-8389A, V-8149C, or V-8160, has failed. Without additional information a distinction between Q_{mass} and Q_{mom} imbalances cannot be made and the precision of the diagnostics cannot be increased.

Between 6 and 8 s into the transient no additional information is available and PRODIAG makes the same diagnoses performed at 5 s. Table 4 shows that the {P,W} time window closes after 7 s (as its time interval Δt is 2.0 s) and remains closed until the end of the diagnostics session.

By 9 s into the transient, TE-127 has increased beyond its secondary threshold ε_s of 0.3% causing the {T,W} time window to open. Since T signals have an infinite time interval, this time window will remain open until the end of the diagnostics session and future changes in any other T signal in the process will be considered in the diagnostics. The increasing trend in TE-127 causes the activation of a CV rule in the PRD knowledge base, namely, rule (H) in Eq. (14), which infers that the mass flow rate through the hot side of the regenerative heat exchanger - represented by the virtual flow transmitter FV-XYZ in Fig. 2 - is increasing. A few rules of the PRD are activated before rule (H) can be applied. First, the enthalpy (temperature) transport CV rule discussed in Sect. 3.1.2 is applied in loops 1 and 2 to transport the unchanging (-) temperature trend measured by TE-125 in loop 1 and TI-026 in loop 2 to the inlet of the hot side and cold side, respectively, of the regenerative heat exchanger. This procedure allows for the inference of a constant inlet temperature T_{in} on both sides of the heat exchanger and accounts for the second term, $T_{in}^{/\uparrow}$, in the condition part of rule (H). The third term in the expression, $T_{out}^{\uparrow}$, is directly accounted for through the measured value of TE-127. Finally, the first term, $Q_{eng}^{/\uparrow}$, is accounted for by inferring that $Q_{eng}^{/\downarrow}$ on the cold side of the heat exchanger and changing the sign of Q_{eng} (to be consistent with the notation) before it is applied to the hot side of the heat exchanger. $Q_{eng}^{/\downarrow}$ is inferred on the cold side of the heat exchanger by applying a Q_{eng} rule in loop 2 that matches W with FT-121, T_{in} with TI-026, and T_{out} with TE-126. Such a rule is obtained by transforming Eq. (12) into a qualitative differential expression valid for $Q_{eng} > 0$ and $T_{out} > T_{in}$, and instantiating W to W^-, T_{in} to T_{in}^- and T_{out} to $T_{out}^{/\downarrow}$.

The use of the heat exchanger CV rule at 9 s to infer the unmeasured trend status of flow upstream of the break narrows down the inference to a Q_{mass} problem only. After the trend status of FV-XYZ has been inferred through the heat exchanger CV rule, the T-H signals are once again provided to the PRD knowledge base. In the PRD, the mass imbalance indicator Q_{mass} rule in Eq. (3) is identified by matching W_{in} with FV-XYZ and W_{out} with FT-132. The activation of this Q_{mass} rule defines the "imbalanced" control volume to be located between the two flow transmitters, i.e., encompassing the two heat exhangers, and sets the mass imbalance indicator of the components within the imbalanced control volume to $Q_{mass}^{\downarrow}$, characterizing the fact that one of the components within the control volume is malfunctioning and causing the mass inventory to decrease. Because pressure information in loop 1 is available through PT-131, rule (D) in Eq. (8) could have also been used resulting in the same $Q_{mass}^{\downarrow}$ inference.

With a mass decrease imbalance detected by the PRD, the CCD is searched to infer that closed valve, junction, or break are the generic component types whose failure could have caused the detected $Q_{mass}^{\downarrow}$ imbalance. Next, given the location of the imbalanced control volume, the P&IDB is searched to determine the existence and identity of these component types between the two heat exchangers. Closed valves and junctions are ruled out for the same reasons they were discarded during the diagnostics between 5 and 8 s. Then, PRODIAG correctly hypothesizes that the detected imbalance is caused by a break between the inlet of the regenerative heat exchanger and the outlet of the letdown heat exchanger. This example illustrates how the use of a CV rule to infer the trend status of unmeasured T-H signals can increase the precision of the diagnostics. To narrow down the location of the break to the letdown heat exchanger would require additional sensor instruments, such as a flow transmitter at the inlet of the letdown heat exchanger.

4.3 The Component Coolant Water System

To confirm PRODIAG's capability to identify unanticipated components faults and to be ported across different processes/plants through modification only of input data files, a different T-H system of the Braidwood PWR, namely the Component Cooling Water system, was selected for a double-blind test. Similar to the CVCS blind test, in this double-blind test the sole information provided to the diagnostic system, through input only, is the simulated transient T-H instrumentation signal data and the CCW P&ID illustrated in Fig. 6. We designate this test as a double-blind test because the identities of the simulated transients were not disclosed to us until after the conclusion of the test and we had no a priori knowledge of which CCW component faults could be modeled with the BNPP simulator.

The purpose of the CCW system is to continuously supply cooling water to plant components during normal, off-normal, and cooldown plant conditions. The component cooling water circulates through partially overlapping parallel closed-loop flow paths that start and end at pump 1B and removes heat from different plant systems, such as the RCS, CVCS, residual heat removal, through various heat exchangers.

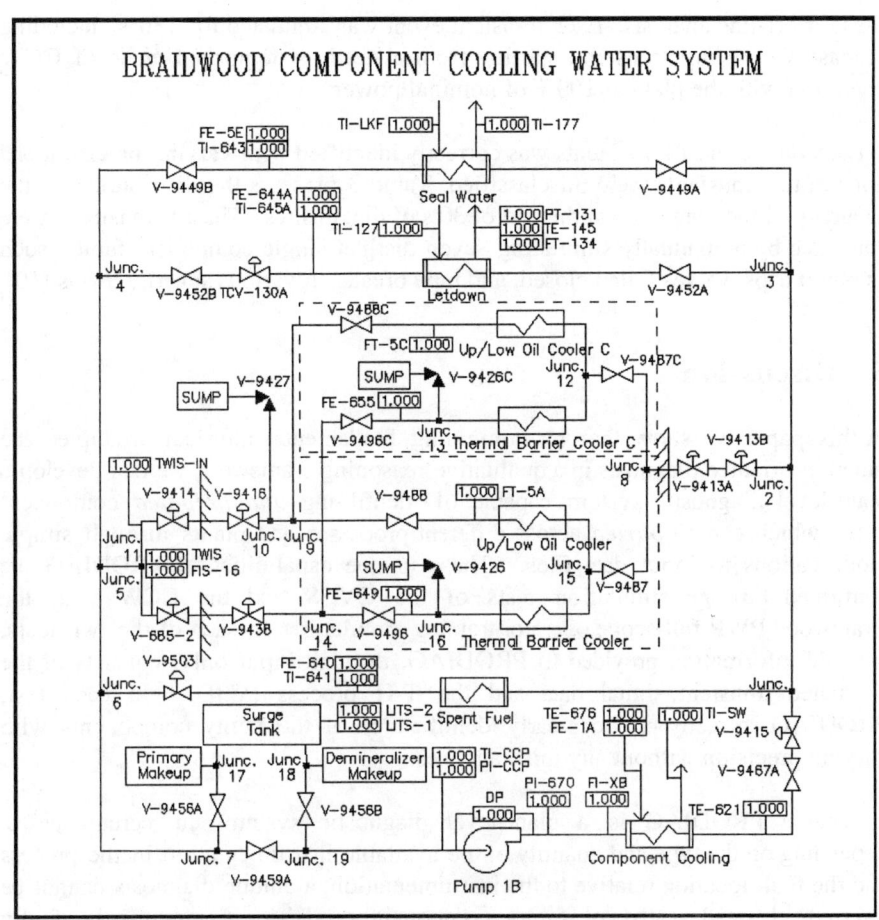

Figure 6: Simplified piping and instrumentation diagram for the component cooling water systems of the Braidwood simulator.

This closed-loop configuration of the CCW system makes it a more challenging diagnostic problem that the CVCS because it tends to decrease the resolution of the diagnosis. Closed-loop configurations require more plant signal information for unique characterization of mass and momentum imbalances than open-loop configurations (see discussion in Sect. 3.1.1.2). In addition, the instantaneous propagation of changes in flow and pressure in an incompressible fluid, such as liquid water, tightly couples all T-H loops of an overlapping closed-loop configuration. Hence, it is sometimes not possible to distinguish between the symptoms of the original faulty components located near the fault and the propagated symptoms spread throughout the system.

A total of 14 CCW single-fault transient events were separately simulated with the BNPP simulator and the transient data for each event were stored in separate files for off-line analysis. Each transient event was simulated for 180 s, including at least 3 s of null transient, starting from a steady-state normal mode of CCW operation with the plant at 100% of nominal power.

Each one of the 14 transients was correctly identified with varying precision and none of the transients were misclassified. Table 3 presents the distribution of the accuracy of the diagnoses at the end of 30 s of diagnostics. The 14 transients were generated by individually simulating seven distinct single-component faults, such as pump trips, valves failed closed, and pipe breaks, at various severity levels [10].

5   Discussion

In this paper we show that by combining fundamental physical principles and function-oriented diagnosis in a qualitative reasoning framework we can develop a plant-level diagnostic system capable of identifying *unanticipated* component faults which can be *ported* across different processes and plants through simple modifications to input data files. These unique capabilities of PRODIAG are confirmed through simulation tests of the CVCS and the CCW with the Braidwood PWR full-scope operator training simulator. In each of the two tests, the sole information provided to PRODIAG, through input only, consists of the simulated transient signal data and the T-H process P&ID. In each test, PRODIAG correctly provides early identification of the faulty components with varying precision without any misdiagnoses.

Because PRODIAG is a plant-level diagnostic system, in certain cases, depending on the type and quantity of the available instrumentation in the process and the fault location relative to the instrumentation, a unique diagnosis cannot be inferred. Instead, a list of faulty component candidates (including the faulty component) is hypothesized. To compensate for the lack of available instrumentation typical of most processes in nuclear plants, PRODIAG infers the trend of unmeasured signals, which, in certain cases, increases the diagnostic precision. Further improvement in the diagnostic precision could be achieved by supplementing the plant-level diagnosis performed by PRODIAG with component-level diagnosis which would test each hypothesized component fault to uniquely identify the fault [1].

Work is under way to remove some of the implementation simplifications of the current version of the system and to expand its range of applicability. For instance, algorithms are being developed to provide the trend of varying T-H signals in the presence of data noise. Future activities will extend the current work to account for processes involving two-phase flow and for the possibility of multiple component faults.

Although not a truly intelligent system as envisioned by the pioneers of machine intelligence, PRODIAG's unique ability to directly apply universal T-H first principles to new and unexpected plant scenarios represents a significant improvement over the current generation of soft computing diagnostic systems. Its flexibility and ability to identify unforeseen events could enhance operator action in difficult-to-assess off-normal plant scenarios.

Acknowledgments

The authors express their gratitude to R. G. Abboud, C. A. Applequist, and W. M. Kuk from Commonwealth Research Corporation for providing their full support. The authors also wish to thank T. M. Chasensky, P. Hippely, and M. Olson from Commonwealth Edison Company for performing the simulation runs.

This work was performed under a Collaborative Research and Development Agreement between Commonwealth Research Corporation and Argonne National Laboratory and was supported by the U.S. Department of Energy, Office of Science, Laboratory Technology Transfer Program, and Office of Nuclear Energy, Science and Technology, under contract W-31-109-ENG-38.

References

1. W. R. Nelson, "REACTOR: An Expert System for Diagnosis and Treatment of Nuclear Reactor Accidents," *Proc. National Conference on Artificial Intelligence,* AAAI, August 18-20, 1982, Pittsburgh, PA, pp. 296-301, (1982).

2. J. Reifman, "Survey of Artificial Intelligence Methods for Detection and Identification of Component Faults in Nuclear Power Plants," *Nucl. Technol.,* Vol. 119, pp. 76-97 (1997).

3. J. De Kleer and J. S. Brown, "A Qualitative Physics Based on Confluences," *AI,* Vol. 24, pp. 7-83 (1984).

4. J. Reifman and T. Y. C. Wei, "PRODIAG - Dynamic Qualitative Analysis for Process Fault Diagnosis," *Proc. Ninth Power Plant Dynamics, Control and Testing Symposium,* Knoxville, Tennessee, May 24-26, 1995, Vol. 1, pp. 40.01-40.15, B. R. Upadhyaya, E. M. Katz, and T. W. Kerlin, Eds., The University of Tennessee, Knoxville, Tennessee (1995).

5. J. Reifman and T. Y. C. Wei, "PRODIAG: A Process-Independent Transient Diagnostic System--I: Theoretical Concepts," *Nucl. Sci. Eng.,* Vol. 131, pp. 329-347 (1999).

6. F. E. FINCH and M. A. KRAMER, "Narrowing Diagnostic Focus Using Functional Decomposition," *AIChE J.,* Vol. 34, 25-36 (1988).

7. J. A. HASSBERGER and J. C. LEE, "Macroscopic Mass and Energy Balance for Nuclear Plant Diagnostics Using Fuzzy Logic," *Proc. Topical Meeting on Artificial Intelligence and Other Innovative Computer Applications in the Nuclear Industry,*

476

Snowbird, Utah, August 31-September 2, 1987, pp. 539-546, M. C. Majumdar, D. Majumdar, and J. I. Sacket, Eds., Plenum Press, New York (1988).

8. I. Bratko, *Prolog Programming for Artificial Intelligence*, Addison-Wesly (1986).
9. *Braidwood Chemical and Volume Control System - System Description*, Commonwealth Edison Company, Braidwood, Illinois (1990).
10. J. Reifman and T. Y. C. Wei, "PRODIAG: A Process-Independent Transient Diagnostic System--II: Validation Tests," *Nucl. Sci. Eng.*, Vol. 131, pp. 348-369 (1999).

Subject Index

active learning, 246
adaptive function, 71
advanced fuzzy control, 69
ALADDIN, 208, 229
artificial intelligence methods, 116
artificial neural networks, 235
aspiration level, 139
ATHLET-code, 3, 4, 14, 28

Bayesian classifier, 416
Boosting, 231
B-spline, 41
BWR, 221

characteristic field, 11, 14, 29, 32
combinatorial explosion, 396
complex system, 364, 368, 369, 391
conservation equations, 450, 464
contingency factors, 154
continued fraction, 397
control rods, 48, 259
core
  geometry, 433
  reload pattern design, *see*
  refuellings
cost uncertainties, 154

decision making under uncertainty,
  414
decoupling scheme, 196
defuzzification, 44, 177
diagnosis system, 339
direct containment heating (DCH),
  377-391
DML-US Software, 388-391
dynamic
  generation, 86, 87
  master logic diagram (DMLD),
  364-365, 367, 372-393
  modelling, 240

early failure detection system, 121
economic calculus, 153
eigen structure based methods, 244
Elman, 218, 226
engineering restriction, 143
ergodic
  theorem, 400
  transformations, 400
expert system, 449, 451

flow measurements, 270
fuel management optimization, 335
FUELCON, 432-448
FUELGEN, 432-448
functional link ANN, 245
function-oriented diagnosis, 449,
  450, 474
fuzzy
  adaptation, 83, 86
  algorithm, 136
  clustering, 211
  C-means, 212
  Control, 67, 86-88, 99-100, 103
  Controller, 3, 9, 30, 54
  Generation, 97
  goal programming, 151
  gradual rules, 162
  hierarchy modeling, 364-365, 372-
  393
  inference system, 153, 176
  logic diagnostic system, 130
  rules, 43
  sets, 175
  system, 40, 45, 48

Gauss measure, 401
genetic algorithms, 182, 315, 335,
  351, 435
global optimization, 317
goal

programming, 135
tree success tree (GTST), 364-371, 377, 391

hierarchy modeling, 364-393
hybrid observer, 3, 8, 10, 22, 28

image processing, 241
inaccuracies in a priori knowledge, 420, 424-425
in-core fuel management, *see* refufuellings
inferential sensing, 286
input fuzzification, 43
inverse problems, 258

Khintchin constant, 405
knowledge base, 449-453, 466, 467, 471

learning process, 236
lexicographic minimum, 140
Lorentz model, 407

market, monopolistic vs. competitive, 434
master logic diagram (MLD), 364-371
material balance, 144
membership function, 175
minimum centroids set method, 343
mixmaster universe, 404
model-based measuring method, 1, 3, 6, 34, 36
modular group, 403

neural networks, 208, 211, 259, 301
neuron-fuzzy control, 174
neutron noise, 263
nodal algorithms, 432
nuclear
   reactor, 65, 83, 97, 351
   reactor design optimization, 317

observer, 7, 25, 34
operating of nuclear power plants, 115
operator control process, 120
optical signal processing, 246
optimal operation, 135

parameter estimation, 259
partial
   least squares, 299
   quotients, 398
physical principles, 449, 450, 452, 474
possibilistic, 211, 220, 227
possibilistic classifier, 418
power
   axial-offset, 53
   distribution control, 193
pressurized water reactor (PWR), 53, 229
process
   diagnosis, 391
   diagnostics, 450
project and technology factors, 155
pulsed neutron activation, 270

qualitative
   classification, 423
   reasoning, 449, 450, 452, 455, 474
quantitative classification, 423

radioactive waste
   management, 153
   processing system, 142
radioactivity, 142
RBF, 211, 222
reactor
   diagnostics, 258, 259
   power, 47, 127
recurrent, 218
refuellings, 432-448
regularization, 285, 290
rigid constraints, 141

rule base 13, 19, 24, 33, 37
rule-based expert system, 435

sensitivity analysis, 150
sensors, 286
shutdown periods, 434
signal processing, 239
simulated
  annealing (SA), 359
  data, 259, 265, 272
simulator, 449, 457, 467, 472
SOM, 213, 209
spline interpolation function, 43-44
state estimation, 9, 20, 25, 34
static state estimation, 126
statistical learning networks, 242
steam generator level control, 185
system constraint, 139

target classification, 414, 416
tolerance limit, 137
total power defect, 47
training method, 180
transient classification, 208, 209
traveling salesman problem, 336

water level measurement, 2, 4, 9

Studies in Fuzziness and Soft Computing

Vol. 25. J. Buckley and Th. Feuring
Fuzzy and Neural: Interactions and Applications, 1999
ISBN 3-7908-1170-X

Vol. 26. A. Yazici and R. George
Fuzzy Database Modeling, 1999
ISBN 3-7908-1171-8

Vol. 27. M. Zaus
Crisp and Soft Computing with Hypercubical Calculus, 1999
ISBN 3-7908-1172-6

Vol. 28. R.A. Ribeiro, H.-J. Zimmermann, R.R. Yager and J. Kacprzyk (Eds.)
Soft Computing in Financial Engineering, 1999
ISBN 3-7908-1173-4

Vol. 29. H. Tanaka and P. Guo
Possibilistic Data Analysis for Operations Research, 1999
ISBN 3-7908-1183-1

Vol. 30. N. Kasabov and R. Kozma (Eds.)
Neuro-Fuzzy Techniques for Intelligent Informations Systems, 1999
ISBN 3-7908-1187-4

Vol. 31. B. Kostek
Soft Computing in Acoustics, 1999
ISBN 3-7908-1190-4

Vol. 32. K. Hirota and T. Fukuda
Soft Computing in Mechatronics, 1999
ISBN 3-7908-1212-9

Vol. 33. L.A. Zadeh and J. Kacprzyk (Eds.)
Computing with Words in Information/ Intelligent Systems 1, 1999
ISBN 3-7908-1217-X

Vol. 34. L.A. Zadeh and J. Kacprzyk (Eds.)
Computing with Words in Information/ Intelligent Systems 2, 1999
ISBN 3-7908-1218-8

Vol. 35. K.T. Atanassov
Intuitionistic Fuzzy Sets, 1999
ISBN 3-7908-1228-5

Vol. 36. L.C. Jain (Ed.)
Innovative Teaching and Learning, 2000
ISBN 3-7908-1246-3

Vol. 37. R. Słowiński and M. Hapke (Eds.)
Scheduling Under Fuzziness, 2000
ISBN 3-7908-1249-8

Druck: Strauss Offsetdruck, Mörlenbach
Verarbeitung: Schäffer, Grünstadt